Elliptic Curves, Modular Forms, and Fermat's Last Theorem

second edition

edited by

John H. Coates and Shing-Tung Yau

International Press
www.intlpress.com

Elliptic Curves, Modular Forms, and Fermat's Last Theorem, 2nd edition

Editors:
John H. Coates, University of Cambridge
Shing-Tung Yau, Harvard University

ISBN 978-1-57146-185-8

Paperback reissue 2010. Previously published in 1997 under ISBN 1-57146-049-7 (clothbound).

Typeset using the LaTeX system.

Table of Contents

Conference on Elliptic Curves and Modular Forms
Hong Kong, December 18-21, 1993

Foreword

A conference, on the general theme of "Elliptic curves and modular forms", was held in the Mathematics Department and the Institute of Mathematical Sciences of The Chinese University of Hong Kong from December 18–21, 1993. The impetus for organizing the conference arose from Andrew Wiles' deep and spectacular work on the celebrated conjecture that every elliptic curve over Q is modular, although only some of the lectures at the conference were specifically related to this theme. At the time of the conference, the difficulties in the last hurdle in Wiles' work (the proof of the conjectural upper bound for the order of the Selmer group attached to the symmetric square of a modular form) had still not been overcome. However, the optimism shared by all at the conference that is was only a matter of time until the proof would become complete has happily been borne out by subsequent events. It is now history that Wiles himself, assisted by R. Taylor, found a beautiful proof of the desired upper bound. As a result, we now know today the remarkable fact that every semi-stable elliptic curve over Q is modular. Not only is this result revolutionary in its own right for the study of the arithmetic of these elliptic curves, but it has the added bonus that it provides at last a proof of Fermat's last theorem, thanks to the earlier work of Frey, Ribet and others. We are grateful to H. Darmon, F. Diamond and R. Taylor for their kind permission to publish their beautiful survey article on Wiles' work as the first article in this second edition. During the conference itself, lectures were given by John Coates, Noam Elkies, Matthias Flach, Jean-Marc Fontaine, Gerhard Frey, Dick Gross, Victor Kolyvagin, Ken Ribet, Karl Rubin, Jean-Pierre Serre, John Tate, Richard Taylor, and Don Zagier. The present short volume is a mixture of the texts of some of these lectures, together with a number of recent articles related to the general theme of the conference. Finally, the organizers of the conference wish to express their warmest thanks to Professor Charles Kao, Vice-Chancellor, and to Professor S.Y. Cheng, Chairman of the Mathematics Department, and Dr. K.W. Leung, of The Chinese University of Hong Kong for their support and assistance throughout the preparation and running of the conference. Finally, and most importantly, the organizers wish to express their gratitude to The Chinese University of Hong Kong, the Ho Sin Hang Education Endowment Fund, the Lee Hysan Foundation Limited, the Pei Hua Education Foundation Limited, and the Sing Ho Yu Foundation for their generous financial support for the conference.

John Coates, Cambridge University
Shing-Tung Yau, Director
Institute of Mathematical Sciences
The Chinese University of Hong Kong

Fermat's Last Theorem

Henri Darmon
Department of Mathematics, McGill University, Montreal, QC
Canada H3A 2K6
Fred Diamond
Departament of Mathematics, MIT, 77 Massachusetts Avenue,
Camridge, MA 02139, USA
Richard Taylor
Departament of Mathematics, Harvard University, 1 Oxford Street,
Cambridge, MA 02138, USA

The authors would like to give special thanks to N. Boston, K. Buzzard, and B. Conrad for providing so much valuable feedback on earlier versions of this paper. They are also grateful to A. Agboola, M. Bertolini, B. Edixhoven, J. Fearnley, B.H. Gross, R. Gross, L. Guo, F. Jarvis, H. Kisilevsky, E. Liverance, J. Manoharmayum, K. Ribet, D. Rohrlich, M. Rosen, R. Schoof, J.-P. Serre, C. Skinner, D. Thakur, J. Tilouine, J. Tunnell, A. Van der Poorten, and L. Washington for their helpful comments.

Darmon thanks the members of CICMA and of the Quebec-Vermont Number Theory Seminar for many stimulating conversations on the topics of this paper, particularly in the Spring of 1995. For the same reason Diamond is grateful to the participants in an informal seminar at Columbia University in 1993-94, and Taylor thanks those attending the Oxford Number Theory Seminar in the Fall of 1995.

Parts of this paper were written while the authors held positions at other institutions: Darmon at Princeton, Diamond at Cambridge and the Institute for Advanced Study, and Taylor at Cambridge and Oxford. During some of the period, Diamond enjoyed the hospitality of Princeton University, and Taylor that of MIT. The writing of this paper was also supported by research grants from NSERC (Darmon), EPSRC # GR/J9461 and NSF # DMS 9304580 (Diamond) and by an advanced fellowship from EPSRC (Taylor).

This article owes everything to the ideas of Wiles, and the arguments presented here are fundamentally his [W3], though they include both the work [TW] and several simplifications to the original arguments, most notably that of Faltings. In the hope of increasing clarity, we have not always stated theorems in the greatest known generality, concentrating instead on what is needed for the proof of the Shimura-Taniyama conjecture for semi-stable elliptic curves. This article can serve as an introduction to the fundamental papers [W3] and [TW], which the reader is encouraged to consult for a different, and often more in-depth, perspective on the topics considered. Another useful more advanced reference is the article [Di2] which strengthens the methods of [W3] and [TW] to prove that every elliptic curve that is semistable at 3 and 5 is modular.

For the reprinting of this paper the only changes we have made are to correct a few typos. We have not updated the material to take account of recent developments in the subject, most notably:

- F. Diamond, *The Taylor-Wiles construction and multiplicity one*, Invent. Math., **128** (1997), 379-391 gives a simplification to the arguments which we would have incorporated into this exposition if we were writing it today.

- Some progress has been made on conjecture 3.13 for small primes. See N. Shepherd-Barron and R. Taylor, *Mod 2 and mod 5 icosahedral representations*, J. Amer. Math. Soc., **10** (1997), 283-298.

- The Shimura-Taniyama conjecture is now known for elliptic curves with conductor not divisible by 27. See B. Conrad, F. Diamond, R. Taylor, *Modularity of certain potentially crystalline Galois representations*, in preparation.

Contents

Introduction

Fermat's Last Theorem

Fermat's Last Theorem states that the equation

$$x^n + y^n = z^n, \quad xyz \neq 0$$

has no integer solutions when n is greater than or equal to 3. Around 1630, Pierre de Fermat claimed that he had found a "truly wonderful" proof of this theorem, but that the margin of his copy of Diophantus' *Arithmetica* was too small to contain it:

> "Cubum autem in duos cubos, aut quadrato quadratum in duos quadrato quadratos, et generaliter nullam in infinitum ultra quadratum potestatem in duos ejusdem nominis fas est dividere; cujus rei demonstrationem mirabile sane detexi. Hanc marginis exiguitas non caperet."

Among the many challenges that Fermat left for posterity, this was to prove the most vexing. A tantalizingly simple problem about whole numbers, it stood unsolved for more than 350 years, until in 1994 Andrew Wiles finally laid it to rest.

Prehistory: The only case of Fermat's Last Theorem for which Fermat actually wrote down a proof is for the case $n = 4$. To do this, Fermat introduced the idea of *infinite descent* which is still one the main tools in the study of Diophantine equations, and was to play a central role in the proof of Fermat's Last Theorem 350 years later. To prove his Last Theorem for exponent 4, Fermat showed something slightly stronger, namely that the equation $x^4 + y^4 = z^2$ has no solutions in relatively prime integers with $xyz \neq 0$. Solutions to such an equation correspond to rational points on the elliptic curve $v^2 = u^3 - 4u$. Since every integer $n \geq 3$ is divisible either by an odd prime or by 4, the result of Fermat allowed one to reduce the study of Fermat's equation to the case where $n = \ell$ is an *odd prime*.

In 1753, Leonhard Euler wrote down a proof of Fermat's Last Theorem for the exponent $\ell = 3$, by performing what in modern language we would call a 3-descent on the curve $x^3 + y^3 = 1$ which is also an elliptic curve. Euler's argument (which seems to have contained a gap) is explained in [Edw], ch. 2, and [Dic1], p. 545.

It took mathematicians almost 100 years after Euler's achievement to handle the case $\ell = 5$; this was settled, more or less simultaneously, by Gustav Peter Lejeune Dirichlet [Dir] and Adrien Marie Legendre [Leg] in 1825. Their elementary arguments are quite involved. (Cf. [Edw], sec. 3.3.)

In 1839, Fermat's equation for exponent 7 also yielded to elementary methods, through the heroic efforts of Gabriel Lamé. Lamé's proof was even more intricate than the proof for exponent 5, and suggested that to go further, new theoretical insights would be needed.

The work of Sophie Germain: Around 1820, in a letter to Gauss, Sophie Germain proved that if ℓ is a prime and $q = 2\ell + 1$ is also prime, then Fermat's equation $x^\ell + y^\ell = z^\ell$ with exponent ℓ has no solutions (x, y, z) with $xyz \neq 0$

(mod ℓ). Germain's theorem was the first really general proposition on Fermat's Last Theorem, unlike the previous results which considered the Fermat equation one exponent at a time.

The case where the solution (x, y, z) to $x^\ell + y^\ell = z^\ell$ satisfies $xyz \not\equiv 0$ (mod ℓ) was called the *first case* of Fermat's Last Theorem, and the case where ℓ divides xyz, the *second case*. It was realized at that time that the first case was generally easier to handle: Germain's theorem was extended, using similar ideas, to cases where $k\ell + 1$ is prime and k is small, and this led to a proof that there were no first case solutions to Fermat's equation with prime exponents $\ell \leq 100$, which in 1830 represented a significant advance. The division between first and second case remained fundamental in much of the later work on the subject. In 1977, Terjanian [Te] proved that if the equation $x^{2\ell} + y^{2\ell} = z^{2\ell}$ has a solution (x, y, z), then 2ℓ divides either x or y, i.e., "the first case of Fermat's Last Theorem is true for even exponents". His simple and elegant proof used only techniques that were available to Germain and her contemporaries.

The work of Kummer: The work of Ernst Eduard Kummer marked the beginning of a new era in the study of Fermat's Last Theorem. For the first time, sophisticated concepts of algebraic number theory and the theory of L-functions were brought to bear on a question that had until then been addressed only with elementary methods. While he fell short of providing a complete solution, Kummer made substantial progress. He showed how Fermat's Last Theorem is intimately tied to deep questions on class numbers of cyclotomic fields which are still an active subject of research. Kummer's approach relied on the factorization

$$(x + y)(x + \zeta_\ell y) \cdots (x + \zeta_\ell^{\ell-1} y) = z^\ell$$

of Fermat's equation over the ring $\mathbb{Z}[\zeta_\ell]$ generated by the ℓth roots of unity. One observes that the greatest common divisor of any two factors in the product on the left divides the element $(1 - \zeta_\ell)$, which is an element of norm ℓ. Since the product of these numbers is a perfect ℓ-th power, one is tempted to conclude that $(x+y), \ldots, (x+\zeta_\ell^{\ell-1}y)$ are each ℓ-th powers in the ring $\mathbb{Z}[\zeta_\ell]$ up to units in this ring, and up to powers of $(1 - \zeta_\ell)$. Such an inference would be valid if one were to replace $\mathbb{Z}[\zeta_\ell]$ by $\mathbb{Z}$, and is a direct consequence of *unique factorization* of integers into products of primes. We say that a ring R has property UF if every non-zero element of R is uniquely a product of primes, up to units. Mathematicians such as Lamé made attempts at proving Fermat's Last Theorem based on the mistaken assumption that the rings $\mathbb{Z}[\zeta_\ell]$ had property UF. Legend even has it that Kummer fell into this trap, although this story now has been discredited; see for example [Edw], sec. 4.1. In fact, property UF is far from being satisfied in general: one now knows that the rings $\mathbb{Z}[\zeta_\ell]$ have property UF only for $\ell < 23$ (cf. [Wa], ch. 1).

It turns out that the full force of property UF is not really needed in the applications to Fermat's Last Theorem. Say that a ring R has property UF_ℓ if the following inference is valid:

$$ab = z^\ell, \text{ and } \gcd(a, b) = 1 \Rightarrow a \text{ and } b \text{ are } \ell\text{th powers up to units of } R.$$

If a ring R has property UF, then it also has property UF_ℓ, but the converse need not be true. Kummer showed that Fermat's last theorem was true for exponent ℓ if $\mathbb{Z}[\zeta_\ell]$ satisfied the property UF_ℓ (cf. [Wa]). The proof is far from trivial, because of difficulties arising from the units in $\mathbb{Z}[\zeta_\ell]$ as well as from the possible failure of property UF. (A number of Kummer's contemporaries, such as Cauchy and Lamé, seem to have overlooked both of these difficulties in their attempts to prove Fermat's Last Theorem.)

Kummer then launched a systematic study of the property UF_ℓ for the rings $\mathbb{Z}[\zeta_\ell]$. He showed that even if $\mathbb{Z}[\zeta_\ell]$ failed to have unique factorization, it still possessed unique factorization into prime ***ideals***. He defined the ***ideal class group*** as the quotient of the group of fractional ideals by its subgroup consisting of principal ideals, and was able to establish the finiteness of this class group. The order of the class group of $\mathbb{Z}[\zeta_\ell]$, denoted h_ℓ, could be taken as a measure of the failure of the ring $\mathbb{Z}[\zeta_\ell]$ to satisfy UF. It was rather straightforward to show that if ℓ did not divide h_ℓ, then $\mathbb{Z}[\zeta_\ell]$ satisfied the property UF_ℓ. In this case, one called ℓ a *regular prime.* Kummer thus showed that Fermat's last theorem is true for exponent ℓ if ℓ is a regular prime.

He did not stop here. For it remained to give an efficient means of computing h_ℓ, or at least an efficient way of checking when ℓ divides h_ℓ. The class number h_ℓ can be factorized as a product

$$h_\ell = h_\ell^+ h_\ell^-,$$

where h_ℓ^+ is the class number of the real subfield $\mathbb{Q}(\zeta_\ell)^+$, and h_ℓ^- is defined as h_ℓ/h_ℓ^+. Essentially because of the units in $\mathbb{Q}(\zeta_\ell)^+$, the factor h_ℓ^+ is somewhat difficult to compute, while, because the units in $\mathbb{Q}(\zeta_\ell)^+$ generate the group of units in $\mathbb{Q}(\zeta_\ell)$ up to finite index, the term h_ℓ^- can be expressed in a simple closed form. Kummer showed that if ℓ divides h_ℓ^+, then ℓ divides h_ℓ^-. Hence, ℓ divides h_ℓ if and only if ℓ divides h_ℓ^-. This allowed one to avoid the difficulties inherent in the calculation of h_ℓ^+. Kummer then gave an elegant formula for h_ℓ^- by considering the Bernoulli numbers B_n, which are rational numbers defined by the formula

$$\frac{x}{e^x - 1} = \sum \frac{B_n}{n!} x^n.$$

He produced an explicit formula for the class number h_ℓ^-, and concluded that if ℓ does not divide the numerator of B_{2i}, for $1 \leq i \leq (\ell - 3)/2$, then ℓ is regular, and conversely.

The conceptual explanation for Kummer's formula for h_ℓ^- lies in the work of Dirichlet on the analytic class number formula, where it is shown that h_ℓ^- can be expressed as a product of special values of certain (abelian) L-series

$$L(s, \chi) = \sum_{n=1}^{\infty} \chi(n) n^{-s}$$

associated to odd Dirichlet characters. Such special values in turn can be expressed in terms of certain generalized Bernoulli numbers $B_{1,\chi}$, which are related to the Bernoulli numbers B_i via congruences mod ℓ. (For more details, see [Wa].)

These considerations led Kummer to initiate a deep study relating congruence properties of special values of L-functions and of class numbers, which was to emerge as a central concern of modern algebraic number theory, and was to reappear – in a surprisingly different guise – at the heart of Wiles' strategy for proving the Shimura-Taniyama conjecture.

Later developments: Kummer's work had multiple ramifications, and led to a very active line of enquiry pursued by many people. His formulae relating Bernoulli numbers to class numbers of cyclotomic fields were refined by Kenneth Ribet [R1], Barry Mazur and Andrew Wiles [MW], using new methods from the theory of modular curves which also play a central role in Wiles' more recent work. (Later Francisco Thaine [Th] reproved some of the results of Mazur and Wiles using techniques inspired directly from a reading of Kummer.) In a development more directly related to Fermat's Last Theorem, Wieferich proved that if ℓ^2 does not divide $2^{\ell-1}-1$, then the first case of Fermat's Last Theorem is true for exponent ℓ. (Cf. [Ri], lecture VIII.)

There were many other refinements of similar criteria for Fermat's Last theorem to be true. Computer calculations based on these criteria led to a verification that Fermat's Last theorem is true for all odd prime exponents less than four million [BCEM], and that the first case is true for all $\ell \leq 8.858 \cdot 10^{20}$ [Su].

The condition that ℓ is a regular prime seems to hold heuristically for about 61% of the primes. (See the discussion on p. 63, and also p. 108, of [Wa], for example.) In spite of the convincing numerical evidence, it is still not known if there are infinitely many regular primes. Ironically, it is not too difficult to show that there are infinitely many irregular primes. (Cf. [Wa].)

Thus the methods introduced by Kummer, after leading to very strong results in the direction of Fermat's Last theorem, seemed to become mired in difficulties, and ultimately fell short of solving Fermat's conundrum[1].

Faltings' proof of the Mordell conjecture: In 1985, Gerd Faltings [Fa] proved the very general statement (which had previously been conjectured by Mordell) that any equation in two variables corresponding to a curve of genus strictly greater than one had (at most) finitely many rational solutions. In the context of Fermat's Last Theorem, this led to the proof that for each exponent $n \geq 3$, the Fermat equation $x^n + y^n = z^n$ has at most finitely many integer solutions (up to the obvious rescaling). Andrew Granville [Gra] and Roger Heath-Brown [HB] remarked that Faltings' result implies Fermat's Last Theorem for a set of exponents of density one.

However, Fermat's Last Theorem was still not known to be true for an infinite set of prime exponents. In fact, the theorem of Faltings seemed ill-equipped for dealing with the finer questions raised by Fermat in his margin, namely of finding a complete list of rational points on *all* of the Fermat curves $x^n + y^n = 1$ simultaneously, and showing that there are no solutions on these curves when $n \geq 3$ except the obvious ones.

Mazur's work on Diophantine properties of modular curves: Although it was not realized at the time, the chain of ideas that was to lead to a proof of Fermat's Last theorem had already been set in motion by Barry Mazur in the mid seventies. The modular curves $X_0(\ell)$ and $X_1(\ell)$ introduced in section 1.2 and 1.5 give rise to another naturally occurring infinite family of Diophantine equations. These equations have certain systematic rational solutions corresponding to the cusps that are defined over $\mathbb{Q}$, and are analogous to the

[1] However, W. McCallum has recently introduced a technique, based on the method of Chabauty and Coleman, which suggests new directions for approaching Fermat's Last Theorem via the cyclotomic theory. An application of McCallum's method to showing the *second* case of Fermat's Last Theorem for regular primes is explained in [Mc].

so-called "trivial solutions" of Fermat's equation. Replacing Fermat curves by modular curves, one could ask for a complete list of all the rational points on the curves $X_0(\ell)$ and $X_1(\ell)$. This problem is perhaps even more compelling than Fermat's Last Theorem: rational points on modular curves correspond to objects with natural geometric and arithmetic interest, namely, elliptic curves with cyclic subgroups or points of order ℓ. In [Maz1] and [Maz2], B. Mazur gave essentially a complete answer to the analogue of Fermat's Last Theorem for modular curves. More precisely, he showed that if $\ell \neq 2, 3, 5$ and 7, (i.e., $X_1(\ell)$ has genus > 0) then the curve $X_1(\ell)$ has no rational points other than the "trivial" ones, namely cusps. He proved analogous results for the curves $X_0(\ell)$ in [Maz2], which implied, in particular, that an elliptic curve over $\mathbb{Q}$ with square-free conductor has no rational cyclic subgroup of order ℓ over $\mathbb{Q}$ if ℓ is a prime which is strictly greater than 7. This result appeared a full ten years before Faltings' proof of the Mordell conjecture.

Frey's strategy: In 1986, Gerhard Frey had the insight that these constructions might provide a precise link between Fermat's Last Theorem and deep questions in the theory of elliptic curves, most notably the Shimura Taniyama conjecture. Given a solution $a^\ell + b^\ell = c^\ell$ to the Fermat equation of prime degree ℓ, we may assume without loss of generality that $a^\ell \equiv -1 \pmod 4$ and that $b^\ell \equiv 0 \pmod{32}$. Frey considered (following Hellegouarch, [He], p. 262; cf. also Kubert-Lang [KL], ch. 8, §2) the elliptic curve

$$E : y^2 = x(x - a^\ell)(x + b^\ell).$$

This curve is *semistable*, i.e., it has square-free conductor. Let $E[\ell]$ denote the group of points of order ℓ on E defined over some (fixed) algebraic closure $\bar{\mathbb{Q}}$ of $\mathbb{Q}$, and let L denote the smallest number field over which these points are defined. This extension appears as a natural generalization of the cyclotomic fields $\mathbb{Q}(\zeta_\ell)$ studied by Kummer. What singles out the field L for special attention is that it has *very little ramification*: using Tate's analytic description of E at the primes dividing abc, it could be shown that L was ramified only at 2 and ℓ, and that the ramification of L at these two primes was rather restricted. (See theorem 2.15 of section 2.2 for a precise statement.) Moreover, the results of Mazur on the curve $X_0(\ell)$ could be used to show that L is *large*, in the following precise sense. The space $E[\ell]$ is a vector space of dimension 2 over the finite field $\mathbb{F}_\ell$ with ℓ elements, and the absolute Galois group $G_\mathbb{Q} = \mathrm{Gal}\,(\bar{\mathbb{Q}}/\mathbb{Q})$ acts $\mathbb{F}_\ell$-linearly on $E[\ell]$. Choosing an $\mathbb{F}_\ell$-basis for $E[\ell]$, the action is described by a representation

$$\bar{\rho}_{E,\ell} : \mathrm{Gal}\,(L/\mathbb{Q}) \hookrightarrow GL_2(\mathbb{F}_\ell).$$

Mazur's results in [Maz1] and [Maz2] imply that $\bar{\rho}_{E,\ell}$ is *irreducible* if $\ell > 7$ (using the fact that E is *semi-stable*). In fact, combined with earlier results of Serre [Se6], Mazur's results imply that for $\ell > 7$, the representation $\bar{\rho}_{E,\ell}$ is surjective, so that $\mathrm{Gal}\,(L/\mathbb{Q})$ is actually isomorphic to $GL_2(\mathbb{F}_\ell)$ in this case.

Serre's conjectures: In [Se7], Jean-Pierre Serre made a careful study of mod ℓ Galois representations $\bar{\rho} : G_\mathbb{Q} \longrightarrow GL_2(\mathbb{F}_\ell)$ (and, more generally, of representations into $GL_2(k)$, where k is any finite field). He was able to make very precise conjectures (see section 3.2) relating these representations to modular forms mod ℓ. In the context of the representations $\bar{\rho}_{E,\ell}$ that occur in Frey's construction, Serre's conjecture predicted that they arose from modular forms (mod ℓ) of weight two and level two. Such modular forms, which correspond to differentials on the modular curve $X_0(2)$, do not exist because $X_0(2)$ has genus 0. Thus Serre's conjecture implied Fermat's Last Theorem. The link between fields with Galois groups contained in $GL_2(\mathbb{F}_\ell)$ and modular forms mod ℓ still appears to be very deep, and Serre's conjecture remains a tantalizing open problem.

Ribet's work: lowering the level: The conjecture of Shimura and Taniyama (cf. section 1.8) provides a direct link between elliptic curves and modular forms. It predicts that the representation $\bar{\rho}_{E,\ell}$ obtained from the ℓ-division points of the Frey curve arises from a modular form of weight 2, albeit a form whose level is quite large. (It is the product of all the primes dividing abc, where $a^\ell + b^\ell = c^\ell$ is the putative solution to Fermat's equation.) Ribet [R5] proved that, if this were the case, then $\bar{\rho}_{E,\ell}$ would *also* be associated with a modular form mod ℓ of weight 2 and level 2, in the way predicted by Serre's conjecture. This deep result allowed him to reduce Fermat's Last Theorem to the Shimura-Taniyama conjecture.

Wiles' work: proof of the Shimura-Taniyama conjecture: In [W3] Wiles proves the Shimura-Taniyama conjecture for semi-stable elliptic curves, providing the final missing step and proving Fermat's Last Theorem. After more than 350 years, the saga of Fermat's Last theorem has come to a spectacular end.

The relation between Wiles' work and Fermat's Last Theorem has been very well documented (see, for example, [R8], and the references contained therein). Hence this article will focus primarily on the breakthrough of Wiles [W3] and Taylor-Wiles [TW] which leads to the proof of the Shimura-Taniyama conjecture for semi-stable elliptic curves.

From elliptic curves to ℓ-adic representations: Wiles' opening gambit for proving the Shimura-Taniyama conjecture is to view it as part of the more general problem of relating two-dimensional Galois representations and modular forms. The Shimura-Taniyama conjecture states that if E is an elliptic curve over $\mathbb{Q}$, then E is modular. One of several equivalent definitions of modularity is that for some integer N there is an eigenform $f = \sum a_n q^n$ of weight two on $\Gamma_0(N)$ such that

$$\#E(\mathbb{F}_p) = p + 1 - a_p$$

for all but finitely primes p. (By an eigenform, here we mean a cusp form which is a normalized eigenform for the Hecke operators; see section 1 for definitions.)

This conjecture acquires a more Galois theoretic flavour when one considers the two dimensional ℓ-adic representation

$$\rho_{E,\ell} : G_{\mathbb{Q}} \longrightarrow GL_2(\mathbb{Z}_\ell)$$

obtained from the action of $G_{\mathbb{Q}}$ on the ℓ-adic Tate module of E: $T_\ell E = \varprojlim E[l^n](\bar{\mathbb{Q}})$. An ℓ-adic representation ρ of $G_{\mathbb{Q}}$ is said to arise from an eigenform $f = \sum a_n q^n$ with integer coefficients a_n if

$$\mathrm{tr}\,(\rho(\mathrm{Frob}_p)) = a_p,$$

for all but finitely many primes p at which ρ is unramified. Here Frob_p is a Frobenius element at p (see section 2), and its image under ρ is a well-defined conjugacy class.

A direct computation shows that $\#E(\mathbb{F}_p) = p + 1 - \operatorname{tr}(\rho_{E,\ell}(\operatorname{Frob}_p))$ for all primes p at which $\rho_{E,\ell}$ is unramified, so that E is modular (in the sense defined above) if and only if for some ℓ, $\rho_{E,\ell}$ arises from an eigenform. In fact the Shimura-Taniyama conjecture can be generalized to a conjecture that every ℓ-adic representation, satisfying suitable local conditions, arises from a modular form. Such a conjecture was proposed by Fontaine and Mazur [FM].

Galois groups and modular forms

Viewed in this way, the Shimura-Taniyama conjecture becomes part of a much larger picture: the emerging, partly conjectural and partly proven correspondence between certain modular forms and two dimensional representations of $G_{\mathbb{Q}}$. This correspondence, which encompasses the Serre conjectures, the Fontaine-Mazur conjecture, and the Langlands program for GL_2, represents a first step toward a higher dimensional, non-abelian generalization of class field theory.

Two-dimensional representations of $G_{\mathbb{Q}}$: In the first part of this century, class field theory gave a complete description of $G_{\mathbb{Q}}^{\mathrm{ab}}$, the maximal (continuous) abelian quotient of $G_{\mathbb{Q}}$. In fact the Kronecker-Weber theorem asserts that $G_{\mathbb{Q}}^{\mathrm{ab}} \cong \prod_p \mathbb{Z}_p^\times$, and one obtains a complete description of all one-dimensional representations of $G_{\mathbb{Q}}$. In the second half of this century much attention has focused on attempts to understand the whole group $G_{\mathbb{Q}}$, or more precisely to describe all its representations. Although there has been a fair degree of success in using modular forms to construct representations of $G_{\mathbb{Q}}$, less is known about how exhaustive these constructions are. The major results in the latter direction along these lines are the work of Langlands [Ll2] and the recent work of Wiles ([W3] completed by [TW]). Both concern two-dimensional representations of $G_{\mathbb{Q}}$ and give significant evidence that these representations are parametrised (in a very precise sense) by certain modular forms. The purpose of this article is to describe both the proven and conjectural parts of this theory, give a fairly detailed exposition of Wiles' recent contribution and explain the application to Fermat's Last theorem. To make this description somewhat more precise let us distinguish three types of representation.

Artin representations and the Langlands-Tunnell theorem: Continuous representations $\rho : G_{\mathbb{Q}} \to GL_2(\mathbb{C})$ are called (two-dimensional) Artin representations. Such representations necessarily have finite image, and are therefore semi-simple. We restrict our attention to those which are irreducible. They are conjectured to be in bijection (in a precise way) with certain newforms (a special class of eigenforms). Those ρ which are odd (i.e. the determinant of complex conjugation is -1), should correspond to weight 1 holomorphic newforms. Those which are even should correspond to certain non-holomorphic (Maass) newforms. Two partial but deep results are known.

(a) (Deligne-Serre) If f is a holomorphic weight one newform then the corresponding Artin representation can be constructed ([DS]).

(b) (Langlands-Tunnell) If ρ is a two dimensional Artin representation with soluble image then the corresponding modular form exists ([Ll2] and [Tu]).

The proof of the latter result is analytic in nature, invoking the trace formula and the theory of L-functions.

ℓ-adic representations and the Fontaine-Mazur conjecture: By an ℓ-adic representation we shall mean any continuous representation $\rho : G_{\mathbb{Q}} \to GL_2(K)$ which is unramified outside a finite set of primes and where K is a finite extension of $\mathbb{Q}_\ell$ (generalizing slightly the notion of ℓ-adic representation that was introduced before). Given a holomorphic newform f one can attach to f a system of ℓ-adic representations, following Eichler, Shimura, Deligne and Serre. These ℓ-adic representations are called modular. The Fontaine-Mazur conjecture (see [FM]) predicts if ρ is an odd, irreducible, ℓ-adic representation whose restriction to the decomposition group at ℓ is well enough behaved, then ρ is modular. (The restriction on the behaviour of the representation on the decomposition group at ℓ is essential in this conjecture; it is not true that all odd, irreducible two dimensional ℓ-adic representation are modular.) Before Wiles' work almost nothing was known about this conjecture, except that certain very special cases could be deduced from the work of Hecke, Langlands and Tunnell.

Mod ℓ representations and Serre's conjecture: A mod ℓ representation is a continuous representation $\bar{\rho} : G_{\mathbb{Q}} \longrightarrow GL_2(\bar{\mathbb{F}}_\ell)$. For example if $E/\mathbb{Q}$ is an elliptic curve then the action of $G_{\mathbb{Q}}$ on the ℓ-division points of E gives rise to a mod ℓ representation $\bar{\rho}_{E,\ell}$ which is just the reduction modulo ℓ of $\rho_{E,\ell}$. One can use the work of Eichler, Shimura, Deligne and Serre to associate to each mod ℓ eigenform a mod ℓ representation of $G_{\mathbb{Q}}$. The mod ℓ representations which arise in this way are called modular. Serre has conjectured [Se7] that every odd (absolutely) irreducible mod ℓ representation is modular and should arise from a mod ℓ eigenform with certain very specific properties. This conjecture can be thought of as having two parts.

The first asserts that every odd irreducible mod ℓ representation is modular. About this very little is known. It is known for $\bar{\rho} : G_{\mathbb{Q}} \to GL_2(\mathbb{F}_2)$ by work of Hecke. It is also known for $\bar{\rho} : G_{\mathbb{Q}} \to GL_2(\mathbb{F}_3)$. This latter result is an application of the Langlands-Tunnell theorem using the two accidents that there is a section to the homomorphism $GL_2(\mathbb{Z}[\sqrt{-2}]) \twoheadrightarrow GL_2(\mathbb{F}_3)$ and that $GL_2(\mathbb{F}_3)$ is soluble. Partial results for $\bar{\rho} : G_{\mathbb{Q}} \to GL_2(\mathbb{F}_5)$ follow from Wiles' work.

Given a mod ℓ representation arising from a mod ℓ eigenform, the second part of Serre's conjecture predicts the minimal weight and level for that mod ℓ eigenform. Here the situation is much better. There has been a lot of work over the last decade (including ideas from Mazur, Ribet, Carayol and Gross) and the problem is nearly completely resolved (see [Di1]). As was pointed out earlier, Ribet's contribution [R5] implies that, if one can show that the Galois representation $\bar{\rho}_{E,\ell}$ arising from the (semi-stable) Frey curve attached to a solution of Fermat's equation with exponent ℓ is modular, then one can show that this representation does not exist—because it would be modular of weight two and level two— and hence one can deduce Fermat's Last Theorem.

However we have seen that to show $\bar{\rho}_{E,\ell}$ is modular it suffices to show that for some ℓ_0, the ℓ_0-adic representation ρ_{E,ℓ_0} is modular. In particular it suffices to verify that either $\rho_{E,3}$ or $\rho_{E,5}$ is modular. Hence the Shimura-Taniyama conjecture can be reduced to (part of) the Fontaine-Mazur conjecture for $\ell = 3$

and 5. We have seen that for these primes part of Serre's conjecture is known, so it turns out it suffices to prove results of the form "Serre's conjecture for ℓ implies the Fontaine-Mazur conjecture for ℓ". This is the direction of Wiles' work, although nothing quite this general has been proven yet.

Deformation theory: Thus the problem Wiles faces is to show that if ρ is an odd ℓ-adic representation which has irreducible modular reduction $\bar{\rho}$ and which is sufficiently well behaved when restricted to the decomposition group at ℓ, then ρ is modular. In fact he only proves a weakened version of such a result, but one which is sufficient to conclude that all semistable elliptic curves are modular.

Wiles approaches the problem by putting it in a more general setting. On the one hand he considers lifts of $\bar{\rho}$ to representations over complete noetherian local $\mathbb{Z}_\ell$-algebras R. For each finite set of primes Σ, one can consider lifts of type Σ; these are lifts which are well-behaved on a decomposition group at ℓ, and whose ramification at primes not in Σ is rather restricted. In particular, such a lift is unramified outside $\Sigma \cup S$ where S is the set of ramified primes of $\bar{\rho}$. A method of Mazur (see [Maz3]) can then be used to show that if $\bar{\rho}$ is absolutely irreducible, then there is a representation

$$\rho_\Sigma^{\text{univ}} : G_\mathbb{Q} \longrightarrow GL_2(R_\Sigma)$$

which is universal in the following sense. If $\rho : G_\mathbb{Q} \to GL_2(R)$ is a lift of $\bar{\rho}$ of type Σ, then there is a unique local homomorphism $R_\Sigma \longrightarrow R$ such that ρ is equivalent to the pushforward of $\rho_\Sigma^{\text{univ}}$. Thus the equivalence classes of type Σ lifts to $GL_2(R)$ can be *identified* with $\text{Hom}(R_\Sigma, R)$. The local ring R_Σ is called the *universal deformation ring* for representations of type Σ.

On the other hand Wiles constructs a candidate for a universal modular lifting of type Σ

$$\rho_\Sigma^{\text{mod}} : G_\mathbb{Q} \longrightarrow GL_2(\mathbb{T}_\Sigma).$$

The ring $\mathbb{T}_\Sigma$ is constructed from the algebra of Hecke operators acting on a certain space of modular forms. The universal property of R_Σ gives a map $R_\Sigma \to \mathbb{T}_\Sigma$. The problem thus becomes: to show that this map is an isomorphism[2]. In fact, it can be shown to be a surjection without great difficulty, and the real challenge is to prove injectivity, i.e., to show, in essence, that R_Σ is not larger than $\mathbb{T}_\Sigma$.

By an ingenious piece of commutative algebra, Wiles found a numerical criterion for this map to be an isomorphism, and for the ring $\mathbb{T}_\Sigma$ to be a local complete intersection. This numerical criterion seems to be very close to a special case of the Bloch-Kato conjecture [BK]. Wiles further showed (by combining arguments from Galois cohomology and from the theory of congruences between modular forms) that this numerical criterion was satisfied if the minimal version $\mathbb{T}_\emptyset$ of this Hecke algebra (obtained by taking $\Sigma = \emptyset$, i.e., allowing the least possible amount of ramification in the deformations) was a complete intersection. Finally in [TW] it was proved that $\mathbb{T}_\emptyset$ is a complete intersection.

---

[2]Maps of this kind were already considered in [Maz3] and [BM], and it is conjectured in [MT] that these maps are isomorphisms in certain cases, though not in exactly the situations considered by Wiles.

Outline of the paper

Chapter 1 recalls some basic notions from the classical theory of elliptic curves and modular forms, such as modular forms and modular curves over $\mathbb{C}$ and $\mathbb{Q}$, Hecke operators and q-expansions, and Eichler-Shimura theory. The Shimura-Taniyama conjecture is stated precisely in section 1.8.

Chapter 2 introduces the basic theory of representations of $G_{\mathbb{Q}}$. We describe Mazur's deformation theory and begin our study of the universal deformation rings using techniques from Galois cohomology and from the theory of finite flat group schemes. We also recall some basic properties of elliptic curves, both to explain Frey's argument precisely and illustrate the uses of ℓ-adic representations.

Chapter 3 explains how to associate Galois representations to modular forms. We then describe what was known and conjectured about associating modular forms to Galois representations before Wiles' work. After introducing the universal modular lifts of certain mod ℓ representations, we give the proof of Wiles' main theorems, taking for granted certain results of a more technical nature that are proved in the last two chapters.

Chapter 4 explains how to prove the necessary results concerning the structure of Hecke algebras: the generalization by Taylor and Wiles of a result of de Shalit, and the generalization by Wiles of a result of Ribet.

Chapter 5 establishes the fundamental results from commutative algebra discovered by Wiles, following modifications of the approach of Wiles and Taylor-Wiles proposed by Faltings and Lenstra.

1 Elliptic curves and modular forms

1.1 Elliptic curves

We begin with a brief review of elliptic curves. A general reference for the results discussed in this section is [Si1] and [Si2].

An elliptic curve E over a field F is a proper smooth curve over F of genus one with a distinguished F-rational point. If E/F is an elliptic curve and if ω is a non-zero holomorphic differential on E/F then E can be realised in the projective plane by an equation (called a Weierstrass equation) of the form

$$(W) \qquad Y^2Z + a_1XYZ + a_3YZ^2 = X^3 + a_2X^2Z + a_4XZ^2 + a_6Z^3$$

such that the distinguished point is $(0:1:0)$ (sometimes denoted ∞ because it corresponds to the "point at infinity" in the affine model obtained by setting $Z = 1$) and $\omega = \frac{dx}{2y+a_1x+a_3}$. We also define the following quantities associated to (W):

$$\begin{aligned} &b_2 = a_1^2 + 4a_2 \qquad b_4 = 2a_4 + a_1a_3 \qquad b_6 = a_3^2 + 4a_6 \\ &b_8 = a_1^2a_6 + 4a_2a_6 - a_1a_3a_4 + a_2a_3^2 - a_4^2 \\ &\Delta = 9b_2b_4b_6 - b_2^2b_8 - 8b_4^3 - 27b_6^2 \\ &j = (b_2^2 - 24b_4)^3/\Delta. \end{aligned}$$

One can check that the equation (W) defines an elliptic curve if and only if Δ is nonzero. One can also check that such equations define elliptic curves which are isomorphic over $\bar{F}$ if and only if they give the same quantity j. Thus j only depends on E so we will denote it j_E. The quantity Δ depends only on the pair (E, ω) so we shall denote it $\Delta(E, \omega)$. If u belongs to $F^\times$ then $u^{12}\Delta(E, u\omega) = \Delta(E, \omega)$.

An elliptic curve E/F has a natural structure of a commutative algebraic group with the distinguished F-rational point as the identity element.

An algebraic map between two elliptic curves which sends the distinguished point of one to the distinguished point of the other is automatically a morphism of algebraic groups. A map between elliptic curves which has finite kernel (and hence, is generically surjective) is called an *isogeny*.

Elliptic curves over $\mathbb{C}$: If $F = \mathbb{C}$, then the curve E is isomorphic as a complex analytic manifold to the complex torus $\mathbb{C}/\Lambda$, where Λ is a lattice in $\mathbb{C}$, i.e., a discrete $\mathbb{Z}$-submodule of $\mathbb{C}$ of rank 2. The group law on $E(\mathbb{C})$ corresponds to the usual addition in $\mathbb{C}/\Lambda$. In terms of Λ, an affine equation for E in $A^2(\mathbb{C})$ is given by

$$y^2 = 4x^3 + g_2x + g_3,$$

where

$$g_2 = -60 \sum_{\lambda \in \Lambda - \{0\}} \frac{1}{z^4}, \qquad g_3 = -140 \sum_{\lambda \in \Lambda - \{0\}} \frac{1}{z^6}.$$

In terms of this equation, the map from $\mathbb{C}/\Lambda$ to $E(\mathbb{C})$ sends z to $(x, y) = (\wp(z), \wp'(z))$, where $\wp(z)$ is the *Weierstrass $\wp$-function* associated to the lattice Λ. (Cf. [Si1], ch. VI.) The inverse map is given by integrating the holomorphic differential ω, i.e., sending $P \in E(\mathbb{C})$ to the image of $\int_\gamma \omega$ in $\mathbb{C}/\Lambda$, where γ is any path on $E(\mathbb{C})$ from ∞ to P, and Λ is the lattice of periods $\int_\gamma \omega$, where γ

ranges over the integral homology $H_1(E(\mathbb{C}), \mathbb{Z})$. Replacing ω by $u\omega$ changes Λ to $u\Lambda$, so that Λ is determined by E only up to homotheties. We scale Λ so that one of its $\mathbb{Z}$-generators is 1, and another, τ, has strictly positive imaginary part. This gives the analytic isomorphism:

$$E(\mathbb{C}) \simeq \mathbb{C}/\langle 1, \tau \rangle.$$

The complex number τ in the complex upper half plane $\mathcal{H}$ is well defined, modulo the natural action of $SL_2(\mathbb{Z})$ on $\mathcal{H}$ by Möbius transformations. (Thus the set of isomorphism classes of elliptic curves over $\mathbb{C}$ can be *identified* with the quotient $\mathcal{H}/SL_2(\mathbb{Z})$.)

The map $z \mapsto e^{2\pi i z}$ identifies $\mathbb{C}/\langle 1, \tau \rangle$ with $\mathbb{C}^\times/q^{\mathbb{Z}}$, where $q = e^{2\pi i \tau}$ is the *multiplicative Tate period.* The analytic isomorphism

$$E(\mathbb{C}) \simeq \mathbb{C}^\times/q^{\mathbb{Z}}$$

has the virtue of generalizing to the p-adic setting in certain cases, as we will see shortly.

Note that $|q| < 1$. The invariant j can be expressed in terms of q by a convergent power series with *integer* coefficients:

$$j = q^{-1} + 744 + 196884q + \cdots . \tag{1.1.1}$$

The following basic facts are a direct consequence of the analytic theory:

Proposition 1.1 *The subgroup $E[n](\mathbb{C})$ of points of order n on $E(\mathbb{C})$ is isomorphic (non-canonically) to $\mathbb{Z}/n\mathbb{Z} \times \mathbb{Z}/n\mathbb{Z}$. More generally, if F is any field of characteristic zero, the subgroup $E[n](F)$ is contained in $\mathbb{Z}/n\mathbb{Z} \times \mathbb{Z}/n\mathbb{Z}$.*

Proof: The analytic theory shows that $E(\mathbb{C})$ is isomorphic as an abstract group to a product of two circle groups, and the first statement follows. The second statement follows from the Lefschetz principle (cf. [Si1], ch. VI, §6). □

Proposition 1.2 *The endomorphism ring* $\mathrm{End}_{\mathbb{C}}(E)$ *of an elliptic curve over $\mathbb{C}$ is isomorphic either to $\mathbb{Z}$ or to an order in a quadratic imaginary field. The same is true if one replaces $\mathbb{C}$ by any field of characteristic* 0.

Proof: An endomorphism of $E(\mathbb{C}) \simeq \mathbb{C}/\Lambda$ induces multiplication by complex number α on the tangent space. Hence $\mathrm{End}_{\mathbb{C}}(E)$ is isomorphic to the ring of $\alpha \in \mathbb{C}$ satisfying $\alpha\Lambda \subset \Lambda$. Such a ring is isomorphic either to $\mathbb{Z}$ or to a quadratic imaginary order. The corresponding statement for fields of characteristic 0 follows as in the proof of proposition 1.1. □

If $\mathrm{End}_{\mathbb{C}}(E) \otimes \mathbb{Q}$ is a quadratic imaginary field, we say that E has *complex multiplication.*

Remark 1.3 It follows from the arithmetic theory of complex multiplication (cf. [Si2], ch. 1) that any elliptic curve E with complex multiplication is defined over an abelian extension of the quadratic imaginary field $K = \mathrm{End}_{\mathbb{C}}(E) \otimes \mathbb{Q}$. If E is defined over $\mathbb{Q}$, then K has class number one. There are only finitely many elliptic curves over $\mathbb{Q}$ with complex multiplication, up to "twists" (i.e., $\mathbb{C}$-isomorphism).

Elliptic curves over $\mathbb{Q}_p$: Now suppose that E is an elliptic curve defined over the p-adic field $\mathbb{Q}_p$. There is an equation

$$(W^{\min}) \qquad Y^2Z + a_1XYZ + a_3YZ^2 = X^3 + a_2X^2Z + a_4XZ^2 + a_6Z^3$$

for E with the property $a_i \in \mathbb{Z}_p$ for all i and $|\Delta|$ is minimal amongst all such equations for E. Although $(W^{\min})$ is not unique, the associated discriminant depends only on E and is denoted $\Delta_E^{\min}$. Moreover the reduction of $(W^{\min})$ modulo the uniformizer p defines a projective curve $\bar{E}$, which is independent of the particular minimal equation chosen. If (W) is any equation for E with coefficients in $\mathbb{Z}_p$ and with discriminant Δ, then $\Delta_E^{\min}$ divides Δ.

If $\bar{E}$ is a smooth curve we say that E has *good reduction* at p. If $\bar{E}$ has a unique singular point which is a node we say that E has *multiplicative reduction* at p. Otherwise $\bar{E}$ has a unique singular point which is a cusp and we say that E has *additive reduction* at p. If E has good or multiplicative reduction we say that it has *semi-stable* reduction at p, or simply that E is *semi-stable.*

If (W) defines a smooth curve mod p then E has good reduction at p and (W) is a minimal equation. If $\Delta \equiv 0 \bmod p$ but $b_2^2 \not\equiv 24b_4 \bmod p$, then modulo p the equation (W) defines a curve with a node. In this case E has multiplicative reduction at p and (W) is a minimal equation.

Curves with good reduction: In that case p does not divide $\Delta_E^{\min}$, and the reduction $\bar{E}$ is an elliptic curve over $\mathbb{F}_p$.

If q is any power of p, and $\mathbb{F}_q$ is the field with q elements, we define the integer N_q to be the number of solutions to the equation $(W^{\min})$ in the projective plane $\mathbb{P}^2(\mathbb{F}_q)$. Thus N_q is the order of the finite group $\bar{E}(\mathbb{F}_q)$. We define the integer a_q by the formula

$$a_q = q + 1 - N_q.$$

The integers a_q are completely determined by a_p: more precisely, we have

$$(1 - a_pp^{-s} + p^{1-2s})^{-1} = 1 + a_pp^{-s} + a_{p^2}p^{-2s} + a_{p^3}p^{-3s} + \cdots . \tag{1.1.2}$$

We call the expression on the left the (local) L-function associated to E over $\mathbb{Q}_p$, and denote it by $L(E/\mathbb{Q}_p, s)$. Concerning the size of a_p we have the following fundamental result of Hasse, whose proof can be found in [Sil], ch. V, §1:

Theorem 1.4 $|a_p| \leq 2\sqrt{p}$.

A further division among curves of good reduction plays a significant role in our later discussion. We say that E has (good) *ordinary* reduction if p does not divide a_p, and that it has *supersingular* reduction if p divides a_p.

When E has good reduction at p, we define its local conductor at p to be $m_p(E) = 0$.

Curves of multiplicative reduction: Elliptic curves over $\mathbb{Q}_p$ which have multiplicative reduction at p can be understood by using the p-adic analytic description discovered by Tate. More precisely, we can formally invert the power series (1.1.1) expressing j in terms of q, to obtain to a power series for q in j^{-1}, having *integer* coefficients:

$$q = j^{-1} + 744j^{-2} + 750420j^{-3} + 872769632j^{-4} + \cdots . \tag{1.1.3}$$

If E has multiplicative reduction, then $j \in \mathbb{Q}_p$ is non-integral, and hence the power series (1.1.3) converges, yielding a well-defined value of q in $p\mathbb{Z}_p$. This is called Tate's p-adic period associated to E over $\mathbb{Q}_p$. Note that we have $v_p(q) = -v_p(j) = v_p(\Delta_E^{min})$.

We say that E has *split* (resp. *non-split*) multiplicative reduction at p if the two tangent lines to the node on $\bar{E}(\mathbb{F}_p)$ have slopes defined over $\mathbb{F}_p$ (resp. $\mathbb{F}_{p^2}$).

Proposition 1.5 (Tate) *There is a p-adic analytic isomorphism*

$$\Phi : \bar{\mathbb{Q}}_p^\times / q^{\mathbb{Z}} \longrightarrow E(\bar{\mathbb{Q}}_p),$$

which has the property that

$$\sigma(\Phi(x)) = \Phi(\sigma x^{\delta(\sigma)}), \quad \forall \sigma \in G_{\mathbb{Q}_p},$$

where $\delta : G_{\mathbb{Q}_p} \longrightarrow \pm 1$ *is*

- *the trivial character, if E has split multiplicative reduction;*
- *the unique unramified quadratic character of $G_{\mathbb{Q}_p}$, if E has non-split multiplicative reduction.*

The proof of this proposition is explained in [Si2], ch. V, for example.

We define the L-function $L(E/\mathbb{Q}_p, s)$ to be

$$L(E/\mathbb{Q}_p, s) = \begin{cases} (1 - p^{-s})^{-1} & \text{if } E \text{ has split reduction,} \\ (1 + p^{-s})^{-1} & \text{if } E \text{ has non-split reduction.} \end{cases} \tag{1.1.4}$$

In both cases the conductor $m_p(E)$ is defined to be 1.

Curves of additive reduction: If E has additive reduction at p, we simply define

$$L(E/\mathbb{Q}_p, s) = 1. \tag{1.1.5}$$

The conductor $m_p(E)$ is defined to be 2, if $p > 3$. When $p = 2$ or 3, it is determined by a somewhat more complicated recipe, given in [Ta].

Elliptic curves over $\mathbb{Q}$: Let E be an elliptic curve defined over $\mathbb{Q}$. In particular E may be viewed as a curve over $\mathbb{Q}_p$ for every p, and we define its (global) conductor by

$$N_E = \prod_p p^{m_p(E)}.$$

The curve E is said to be *semi-stable* if it is semi-stable over all p-adic fields $\mathbb{Q}_p$. Note that E is semi-stable if and only if its conductor N_E is square-free.

Using the fact that $\mathbb{Q}$ has class number 1, one can show that E has a *global minimal Weierstrass model* ($W^{\min}$) which gives the equation of a minimal Weierstrass model over each $\mathbb{Q}_p$. The associated discriminant, denoted $\Delta_E^{\min}$, depends only on E. The associated differential, denoted $\omega_E^{\rm Neron}$, is called the *Néron differential*. It is well-defined up to sign.

The following, known as the Mordell-Weil theorem, is the fundamental result about the structure of the group of rational points $E(\mathbb{Q})$. (Cf. For example [Sil].)

Theorem 1.6 *The group $E(\mathbb{Q})$ is a finitely generated abelian group. Hence*

$$E(\mathbb{Q}) \simeq T \oplus \mathbb{Z}^r,$$

where T is the (finite) torsion subgroup of $E(\mathbb{Q})$, and $r \geq 0$ is the rank *of E over $\mathbb{Q}$.*

Concerning the possible structure of T, there is the following deep result of Mazur, a variant of which also plays a crucial role in the proof of Fermat's Last Theorem:

Theorem 1.7 *If $E/\mathbb{Q}$ is an elliptic curve, then its torsion subgroup is isomorphic to one of the following possibilities:*

$$\mathbb{Z}/n\mathbb{Z}, \quad 1 \leq n \leq 10, \quad n = 12, \qquad \mathbb{Z}/2n\mathbb{Z} \times \mathbb{Z}/2\mathbb{Z}, \quad 1 \leq n \leq 4.$$

The proof is given in [Maz1] (see also [Maz2]). Thanks to this result, the structure of the torsion subgroup T is well understood. (Recently, the techniques of Mazur have been extended by Kamienny [Kam] and Merel [Mer] to prove uniform boundedness results on the torsion of elliptic curves over general number fields.)

Much more mysterious is the behaviour of the rank r. It is not known if r can be arbitrarily large, although results of Mestre [Mes] and Nagao [Na] show that it is greater or equal to 13 for infinitely many elliptic curves over $\mathbb{Q}$. It turns out that many of the deep results on $E(\mathbb{Q})$ and on r are based on the relation with L-functions.

We define the global L-function of the complex variable s by:

$$L(E/\mathbb{Q}, s) = \prod_p L(E/\mathbb{Q}_p, s). \tag{1.1.6}$$

Exercise 1.8 Using theorem 1.4, show that the infinite product defining the L-function $L(E/\mathbb{Q}, s)$ converges absolutely on the right half plane $\mathrm{Real}(s) > 3/2$.

Conjecture 1.9 (Birch-Swinnerton-Dyer) *The L-function $L(E/\mathbb{Q}, s)$ has an analytic continuation to the entire complex plane,and in particular is analytic at $s = 1$. Furthermore:*

$$\mathrm{ord}_{s=1} L(E/\mathbb{Q}, s) = r.$$

There is also a more precise form of this conjecture, which expresses the leading coefficient of $L(E/\mathbb{Q}, s)$ at $s = 1$ in terms of certain arithmetic invariants of $E/\mathbb{Q}$. For more details, see [Sil], conj. 16.5.

As we will explain in more detail in section 1.8, the analytic continuation of $L(E/\mathbb{Q}, s)$ now follows from the work of Wiles and Taylor-Wiles and a strengthening by Diamond [Di2], for a very large class of elliptic curves over $\mathbb{Q}$, which includes all the semi-stable ones.

Abelian varieties: Elliptic curves admit higher-dimensional analogues, called *abelian varieties*, which also play a role in our discussion. Analytically, the set of complex points on an abelian variety is isomorphic to a quotient $\mathbb{C}^g/\Lambda$, where Λ is a lattice in $\mathbb{C}^g$ of rank $2g$, satisfying the so-called Riemann period relations. A good introduction to the basic theory of abelian varieties can be found in [CS] and [Wel].

1.2 Modular curves and modular forms over $\mathbb{C}$

Modular curves: The group $SL_2(\mathbb{Z})$ of two by two integer matrices of determinant one acts by fractional linear (Möbius) transformations on the complex upper half plane

$$\mathcal{H} = \{z \in \mathbb{C} \mid \mathrm{Im}\,(z) > 0\},$$

equipped with its standard complex analytic structure. The *principal congruence group* $\Gamma(N)$ of level N is the subgroup of matrices in $SL_2(\mathbb{Z})$ which reduce to the identity matrix modulo the positive integer N. A subgroup Γ of $SL_2(\mathbb{Z})$ is called a *congruence group* if it contains $\Gamma(N)$ for some N. The level of Γ is the smallest N for which this is true. The most important examples of congruence groups are:

- The group $\Gamma_0(N)$ consisting of all matrices that reduce modulo N to an upper triangular matrix.
- The group $\Gamma_1(N)$ consisting of all matrices that reduce modulo N to a matrix of the form $\begin{pmatrix} 1 & * \\ 0 & 1 \end{pmatrix}$.
- The principal congruence group $\Gamma(N)$ of level N consisting of all matrices that reduce modulo N to the identity.

Notice the natural inclusions of normal subgroups $\Gamma(N) \subset \Gamma_1(N) \subset \Gamma_0(N)$. The quotient $\Gamma_0(N)/\Gamma_1(N)$ is canonically isomorphic to $(\mathbb{Z}/N\mathbb{Z})^\times$ via

$$\begin{pmatrix} a & b \\ c & d \end{pmatrix} \mapsto d \bmod N.$$

For any subgroup H of $(\mathbb{Z}/N\mathbb{Z})^\times$, we denote by $\Gamma_H(N)$ the group of matrices in $\Gamma_0(N)$ whose image in $\Gamma_0(N)/\Gamma_1(N)$ belongs to H.

If Γ is a congruence subgroup of $SL_2(\mathbb{Z})$, define Y_Γ to be the quotient of the upper half plane $\mathcal{H}$ by the action of Γ. One equips Y_Γ with the analytic structure coming from the projection map $\pi : \mathcal{H} \longrightarrow Y_\Gamma$. (More precisely, if $y = \pi(\tau)$, and $G_\tau \subset \Gamma$ is the stabilizer of τ in Γ, then the local ring $\mathcal{O}_{Y_\Gamma,y}$ is identified with the local ring of germs of holomorphic functions at τ which are invariant under the action of G_τ.) This makes Y_Γ into a connected complex analytic manifold of dimension one, i.e., a Riemann surface. If Γ is $\Gamma_0(N)$ (resp. $\Gamma_1(N)$, or $\Gamma(N)$), we will also denote Y_Γ by $Y_0(N)$ (resp. $Y_1(N)$, or $Y(N)$). One compactifies Y_Γ by adjoining a finite set of *cusps* which correspond to orbits of $\mathbb{P}^1(\mathbb{Q}) = \mathbb{Q} \cup \{\infty\}$ under Γ. Call X_Γ the corresponding compact Riemann surface. (For more details, notably on the definition of the analytic structure on X_Γ at the cusps, see for example [Kn], p. 311, or [Shi2], ch. 1.) It follows from the definition of this analytic structure that the field K_Γ of meromorphic functions on X_Γ is equal to the set of meromorphic functions on $\mathcal{H}$ satisfying

- (Transformation property): $f(\gamma\tau) = f(\tau)$, for all $\gamma \in \Gamma$;
- (Behaviour at the cusps): For all $\gamma \in SL_2(\mathbb{Z})$, the function $f(\gamma\tau)$ has a Puiseux series expansion $\sum_{-m}^{\infty} a_n q^{n/h}$ in fractional powers of $q = e^{2\pi i\tau}$.

Riemann's existence theorem (cf. for example [For], ch. 2) asserts that the analytic structure on X_Γ comes from an algebraic one, i.e., the field K_Γ is a finitely generated extension of $\mathbb{C}$ of transcendence degree 1. Thus we can, and will, view X_Γ as a complex algebraic curve over $\mathbb{C}$. If Γ is $\Gamma_0(N)$ (resp. $\Gamma_1(N)$, or $\Gamma(N)$), we will also denote X_Γ by $X_0(N)$ (resp. $X_1(N)$, or $X(N)$).

Examples and exercises:
1. For $N = 1$, the curve $X_0(N) = X_1(N) = X(N)$ is a curve of genus 0, and its field of functions is the ring $\mathbb{C}(j)$, where j is the classical modular function,

$$j(\tau) = q^{-1} + 744 + 196884q + \cdots, \quad q = e^{2\pi i\tau}.$$

(Cf., for example, [Se4], ch. 7.)

2. For $\tau \in \mathcal{H} \cup \mathbb{P}^1(\mathbb{Q}) = \bar{\mathcal{H}}$, define G_τ to be the stabilizer of τ in $PSL_2(\mathbb{Z})$, and let $e_\tau = \#(G_\tau/(G_\tau \cap \Gamma))$. Show that e_τ depends only on the Γ-orbit of τ in $\bar{\mathcal{H}}$, and that $e_\tau = 1$ for all but finitely many τ in $\bar{\mathcal{H}}/\Gamma$. Using the Riemann-Hurwitz formula (cf. [Ki], sec. 4.3) show that the genus of X_Γ is given by

$$g(\Gamma) = 1 - [PSL_2(\mathbb{Z}) : \Gamma] + \frac{1}{2} \sum_{\tau \in \bar{\mathcal{H}}/\Gamma} (e_\tau - 1).$$

Use this to compute the genus of $X_0(p)$, $X_1(p)$, and $X(p)$ for p prime. For details, see [Shi2], sec. 1.6 or [Ogg].

3. For $\Gamma = \Gamma(2)$, show that X_Γ is isomorphic to $\mathbb{P}^1$, and that Y_Γ is isomorphic to $\mathbb{P}^1 - \{0, 1, \infty\}$. Show that $\Gamma/\langle\pm 1\rangle$ is the free group on the two generators $g_1 = \begin{pmatrix} 1 & 2 \\ 0 & 1 \end{pmatrix}$ and $g_2 = \begin{pmatrix} 1 & 0 \\ 2 & 1 \end{pmatrix}$.

4. Define a homomorphism $\Gamma(2) \longrightarrow \mathbb{Z}/n\mathbb{Z} \times \mathbb{Z}/n\mathbb{Z}$, by sending g_1 to $(1,0)$ and g_2 to $(0,1)$, and let Γ denote its kernel. Show that Γ is not in general a congruence subgroup and that the curve $Y_\Gamma := \mathcal{H}/\Gamma$ is birationally isomorphic to the Fermat curve of degree n with affine equation $x^n + y^n = 1$.

Moduli interpretations: The points in $Y_\Gamma = \mathcal{H}/\Gamma$ can be interpreted as elliptic curves over $\mathbb{C}$ with some extra "level N" structure. More precisely,

- If $\Gamma = \Gamma_0(N)$, then the Γ-orbit of $\tau \in \mathcal{H}$ corresponds to the complex torus $E = \mathbb{C}/\langle 1, \tau\rangle$ with the distinguished cyclic subgroup of order N generated by $\frac{1}{N}$. Thus, points on $Y_0(N)$ parametrize isomorphism classes of pairs (E, C) where E is an elliptic curve over $\mathbb{C}$ and C is a cyclic subgroup of E of order N.

- If $\Gamma = \Gamma_1(N)$, then the Γ-orbit of τ corresponds to the complex torus $E = \mathbb{C}/\langle 1, \tau\rangle$ with the distinguished point of order N given by $\frac{1}{N}$. Hence, points on $Y_1(N)$ parametrize isomorphism classes of pairs (E, P) where now P is a point on E of exact order N.

Remark 1.10 One checks that the above rules set up a bijection between points on Y_Γ and elliptic curves with the appropriate structures, and that the projection $Y_1(N) \longrightarrow Y_0(N)$ sending $\Gamma_1(N)\tau$ to $\Gamma_0(N)\tau$ becomes the "forgetful" map sending (E, P) to $(E, \langle P\rangle)$.

Remark 1.11 (This remark will be used in section 1.3 when discussing Hecke operators.) Define an n-isogeny of Γ-structures to be an n-isogeny of the underlying elliptic curves which sends one Γ-structure to the other. If p is a prime not dividing N, then there are exactly $p+1$ distinct p-isogenies from $(\mathbb{C}/\langle\tau, 1\rangle, \frac{1}{N})$, whose images are the pairs:

$$\left(\mathbb{C}/\left\langle\frac{\tau+i}{p}, 1\right\rangle, \frac{1}{N}\right) \quad (i = 0, \dots, p-1), \qquad \left(\mathbb{C}/\langle p\tau, 1\rangle, \frac{p}{N}\right).$$

If p divides N, then there are only p distinct p-isogenies from $(\mathbb{C}/\langle\tau, 1\rangle, \frac{1}{N})$, since $(\mathbb{C}/\langle p\tau, 1\rangle, \frac{p}{N})$ is not a $\Gamma_1(N)$-structure (the point p/N not being of exact order N on the complex torus $\mathbb{C}/\langle p\tau, 1\rangle$).

Modular forms: Let k be an even positive integer. A *modular form* of weight k on Γ is a holomorphic function f on $\mathcal{H}$ satisfying:

- (Transformation property): $f(\gamma\tau) = (c\tau + d)^k f(\tau)$, for all $\gamma = \begin{pmatrix} a & b \\ c & d \end{pmatrix} \in \Gamma$.
- (Behaviour at the cusps): For all $\gamma \in PSL_2(\mathbb{Z})$, the function $(c\tau + d)^{-k} f(\gamma\tau)$ has a Puiseux series expansion $\sum_0^\infty a_n q^{n/h}$ in fractional powers of $q = e^{2\pi i\tau}$. We call $\sum a_n q^{n/h}$ the Fourier expansion of f at the cusp $\gamma^{-1}(i\infty)$.

A modular form which satisfies the stronger property that the constant coefficient of its Fourier expansion at each cusp vanishes is called a *cusp form*. We denote by $M_k(\Gamma)$ the complex vector space of modular forms of weight k on Γ, and by $S_k(\Gamma)$ the space of cusp forms on Γ. (For examples, see [DI], sec. 2.2 and the references therein, especially, [Shi2], ch. 2.)

This article is mainly concerned with modular forms of weight 2, and hence we will focus our attention on these from now on. A pleasant feature of the case $k = 2$ is that the cusp forms in $S_2(\Gamma)$ admit a direct geometric interpretation as holomorphic differentials on the curve X_Γ.

Lemma 1.12 *The map $f(\tau) \mapsto \omega_f := 2\pi i f(\tau)d\tau$ is an isomorphism between the space $S_2(\Gamma)$ and the space $\Omega^1(X_\Gamma)$ of holomorphic differentials on the curve X_Γ.*

Sketch of proof: One checks that the transformation property satisfied by $f(\tau)$ under Γ causes the expression $f(\tau)d\tau$ to be Γ-invariant, and that the condition of vanishing at the cusps translates into holomorphicity of $f(\tau)d\tau$. (Note, for example, that $2\pi i d\tau = dq/q$, so that ω_f is holomorphic at $i\infty$ precisely when $f(q)$ vanishes at $q = 0$.) □

As a corollary, we find:

Corollary 1.13 *The space $S_2(\Gamma)$ is finite-dimensional, and its dimension is equal to the genus g of X_Γ.*

Proof: This follows directly from the Riemann-Roch theorem, cf. [Ki], sec. 6.3. □

To narrow still further the focus of our interest, we will be mostly concerned with the cases $\Gamma = \Gamma_0(N)$ and $\Gamma_1(N)$. A slightly more general framework is sometimes convenient, so we suppose from now on that Γ satisfies

$$\Gamma_1(N) \subset \Gamma \subset \Gamma_0(N).$$

Such a group Γ is necessarily of the form $\Gamma_H(N)$ for some subgroup H of $(\mathbb{Z}/N\mathbb{Z})^\times$. Because the transformation $\tau \mapsto \tau + 1$ belongs to Γ the forms in $S_2(\Gamma)$ are periodic functions on $\mathcal{H}$ of period 1, and hence their Fourier expansions at $i\infty$ are of the form

$$f(\tau) = \sum_{n>0} a_n q^n, \quad q = e^{2\pi i\tau}, a_n \in \mathbb{C}.$$

The Petersson inner product: The spaces $S_2(\Gamma)$ are also equipped with a natural Hermitian inner product given by

$$\langle f, g\rangle = \frac{i}{8\pi^2}\int_{X_\Gamma} \omega_f \wedge \bar{\omega}_g = \int_{\mathcal{H}/\Gamma} f(\tau)\bar{g}(\tau)dxdy,$$

where $\tau = x + iy$. This is called the Petersson inner product.

The diamond operators: Suppose now that $\Gamma = \Gamma_1(N)$ and let d be an element of $(\mathbb{Z}/N\mathbb{Z})^\times$. The map $\langle d\rangle$ which sends an elliptic curve with Γ-structure (E, P) to the pair (E, dP) gives an automorphism of Y_Γ which extends to X_Γ. It is called the *diamond operator*. For τ in $\mathcal{H}$ and $\gamma = \left(\begin{smallmatrix} a & b \\ c & d\end{smallmatrix}\right)$ in $\Gamma_0(N)$, we have

$$\langle d\rangle(\Gamma\tau) = \Gamma(\gamma\tau).$$

Hence $\langle d\rangle$ acts on $S_2(\Gamma)$, identified with the holomorphic differentials on X_Γ, by the rule

$$\langle d\rangle f(\tau) = (c\tau + d)^{-2} f\left(\frac{a\tau + b}{c\tau + d}\right).$$

In geometric terms, the diamond operators are the Galois automorphisms of the natural (branched) covering $X_1(N) \longrightarrow X_0(N)$ whose Galois group is isomorphic to $\Gamma_0(N)/\langle \pm\Gamma_1(N)\rangle = (\mathbb{Z}/N\mathbb{Z})^\times/\langle\pm 1\rangle$. Given an even Dirichlet character $\chi : (\mathbb{Z}/N\mathbb{Z})^\times \longrightarrow \mathbb{C}^\times$, say that f is a modular form of level N and character χ if it belongs to the χ-eigenspace in $S_2(\Gamma_1(N))$ under this action. Let $S_2(N, \chi)$ denote the space of all such forms. Thus a function f in $S_2(N, \chi)$ is a cusp form on $\Gamma_1(N)$ which satisfies the stronger transformation property:

$$f\left(\frac{a\tau + b}{c\tau + d}\right) = \chi(d)(c\tau + d)^2 f(\tau), \text{ for all } \left(\begin{smallmatrix} a & b \\ c & d\end{smallmatrix}\right) \in \Gamma_0(N).$$

Note that if $\mathbf{1}$ is the trivial character, then $S_2(N, \mathbf{1})$ is canonically identified with $S_2(\Gamma_0(N))$, which we will also denote by $S_2(N)$. Finally note the direct sum decomposition:

$$S_2(\Gamma_1(N)) = \oplus_\chi S_2(N, \chi),$$

where the sum ranges over all the even Dirichlet characters modulo N.

Exercise 1.14 Show that if $f(\tau)$ belongs to $S_2(N)$, then $f(a\tau)$ belongs to $S_2(mN)$, for each integer a dividing m.

Jacobians of modular curves: Let V be the dual space

$$V = S_2(\Gamma)^\vee := \operatorname{Hom}(S_2(\Gamma), \mathbb{C}).$$

It is a complex vector space of dimension $g = \text{genus}(X_\Gamma)$. The integral homology $\Lambda = H_1(X_\Gamma, \mathbb{Z})$ maps naturally to V by sending a homology cycle c to the functional ϕ_c defined by $\phi_c(f) = \int_c \omega_f$. The image of Λ is a lattice in V, i.e., a $\mathbb{Z}$-module of rank $2g$ which is discrete (cf. [Mul], cor. 3.8). Fix a base point $\tau_0 \in \mathcal{H}$, and define the Abel-Jacobi map $\Phi_{\text{AJ}} : X_\Gamma(\mathbb{C}) \longrightarrow V/\Lambda$ by $\Phi_{\text{AJ}}(P)(f) = \int_{\tau_0}^P \omega_f$. Note that this is well-defined, i.e., it does not depend on the choice of path on X_Γ from τ_0 to P, up to elements in Λ.

We extend the map Φ_{AJ} by linearity to the group $\operatorname{Div}(X_\Gamma)$ of divisors on X_Γ, and observe that the restriction of Φ_{AJ} to the group $\operatorname{Div}^0(X_\Gamma)$ of degree 0 divisors does not depend on the choice of base-point τ_0. Moreover we have the Abel-Jacobi theorem:

Theorem 1.15 *The map*

$$\Phi_{\text{AJ}} : \operatorname{Div}^0(X_\Gamma) \longrightarrow V/\Lambda$$

has a kernel consisting precisely of the group $P(X_\Gamma)$ of principal divisors on X_Γ. Hence Φ_{AJ} induces an isomorphism from $\operatorname{Pic}^0(X_\Gamma) := \operatorname{Div}^0(X_\Gamma)/P(X_\Gamma)$ to V/Λ.

For the proof, see [Mul], ch. 3. The quotient V/Λ is a complex torus, and is equal to the group of complex points of an abelian variety. We denote this abelian variety by J_Γ, the *Jacobian variety* of X_Γ over $\mathbb{C}$. If $\Gamma = \Gamma_0(N)$ or $\Gamma_1(N)$, we will also write $J_0(N)$ or $J_1(N)$ respectively for the Jacobian J_Γ.

1.3 Hecke operators and Hecke theory

We maintain our running assumption that Γ satisfies

$$\Gamma_1(N) \subset \Gamma \subset \Gamma_0(N).$$

If p is a prime not dividing the level N, we define the Hecke operator T_p on $S_2(\Gamma)$ by the formula

$$T_p(f) = \frac{1}{p}\sum_{i=0}^{p-1} f\left(\frac{\tau+i}{p}\right) + p\langle p\rangle f(p\tau).$$

We give a more conceptual description of T_p in terms of remark 1.11 in the case $\Gamma = \Gamma_1(N)$. We have

$$\omega_{T_p(f)} = \sum \phi_i^*(\omega_f),$$

where $\phi_i(\tau) = \frac{\tau+i}{p}$, and $\phi_\infty(\tau) = \langle p\rangle p\tau$ represent the $p+1$ curves with Γ-structure that are images of $(\mathbb{C}/\langle\tau, 1\rangle, \frac{1}{N})$ by a p-isogeny, and the ϕ_i^* are the pull-back maps on differential forms on $\mathcal{H}$. (An isogeny of elliptic curves with Γ-structure is simply an isogeny between the underlying curves which sends one Γ-structure to the other.) Such a description makes it evident that $T_p(f)$ belongs to $S_2(\Gamma)$, if f does. In terms of the Fourier expansion of $f = \sum a_n q^n$, the formula for the operator T_p on $S_2(N, \chi)$ is given by:

$$T_p(f) = \sum_{p|n} a_n q^{n/p} + p\chi(p) \sum a_n q^{pn}.$$

If p divides N, then we define the Hecke operator U_p analogously, by summing again over all the cyclic p-isogenies of Γ-structures. Since there are only p of them, the formula becomes simpler:

$$U_p(f) = \frac{1}{p} \sum_{i=0}^{p-1} f\left(\frac{\tau + i}{p}\right) = \sum_{p|n} a_n q^{n/p}.$$

The reader is invited to check that the Hecke operators of the form T_p or U_q commute with each other, and also that they commute with the diamond operators introduced in the previous section.

We extend the definition of the Hecke operators to operators T_{p^n}, with $n > 1$, by the inductive formulae

$$T_{p^{n+1}} = T_p T_{p^n} - \langle p \rangle p T_{p^{n-1}}, \quad \text{if } (p, N) = 1,$$

and $T_{p^n} = U_p^n$ otherwise. We then define the operator T_n, where $n = \prod p_i^{e_i}$ is written as a product of powers of distinct primes p_i, by

$$T_n = \prod_i T_{p_i^{e_i}}.$$

This definition makes the Hecke operators multiplicative, i.e., $T_m T_n = T_{mn}$ if $(m, n) = 1$. (A more conceptual definition of the Hecke operator T_n is that $T_n(f)$ is obtained by summing the pullback of ω_f over the maps describing all the cyclic n-isogenies of Γ-structures.) The relations among the different Hecke operators can be stated succinctly by saying that they obey the following (formal) identity:

$$\prod_{p \nmid N} (1 - T_p p^{-s} + \langle p \rangle p^{1-2s})^{-1} \prod_{p|N} (1 - U_p p^{-s})^{-1} = \sum_n T_n n^{-s}. \tag{1.3.1}$$

The reader can consult [DI], sec. 3 and the references therein (especially [Shi2], ch. 3 or [Kn]) for more details and different points of view on Hecke operators. Let $\mathbb{T}$ be the subring of $\mathrm{End}_{\mathbb{C}}(S_2(\Gamma))$ generated over $\mathbb{C}$ by all the Hecke operators T_p for $p \nmid N$, U_q for $q|N$, and $\langle d \rangle$ acting on $S_2(\Gamma)$.

Definition 1.16 A modular form f is an *eigenform* if it is a simultaneous eigenvector for *all* the Hecke operators in $\mathbb{T}$, i.e., if there exists a $\mathbb{C}$-algebra homomorphism $\lambda : \mathbb{T} \longrightarrow \mathbb{C}$ such that $Tf = \lambda(T)f$, for all $T \in \mathbb{T}$.

A direct calculation shows that the coefficients a_n of an eigenform f can be recovered from the homomorphism λ by the formula:

$$a_n(f) = a_1(f)\lambda(T_n).$$

It follows that the first Fourier coefficient a_1 of a non-zero eigenform is always non-zero, and that the non-trivial eigenspaces for $\mathbb{T}$ are all one-dimensional:

Proposition 1.17 *Given a non-zero algebra homomorphism $\lambda : \mathbb{T} \longrightarrow \mathbb{C}$, there is exactly one eigenform f up to scaling, which satisfies $Tf = \lambda(T)f$, for all $T \in \mathbb{T}$.*

Sketch of proof: The proof of the existence of f is an exercise in commutative algebra (localize $S_2(\Gamma)$ at the kernel of λ), and the uniqueness is clear from the formula above. □

We call an eigenform satisfying $a_1 = 1$ a *normalized* eigenform.

Atkin-Lehner theory: It is natural to ask whether $S_2(\Gamma)$ can be decomposed into a basis consisting of distinct normalized eigenforms. Unfortunately, this is not always possible, as the following exercise illustrates.

Exercise 1.18 Suppose that p^3 divides N exactly. Let $\mathbb{T}'$ be the algebra of Hecke operators (generated by the operators T_q with $q \nmid N/p^3$, and U_q with $q|N/p^3$) acting on $S_2(N/p^3)$. Let $f = \sum_{n=1}^{\infty} a_n q^n$ be a $\mathbb{T}'$-eigenform of level N/p^3 in $S_2(N/p^3)$. Show that the space S_f spanned by the forms $f(\tau)$, $f(p\tau)$, $f(p^2\tau)$, and $f(p^3\tau)$ is contained in $S_2(N)$, and is stable for the action of the Hecke operators T_q, $q \nmid N$, and U_q, $q|N$. Show that S_f has no basis of simultaneous eigenforms for the Hecke algebra $\mathbb{T}$ of level N, so that the action of $\mathbb{T}$ on S_f is not semi-simple.

Let $\mathbb{T}^0$ denote the subalgebra of $\mathbb{T}$ generated only by the "good" Hecke operators T_q with $q \nmid N$, and $\langle d \rangle$.

Proposition 1.19 *If q does not divide N, the adjoint of the Hecke operator T_q with respect to the Petersson scalar product is the operator $\langle q \rangle^{-1} T_q$, and the adjoint of $\langle q \rangle$ is $\langle q \rangle^{-1}$. In particular, the Hecke operators commute with their adjoints.*

Proof: See [Kn], th. 9.18 and 8.22, or [Ogg]. □

Proposition 1.19 implies, by the spectral theorem for commuting operators that commute with their adjoints:

Proposition 1.20 *The algebra $\mathbb{T}^0$ is semi-simple (i.e, it is isomorphic to a product $\mathbb{C} \times \cdots \times \mathbb{C}$ of a certain number of copies of $\mathbb{C}$), and there is a basis of $S_2(\Gamma)$ consisting of simultaneous eigenvectors for the operators T_q.*

Thus, $\mathbb{T}^0$ has the merit of being semi-simple, while $\mathbb{T}$ is not in general. The cost of replacing $\mathbb{T}$ by $\mathbb{T}^0$, however, is that one loses "multiplicity one", i.e., the eigenspaces for $\mathbb{T}^0$ need not be one-dimensional. For example, the space S_f defined in the previous exercise is contained in a single eigenspace for $\mathbb{T}^0$.

The theory of Atkin-Lehner [AL] gives essentially a complete understanding of the structure of the algebra $\mathbb{T}$, and the structure of the space of eigenforms. To motivate the main result, observe that the problem in the exercise above seems to be caused by forms of level N that are coming from forms of lower level N/p^3 by a straightforward operation, and are therefore not "genuinely" of level N. They are the analogues, in the context of modular forms, of non-primitive Dirichlet characters.

Definition 1.21 We define the *old* subspace of $S_2(\Gamma)$ to be the space spanned by those functions which are of the form $g(az)$, where g is in $S_2(\Gamma_1(M))$ for some $M < N$ and aM divides N. We define the *new* subspace of $S_2(\Gamma)$ to be the orthogonal complement of the old subspace with respect to the Petersson scalar product. A normalized eigenform in the new subspace is called a *newform of level N*.

The following result is the main consequence of the theory of Atkin-Lehner. It gives a complete answer to the question of what is the structure of the algebra $\mathbb{T}$ acting on $S_2(\Gamma)$.

Theorem 1.22 *If f is in the new subspace of $S_2(\Gamma)$ and is an eigenvector for all the operators in $\mathbb{T}^0$, then it is also an eigenform for $\mathbb{T}$, and hence is unique up to scaling. More generally, if f is a newform of level $N_f|N$, then the space S_f defined by*

$$S_f = \{g \in S_2(\Gamma) \text{ such that } Tg = \lambda_f(T)g, \text{ for all } T \in \mathbb{T}^0\}$$

is stable under the action of all the Hecke operators in $\mathbb{T}$. It is spanned by the linearly independent forms $f(az)$ where a ranges over the divisors of N/N_f. Furthermore, we have

$$S_2(\Gamma) = \oplus_f S_f,$$

where the sum is taken over all newforms f of some level N_f dividing N.

See [AL] for the proof in the case $\Gamma = \Gamma_0(N)$, and [La2], ch. VIII for the general case. (See also [DI], sec. 6 for an overview.)

Exercise 1.23 Consider the case where $\Gamma = \Gamma_0(22)$. Show that $X_0(22)$ is of genus 2, and hence that $S_2(22)$ has dimension 2. Show that $S_2(22)$ is equal to S_f, where $f = (\eta(\tau)\eta(11\tau))^2$ is a newform of level 11, so that in particular there are no newforms of level 22 on Γ. Show that $\mathbb{T}^0$ is isomorphic to $\mathbb{C}$ in this case, and that $\mathbb{T}$ is isomorphic to the semisimple algebra $\mathbb{C} \times \mathbb{C}$.

Action on homology and Jacobians: Note that the Hecke operators act on $V = S_2(\Gamma)^\vee$ by duality. One checks (cf. [Kn], props. 11.23, 11.24) that the sublattice Λ of V is stable under the action of all the Hecke operators T_n, and of the diamond operators $\langle d \rangle$. Therefore the operators T_n and $\langle d \rangle$ give rise to endomorphisms of the torus V/Λ, and hence the Jacobian variety J_Γ, in a natural way. The involution $\tau \mapsto -\bar{\tau}$ gives rise to an involution on $X_\Gamma(\mathbb{C})$ (which is complex conjugation on the model of X_Γ over $\mathbb{R}$ deduced from the $\mathbb{Q}$- model defined in section 1.5). Since complex conjugation is continuous it also acts on the integral homology $\Lambda = H_1(X_\Gamma(\mathbb{C}), \mathbb{Z})$. Let Λ^+ and Λ^- be the sublattices of Λ on which complex conjugation acts by $+1$ and -1. These are sublattices of Λ of rank g which are stable under the Hecke operators, since complex conjugation commutes with the Hecke action.

A more algebraic description of the action of T_p on J_Γ is given via the notion of an *algebraic correspondence.* A correspondence on a curve X is a divisor C on $X \times X$ which is taken modulo divisors of the form $\{P\} \times X$ and $X \times \{Q\}$. Let π_1 and π_2 denote the projections of $X \times X$ onto each factor. Then the correspondence C induces a map on divisors of X, by setting

$$C(D) = \pi_2(\pi_1^{-1}(D) \cdot C).$$

(For the definition of the intersection $D_1 \cdot D_2$ of two divisors, see [Wel].) The map C preserves divisors of degree 0, and sends principal divisors on X to principal divisors. It gives a well defined algebraic endomorphism of the Jacobian variety Jac (X). Given a correspondence C, its transpose $C^\vee$ is defined to be the divisor of $X \times X$ obtained by interchanging the two factors of $X \times X$. One can define a natural notion of composition for correspondences, and the set of correspondences forms a ring. The general theory of correspondences and the proofs of the above facts are given in [Wel], particularly the second chapter.

The Hecke correspondence T_n is defined as the closure in $X_\Gamma \times X_\Gamma$ of the locus of points (A, B) in $Y_\Gamma \times Y_\Gamma$ such that there is a degree n cyclic isogeny of elliptic curves with Γ-structure from A to B. For example, if p is a prime not dividing N, then T_p is an algebraic curve in $X_1(N) \times X_1(N)$ which is birational to $X_{\Gamma_1(N)\cap\Gamma_0(p)}$. The induced map on divisors in this case satisfies

$$T_p((E,P)) = \sum (E/C, P \bmod C)$$

where the sum runs over the subgroups C of E having order p. Note also that if (A, B) belongs to T_p, then the isogeny dual to $A \longrightarrow B$ gives a p-isogeny from B to $\langle p \rangle A$, so that

$$T_p^\vee = \langle p \rangle^{-1} T_p.$$

1.4 The L-function associated to a cusp form

For this section, let f in $S_2(\Gamma_1(N))$ be a cusp form with Fourier expansion at $i\infty$ given by $\sum_n a_n q^n$. One has the following estimate for the size of the Fourier coefficients a_n:

Theorem 1.24 *The coefficients $a_n \in \mathbb{C}$ satisfy the inequality*

$$|a_n| \le c(f)\sigma_0(n)\sqrt{n},$$

where $c(f)$ is a constant depending only on f, and $\sigma_0(n)$ denotes the number of positive divisors of n.

Sketch of proof: This follows from proposition 1.51 of section 1.7 which relates the p-th Fourier coefficients of eigenforms, for p a prime not dividing the level of Γ, to the number of points on certain abelian varieties over the finite field $\mathbb{F}_p$. The estimates of Hasse and Weil for the number of points on abelian varieties over finite fields (stated in theorem 1.4 of section 1.1 for the special case of elliptic curves; see [Wel], §IV for the general case) thus translate into asymptotic bounds for the Fourier coefficients of these eigenforms. We note that the cruder estimate $|a_n| = O(n)$, which is enough for the purposes of this section, can be derived by a more elementary, purely analytic argument; cf. [Ogg], ch. IV, prop. 16. □

The L-function associated to f is defined by the formula:

$$L(f,s) = \sum a_n n^{-s}.$$

As in exercise 1.8, one can show that the infinite sum defining $L(f, s)$ converges absolutely in the right half-plane Re $(s) > \frac{3}{2}$. A much better insight is gained into the function $L(f, s)$ by noting that it is essentially the Mellin transform of the modular form f. More precisely, if we set $\Lambda(f,s) = N^{s/2}(2\pi)^{-s}\Gamma(s)L(f,s)$, then we have

$$\Lambda(f,s) = N^{s/2}\int_0^\infty f(iy)y^s dy/y \tag{1.4.1}$$

Exercise 1.25 Check the formula above.

This integral representation for $L(f,s)$ gives the analytic continuation of $L(f,s)$ to the entire complex plane. The modular invariance of f translates into a functional equation for $L(f,s)$: more precisely, let w_N be the Atkin-Lehner involution defined by $w_N(\tau) = -1/N\tau$. The reader may check that w_N induces an involution of X_Γ, and hence of $\Omega^1(X_\Gamma) = S_2(\Gamma)$. One finds that $L(f,s)$ satisfies the functional equation:

$$\Lambda(f,s) = -\Lambda(w_N(f), 2-s).$$

For a proof of this, see [Ogg], ch. V, lemma 1. Eigenforms for $\mathbb{T}$ in $S_2(N,\chi)$ have a great importance in the theory because their associated L-functions have an Euler product expansion, in addition to an analytic continuation and functional equation:

Theorem 1.26 *If $f = \sum a_n q^n$ is a normalized eigenform in $S_2(N,\chi)$ for all the Hecke operators, then the L-function $L(f,s) = \sum a_n n^{-s}$ has the Euler product expansion*

$$\prod_{p\nmid N}(1 - a_p p^{-s} + \chi(p)p^{1-2s})^{-1}\prod_{p|N}(1-a_p p^{-s})^{-1}.$$

Proof: This follows directly from equation (1.3.1) of section 1.3. □

If f is a newform of level N, then it is also an eigenform for w_N, so that the functional equation may be viewed as relating $L(f,s)$ and $L(f,2-s)$. We can also state the following more precise version of theorem 1.24 (see lemma 3.2 of [Hi2] for example for parts (b), (c) and (d)).

Theorem 1.27 *Suppose that f is a newform of level N_f and let N_χ denote the conductor of its character χ.*

(a) If p does not divide N_f then $|a_p| \le 2\sqrt{p}$.

(b) If $p||N_f$ and p does not divide N_χ then $a_p^2 = \chi_0(p)$ where χ_0 is the primitive character associated to χ.

(c) If p divides N_f and p does not divide N_f/N_χ then $|a_p| = \sqrt{p}$.

(d) If p^2 divides N_f and p divides N_f/N_χ then $a_p = 0$.

1.5 Modular curves and modular forms over $\mathbb{Q}$

Modular curves: For Γ between $\Gamma_0(N)$ and $\Gamma_1(N)$, the modular curve X_Γ has a model over $\mathbb{Q}$. We describe such a model in the case of $\Gamma = \Gamma_0(N)$; the construction for general Γ follows from similar considerations.

The key remark here is that, as was noted in section 1.2, the complex points on the curve $Y_0(N)$ have a natural interpretation as moduli of elliptic curves together with a cyclic subgroup of order N, given by sending the point $\tau \in \mathcal{H}/\Gamma_0(N)$ to the pair $(\mathbb{C}/\langle 1,\tau\rangle, \langle 1/N\rangle)$.

Consider the "universal elliptic curve"

$$E_j : y^2 + xy = x^3 - \frac{36}{j-1728}x - \frac{1}{j-1728}.$$

It is an elliptic curve over the function field $\mathbb{Q}(j)$, with j-invariant j. Let d be the order of $\mathbb{P}^1(\mathbb{Z}/N\mathbb{Z})$, and let $C_1, \ldots, C_d$ denote the set of all cyclic subgroups of E_j of order N, defined over $\overline{\mathbb{Q}(j)}$, an algebraic closure of $\mathbb{Q}(j)$. Fix one of these subgroups, C. The Galois group $\mathrm{Gal}(\overline{\mathbb{Q}(j)}/\mathbb{Q}(j))$ permutes the C_i in a natural way. Let F_N be the smallest extension of $\mathbb{Q}(j)$ (viewed as embedded in $\overline{\mathbb{Q}(j)}$) with the property that $\sigma(C) = C$, for all $\sigma \in \mathrm{Gal}(\overline{\mathbb{Q}(j)}/F_N)$. It can be seen (cf. [Shi2], thm. 6.6) that the Galois action on the C_i is transitive so that $F_N/\mathbb{Q}(j)$ is of degree d, and that it is a *regular* extension, i.e., $F_N \cap \bar{\mathbb{Q}} = \mathbb{Q}$. Geometrically, F_N can be viewed as the function field of a curve $X_{/\mathbb{Q}}$ over $\mathbb{Q}$, with the inclusion of $\mathbb{Q}(j)$ corresponding to a map from $X_{/\mathbb{Q}}$ to the projective "j-line" over $\mathbb{Q}$. The pair (E_j, C) is an elliptic curve over $\mathbb{Q}(j)$ with a subgroup of order N defined over F_N. Using (E_j, C), each complex point of X gives an elliptic curve over $\mathbb{C}$ with a subgroup of order N, provided the point does not lie over $j = 0$, 1728 or ∞. The resulting map to $X_0(N)$ extends to an isomorphism from X to $X_0(N)$. The curve X thus constructed, together with this identification, is the desired model of $X_0(N)$ over $\mathbb{Q}$.

More concretely, the functions $j = j(\tau)$ and $j_N = j(N\tau)$ are related by a polynomial equation $\Phi_N(j, j_N) = 0$ with coefficients in $\mathbb{Q}$, of bidegree d. The field F_N is the function field of the affine curve $\Phi_N(X, Y) = 0$, and the mapping $\tau \mapsto (j(\tau), j(N\tau))$ gives a birational equivalence between $\mathcal{H}/\Gamma_0(N)$ and the complex curve defined by the equation Φ_N. In practice it is not feasible to write down the polynomial Φ_N, except for certain very small values of N. To study the models over $\mathbb{Q}$ of $X_0(N)$, more indirect methods are needed, which rely crucially on the moduli interpretation of $X_0(N)$. Similar remarks hold for $X_1(N)$.

Models over $\mathbb{Z}$: The work of Igusa [Ig], Deligne-Rapoport [DR], Drinfeld [Dr], and Katz-Mazur [KM] uses the moduli-theoretic interpretation to describe a canonical proper model for X_Γ over Spec $\mathbb{Z}$. These models allow us to talk about the reduction of X_Γ over finite fields $\mathbb{F}_p$, for p prime. The curve has good reduction at primes p not dividing N, with the "non-cuspidal" points of $X_{\Gamma/\mathbb{F}_p}$ corresponding to elliptic curves over $\bar{\mathbb{F}}_p$ with Γ-structure. The singular fibers at primes p dividing N can also be described precisely; an important special case (see [DR]) is that of $\Gamma_0(N)$ with p exactly dividing N. For further discussion and references, see [DI], sec. 8, 9.

From now on, when we write X_Γ, $X_0(N)$, or $X_1(N)$, we will mean the curve over $\mathbb{Q}$ which are the models described above for the complex curves defined in section 1.2.

Remark 1.28 When considering q-expansions, it is more convenient to use a different set of models over $\mathbb{Z}$ for these complex curves. We define $X_1^\mu(N)$ in the case of $\Gamma_1(N)$ as a model over $\mathbb{Z}$ which parametrizes pairs (E, i) where i

is an embedding of μ_N in the (generalized) elliptic curve E. (So $X_1^\mu(N)$ is the model denoted $X_\mu(N)$ in [DI], sec. 9.3, assuming $N > 4$.) For Γ between $\Gamma_0(N)$ and $\Gamma_1(N)$, we define X_Γ^μ as the corresponding quotient of $X_1^\mu(N)$. This model has good reduction at primes p not dividing N, but unlike the models mentioned above, its fibers at primes p dividing N are smooth and irreducible, but not proper. In the case of $\Gamma = \Gamma_0(N)$, the curve $X_{\Gamma,\mathbb{Q}}^\mu$ can be identified with $X_0(N)$. However, this is not the case in general: the cusp ∞ is a rational point of $X_{\Gamma,\mathbb{Q}}^\mu$ but not necessarily of X_Γ.

Jacobians: Weil's theory [Wel] of the Jacobian shows that the Jacobians J_Γ defined in section 1.2 as complex tori also admit models over $\mathbb{Q}$. When we speak of J_Γ, $J_0(N)$ and $J_1(N)$ from now on, we will refer to these as abelian varieties defined over $\mathbb{Q}$. Thus, the points in $J_\Gamma(K)$, for any $\mathbb{Q}$-algebra K, are identified with the divisor classes on X_Γ of degree 0, defined over K.

We let $J_{\Gamma/\mathbb{Z}}$, denote the Néron model of the Jacobian J_Γ over Spec$(\mathbb{Z})$. Using this model we define $J_{\Gamma/A}$ for arbitrary rings A. In particular we can consider $J_{\Gamma/\mathbb{F}_p}$, the reduction of the Jacobian in characteristic p, which is closely related to the reduction of the integral model of the curve X_Γ mentioned above. In particular, if p does not divide the level of Γ, then $J_{\Gamma/\mathbb{F}_p}$ can be identified with the Jacobian of $X_{\Gamma/\mathbb{F}_p}$. For a treatment of the case $\Gamma = \Gamma_0(N)$ with p exactly dividing N, see the appendix of [Maz1]; for more general discussion and references, see [DI], sec. 10, especially sec. 10.3.

Hecke operators: The Hecke operators have a natural moduli interpretation, which was already touched upon in section 1.3. In particular, one finds that the operator T_n arises from a correspondence which is defined over $\mathbb{Q}$, and gives rise to an endomorphism of the Jacobian J_Γ which is defined over $\mathbb{Q}$. This in turn gives rise to an endomorphism of the Néron model $J_{\Gamma/\mathbb{Z}}$, and we can then consider the endomorphism T_n on the reduction of the Jacobian in characteristic p. Recall that if p is a prime not dividing N, we may identify this reduction with the Jacobian of $X_{\Gamma/\mathbb{F}_p}$. (Cf. [MW], ch. 2, sec. 1, prop. 2.) Furthermore, one can show that the moduli-theoretic interpretation of the Hecke operator remains valid in characteristic p; i.e., the endomorphism T_n of $J_{\Gamma/\mathbb{F}_p}$ is induced by a map on divisors satisfying, for all ordinary elliptic curves A with Γ-structure:

$$T_n(A) = \sum_i i(A),$$

where the sum is taken over all cyclic isogenies of degree n. (See [DI], sec. 10.2 for further discussion and references.)

This description allows one to analyse, for example, the Hecke operator T_p over $\mathbb{F}_p$, when $(p, N) = 1$. Let us work with $\Gamma = \Gamma_1(N)$, to illustrate the idea. For a variety X over $\bar{\mathbb{F}}_p$, let ϕ_X be the Frobenius morphism on X defined by raising coordinates to the pth power. Thus if (E, P) corresponds to a point of $X_1(N)_{/\mathbb{F}_p}$, then ϕ_E is an isogeny of degree p from (E, P) to the pair $(E_\infty, P_\infty) = \phi_{X_1(N)}(E, P)$. The graph of $\phi_{X_1(N)}$ in $(X_1(N) \times X_1(N))_{/\mathbb{F}_p}$ is a correspondence of degree p, which we call F. Let F' be the transpose of this correspondence. The endomorphism F of J_Γ induced by F is the Frobenius endomorphism ϕ_{J_Γ}, and the endomorphism F' is the dual endomorphism (in the sense of duality of abelian varieties). Now consider the divisor

$$F'((E, P)) = (E_1, P_1) + \cdots + (E_p, P_p),$$

where the (E_i, P_i) are elliptic curves with Γ-structure in characteristic p. Since ϕ_{E_i} is an isogeny of degree p from (E_i, P_i) to (E, P), we also have the dual isogeny from (E, P) to (E_i, pP_i). If E is ordinary at p, then the points (E_∞, P_∞), (E_1, pP_1), ..., (E_p, pP_p) are a complete list of the distinct curves with Γ-structure which are p-isogenous to (E, P). Hence one has the equality of divisors on $X_1(N)_{/\mathbb{F}_p}$:

$$T_p((E,P)) = (E_\infty, P_\infty) + (E_1, pP_1) + \cdots + (E_p, pP_p) = (F + \langle p \rangle F')((E,P)).$$

Since the ordinary points are dense on $X_1(N)_{/\mathbb{F}_p}$, we deduce that $T_p = (F + \langle p \rangle F')$ as endomorphisms of $J_1(N)_{/\mathbb{F}_p}$. This equation, known as Eichler-Shimura congruence relation, plays a central role in the theory. (For more details, see [DI], sec. 10.2, 10.3.)

Theorem 1.29 *If $p \nmid N$ then the endomorphism T_p of $J_{\Gamma/\mathbb{F}_p}$ satisfies*

$$T_p = F + \langle p \rangle F'.$$

Remark 1.30 This was proved by Eichler [Ei] to hold for all but finitely many p in the case of $\Gamma_0(N)$, and by Shimura ([Shi1], see also [Shi2], ch. 7) in the case of $\Gamma_1(N)$. The fact that it holds for all p not dividing N follows from work of Igusa [Ig].

Modular forms: In the same way that modular curves have models over $\mathbb{Q}$ and over $\mathbb{Z}$, the Fourier coefficients of modular forms also have natural rationality and integrality properties. We start by sketching the proof of:

Theorem 1.31 *The space $S_2(\Gamma)$ has a basis consisting of modular forms with integer Fourier coefficients.*

Proof: The Hecke operators act on the integral homology Λ^+ in a way that is compatible with the action on $S_2(\Gamma)$ and respects the natural (Poincaré) duality between these two spaces. Hence, if $\{\lambda_n\}_{n \in \mathbb{N}}$ is a system of eigenvalues for the T_n, then the λ_n are algebraic integers in some finite extension K of $\mathbb{Q}$, and the system $\{\lambda_n^\sigma\}_{n \in \mathbb{N}}$ is a system of eigenvalues for the T_n for any Galois automorphism σ of $\bar{\mathbb{Q}}/\mathbb{Q}$. Hence, we have shown:

Proposition 1.32 *If $f \in S_2(M, \chi)$ is a newform of some level M dividing N, then its Fourier coefficients lie in a finite extension K of $\mathbb{Q}$. Moreover, if $\sigma \in \mathrm{Gal}(\bar{\mathbb{Q}}/\mathbb{Q})$ is any Galois automorphism, then the Fourier series f^σ obtained by applying σ to the Fourier coefficients is a newform in $S_2(M, \chi^\sigma)$.*

The explicit description of $S_2(\Gamma)$ given in section 1.3 implies that $S_2(\Gamma)$ is spanned by forms having Fourier coefficients which are algebraic integers in some finite (Galois) extension K of $\mathbb{Q}$, and that the space of forms with Fourier coefficients in K is stable under the natural action of $\mathrm{Gal}(K/\mathbb{Q})$ on Fourier expansions. An application of Hilbert's theorem 90 shows that $S_2(\Gamma)$ has a basis consisting of forms with rational Fourier expansions, and the integrality of the Fourier coefficients of eigenforms yields the integrality statement of theorem 1.31. □

We define $S_2(\Gamma, \mathbb{Z})$ to be the space of modular forms with integral Fourier coefficients in $S_2(\Gamma)$. Theorem 1.31 states that $S_2(\Gamma, \mathbb{Z}) \otimes \mathbb{C} = S_2(\Gamma)$. Given any ring A, we define

$$S_2(\Gamma, A) = S_2(\Gamma, \mathbb{Z}) \otimes A,$$

and define $S_2(N, A)$ and $S_2(N, \chi, A)$ (where χ now is a character with values in $A^\times$) in the obvious way. If A is contained in $\mathbb{C}$, the q-expansion principle below allows us to identify $S_2(\Gamma, A)$ with the set of modular forms in $S_2(\Gamma)$ with Fourier coefficients in A.

The q-expansion principle: Because the modular curve $X_0(N)$ has a model over $\mathbb{Q}$, the space of modular forms $S_2(N) = \Omega^1(X_0(N))$ has a natural rational structure, given by considering the differential forms on $X_0(N)$ defined over $\mathbb{Q}$. The fundamental q-expansion principle (see [DR], ch. 7, or [Kat], sec. 1.6) says that these algebraic structures are the same as those obtained analytically by considering q-expansions at ∞. More generally, using the model X_Γ^μ, we obtain the q-expansion principle over $\mathbb{Z}$ for cusp forms on Γ (cf. [DI], sec. 12.3).

Theorem 1.33 *The map $S_2(\Gamma) \longrightarrow \Omega^1(X_\Gamma)$ defined by $f \mapsto \omega_f$ induces an isomorphism from $S_2(\Gamma, \mathbb{Z})$ to $\Omega^1(X_\Gamma^\mu)$. Furthermore, if A is flat over $\mathbb{Z}$ or N is invertible in A, then the induced map $S_2(\Gamma, A) \to \Omega^1(X_{\Gamma,A}^\mu)$ is an isomorphism. Furthermore, if A is a subring of $\mathbb{C}$, then this isomorphism identifies $S_2(\Gamma, A)$ with set of modular forms in $S_2(\Gamma)$ having coefficients in A.*

1.6 The Hecke algebra

It follows directly from the formulas for the Hecke operators acting on q-expansions that the T_n leave $S_2(\Gamma_0(N), \mathbb{Z})$ stable, as well as the subspace of $S_2(N, \chi)$ with coefficients in $\mathbb{Z}[\chi]$. Using the q-expansion principle (theorem 1.33), one can also show ([DI], sec. 12.4) that the diamond operators preserve the spaces of cusp forms on Γ with integral Fourier expansions, and hence that the space $S_2(\Gamma, \mathbb{Z})$ is preserved by all the Hecke operators.

We define $\mathbb{T}_\mathbb{Z}$ to be the ring generated over $\mathbb{Z}$ by the Hecke operators T_n and $\langle d \rangle$ acting on the space $S_2(\Gamma, \mathbb{Z})$. More generally, if A is any ring, we define $\mathbb{T}_A$ to be the A-algebra $\mathbb{T}_\mathbb{Z} \otimes A$. This Hecke ring acts on the space $S_2(\Gamma, A)$ in a natural way. Before studying the structure of the Hecke rings $\mathbb{T}_A$ as we vary the rings A, we note the following general result (Cf. [Shi2], ch. 3.):

Lemma 1.34 *The space $S_2(\Gamma, A)^\vee = \mathrm{Hom}_A(S_2(\Gamma, A), A)$ is a free $\mathbb{T}_A$-module of rank one.*

Sketch of proof: One checks that the pairing $\mathbb{T}_\mathbb{Z} \times S_2(\Gamma, \mathbb{Z}) \longrightarrow \mathbb{Z}$ defined by $(T, f) \mapsto a_1(Tf)$ sets up a perfect, $\mathbb{T}_\mathbb{Z}$-equivariant duality between $\mathbb{T}_\mathbb{Z}$ and $S_2(\Gamma, \mathbb{Z})$. The result for arbitrary A follows. □

Hecke rings over $\mathbb{C}$: If $A = \mathbb{C}$, then the structure of the ring $\mathbb{T} = \mathbb{T}_\mathbb{C}$ is completely described by theorem 1.22. More precisely, if $\mathbb{T}_{\mathbb{C},f}$ denotes the image of the Hecke algebra acting on the space S_f defined in section 1.3, then

$$\mathbb{T}_\mathbb{C} = \oplus_f \mathbb{T}_{\mathbb{C},f},$$

where the direct sum ranges over all distinct newforms f of some level N_f dividing N. Furthermore, the algebra $\mathbb{T}_{\mathbb{C},f}$ can be described explicitly. In particular, if f is a newform of level N then $\mathbb{T}_{\mathbb{C},f}$ is isomorphic to $\mathbb{C}$, but if N_f is not equal to N then the ring $\mathbb{T}_{\mathbb{C},f}$ need not be a semi-simple algebra over $\mathbb{C}$.

Lemma 1.34 in the case $A = \mathbb{C}$ says that $V = S_2(\Gamma)^\vee$ is a free $\mathbb{T}_\mathbb{C}$-module of rank one, but we also have:

Lemma 1.35 *The module $S_2(\Gamma)$ is a free $\mathbb{T}_\mathbb{C}$-module of rank one.*

Proof: Let $g_1, \dots, g_t$ be a complete system of newforms of levels $N_1, \dots, N_t$ dividing N. One can check that the form

$$g = g_1(N/N_1\tau) + \cdots + g_t(N/N_t\tau)$$

generates $S_2(\Gamma)$ as a $\mathbb{T}_\mathbb{C}$-module. The map $\mathbb{T}_\mathbb{C} \longrightarrow S_2(\Gamma)$ defined by $T \mapsto Tg$ gives an isomorphism from $\mathbb{T}_\mathbb{C}$ to $S_2(\Gamma)$, as $\mathbb{T}_\mathbb{C}$-modules. □

Remark 1.36 Lemmas 1.34 and 1.35 imply that $\mathbb{T}_\mathbb{C}$ is a *Gorenstein* $\mathbb{C}$-algebra (of finite rank), i.e., $\mathrm{Hom}_\mathbb{C}(\mathbb{T}_\mathbb{C}, \mathbb{C})$ is isomorphic to $\mathbb{T}_\mathbb{C}$ as a $\mathbb{T}_\mathbb{C}$-module.

Hecke rings over $\mathbb{Q}$: Let $[f]$ be the Galois orbit (under the action of $G_\mathbb{Q} = \mathrm{Gal}\,(\bar{\mathbb{Q}}/\mathbb{Q})$) of a normalized newform f of some level N_f dividing N, and let K_f be the field extension of $\mathbb{Q}$ generated by the Fourier coefficients of f. The space $\oplus_{g\in[f]} S_g$ is a vector space of dimension $[K_f : \mathbb{Q}]\sigma_0(N/N_f)$, which is spanned by modular forms with rational Fourier coefficients. Let $S_{[f]}$ be the $\mathbb{Q}$-subspace of forms in $\oplus_{g\in[f]} S_g$ with rational Fourier coefficients. The space $S_{[f]}$ is stable under the action of $\mathbb{T}_\mathbb{Q}$, and letting $\mathbb{T}_{\mathbb{Q},[f]}$ be the image of $\mathbb{T}_\mathbb{Q}$ acting on $S_{[f]}$, we obtain the direct sum decomposition

$$\mathbb{T}_\mathbb{Q} = \oplus_{[f]} \mathbb{T}_{\mathbb{Q},[f]},$$

where the sum is taken over the distinct $G_\mathbb{Q}$-orbits of normalized newforms f of some level N_f dividing N. If N_f is equal to N, then the algebra $\mathbb{T}_{\mathbb{Q},[f]}$ is isomorphic to the field K_f. If N_f is a proper divisor of N, then, as in the complex case, the algebra $\mathbb{T}_{\mathbb{Q},[f]}$ is a (not necessarily semi-simple) algebra over $\mathbb{Q}$ of rank $\sigma_0(N/N_f)[K_f : \mathbb{Q}]$. The nature of the fields K_f, and in general the structure of $\mathbb{T}_\mathbb{Q}$, is very poorly understood at this stage; for example, one does not know how to characterize the number fields that occur as a K_f for some f (but they are all known to be totally real or CM fields).

The ring $\mathbb{T}_\mathbb{Q}$ acts naturally on the rational homology $H_1(X_\Gamma, \mathbb{Q}) = \Lambda \otimes \mathbb{Q}$, and we have

Lemma 1.37 *The module $\Lambda \otimes \mathbb{Q}$ is free of rank two over $\mathbb{T}_\mathbb{Q}$.*

Sketch of proof: The modules $\Lambda^+ \otimes \mathbb{C} \simeq V$ and $\Lambda^- \otimes \mathbb{C} \simeq V$ are free of rank one over $\mathbb{T}_\mathbb{C}$, by lemma 1.34. This implies that $\Lambda^+ \otimes \mathbb{Q}$ and $\Lambda^- \otimes \mathbb{Q}$ are both free of rank one over $\mathbb{T}_\mathbb{Q}$. □

Hecke rings over $\mathbb{Z}$: The ring $\mathbb{T}_\mathbb{Z}$ is a certain (not necessarily maximal) order in $\mathbb{T}_\mathbb{Q}$. One still has an injection

$$\mathbb{T}_\mathbb{Z} \hookrightarrow \oplus_{[f]} \mathbb{T}_{\mathbb{Z},[f]},$$

where now $\mathbb{T}_{\mathbb{Z},[f]}$ denotes the ring generated over $\mathbb{Z}$ by the Hecke operators acting on $S_{[f]}$. Of course the structure of $\mathbb{T}_\mathbb{Z}$ is even more mysterious than that of $\mathbb{T}_{\mathbf{Q}}$! The ring $\mathbb{T}_{\mathbf{Z}}$ acts naturally on Λ, but it is not the case in general that Λ is free of rank two over $\mathbb{T}_{\mathbf{Z}}$, i.e., that the integral analogue of lemma 1.37 is true. (See remark 1.42 below.)

Hecke rings over $\mathbf{Q}_\ell$: The study of the algebras $\mathbb{T}_{\mathbf{Z}_\ell}$ and $\mathbb{T}_{\mathbf{Q}_\ell}$ arises naturally because of the Hecke action on the Tate module $\mathcal{T}_\ell(J_\Gamma)$,

$$\mathcal{T}_\ell(J_\Gamma) := \varprojlim (J_\Gamma)[\ell^n],$$

where the inverse limit is taken with respect to the multiplication by ℓ maps. The action of $\mathbb{T}_{\mathbf{Z}_\ell}$ on $\mathcal{T}_\ell(J_\Gamma)$ is compatible with that of $G_{\mathbf{Q}}$, and it is this pair of actions on the Tate module which is used to associate two-dimensional Galois representations to modular forms.

It will sometimes be more convenient to consider the ring $\mathbb{T}_{\mathbf{Q}_\ell}$ and its action on

$$\mathcal{V} = \mathcal{T}_\ell(J_\Gamma) \otimes_{\mathbf{Z}_\ell} \mathbf{Q}_\ell.$$

We first record a useful duality property enjoyed by the Tate modules. The Weil pairings on the groups $J_\Gamma[\ell^n]$ for $n \geq 1$ induce a perfect pairing

$$\langle\ ,\ \rangle : \mathcal{T}_\ell(J_\Gamma) \times \mathcal{T}_\ell(J_\Gamma) \to \mathbf{Z}_\ell.$$

Since each Hecke operator T is adjoint to wTw where $w = w_N$ is the Atkin-Lehner involution, we have the following lemma.

Lemma 1.38 *The map $x \mapsto \phi_x$ where $\phi_x(y) = \langle x, wy\rangle$ defines an isomorphism of $\mathbb{T}_{\mathbf{Z}_\ell}$-modules,*

$$\mathcal{T}_\ell(J_\Gamma) \cong \mathcal{T}_\ell(J_\Gamma)^\vee = \mathrm{Hom}_{\mathbf{Z}_\ell}(\mathcal{T}_\ell(J_\Gamma), \mathbf{Z}_\ell),$$

and hence an isomorphism of $\mathbb{T}_{\mathbf{Q}_\ell}$-modules

$$\mathcal{V} \cong \mathcal{V}^\vee = \mathrm{Hom}_{\mathbf{Q}_\ell}(\mathcal{V}, \mathbf{Q}_\ell).$$

The following lemma allows us to regard $\mathbb{T}_{\mathbf{Q}_\ell}$ as a coefficient ring for a two-dimensional Galois representation

$$\mathrm{Gal}\,(\bar{\mathbf{Q}}/\mathbf{Q}) \to \mathrm{Aut}_{\,\mathbb{T}_{\mathbf{Q}_\ell}}(\mathcal{V}) \cong GL_2(\mathbb{T}_{\mathbf{Q}_\ell}).$$

Lemma 1.39 *The module $\mathcal{V}$ is free of rank 2 over $\mathbb{T}_{\mathbf{Q}_\ell}$.*

Proof. Lemma 1.37 implies directly that the module $\mathcal{V} = \Lambda \otimes \mathbf{Q}_\ell$ is free of rank two over $\mathbb{T}_{\mathbf{Q}_\ell}$. □

Corollary 1.40 *The ring $\mathbb{T}_{\mathbf{Q}_\ell}$ is a Gorenstein $\mathbf{Q}_\ell$-algebra; i.e.,*

$$\mathbb{T}_{\mathbf{Q}_\ell}^\vee = \mathrm{Hom}_{\mathbf{Q}_\ell}(\mathbb{T}_{\mathbf{Q}_\ell}, \mathbf{Q}_\ell)$$

is free of rank one over $\mathbb{T}_{\mathbf{Q}_\ell}$.

Proof. Choosing a basis for $\mathcal{V}$ over $\mathbb{T}_{\mathbf{Q}_\ell}$, we obtain an isomorphism

$$\mathbb{T}_{\mathbf{Q}_\ell} \oplus \mathbb{T}_{\mathbf{Q}_\ell} \cong \mathbb{T}_{\mathbf{Q}_\ell}^\vee \oplus \mathbb{T}_{\mathbf{Q}_\ell}^\vee.$$

Decomposing $\mathbb{T}_{\mathbb{Q}_\ell}$ as $\prod_i R_i$ where each factor R_i is a finite-dimensional local $\mathbb{Q}_\ell$-algebra, we obtain an isomorphism

$$R_i \oplus R_i \cong R_i^\vee \oplus R_i^\vee$$

for each i. At least one of the four maps $R_i^\vee \longrightarrow R_i$ deduced from this isomorphism must be surjective, and by counting dimensions, we see that it must be injective as well. It follows that $\mathbb{T}_{\mathbb{Q}_\ell}$ is isomorphic to $\mathbb{T}_{\mathbb{Q}_\ell}^\vee$. □

Recall that for primes p not dividing N, the Jacobian J_Γ has good reduction mod p, and the Eichler-Shimura relation, theorem 1.29, states that on $J_{\Gamma/\mathbb{F}_p}$, we have

$$T_p = F + \langle p \rangle F'.$$

For primes p not dividing $N\ell$, we may identify $\mathcal{T}_\ell(J_\Gamma)$ with the ℓ-adic Tate module of the reduction (see [ST]) and consider the Frobenius endomorphism F on the free rank two $\mathbb{T}_{\mathbb{Q}_\ell}$-module $\mathcal{V}$. As a consequence of the Eichler-Shimura relation, we find:

Theorem 1.41 *For p not dividing $N\ell$, the characteristic polynomial of F on the $\mathbb{T}_{\mathbb{Q}_\ell}$-module $\mathcal{V}$ is*

$$X^2 - T_p X + \langle p \rangle p.$$

Proof: (We are grateful to Brian Conrad for showing us this argument.) Since $FF' = p$, it follows from the Eichler-Shimura relation that

$$F^2 - T_p F + \langle p \rangle p = 0.$$

To conclude that this is in fact the characteristic polynomial, it suffices to compute the trace of F. To do so, we use the $\mathbb{T}_{\mathbb{Q}_\ell}$ isomorphism

$$\mathcal{V} \to \mathcal{V}^\vee$$

defined by the modified pairing $\langle \cdot, w \cdot \rangle$. Under this modified pairing, F is adjoint to $wF'w = \langle p \rangle F'$, so the trace of F on $\mathcal{V}$ is the same as that of $\phi \mapsto \phi \circ (\langle p \rangle F')$ on $\mathcal{V}^\vee$. Choosing bases for $\mathcal{V}$ and $\mathbb{T}_{\mathbb{Q}_\ell}^\vee$, one sees that this is the same as the trace of $\langle p \rangle F'$. Hence

$$2 \operatorname{tr} F = \operatorname{tr} F + \operatorname{tr} (\langle p \rangle F') = \operatorname{tr} (T_p) = 2 T_p.$$

□

Hecke rings over $\mathbb{Z}_\ell$: The ring $\mathbb{T}_{\mathbb{Z}_\ell}$ is free of finite rank over $\mathbb{Z}_\ell$. It therefore decomposes as

$$\mathbb{T}_{\mathbb{Z}_\ell} = \prod \mathbb{T}_\mathfrak{m},$$

where the product runs over the maximal ideals $\mathfrak{m}$ of $\mathbb{T}_{\mathbb{Z}_\ell}$, and $\mathbb{T}_\mathfrak{m}$ is the localization of $\mathbb{T}_{\mathbb{Z}_\ell}$ at $\mathfrak{m}$. For each $\mathfrak{m}$, $\mathbb{T}_\mathfrak{m}$ is a complete local $\mathbb{Z}_\ell$-algebra, free of finite rank as a $\mathbb{Z}_\ell$-module.

Remark 1.42 While the analogue of lemma 1.39 does not always hold for $\mathbb{T}_{\mathbb{Z}_\ell}$ (see [MRi], sec. 13) we shall see that it holds for certain localizations $\mathbb{T}_\mathfrak{m}$. Results of this type are much deeper than lemma 1.39 and were first obtained by Mazur [Maz1], sec. 14, 15. We shall return in chapter 4 to explain Mazur's result and its generalizations, which play a role in the arguments of [W3] and [TW].

Example 1.43 The curve $X_0(19)$ has genus 1, and $X_0(57)$ has genus 5. By consulting the tables in the Antwerp volume [Ant4] or Cremona's book [Cr], one finds that there is exactly one newform of level 19, and that there are three newforms of level 57, which all have rational Fourier coefficients. Their Fourier coefficients a_p, for the first few primes p, are listed in the following table:

	2	3	5	7	11	13	17	19	23	29	31
$19A$	0	−2	3	−1	3	−4	−3	1	0	6	−4
$57A$	−2	−1	−3	−5	1	2	−1	−1	−4	−2	−6
$57B$	1	1	−2	0	0	6	−6	−1	4	2	8
$57C$	−2	1	1	3	−3	−6	3	−1	4	−10	2

Setting $f = 19A$, we find that a basis of simultaneous eigenforms in $S_2(57, \mathbb{C})$ for the Hecke operators T_p, $l \neq 3, 19$, and U_3, U_{19} is:

$$f(\tau) + (1 + \sqrt{-2})f(3\tau), \quad f(\tau) + (1 - \sqrt{-2})f(3\tau), \quad 57A, \quad 57B, \quad 57C.$$

It appears from the table that the Fourier coefficients corresponding to the forms $57B$ and $57C$ are congruent modulo 3. This is in fact the case. One finds that the Hecke ring $\mathbb{T}_{\mathbb{Z}_3}$ generated over $\mathbb{Z}_3$ by the Hecke operators acting on $S_2(57, \mathbb{Z}_3)$ is isomorphic to the subalgebra of $\mathbb{Z}_3^5$:

$$\{(x, y, z, t, w) \text{ such that } t \equiv w \pmod 3\}.$$

The isomorphism sends T_p (for $p \neq 3, 19$) to the element

$$(a_p(19A), a_p(19A), a_p(57A), a_p(57B), a_p(57C)).$$

It sends U_3 to $(-1 + \sqrt{-2}, -1 - \sqrt{-2}, -1, 1, 1)$ and U_{19} to $(1, 1, -1, -1, -1)$. Thus there are exactly 4 distinct maximal ideals of $\mathbb{T}_{\mathbb{Z}_3}$, and the localizations at these maximal ideals are the rings $\mathbb{Z}_3$, $\mathbb{Z}_3$, $\mathbb{Z}_3$, and

$$\mathbb{T}_{\mathfrak{m}} = \{(t, w) \text{ such that } t \equiv w \pmod 3\},$$

where $\mathfrak{m}$ is the ideal generated by 3 and $T_n - a_n(57B)$ for all n.

1.7 The Shimura construction

Let $f = \sum a_n q^n$ be an eigenform on Γ with (not necessarily rational) Fourier coefficients, corresponding to a surjective algebra homomorphism $\lambda_f : \mathbb{T}_{\mathbb{Q}} \longrightarrow K_f$, where K_f is the field generated over $\mathbb{Q}$ by the Fourier coefficients of f. We briefly review in this section a construction of Shimura ([Shi2], ch. 7) which associates to f (or rather, to the orbit $[f]$ of f under $G_{\mathbb{Q}}$) an abelian variety A_f defined over $\mathbb{Q}$ and of dimension $[K_f : \mathbb{Q}]$.

Let $I_f \subset \mathbb{T}_{\mathbb{Z}}$ be the ideal $\ker(\lambda_f) \cap \mathbb{T}_{\mathbb{Z}}$. The image $I_f(J_\Gamma)$ is a (connected) subabelian variety of J_Γ which is stable under $\mathbb{T}_{\mathbb{Z}}$ and is defined over $\mathbb{Q}$.

Definition 1.44 The abelian variety A_f associated to f is the quotient

$$A_f = J_\Gamma / I_f J_\Gamma$$

From this definition one sees that A_f is defined over $\mathbb{Q}$ and depends only on $[f]$, and that its endomorphism ring contains $\mathbb{T}_{\mathbb{Z}}/I_f$ which is isomorphic to an order in K_f.

Remark 1.45 Using theorem 1.22, one can show that $J_0(N)$ is isogenous to $\prod_{[f]} A_{[f]}^{\sigma_0(N/N_f)}$.

We now describe the abelian variety A_f as a complex torus. Let V_f be the subspace of $V = S_2(\Gamma)^\vee$ on which $\mathbb{T}$ acts by λ_f. Theorem 1.22 and lemma 1.34 show that V_f is a one-dimensional complex vector space. Let π_f be the orthogonal projection of V to V_f, relative to the Petersson scalar product. The projector π_f belongs naturally to $\mathbb{T}_{K_f}$.

Let $[f]$ be the set of all eigenforms whose Fourier coefficients are Galois conjugate to those of f. The number of forms in $[f]$ is equal to the degree d of K_f over $\mathbb{Q}$. Now, we set

$$V_{[f]} = \oplus_{g\in[f]} V_g, \quad \pi_{[f]} = \sum_{g\in[f]} \pi_g.$$

Note that $\pi_{[f]}$ is simply the orthogonal projection of V to $V_{[f]}$. Note also that $\pi_{[f]}$ belongs to the Hecke algebra $\mathbb{T}_{\mathbb{Q}}$.

Lemma 1.46 *The abelian variety A_f is isomorphic over $\mathbb{C}$ to the complex torus $V_{[f]}/\pi_{[f]}(\Lambda)$, with the map $\pi_{[f]} : V/\Lambda \longrightarrow V_{[f]}/\pi_{[f]}(\Lambda)$ corresponding to the natural projection from J_Γ to A_f.*

In particular, one sees that A_f is an abelian variety of dimension $d = [K_f : \mathbb{Q}]$. Hence if f has rational Fourier coefficients, then the abelian variety A_f is an elliptic curve. This elliptic curve is called the *strong modular elliptic curve* associated to f if also f is a newform of level N and $\Gamma = \Gamma_0(N)$.

Example 1.47 If $\Gamma = \Gamma_0(26)$, one checks that the genus of X_Γ is two, and that there are two distinct normalized eigenforms f_1 and f_2 in $S_2(26)$. From the tables in [Ant4] or [Cr], one sees that f_1 and f_2 have integral Fourier coefficients, whose values for the primes ≤ 31 are:

	2	3	5	7	11	13	17	19	23	29	31
f_1	−1	1	−3	−1	6	1	−3	2	0	6	−4
f_2	1	−3	−1	1	−2	−1	−3	6	−4	2	4

Hence the abelian varieties A_{f_1} and A_{f_2} are elliptic curves. The above table suggests (and this can be checked directly by looking at the equations for these curves given in [Ant4] or [Cr], or by using the discussion in [DO], lemma 2.1) that the Fourier coefficients of f_1 and f_2 are congruent modulo 2. The natural projection $J_0(26) \longrightarrow A_{f_1} \oplus A_{f_2}$ is an isogeny whose kernel is isomorphic to $\mathbb{Z}/2\mathbb{Z} \times \mathbb{Z}/2\mathbb{Z}$, and $J_0(26)$ is *not* isomorphic to $A_{f_1} \oplus A_{f_2}$. More generally, one knows that a Jacobian can never decompose as a non-trivial direct product of two principally polarized abelian varieties, cf. [Maz1], prop. 10.6. (There are no non-zero homomorphisms from A_{f_1} to A_{f_2}, and so a non-trivial product decomposition of J_Γ would have to induce a decomposition as a product of principally polarized abelian varieties.)

Let $T_\ell(A_f)$ be the Tate module of the abelian variety A_f,

$$T_\ell(A_f) := \varprojlim (A_f)[\ell^n],$$

where the inverse limit is taken with respect to the multiplication by ℓ maps. This module is naturally a module for $\mathbb{T}_f \otimes \mathbb{Z}_\ell$, and $\mathcal{T}_\ell(A_f) \otimes \mathbb{Q}_\ell$ is a module for $K_f \otimes \mathbb{Q}_\ell$ as well.

Lemma 1.48 *The module $\mathcal{T}_\ell(A_f) \otimes \mathbb{Q}_\ell$ is a free module of rank 2 over $K_f \otimes \mathbb{Q}_\ell$.*

Proof: This follows directly from lemma 1.39. □

Proposition 1.49 *The algebra* $\mathrm{End}_{\mathbb{Q}}(A_f) \otimes \mathbb{Q}$ *is isomorphic to K_f. In particular A_f is simple over $\mathbb{Q}$.*

The proof is given in [R2], cor. 4.2. The main ingredient is the *irreducibility* of the Galois representation attached to A_f as in section 3.1. (Cf. theorem 3.1.)

Properties of A_f: good reduction:

Theorem 1.50 *If f is an eigenform of level N, and p is a prime that does not divide N, then the abelian variety A_f has good reduction at p.*

Proof: This follows from the fact that J_Γ has good reduction at such primes, which in turn is a consequence of the good reduction of X_Γ. □

If p is a prime not dividing N, it then becomes natural to study the number $N_{f,p}$ of points on the abelian variety A_f over the finite field $\mathbb{F}_p$. It is given by the following formula:

Proposition 1.51 *The number of points $N_{f,p}$ is given by the formula*

$$N_{f,p} = \mathrm{Norm}_{K_f/\mathbb{Q}}(\lambda_f(1 - a_p(f) + \langle p \rangle p)),$$

where $a_f(p) \in K_f$ is the p-th Fourier coefficient of f.

Proof: By Weil's theory [Wel], this number is given by the determinant

$$N_{f,p} = \det(1 - F),$$

where F is the Frobenius endomorphism acting on the ℓ-adic Tate module $\mathcal{T}_\ell(A_f)$. So theorem 1.41 implies

$$\det(1 - F) = \mathrm{Norm}_{K_f/\mathbb{Q}}(\lambda_f(1 - T_p + \langle p \rangle p)),$$

and proposition 1.51 follows. □

Defining the local Hasse-Weil L-function of A_f over $\mathbb{F}_p$ by the formula

$$L(A_f/\mathbb{F}_p, s) = \det(1 - Fp^{-s})^{-1},$$

the proof above gives the formula:

$$L(A_f/\mathbb{F}_p, s) = \prod_\sigma L_p(f^\sigma, s),$$

where the product is taken over all complex embeddings $\sigma : K_f \hookrightarrow \mathbb{C}$, and $L_p(f^\sigma, s)$ is the Euler factor at p in the L-function that was associated to f^σ in section 1.4.

In particular, if $A_f = E$ is an elliptic curve, i.e., f has rational Fourier coefficients and f is on $\Gamma_0(N)$, then the number of points on E over $\mathbb{F}_p$ is given by the formula

$$\#E(\mathbb{F}_p) = p + 1 - a_p(f). \tag{1.7.1}$$

Properties of A_f: bad reduction: A fundamental quantity associated to any abelian variety A over $\mathbb{Q}$ is its *arithmetic conductor*, which measures the amount of bad reduction of A. If we factor this conductor as a product of prime powers, $\prod_p p^{m_p}$, then the exponents m_p are equal to 0 precisely when A has good reduction at p. The definition of $m_p(A)$ was already given in section 1.1 when A is an abelian variety of dimension 1, i.e., an elliptic curve. In general, the exponent m_p coincides with $m_p(\rho_{A,\ell})$ (see section 2.1 below), where $\rho_{A,\ell}$ is the Galois representation on the ℓ-adic Tate module of A.

Regarding the bad reduction of the abelian variety A_f, one has the following consequence of the results of Langlands, Deligne and Carayol [Ca3] discussed below in section 3.1.

Theorem 1.52 *If f is a newform of level N, then the conductor of the abelian variety A_f is equal to N^g.*

1.8 The Shimura-Taniyama conjecture

Let E be any elliptic curve defined over $\mathbb{Q}$, and let N denote its arithmetic conductor, defined as in section 1.1. Then we have:

Proposition 1.53 *The following are equivalent:*

(a) The curve E is isogenous over $\mathbb{Q}$ to A_f, for some newform f on some congruence group Γ.

(b) The curve E is isogenous over $\mathbb{Q}$ to A_f, for a newform f on $\Gamma_0(N)$.

(c) There is a non-constant morphism defined over $\mathbb{Q}$, from $X_0(N)$ to E.

Sketch of proof: The implication $(b) \Rightarrow (a)$ is immediate, and $(a) \Rightarrow (b)$ follows from the work of Carayol [Ca3] (cf. proposition 3.20 below). If (b) holds, then there is a surjective map $J_0(N) \longrightarrow E$. Composing with the Abel-Jacobi map $X_0(N) \longrightarrow J_0(N)$ gives a map to E which is not constant, by the definition of the Jacobian, and hence (c) holds. Conversely, a non-constant map of curves $\pi : X_0(N) \longrightarrow E$ induces, by Albanese (covariant) functoriality, a surjective map of Jacobians $\pi_* : J_0(N) \longrightarrow E$, where we have identified E with its own Jacobian in the natural way. Since $J_0(N)$ is isogenous to a product (possibly with multiplicities) of abelian varieties A_f where f run over newforms of level N_f dividing N (cf. remark 1.45), it follows that there is a surjective map $A_f \longrightarrow E$ for some A_f. Since A_f is simple (proposition 1.49), this map is an isogeny, and part (b) follows. □

We call an elliptic curve over $\mathbb{Q}$ satisfying the equivalent properties above a *modular elliptic curve.* A startling conjecture that was first proposed by Taniyama in the 1950's and made more precise by Shimura predicts that the Shimura construction explained in the previous section is surjective:

Conjecture 1.54 *All elliptic curves defined over $\mathbb{Q}$ are modular.*

This conjecture is now known to be true for a very wide class of elliptic curves. The results of Wiles [W3] and Taylor-Wiles [TW] imply it is true for all semi-stable elliptic curves, and a strengthening of the method, [Di2], implies the conjecture for all elliptic curves which have semi-stable reduction at 3 and 5.

Remark 1.55 See [DI], sec. 13 for a more thorough list of equivalent forms of the conjecture.

Relation with L-functions: Define numbers $a_p = a_p(E)$ as in section 1.1. Recall the (global) Hasse-Weil L-function defined in section 1.1 by equations (1.1.2), (1.1.4), (1.1.5), and (1.1.6). The following proposition reveals some of the importance of the Shimura-Taniyama conjecture:

Proposition 1.56 *If E is modular, i.e., is associated to a newform f by the Shimura construction, then*

$$L(E/\mathbb{Q}, s) = L(f, s).$$

In particular, $L(E/\mathbb{Q}, s)$ has an analytic continuation to the entire complex plane, and a functional equation.

Sketch of proof: If E is isogenous to an elliptic curve A_f associated to a newform f on $\Gamma_0(N)$ by the Shimura construction, then the two L-functions $L(A_f/\mathbb{Q}, s)$ and $L(E/\mathbb{Q}, s)$ are equal. On the other hand, formula (1.7.1) directly implies that $L(A_f/\mathbb{Q}, s)$ is equal to $L(f, s)$, at least up to finitely many Euler factors (corresponding precisely, by the work of Igusa, to the primes dividing N). The equality of all the Euler factors follows from the work of Deligne, Langlands, and Carayol (cf. [Ca1]). □

Knowing the analytic continuation and functional equation of the Hasse-Weil L-function of an elliptic curve is of great importance in the theory. For example, the conjecture of Birch and Swinnerton-Dyer (which was stated in a weak form in section 1.1, conjecture 1.9) relates arithmetic invariants of E such as the rank of its Mordell-Weil group and the order of its Shafarevich-Tate group to the behaviour at $s = 1$ of $L(E/\mathbb{Q}, s)$. (Note that $s = 1$ is outside the domain of absolute convergence of the infinite product used to define $L(E/\mathbb{Q}, s)$.)

One consequence of this conjecture is that $E(\mathbb{Q})$ is finite if $L(E/\mathbb{Q}, 1)$ is non-zero. This was proved by Coates and Wiles in [CW] for elliptic curves with complex multiplication, and by Kolyvagin [Kol] for modular elliptic curves. (Recently, a different proof for modular elliptic curves has been announced by K. Kato.) Thanks to the breakthroughs of [W3], [TW], and [Di2], the results of Kolyvagin and Kato are now unconditional for a very large class of elliptic curves.

The Shimura-Taniyama conjecture for abelian varieties: We may view conjecture 1.54 as asserting that the map

$$\left\{\begin{array}{c}\text{Newforms of weight 2 on } X_0(N) \\ \text{with rational Fourier coefficients}\end{array}\right\} \longrightarrow \left\{\begin{array}{c}\text{Isogeny classes of elliptic curves} \\ \text{over } \mathbb{Q} \text{ of conductor N}\end{array}\right\}$$

provided by the Shimura construction is a bijection. It is natural to extend this result to all eigenforms, not just those having rational Fourier expansions. We say that an abelian variety defined over $\mathbb{Q}$ is *modular* if it is isogenous to an abelian variety of the form A_f for some newform f on $\Gamma_1(N)$. It would not be reasonable to expect that all abelian varieties over $\mathbb{Q}$ are modular: the abelian varieties arising from the Shimura construction are very special, in many respects. Most importantly, they have a very large ring of endomorphisms. Following Ribet [R7], we make the definition:

Definition 1.57 An abelian variety over $\mathbb{Q}$ of dimension g is said to be of GL_2-type if its endomorphism ring over $\mathbb{Q}$ contains an order in a field of degree g over $\mathbb{Q}$.

Then we have the following generalized Shimura-Taniyama conjecture:

Conjecture 1.58 *Every simple abelian variety A over $\mathbb{Q}$ which is of GL_2-type is modular.*

This generalization of the Shimura-Taniyama conjecture is far from being proved, and tackling it still seems to require some major new ideas, even after the ideas introduced by Wiles.

2 Galois theory

2.1 Galois representations

If F is a perfect field, we will let $\bar{F}$ denote an algebraic closure of F and $G_F = \mathrm{Gal}(\bar{F}/F)$ its absolute Galois group. Recall that G_F is a profinite group: more precisely, $G_F = \varprojlim \mathrm{Gal}(L/F)$ as L runs over finite Galois extensions of F contained in $\bar{F}$. Hence G_F carries a natural topology. If ℓ is a prime different from the characteristic of F we will let $\epsilon_\ell : G_F \to \mathbb{Z}_\ell^\times$ denote the ℓ-adic cyclotomic character, i.e. if ζ is an ℓ-power root of unity in $\bar{F}$ then $\sigma(\zeta) = \zeta^{\epsilon_\ell(\sigma)}$ for all $\sigma \in G_F$. We will simply write ϵ if the choice of ℓ is clear from the context.

The finite fields $\mathbb{F}_p$: The group $G_{\mathbb{F}_p}$ is isomorphic to $\hat{\mathbb{Z}}$, the profinite completion of $\mathbb{Z}$. A natural topological generator is provided by the Frobenius element Frob_p, the automorphism of $\bar{\mathbb{F}}_p$ which raises elements of $\bar{\mathbb{F}}_p$ to their p^{th} power. If ℓ is different from p, then $\epsilon_\ell(\mathrm{Frob}_p) = p$.

The p-adic fields $\mathbb{Q}_p$: The p-adic valuation $v_p : \mathbb{Q}_p^\times \twoheadrightarrow \mathbb{Z}$ extends uniquely to a valuation $v_p : \bar{\mathbb{Q}}_p^\times \twoheadrightarrow \mathbb{Q}$. Let $\mathcal{O}_{\bar{\mathbb{Q}}_p}$ denote the ring of integers of $\bar{\mathbb{Q}}_p$ and $\mathfrak{m}_{\bar{\mathbb{Q}}_p}$ its maximal ideal. The residue field $\mathcal{O}_{\bar{\mathbb{Q}}_p}/\mathfrak{m}_{\bar{\mathbb{Q}}_p}$ is an algebraic closure of $\mathbb{F}_p$, which we will identify with $\bar{\mathbb{F}}_p$. The valuation v_p is compatible with the action of $G_{\mathbb{Q}_p}$, and hence $\mathcal{O}_{\bar{\mathbb{Q}}_p}$ and $\mathfrak{m}_{\bar{\mathbb{Q}}_p}$ are stable under $G_{\mathbb{Q}_p}$. In particular $G_{\mathbb{Q}_p}$ acts on $\mathcal{O}_{\bar{\mathbb{Q}}_p}/\mathfrak{m}_{\bar{\mathbb{Q}}_p}$ and we obtain a map $\varrho : G_{\mathbb{Q}_p} \to G_{\mathbb{F}_p}$, which is in fact a surjection. We call the kernel the *inertia group* and denote it I_p. It has a unique maximal pro-p subgroup: the *wild inertia group*, denoted P_p. There is a canonical isomorphism

$$I_p/P_p \cong \prod_{\ell \neq p} \mathbb{Z}_\ell(1),$$

where the (1) indicates that if $f \in G_{\mathbb{Q}_p}/P_p$ is a lifting of Frobenius and if $\sigma \in I_p/P_p$ then $f\sigma f^{-1} = \sigma^p$.

The inertia group I_p is filtered by a series of subgroups called the higher ramification groups. More precisely, there are closed normal subgroups I_p^u in $G_{\mathbb{Q}_p}$ for any $u \in [-1, \infty]$, with the following properties:

- If $u \leq v$ then $I_p^u \supset I_p^v$, i.e., the I_p^u form a *decreasing filtration.*
- For $u \leq 0$, $I_p^u = I_p$ while $I_p^\infty = \{1\}$.
- $P_p = \bigcup_{u>0} I_p^u$.
- $I_p^u = \bigcap_{v<u} I_p^v$.

These groups are defined in section 3 of chapter IV of [Se2], where they are denoted $G_{\mathbb{Q}_p}^u$ (except that there, $G_{\mathbb{Q}_p}^{-1} = G_{\mathbb{Q}_p}$ whereas we have $I_p^{-1} = I_p$).

The local Kronecker-Weber theorem ([Se2], ch. 14, sec. 7, for example) asserts that the map

$$\varrho \times \epsilon_p : G_{\mathbb{Q}_p}^{\mathrm{ab}} \longrightarrow G_{\mathbb{F}_p} \times \mathbb{Z}_p^\times.$$

is an isomorphism. (We will use G^{ab} to denote the abelianisation of a profinite group G, i.e. the unique maximal abelian continuous image.) Under it I_p goes to $\mathbb{Z}_p^\times$ and for $u > 0$, I_p^u goes to $(1 + p^{\lceil u \rceil}\mathbb{Z}_p) \subset \mathbb{Z}_p^\times$, where $\lceil u \rceil$ denotes the least integer greater than or equal to u. If $\ell \neq p$ then ϵ_ℓ is trivial on I_p and takes Frob_p to $p \in \mathbb{Z}_\ell^\times$.

The field $\mathbb{Q}$: If p is a rational prime, the usual p-adic valuation $v_p : \mathbb{Q}^\times \to \mathbb{Z}$ extends to a valuation $v : \bar{\mathbb{Q}}^\times \to \mathbb{Q}$. This extension is not unique, but all extensions are permuted transitively by $G_\mathbb{Q}$. Fix one such extension v_p and let G_p denote its stabiliser in $G_\mathbb{Q}$. Then G_p acts on the completion $(\bar{\mathbb{Q}})_{v_p}$ and preserves the subfield of elements algebraic over $\mathbb{Q}_p$. This field is an algebraic closure of $\mathbb{Q}_p$, which we shall identify with $\bar{\mathbb{Q}}_p$. One can check that the resulting map $G_p \to G_{\mathbb{Q}_p}$ is an isomorphism. Thus we obtain subgroups $G_p \supset I_p \supset I_p^u$ for $u \in [-1, \infty]$ and a distinguished element $\mathrm{Frob}_p \in G_p/I_p$. These objects depend on the choice of v_p and vary by conjugation in $G_\mathbb{Q}$ as this choice is varied. We call an algebraic extension $F/\mathbb{Q}$ *unramified at p* if all conjugates of I_p lie in $G_F \subset G_\mathbb{Q}$; otherwise we say that $F/\mathbb{Q}$ is *ramified*. If $F/\mathbb{Q}$ is Galois and unramified at p, then there is a well-defined conjugacy class $[\mathrm{Frob}_p] \subset \mathrm{Gal}\,(F/\mathbb{Q})$.

By replacing the p-adic completion by an Archimedean completion, one has a well-defined conjugacy class $[c]$ in $G_\mathbb{Q}$ consisting of those elements that arise as complex conjugation for some embedding $\bar{\mathbb{Q}} \hookrightarrow \mathbb{C}$. We will denote by G_∞ the subgroup $\{1, c\}$ for one such element c.

We have the following fundamental results concerning the structure of $G_\mathbb{Q}$ (see for instance [La1]).

Theorem 2.1 *If $F/\mathbb{Q}$ is a finite extension then F is only ramified at finitely many primes (those dividing the discriminant of $F/\mathbb{Q}$).*

Theorem 2.2 (Hermite-Minkowski) *If S is a finite set of primes and if $d \in \mathbb{Z}_{>0}$ then there are only finitely many extensions $F/\mathbb{Q}$ of degree d which are unramified outside S (contained in a fixed algebraic closure $\bar{\mathbb{Q}}$).*

Theorem 2.3 (Chebotarev) *If $F/\mathbb{Q}$ is a Galois extension unramified outside a finite set S of primes then $\bigcup_{p \notin S}[\mathrm{Frob}_p]$ is dense in* $\mathrm{Gal}\,(F/\mathbb{Q})$.

Theorem 2.4 (Kronecker-Weber) *The product of the p-adic cyclotomic characters gives an isomorphism*

$$\prod_p \epsilon_p : G_\mathbb{Q}^{\mathrm{ab}} \xrightarrow{\sim} \prod_p \mathbb{Z}_p^\times.$$

We remark that the Chebotarev density theorem (theorem 2.3) applied to the extension of $\mathbb{Q}$ generated by the mth roots of unity implies Dirichlet's theorem that if n and m are coprime integers then there are infinitely many primes p with $p \equiv n \bmod m$. We also remark that the Kronecker-Weber theorem is equivalent to the fact that the maximal abelian extension of $\mathbb{Q}$ is obtained by adjoining all roots of unity to $\mathbb{Q}$.

Representations: A d-dimensional representation of $G_\mathbb{Q}$ is a homomorphism

$$G_\mathbb{Q} \longrightarrow GL_d(K),$$

where K is any field. (In our later discussion, we will also consider representations with coefficients in a ring.) Often K (and hence $GL_d(K)$) comes equipped with a natural topology, and it is then customary to restrict one's attention to continuous homomorphisms $G_{\mathbb{Q}} \longrightarrow GL_d(K)$.

Since any one-dimensional representation has abelian image, theorem 2.4 allows one to give a complete description of the one-dimensional representations of $G_{\mathbb{Q}}$ together with the behaviour of these representations on the decomposition groups at all primes. The aim of this article is to discuss attempts to give a similar theory for the two-dimensional representations of $G_{\mathbb{Q}}$.

We will call a representation ρ of $G_{\mathbb{Q}}$ unramified at p if it is trivial on the inertia group I_p. Otherwise we say that it is ramified at p. If ρ is unramified at p, then $\rho(\mathrm{Frob}_p)$ is well-defined (and its conjugacy class is independent of the choice of v_p).

If ρ is a representation of a group G and i is a non-negative integer, then we let $\wedge^i \rho$ denote the representation of G on the ith exterior power of the underlying module of ρ. If H is a subgroup of G, we will let ρ^H (resp. ρ_H) denote the H-invariants (resp. H-coinvariants) of the underlying module of ρ. If H is normal in G, then we shall also use ρ^H and ρ_H to denote the corresponding representation of G/H. In particular if p is a prime and ρ is a representation of $G_{\mathbb{Q}}$ or $G_{\mathbb{Q}_p}$, then $\rho^{I_p}(\mathrm{Frob}_p)$ and $\rho_{I_p}(\mathrm{Frob}_p)$ are well-defined.

We will be primarily interested in three types of representation of $G_{\mathbb{Q}}$.

- **Artin representations**, i.e. continuous representations $G_{\mathbb{Q}} \to GL_d(\mathbb{C})$. Since all compact totally disconnected subgroups of $GL_d(\mathbb{C})$ are finite, Artin representations have finite image. Hence they are semi-simple and are unramified at all but finitely many primes, by theorem 2.1.

- **Mod ℓ representations**, i.e. continuous representations $G_{\mathbb{Q}} \to GL_d(k)$, where k is a finite field of characteristic ℓ. These always have finite image and hence, like Artin representations, are unramified at all but finitely many primes.

- **ℓ-adic representations**, i.e. continuous representations $G_{\mathbb{Q}} \to GL_d(K)$, where K is a finite extension of $\mathbb{Q}_\ell$. We require that an ℓ-adic representation be unramified at all but finitely many primes.

Remark 2.5 (a) Continuous representations $G_{\mathbb{Q}} \to GL_d(K)$, unlike those to $GL_d(\mathbb{C})$ or $GL_d(k)$, may be ramified at infinitely many primes. We shall not need to consider such representations. (One rarely if ever does.) However, the term "ℓ-adic representation" is often used in the more general sense elsewhere in the literature.

(b) Note that the image of an ℓ-adic representation can very well be infinite. For instance this is the case if $d = 1$ and the representation is the cyclotomic character.

(c) Note that in the case of mod ℓ and ℓ-adic representations, the representations need not be semi-simple.

Proposition 2.6 *Let S be any finite set of primes.*

(a) An Artin representation $\rho : G_{\mathbb{Q}} \to GL_d(\mathbb{C})$ is determined by the values of $\mathrm{tr}\,\rho(\mathrm{Frob}_p)$ on the primes $p \notin S$ at which ρ is unramified.

(b) *A semi-simple mod ℓ representation $\rho : G_{\mathbb{Q}} \to GL_d(k)$ is determined by the values of* $\operatorname{tr} \wedge^i \rho(\operatorname{Frob}_p)$ *$(i = 1, ..., d)$ on the primes $p \notin S$ at which ρ is unramified. If $\ell > d$ then it is determined by* $\operatorname{tr} \rho(\operatorname{Frob}_p)$ *on the primes $p \notin S$ at which ρ is unramified.*

(c) *A semi-simple ℓ-adic representation $\rho : G_{\mathbb{Q}} \to GL_d(K)$ is determined by the values of* $\operatorname{tr} \rho(\operatorname{Frob}_p)$ *on the primes $p \notin S$ at which ρ is unramified.*

Proof: Combining the Chebotarev density theorem (theorem 2.3) with the continuity of ρ we see that we must show that $\operatorname{tr} \rho$ (resp. $\operatorname{tr} \wedge^i \rho$) determines ρ up to conjugacy in the various settings. For the characteristic zero representations, see [Bour], ch. 8, sec. 12.1, prop. 3; for mod ℓ representations, this is the Brauer-Nesbitt theorem ([CR], (30.16)). □

Let K denote a finite extension of $\mathbb{Q}_\ell$, let $\mathcal{O}$ be its ring of integers, λ the maximal ideal of $\mathcal{O}$ and k the residue field. If $\rho : G_{\mathbb{Q}} \to GL_d(K)$ is an ℓ-adic representation, then the image of ρ is compact, and hence ρ can be conjugated to a homomorphism $G_{\mathbb{Q}} \to GL_d(\mathcal{O})$. Reducing modulo the maximal ideal λ gives a *residual representation* $\bar{\rho} : G_{\mathbb{Q}} \to GL_d(k)$. This representation may depend on the particular $GL_d(K)$-conjugate of ρ chosen, but its semi-simplification $\bar{\rho}^{ss}$ (i.e. the unique semi-simple representation with the same Jordan-Hölder factors) is uniquely determined by ρ, by proposition 2.6 (b).

Note that the kernel of the reduction map $GL_d(\mathcal{O}) \to GL_d(k)$ is a pro-ℓ-group. In particular we see that if $p \neq \ell$ then $\rho(P_p)$ is finite and $\rho(P_p)$ is isomorphic to $\bar{\rho}(P_p)$.

Suppose ρ is a representation of $G_{\mathbb{Q}}$ of one of the three sorts above. If $p \neq \ell$ then we define the conductor, $m_p(\rho)$, of ρ at p by

$$m_p(\rho) = \int_{-1}^{\infty} \operatorname{codim} \rho^{I_p^u} \, du = \operatorname{codim} \rho^{I_p} + \int_0^{\infty} \operatorname{codim} \rho^{I_p^u} \, du.$$

This is well-defined as $\rho(P_p)$ is a finite group. If ρ is an Artin representation this makes sense also for $p = \ell$. (It would even make sense for a mod ℓ representation if $p = \ell$, but in this case it does not seem to be a useful notion.) It is known that $m_p(\rho)$ is an integer (see chapter VI of [Se2]). Moreover it is easily seen that $m_p(\rho) = 0$ if and only if ρ is unramified at p. We define the conductor $N(\rho)$ of ρ to be

$$\prod_p p^{m_p(\rho)},$$

where the product is over all primes $p \neq \ell$ in the case of an ℓ-adic or mod ℓ representation and over all primes in the case of an Artin representation. This makes sense because ρ is unramified almost everywhere.

The following lemma is an exercise.

Lemma 2.7 *Suppose that $\rho : G_{\mathbb{Q}} \to GL_d(K)$ is an ℓ-adic representation and that $\bar{\rho} : G_{\mathbb{Q}} \to GL_d(k)$ is a reduction of ρ. Then for each prime $p \neq \ell$ we have* $\int_0^{\infty} \operatorname{codim} \rho^{I_p^u} \, du = \int_0^{\infty} \operatorname{codim} \bar{\rho}^{I_p^u} \, du$. *Thus*

$$N(\rho) = N(\bar{\rho}) \prod_{p \neq \ell} p^{(\dim \bar{\rho}^{I_p} - \dim \rho^{I_p})},$$

and in particular $N(\bar{\rho})$ divides $N(\rho)$.

We take this opportunity to introduce some important notation. Let G denote a group and R a ring. If $\rho : G \to GL_d(R)$ is a representation we shall let M_ρ denote the underlying $R[G]$-module (so that $M_\rho \cong R^d$ as an R-module). If M is an $R[G]$-module we shall let $\mathrm{End}\,(M)$ denote the module of R-linear endomorphisms of M. It is also an $R[G]$-module with the G-action being defined by

$$(g(\phi))(m) = g(\phi(g^{-1}m)).$$

If M is a finitely generated free R-module we will let $\mathrm{End}^0(M)$ denote the sub-$R[G]$-module of $\mathrm{End}\,(M)$ consisting of endomorphisms of trace zero. We will use $\mathrm{ad}\rho$ to denote $\mathrm{End}\,(M_\rho)$ and $\mathrm{ad}^0\rho$ to denote $\mathrm{End}^0(M_\rho)$. Note that if d is invertible in R then $\mathrm{ad}\rho \cong \mathrm{ad}^0\rho \oplus R$ as $R[G]$-modules, where R has the trivial action of G. Note also then that the kernel of $\mathrm{ad}^0\rho$ is the same as the kernel of the composite map $G \xrightarrow{\rho} GL_d(R) \to PGL_d(R)$.

We also remark that if R is an algebraically closed field of characteristic other than 2 and $\rho : G \to GL_2(R)$ is irreducible, then either $\mathrm{ad}^0\rho$ is irreducible or there is a subgroup $H \subset G$ of index 2 and a character $\chi : H \to R^\times$ such that $\rho \cong \mathrm{Ind}\,_H^G\chi$. In the latter case $\mathrm{ad}^0\rho \cong \delta \oplus \mathrm{Ind}\,_H^G(\chi/\chi')$, where δ is the nontrivial character of G/H and χ' is the composite of χ and conjugation by an element of $G - H$. Moreover, either $\mathrm{Ind}\,_H^G(\chi/\chi')$ is irreducible or $\mathrm{ad}^0\rho \cong \delta_1 \oplus \delta_2 \oplus \delta_3$ where the δ_i are distinct characters of G of order 2. In the latter case, ρ is induced from a character from each of the subgroups $H_i = \ker \delta_i$ of index 2 in G. We leave the verification of these facts as an exercise to the reader.

2.2 Representations associated to elliptic curves

Perhaps the simplest examples of non-abelian ℓ-adic representations arise from elliptic curves defined over $\mathbb{Q}$.

We will let $E[n](\bar{\mathbb{Q}})$ denote the group of n-torsion points on $E(\bar{\mathbb{Q}})$. By proposition 1.1 there is a non-canonical isomorphism $E[n](\bar{\mathbb{Q}}) \cong (\mathbb{Z}/n\mathbb{Z})^2$. Furthermore, $E[n](\bar{\mathbb{Q}})$ carries a natural action of $G_\mathbb{Q}$ and so we get a representation (defined up to conjugation)

$$\bar{\rho}_{E,n} : G_\mathbb{Q} \to GL_2(\mathbb{Z}/n\mathbb{Z}).$$

If ℓ is a prime different from the characteristic of $\mathbb{Q}$ then we set $T_\ell E = \varprojlim E[\ell^n](\bar{\mathbb{Q}})$, which is non-canonically isomorphic to $\mathbb{Z}_\ell^2$. Again it has a natural continuous action of $G_\mathbb{Q}$, and so we get a representation (defined up to conjugation)

$$\rho_{E,\ell} : G_\mathbb{Q} \to GL_2(\mathbb{Z}_\ell).$$

Note that $\bar{\rho}_{E,\ell^n} \cong (\rho_{E,\ell} \bmod \ell^n)$.

Global properties: We have the following basic properties of the representations $\bar{\rho}_{E,\ell^n}$ and $\rho_{E,\ell}$:

Proposition 2.8 *(a) The determinant of $\rho_{E,\ell}$ is ϵ_ℓ.*

(b) The representation $\rho_{E,\ell}$ is absolutely irreducible for all ℓ. For fixed E, $\bar{\rho}_{E,\ell}$ is absolutely irreducible for all but finitely many ℓ.

(c) If E does not have complex multiplication then $\rho_{E,\ell}$ (and hence $\bar{\rho}_{E,\ell}$) is surjective for all but finitely many ℓ.

Proof: Part (a) follows from the existence of the non-degenerate alternating Galois-equivariant Weil pairing

$$\mathcal{T}_\ell E \times \mathcal{T}_\ell E \longrightarrow \mathbb{Z}_\ell(1) := \lim_{\leftarrow} \mu_{\ell^n}.$$

Part (b) is proved in [Se5], ch. IV. Part (c) is the main result of [Se6]. □

The following deep result, which is stronger than the second assertion of part (b) of proposition 2.8, is due to Mazur [Maz2], thms. 1 and 2. (Parts (b) and (c) can actually be deduced directly from theorem 1.7, using the fact that a semi-stable curve with a rational subgroup of order ℓ is necessarily isogenous to a curve with a rational point of order ℓ.)

Theorem 2.9 *Suppose that $E/\mathbb{Q}$ is an elliptic curve.*

(a) If $\ell > 163$ is a prime then $\bar{\rho}_{E,\ell}$ is irreducible.

(b) If E is semistable then $\bar{\rho}_{E,\ell}$ is irreducible for $\ell > 7$.

(c) If E is semistable and $\bar{\rho}_{E,2}$ is trivial then $\bar{\rho}_{E,\ell}$ is irreducible for $\ell > 3$.

Remark 2.10 Combined with the results of Serre [Se6], Mazur's results imply that if E is semistable everywhere then $\bar{\rho}_{E,\ell}$ is surjective for $\ell > 7$ ([Maz2], thm. 4).

Local behaviour of $\rho_{E,\ell}$ and $\bar{\rho}_{E,\ell}$:

Proposition 2.11 *Suppose E has good reduction at p.*

(a) If $\ell \neq p$, then $\rho_{E,\ell}$ is unramified at p, and we have the formula

$$\operatorname{tr} \rho_{E,\ell}(\operatorname{Frob}_p) = p + 1 - \#\tilde{E}_p(\mathbb{F}_p).$$

In particular $\operatorname{tr} \rho_{E,\ell}(\operatorname{Frob}_p)$ *belongs to $\mathbb{Z}$ and is independent of $\ell \neq p$.*

(b) For all $n \geq 1$ there is a finite flat group scheme $\mathcal{F}_n/\mathbb{Z}_p$ such that

$$E[p^n](\bar{\mathbb{Q}}_p) \cong \mathcal{F}_n(\bar{\mathbb{Q}}_p)$$

as G_p-modules.

(c)
- *If E has good ordinary reduction at p (i.e., a_p is not divisible by p) then $\tilde{E}_p[p](\bar{\mathbb{F}}_p)$ has order p and*

$$\rho_{E,p}|_{I_p} \sim \begin{pmatrix} \epsilon_p & * \\ 0 & 1 \end{pmatrix};$$

- *If E has supersingular reduction at p, then $\tilde{E}_p[p](\bar{\mathbb{F}}_p)$ is trivial and $\bar{\rho}_{E,p}|_{G_p}$ is irreducible.*

Proofs: The proof of part (a) can be found in chapters V and VII of [Si1]. For part (b) one considers the finite flat group scheme $\mathcal{F}_n = \mathcal{E}[p^n]$, where $\mathcal{E}$ is the model for E over $\mathbb{Z}_p$ defined by $(W^{\min})$. (See [Sha] for an introduction to theory of finite flat group schemes.) Part (c) follows from the results in chapters V and VII of [Si1] together with basic results on finite flat group schemes.

Proposition 2.12 *Suppose that E has multiplicative reduction at p. Let $q = q_{E,p} \in p\mathbb{Z}_p$ be the multiplicative Tate period attached to E, defined as in section 1.1. Let $\delta : G_p/I_p \to \{\pm 1\}$ be the unique non-trivial unramified quadratic character of G_p if E has non-split multiplicative reduction, and let δ be the trivial character if E has split multiplicative reduction. Then*

(a)

$$\rho_{E,\ell}|_{G_p} \sim \begin{pmatrix} \epsilon_\ell & * \\ 0 & 1 \end{pmatrix} \otimes \delta,$$

and if $\ell \neq p$ then $m_p(\rho_{E,\ell}) = 1$.

(b)

$$\bar{\rho}_{E,\ell}|_{G_p} \sim \begin{pmatrix} \epsilon_\ell & \Psi \\ 0 & 1 \end{pmatrix} \otimes \delta,$$

and $\Psi \in H^1(G_{\mathbb{Q}_p}, \mathbb{Z}/\ell\mathbb{Z}(1)) \cong \mathbb{Q}_p^\times/(\mathbb{Q}_p^\times)^\ell$ corresponds to the Tate period $q_{E,p} \in \mathbb{Q}_p^\times/(\mathbb{Q}_p^\times)^\ell$.

(c) If $\ell \neq p$ then $m_p(\bar{\rho}_{E,\ell}) = 0$ (i.e., $\bar{\rho}_{E,\ell}$ is unramified at p) if and only if $\ell | v_p(\Delta_E^{min}) = -v_p(j_E)$.

(d) There is a finite flat group scheme $\mathcal{F}/\mathbb{Z}_p$ such that $E[p](\bar{\mathbb{Q}}_p) \cong \mathcal{F}(\bar{\mathbb{Q}}_p)$ if and only if $p | v_p(\Delta_E^{min}) = -v_p(j_E)$.

Sketch of proof: By Tate's proposition 1.5 of section 1.1, we have

$$E(\bar{\mathbb{Q}}_p) \cong (\bar{\mathbb{Q}}_p^\times/q^{\mathbb{Z}})(\delta)$$

as G_p-modules. (The (δ) indicates that G_p acts on $(\bar{\mathbb{Q}}_p^\times/q^{\mathbb{Z}})(\delta)$ by $\sigma : x \mapsto \sigma(x)^{\delta(\sigma)}$.) The proposition follows directly from this, together with [Edi], prop. 8.2. for the last part. (For further details, see ch. VII, prop. 5.1 of [Si1], ch. IV, thm. 10.2. and ch. V of [Si2], and sec. 2.9 of [Se7].) □

Proposition 2.13 *Suppose that E has additive reduction at p. Then for any $\ell \neq p$ the conductor of $\rho_{E,\ell}|_{G_p}$ is at least 2, and if $p > 3$ then it is equal to 2.*

Remark 2.14 For an elliptic curve E with any type of reduction at p, the conductor of $\rho_{E,\ell}|_{G_p}$ (for $\ell \neq p$) coincides with the local conductor of E at p given by a formula of Ogg (proved by Saito [Sa] in the case $p = 2$). In particular the conductor of $\rho_{E,\ell}|_{G_p}$ is independent of $\ell \neq p$. For further discussion and references, see ch. IV, secs. 10 and 11 of [Si2].

The Frey curve: We now consider the elliptic curve

$$E_{A,B} : Y^2Z = X(X - AZ)(X + BZ),$$

where A and B are non-zero coprime integers with $A + B \neq 0$. Set $C = A + B$. This equation has discriminant $\Delta = 16(ABC)^2$. If p is an odd prime then this curve has good reduction at p if $p \nmid (ABC)$ and has multiplicative reduction at p if $p | ABC$. Thus $\Delta_{E_{A,B}}^{\min} = 2^{4-12n}(ABC)^2$ for some $n \in \mathbb{Z}_{\geq 0}$. If for instance $A \equiv -1 \bmod 4$ and $B \equiv 0 \bmod 16$ then $E_{A,B}$ has a minimal Weierstrass equation

$$Y^2Z + XYZ = X^3 + \frac{B - A - 1}{4}X^2Z - (AB/16)XZ^2.$$

In this special case $E_{A,B}$ has semi-stable reduction everywhere and $\Delta^{\min}_{E_{A,B}} = 2^{-8}(ABC)^2$.

The connection with Fermat's Last Theorem is provided by the elliptic curve

$$E : Y^2Z = X(X - a^\ell Z)(x + b^\ell Z)$$

considered by Hellegouarch [He] and Frey [Fr], where $a^\ell + b^\ell = c^\ell$ is a hypothetical non-trivial solution to Fermat's Last theorem, and where we have supposed without loss of generality that a, b and c are pairwise coprime, that b is even and that $a \equiv -1 \bmod 4$. Frey suggested that properties of this hypothetical elliptic curve E could lead to a contradiction if the curve were also known to be modular. This suggestion was made precise by Serre [Se7], who observed that the following list of properties would yield the desired contradiction, when combined with his conjectures on Galois representations associated to modular forms (to be discussed in section 3.2).

Theorem 2.15 *Suppose that $\ell > 3$ is prime. The representation $\bar{\rho}_{E,\ell}$ has the following properties.*

(a) $\bar{\rho}_{E,\ell}$ is irreducible.

(b) $\bar{\rho}_{E,\ell}$ is unramified outside ℓ and 2.

(c) There is a finite flat group scheme $\mathcal{F}/\mathbb{Z}_\ell$ such that $\mathcal{F}(\bar{\mathbb{Q}}_\ell) \cong E[\ell](\bar{\mathbb{Q}}_\ell)$ as G_ℓ-modules.

(d) $\#\bar{\rho}_{E,\ell}(I_2) = \ell$.

Proof: Part (a) follows from Mazur's theorem (theorem 2.9, (c)). To prove (b), (c), and (d), we note the key fact that $\Delta^{\min}_E = 2^{-8}(abc)^{2\ell}$ is a perfect ℓ-th power, up to powers of 2. The fact that $\bar{\rho}_{E,\ell}$ is unramified at $p \neq \ell, 2$ (part (b)) follows from this, from proposition 2.11, (a) (for primes $p \neq \ell$ of good reduction for E) and from proposition 2.12 (c) (for primes $p \neq \ell, 2$ of multiplicative reduction for E). Part (c) is a consequence of proposition 2.12 (d), and part (d) follows from proposition 2.12 (c). □

Remark 2.16 Part (a) implies that the image of $\bar{\rho}_{E,\ell}$ is large, while parts (b) and (c) state that the image has very limited ramification. We remark that the conclusion of the theorem ensures that $m_p(\bar{\rho}_{E,\ell}) = 0$ for $p \neq 2, \ell$ and that $m_2(\bar{\rho}_{E,\ell}) = 1$.

2.3 Galois cohomology

In this section M will denote a continuous discrete $G_{\mathbb{Q}}$-module of finite cardinality. If M and N are two such modules we will endow the space $\mathrm{Hom}(M, N)$ of homomorphisms of abelian groups with an action of $G_{\mathbb{Q}}$ via the formula

$$(g(\phi))(m) = g(\phi(g^{-1}m)).$$

We will use M^* to denote $\mathrm{Hom}(M, \mu_n(\bar{\mathbb{Q}}))$, where n is an integer such that $nM = (0)$ and where $\mu_n(\bar{\mathbb{Q}})$ denotes the group of nth roots of unity in $\bar{\mathbb{Q}}$. We will use $\vee$ to denote Pontryagin duals. All cohomology groups of profinite groups will mean continuous cohomology. The general references for this section are Milne [Mi] and Serre [Se1].

We start by recalling Tate's local duality theorem (see chapter 1, section 2 of [Mi]).

Theorem 2.17 *Let v be a place of $\mathbb{Q}$ (i.e. either a prime number or ∞).*

(a) $H^i(G_v, M)$ is finite for all i.

(b) For all integers n there are compatible embeddings $H^2(G_v, \mu_n(\bar{\mathbb{Q}})) \hookrightarrow \mathbb{Q}/\mathbb{Z}$ (i.e. compatible with the maps coming from $\mu_n(\bar{\mathbb{Q}}) \hookrightarrow \mu_m(\bar{\mathbb{Q}})$ if $n|m$).

(c) For $v \neq \infty$ and $i = 0, 1, 2$ the cup product and the above embeddings give rise to a perfect pairing

$$H^i(G_v, M) \times H^{2-i}(G_v, M^*) \to H^2(G_v, \mu_n(\bar{\mathbb{Q}})) \hookrightarrow \mathbb{Q}/\mathbb{Z}.$$

(This also holds when $v = \infty$, if one replaces $H^0(G_v, M) = H^0(\mathbb{R}, M)$ by Tate's modified cohomology group: $\hat{H}^0(\mathbb{R}, M) := M^{\tau=1}/(1+\tau)M$, where τ is complex conjugation.)

(d) If $v \neq \infty$ is a prime, then $H^i(G_v, M) = (0)$ for $i > 2$ and

$$\#H^1(G_v, M) = \#H^0(G_v, M)\#H^2(G_v, M)\#(M \otimes \mathbb{Z}_v).$$

(e) If $v \neq \infty$ and $M \otimes \mathbb{Z}_v = 0$ then $H^1(G_v/I_v, M^{I_v})$ and $H^1(G_v/I_v, M^{*I_v})$ are the annihilators of each other under the pairing

$$H^1(G_v, M) \times H^1(G_v, M^*) \to \mathbb{Q}/\mathbb{Z}.$$

By a collection of local conditions for M we shall mean a collection $\mathcal{L} = \{L_v\}$ of subgroups $L_v \subset H^1(G_v, M)$ as v runs over all places of $\mathbb{Q}$, such that $L_v = H^1(G_v/I_v, M^{I_v})$ for all but finitely many v. Note that if $\mathcal{L}$ is a collection of local conditions for M then $\mathcal{L}^* = \{L_v^\perp\}$ is a collection of local conditions for M^*. If $\mathcal{L}$ is a collection of local conditions for M we define the corresponding Selmer group, $H^1_{\mathcal{L}}(\mathbb{Q}, M)$, to be the subgroup of $x \in H^1(G_{\mathbb{Q}}, M)$ such that for all places v of $\mathbb{Q}$ we have

$$\mathrm{res}_v x \in L_v \subset H^1(G_v, M).$$

Suppose that v is a finite place of $\mathbb{Q}$. We have an exact sequence

$$(0) \to H^0(G_v, M) \to M^{I_v} \xrightarrow{\mathrm{Frob}_v - 1} M^{I_v} \to H^1(G_v/I_v, M^{I_v}) \to (0),$$

and hence we see that $\#H^1(G_v/I_v, M^{I_v}) = \#H^0(G_v, M)$.

We have the following important observation of Wiles [W3], inspired by a formula of Greenberg [Gre]. An important theme in number theory has been the calculation of Selmer groups. This result, although it does not allow the absolute calculation of Selmer groups, allows the comparison of two (dual) Selmer groups. In the applications in this paper we shall apply the theorem where the various data have been chosen to make one of the Selmer groups trivial. In such a situation it allows the exact calculation of the order of the non-trivial Selmer group.

Theorem 2.18 *If $\mathcal{L}$ is a collection of local conditions for M then the Selmer group $H^1_{\mathcal{L}}(\mathbb{Q}, M)$ is finite. Moreover we have the formula*

$$\frac{\#H^1_{\mathcal{L}}(\mathbb{Q}, M)}{\#H^1_{\mathcal{L}^*}(\mathbb{Q}, M^*)} = \frac{\#H^0(G_{\mathbb{Q}}, M)}{\#H^0(G_{\mathbb{Q}}, M^*)} \prod_v \frac{\#L_v}{\#H^0(G_v, M)}.$$

We note that all but finitely many terms in the product are 1 so that it makes sense.

Proof of theorem 2.18: Choose a finite set, S, of places of $\mathbb{Q}$, which contains ∞, all the places whose residue characteristic divides the order of M, all places at which M is ramified and all places at which $L_v \neq H^1(G_v/I_v, M^{I_v})$. Let $\mathbb{Q}_S$ denote the maximal extension of $\mathbb{Q}$ unramified outside S and let $G_S = \mathrm{Gal}(\mathbb{Q}_S/\mathbb{Q})$. Then we have an exact sequence

$$(0) \to H^1_{\mathcal{L}}(\mathbb{Q}, M) \to H^1(G_S, M) \to \bigoplus_{v\in S} H^1(G_v, M)/L_v.$$

Because $H^1(G_S, M)$ is finite (see cor. 4.15 of ch. 1 of [Mi]), we see that $H^1_{\mathcal{L}}(\mathbb{Q}, M)$ is finite.

Dualising this sequence for M^* and using theorem 2.17, we see that we have an exact sequence

$$\bigoplus_{v\in S} L_v \to H^1(G_S, M^*)^\vee \to H^1_{\mathcal{L}^*}(\mathbb{Q}, M^*)^\vee \to (0).$$

On the other hand we have the following part of the Poitou-Tate nine term sequence (see thm. 4.10 of ch. 1 of [Mi]):

$$\begin{array}{ccccc} (0)\to H^0(G_S, M) & \to & \left(\bigoplus_{v\in S} H^0(G_v, M)\right)/(1+c)M & \to & H^2(G_S, M^*)^\vee \\ & & & & \downarrow \\ H^1(G_S, M^*)^\vee \leftarrow & & \bigoplus_{v\in S} H^1(G_v, M) & \leftarrow & H^1(G_S, M), \end{array}$$

where we regard $(1+c)M$ as contained in $H^0(G_\infty, M)$. Replacing $H^1(G_v, M)$ by L_v for $v \in S$ and combining this with the previous exact sequence we get

$$\begin{array}{ccccc} (0) & \to & H^0(G_S, M) & \to & \left(\bigoplus_{v\in S} H^0(G_v, M)\right)/(1+c)M \\ & & & & \downarrow \\ \bigoplus_{v\in S} L_v & \leftarrow & H^1_{\mathcal{L}}(\mathbb{Q}, M) & \leftarrow & H^2(G_S, M^*)^\vee \\ \downarrow & & & & \\ H^1(G_S, M^*)^\vee & \to & H^1_{\mathcal{L}^*}(\mathbb{Q}, M^*)^\vee & \to & (0). \end{array}$$

Theorem 2.18 follows from this and the global Euler characteristic formula (see thm. 5.1 of ch. 1 of [Mi]):

$$\frac{\#H^1(G_S, M^*)}{\#H^0(G_S, M^*)\#H^2(G_S, M^*)} = \frac{(\#M^*)}{\#H^0(G_\infty, M^*)} = \#(1+c)M.$$

□

We remark, but will not use elsewhere in this article, that the above argument works over any number field:

Theorem 2.19 *Let F be a number field and M a discrete G_F module of finite cardinality. For each place v of F let G_v denote a decomposition group at v*

and if v is finite let $I_v \subset G_v$ denote the inertia group. Fix subgroups $L_v \subset H^1(G_v, M)$, and such that for all but finitely many v, $L_v = H^1(G_v/I_v, M^{I_v})$. Let $H^1_{\mathcal{L}}(F, M)$ (respectively, $H^1_{\mathcal{L}^*}(F, M^*)$) denote the inverse image of $\prod_v L_v$ (respectively, $\prod_v L_v^{\perp}$) under the map $H^1(G_F, M) \to \prod_v H^1(G_v, M)$ (respectively, $H^1(G_F, M^*) \to \prod_v H^1(G_v, M^*)$). Then $H^1_{\mathcal{L}}(G, M)$ and $H^1_{\mathcal{L}^*}(G, M^*)$ are finite and we have the formula

$$\frac{\# H^1_{\mathcal{L}}(F, M)}{\# H^1_{\mathcal{L}^*}(F, M^*)} = \frac{\# H^0(G_F, M)}{\# H^0(G_F, M^*)} \prod_v \frac{\# L_v}{\# H^0(G_v, M)}.$$

2.4 Representations of $G_{\mathbb{Q}_\ell}$

In this section we assume that ℓ is an **odd** prime. If G is any topological group then by a finite $\mathcal{O}[G]$-module we shall mean a discrete $\mathcal{O}$-module of finite cardinality with a continuous action of G. By a profinite $\mathcal{O}[G]$-module we shall mean an inverse limit of finite $\mathcal{O}[G]$-modules.

If M is a profinite $\mathcal{O}[G_\ell]$-module then we will call M

- **good** if for every discrete quotient M' of M there is a finite flat group scheme $\mathcal{F}/\mathbb{Z}_\ell$ such that $M' \cong F(\bar{\mathbb{Q}}_\ell)$ as $\mathbb{Z}_\ell[G_\ell]$-modules;
- **ordinary** if there is an exact sequence

$$(0) \to M^{(-1)} \to M \to M^{(0)} \to (0)$$

  of profinite $\mathcal{O}[G_\ell]$-modules such that I_ℓ acts trivially on $M^{(0)}$ and by ϵ on $M^{(-1)}$ (equivalently, if for all $\sigma, \tau \in I_\ell$ we have $(\sigma - \epsilon(\sigma))(\tau - 1) = 0$ on M);
- **semi-stable** if M is either good or ordinary.

Suppose that R is a complete Noetherian local $\mathcal{O}$-algebra with residue field k. We will call a continuous representation $\rho : G_\ell \to GL_2(R)$ good, ordinary or semistable, if

$$\det \rho|_{I_\ell} = \epsilon \tag{2.4.1}$$

and if the underlying profinite $\mathcal{O}[G_\ell]$-module, M_ρ is good, ordinary or semistable. We write $\bar{\rho}$ for $\rho \bmod \mathfrak{m}_R$. We record the following consequence of Nakayama's lemma.

Lemma 2.20 *If M_ρ and $\bar{\rho}$ are ordinary, then $M_\rho^{(-1)}$ and $M_\rho^{(0)}$ are each free of rank one over R and ρ is ordinary.*

Remark 2.21 (a) These definitions are somewhat ad hoc, but at the moment that is all that seems to be available (though the work of Fontaine and Laffaille [FL] and its generalisations may well provide a more systematic setting).

(b) For part of the motivation for our definitions, see proposition 2.23 and remark 2.24. For further motivation, we shall see in the next chapter that representations arising from certain modular forms are semistable.

Moreover the Fontaine-Mazur conjecture predicts that, conversely, any representation of $G_{\mathbb{Q}}$ with semistable restriction to G_ℓ arises from such a modular form. We shall state a weak form of the Fontaine-Mazur conjecture below (conjecture 3.17).

(c) The terminology in [W3] to describe representations of G_ℓ is slightly different. In particular, we impose the condition (2.4.1) in our definitions of good and ordinary, as this is all we shall need here. Assuming ρ satisfies this condition, the notion of ordinary in [W3] coincides with the one here, and ρ is flat in the sense of [W3] if and only if it is good but not ordinary. (Note that a representation may be both good and ordinary, for instance $\rho_{E,\ell}$ for an elliptic curve with good, ordinary reduction; see proposition 2.23.)

The following assertions, in the good case, are consequences of results of Raynaud [Ray] (in particular, see sec. 2.1, prop. 2.3.1 and cor. 3.3.6 of [Ray]). In the ordinary case they are elementary.

Lemma 2.22 *(a) Stable profinite $\mathcal{O}[G_\ell]$-modules are closed under taking sub-objects, quotients and direct products. The same is true for ordinary profinite $\mathcal{O}[G_\ell]$-modules.*

(b) Suppose that M is a profinite $\mathcal{O}[G_\ell]$-module and that $\{M_i\}$ is a family of sub-objects with trivial intersection, such that each M/M_i is good (resp. ordinary). Then M is good (resp. ordinary).

(c) If M is a finite $\mathcal{O}[G_\ell]$-module then M is good if and only if there is a finite flat group scheme $\mathcal{F}/\mathbb{Z}_\ell$ such that $M \cong \mathcal{F}(\bar{\mathbb{Q}}_\ell)$ as $\mathbb{Z}_\ell[G_\ell]$-modules.

(d) If M and M' are profinite $\mathcal{O}[G_\ell]$-modules with $M' \cong M$ as $\mathbb{Z}_\ell[I_\ell]$-modules then M is good (resp. ordinary) if and only if M' is good (resp. ordinary).

(e) Suppose that M is a profinite $\mathcal{O}[G_\ell]$-module which is finite and free over $\mathcal{O}$. Then M is good if and only if there exists an ℓ-divisible group $\mathcal{F}/\mathbb{Z}_\ell$ such that M is isomorphic to the Tate module of $\mathcal{F}$ as a $\mathbb{Z}_\ell[G_\ell]$-module.

Together with results stated in section 2.2, we have the following.

Proposition 2.23 *Suppose that E is an elliptic curve over $\mathbb{Q}$ and $\mathcal{O} = \mathbb{Z}_\ell$.*

- *if E has good (resp. semistable) reduction at ℓ, then $\rho_{E,\ell}$ and $\bar{\rho}_{E,\ell}$ are good (resp. semistable);*
- *if E has semistable reduction at ℓ, then $\rho_{E,\ell}$ is ordinary if and only if $\bar{\rho}_{E,\ell}$ is ordinary if and only if either E has multiplicative reduction or E has good ordinary reduction.*

Remark 2.24 It is also true that if $\rho_{E,\ell}$ is good (resp. semistable), then E has good (resp. semistable) reduction at ℓ, but the result is more difficult and we shall not need it.

We will need a few more definitions. We will let $\psi : I_\ell \to \bar{\mathbb{F}}_\ell^\times$ denote the character $\sigma \mapsto (\sigma\varpi)/\varpi \bmod \varpi$ where $\varpi = \sqrt[\ell^2-1]{\ell}$. If F is a field of characteristic other than ℓ and if M is a profinite $\mathbb{Z}_\ell[G_\ell]$-module then we set $M(1) = \varprojlim M \otimes_{\mathbb{Z}_\ell} \mu_{\ell^n}(\bar{F})$. If $\rho : G_\ell \to GL_2(R)$ is ordinary the extension

$$(0) \to R(1) \to M_\rho \to R \to (0)$$

of $R[I_\ell]$-modules gives rise to a class $c_\rho \in H^1(I_\ell, R(1))$. Kummer theory and the valuation on $(\bar{\mathbb{Q}}_\ell^{I_\ell})^\times$ give rise to a map $v : H^1(I_\ell, R(1)) \to H \otimes_{\mathbb{Z}_\ell} R \to R$, where H denotes the ℓ-adic completion of $(\bar{\mathbb{Q}}_\ell^{I_\ell})^\times$. Then we also have the following lemma.

Lemma 2.25 *(a) If $\bar{\rho} : G_\ell \to GL_2(k)$ is good then either $\bar{\rho}$ is ordinary or $\bar{\rho}|_{I_\ell} \otimes \bar{k} = \psi \oplus \psi^\ell$.*

(b) If $\rho : G_\ell \to GL_2(R)$ is such that M_ρ is good and $\bar{\rho}$ is ordinary, then ρ is good and ordinary.

(c) If $\rho : G_\ell \to GL_2(R)$ is ordinary then ρ is good if and only if $v(c_\rho) = 0$.

Note that we need only consider the case that R has finite cardinality to prove the lemma. Parts (a) and (b) again follow from Raynaud's results [Ray] (for part (b) consider the connected-étale sequences for M_ρ and $M_{\bar{\rho}}$).

We sketch the proof of part (c) using an argument suggested to us by Edixhoven. As in [Edi], prop. 8.2 it suffices to determine which extensions

$$(0) \longrightarrow R(1) \longrightarrow M \longrightarrow R \longrightarrow (0) \qquad (2.4.2)$$

of $R[I_\ell]$-modules arise from finite flat group schemes over $\mathbb{Z}_\ell^{\rm nr}$, the ring of integers of the maximal unramified extension of $\mathbb{Z}_\ell$. By prop. 17.4 of [Oo], the extension (2.4.2) arises from a finite flat group scheme if and only if it corresponds to an extension of sheaves of R-modules in the fpqc topology over $\mathbb{Z}_\ell^{\rm nr}$. Therefore we must compute the image of

$$\mathrm{Ext}^1_{R\text{-mod/fpqc}}(R, R(1)) \to \mathrm{Ext}^1_{R[I_\ell]}(R, R(1)).$$

Since the sheaf Ext^1 of R by $R(1)$ vanishes, this is equivalent to computing the image of

$$H^1_{\rm fpqc}(\mathbb{Z}_\ell^{\rm nr}, R(1)) \to H^1(I_\ell, R(1)).$$

Part (c) follows from the fact this is precisely the kernel of v.

Remark 2.26 The authors expect that the theory of Fontaine and Laffaille [FL] discussed in the next section could be used to prove that if $\rho : G_\ell \to GL_2(R)$ is such that M_ρ is good and $\bar{\rho}$ is good, then $\det \rho|_{I_\ell}$ is cyclotomic, i.e., ρ is good.

For semistable representations $\rho : G_\ell \to GL_2(\mathcal{O}/\lambda^n)$ we shall define $\mathcal{O}$-submodules

$$H^1_{\rm f}(G_\ell, \mathrm{ad}\rho) \subset H^1_{\rm ss}(G_\ell, \mathrm{ad}\rho) \subset H^1(G_\ell, \mathrm{ad}\rho).$$

Before doing so, let us consider more generally a continuous representation ρ of a profinite group G with values in $GL_d(\mathcal{O}/\lambda^n)$. For each continuous cocycle $\xi : G \to \mathrm{ad}\rho$, we define a representation

$$\begin{aligned} \rho_\xi : G \to &\quad GL_2(R_n) \\ g \mapsto &\ (1+\varepsilon\xi(g))\rho(g) \end{aligned}$$

where $R_n = (\mathcal{O}/\lambda^n)[\varepsilon]/(\varepsilon^2)$. We find that the map $\xi \to \rho_\xi$ induces a bijection between $H^1(G, \mathrm{ad}\rho)$ and equivalence classes of representations $\rho' : G \to GL_2(R_n)$ such that $\rho = \rho' \bmod \varepsilon R_n$, where ρ_1' and ρ_2' are deemed equivalent if they are conjugate by an element of $1+\varepsilon M_2(R_n)$. Now let M_ρ denote the $(\mathcal{O}/\lambda^n)[G]$-module corresponding to ρ, and for each continuous cocycle ξ, let E_ξ denote the $R_n[G]$-module corresponding to ρ_ξ. Note that multiplication by ε defines an isomorphism from $M_\rho = E_\xi/\varepsilon E_\xi$ to εE_ξ of $(\mathcal{O}/\lambda^n)[G]$-modules. We thus obtain an extension

$$0 \to M_\rho \to E_\xi \to M_\rho \to 0$$

and hence a class in $\mathrm{Ext}^1(M_\rho, M_\rho)$ in the category of profinite $(\mathcal{O}/\lambda^n)[G]$-modules. Moreover if ρ_{ξ_1} and ρ_{ξ_2} are equivalent, then E_{ξ_1} and E_{ξ_2} define the same extension class. We thus obtain a map

$$H^1(G, \mathrm{ad}\rho) \to \mathrm{Ext}^1(M_\rho, M_\rho),$$

which the reader can check is an $(\mathcal{O}/\lambda^n)$-linear isomorphism. Classes in $H^1(G, \mathrm{ad}^0\rho)$ correspond to the equivalence classes of representations ρ' satisfying $\det\rho' = \det\rho$, and ϕ in $\mathrm{Hom}(G, \mathcal{O}/\lambda^n)$ corresponds to the twist of ρ by the character $g \mapsto 1+\varepsilon\phi(g)$. In the case $\ell \not| d$ where we have the decomposition

$$H^1(G, \mathrm{ad}\rho) = H^1(G, \mathrm{ad}^0\rho) \oplus \mathrm{Hom}(G, \mathcal{O}/\lambda^n),$$

we see that if ρ' corresponds to a class in $H^1(G, \mathrm{ad}\rho)$, then the projection to $\mathrm{Hom}(G, \mathcal{O}/\lambda^n)$ is the homomorphism ϕ such that $\det\rho' = (1+\varepsilon d\phi)\det\rho$.

We now return to the case of a semistable representation

$$\rho : G_\ell \to GL_2(\mathcal{O}/\lambda^n)$$

and define the cohomology groups $H^1_{\mathrm{f}}(G_\ell, \mathrm{ad}\rho)$ and $H^1_{\mathrm{ss}}(G_\ell, \mathrm{ad}\rho)$ as follows. Let $H^1_{\mathrm{ss}}(G_\ell, \mathrm{ad}\rho)$ denote the natural image in $H^1(G_\ell, \mathrm{ad}\rho)$ of $\mathrm{Ext}^1(M_\rho, M_\rho)$ taken in the category of semi-stable profinite $\mathcal{O}/\lambda^n[G_\ell]$-modules. If ρ is not good then $H^1_{\mathrm{f}}(G_\ell, \mathrm{ad}\rho)$ is defined to be $H^1_{\mathrm{ss}}(G_\ell, \mathrm{ad}\rho)$. If, however, ρ is good then $H^1_{\mathrm{f}}(G_\ell, \mathrm{ad}\rho)$ will denote the natural image in $H^1(G_\ell, \mathrm{ad}\rho)$ of the group $\mathrm{Ext}^1(M_\rho, M_\rho)$ taken in the category of good profinite $\mathcal{O}/\lambda^n[G_\ell]$-modules. We define $H^1_{\mathrm{f}}(G_\ell, \mathrm{ad}^0\rho)$ (resp. $H^1_{\mathrm{ss}}(G_\ell, \mathrm{ad}^0\rho)$) as the intersection of $H^1_{\mathrm{f}}(G_\ell, \mathrm{ad}\rho)$ (resp. $H^1_{\mathrm{ss}}(G_\ell, \mathrm{ad}\rho)$) and $H^1(G_\ell, \mathrm{ad}^0\rho)$. Note that if ρ is good (resp. semistable) and $\xi : G_\ell \to \mathrm{ad}\rho$ is a cocycle, then E_ξ is good (resp. semistable) if and only if ρ_ξ is.

Suppose that $\rho : G_\ell \to GL_2(\mathcal{O}/\lambda^n)$ is ordinary. Consider the exact sequence

$$0 \to M_\rho^{(-1)} \to M_\rho \to M_\rho^{(0)} \to 0,$$

where $M_\rho^{(-1)}$ denote the maximal submodule of M_ρ where I_ℓ acts by ϵ. Let $\mathrm{ad}^{(-1)}\rho$ denote the submodule $\mathrm{Hom}(M_\rho^{(0)}, M_\rho^{(-1)})$ of $\mathrm{ad}^0\rho$. Then

$$H^1_{\mathrm{ss}}(G_\ell, \mathrm{ad}\rho) = \ker\left(H^1(G_\ell, \mathrm{ad}\rho) \to H^1(I_\ell, \mathrm{ad}\rho/\mathrm{ad}^{(-1)}\rho)\right).$$

The same is true with ad^0 replacing ad.

If $\rho : G_\ell \to GL_2(\mathcal{O})$ is semi-stable then we set

$$H^1_{\mathrm{f}}(G_\ell, \mathrm{ad}\rho \otimes K/\mathcal{O}) = \lim_{\rightarrow} H^1_{\mathrm{f}}(G_\ell, \mathrm{ad}\rho \otimes (\lambda^{-n}/\mathcal{O})) \subset H^1(G_\ell, \mathrm{ad}\rho \otimes K/\mathcal{O}).$$

We define $H^1_{\mathrm{f}}(G_\ell, \mathrm{ad}^0\rho \otimes K/\mathcal{O})$ as

$$H^1_{\mathrm{f}}(G_\ell, \mathrm{ad}\rho \otimes K/\mathcal{O}) \cap H^1(G_\ell, \mathrm{ad}^0\rho \otimes K/\mathcal{O}).$$

We make similar definitions for $H^1_{\mathrm{ss}}(G_\ell, \mathrm{ad}\rho \otimes K/\mathcal{O})$ and $H^1_{\mathrm{ss}}(G_\ell, \mathrm{ad}^0\rho \otimes K/\mathcal{O})$. We will need the following calculations.

Proposition 2.27 *(a) Suppose that $\bar\rho : G_\ell \to GL_2(k)$ is semi-stable. Then*

$$\#H^1_{\mathrm{f}}(G_\ell, \mathrm{ad}^0\bar\rho) \leq \#H^0(G_\ell, \mathrm{ad}^0\bar\rho)\#k$$

and equality holds if $\bar\rho$ is ordinary[3].

(b) Suppose that $\rho : G_\ell \to GL_2(\mathcal{O})$ is both good and ordinary. Let χ_1 and χ_2 be the unramified characters such that $\rho \sim \begin{pmatrix} \chi_1\epsilon & * \\ 0 & \chi_2 \end{pmatrix}$. Let

$$c_\ell = (\chi_1/\chi_2)(\mathrm{Frob}\,_\ell) - 1.$$

If $c_\ell \neq 0$, then

$$H^1_{\mathrm{ss}}(G_\ell, \mathrm{ad}^0\rho \otimes K/\mathcal{O})/H^1_{\mathrm{f}}(G_\ell, \mathrm{ad}^0\rho \otimes K/\mathcal{O})$$

is finite of order $\#(\mathcal{O}/c_\ell\mathcal{O})$.

(c) If $\rho \bmod \lambda$ is either not good or not ordinary, then

$$H^1_{\mathrm{ss}}(G_\ell, \mathrm{ad}^0\rho \otimes K/\mathcal{O}) = H^1_{\mathrm{f}}(G_\ell, \mathrm{ad}^0\rho \otimes K/\mathcal{O}).$$

Proof. Part (c) is clear, and for parts (a) and (b) it suffices to prove the following two results.

Proposition 2.28 *Suppose that $\rho : G_\ell \to GL_2(\mathcal{O}/\lambda^n)$ is good. Then*

$$\#H^1_{\mathrm{f}}(G_\ell, \mathrm{ad}\rho) = \#H^0(G_\ell, \mathrm{ad}^0\rho)(\#\mathcal{O}/\lambda^n)^2.$$

Lemma 2.29 *Suppose that $\rho : G_\ell \to GL_2(\mathcal{O}/\lambda^n)$ is ordinary, so that we have $\rho \sim \begin{pmatrix} \chi_1\epsilon & * \\ 0 & \chi_2 \end{pmatrix}$, for some unramified characters χ_1 and χ_2.*

(a) We have

$$\#H^1_{\mathrm{ss}}(G_\ell, \mathrm{ad}^0\rho) \leq \#H^0(G_\ell, \mathrm{ad}^0\rho)\#(\mathcal{O}/\lambda^n)\#(\mathcal{O}/(\lambda^n, c_\ell))$$

where $c_\ell = (\chi_1/\chi_2)(\mathrm{Frob}\,_\ell) - 1$. Moreover if ρ is good, then equality holds.

(b) If $n = 1$ and ρ is not good then $\#H^1_{\mathrm{ss}}(G_\ell, \mathrm{ad}^0\rho) = \#k$.

[3] The authors expect equality to hold without this hypothesis; cf. remark 2.26.

End of proof of proposition 2.27: To deduce proposition 2.27 (in the good case) from proposition 2.28 and lemma 2.29, note that if $\rho : G_\ell \to GL_2(\mathcal{O}/\lambda^n)$ is good, then

$$H^1(G_\ell/I_\ell, \mathcal{O}/\lambda^n) \subset H^1_{\mathrm{f}}(G_\ell, \mathrm{ad}\rho) \cap H^1(G_\ell, \mathcal{O}/\lambda^n)$$

by lemma 2.22 (d), and this gives the inequality in proposition 2.27(a). Furthermore if ρ is also ordinary, then the above group is contained in

$$\mathrm{Im}\,(H^1_{\mathrm{f}}(G_\ell, \mathrm{ad}\rho) \to H^1(G_\ell, \mathcal{O}/\lambda^n)) \subset H^1(G_\ell/I_\ell, \mathcal{O}/\lambda^n)$$

where the last inclusion comes from lemma 2.25 (b). Therefore

$$\#H^1_{\mathrm{f}}(G_\ell, \mathrm{ad}^0\rho) = \#H^0(G_\ell, \mathrm{ad}^0\rho)\#(\mathcal{O}/\lambda^n).$$

□

We will prove proposition 2.28 in the next section using the theory of Fontaine and Laffaille. The proof of lemma 2.29 is a somewhat technical exercise in the Galois cohomology of local fields, for which we refer the reader to ch. 1 of [W3], prop. 1.9, parts (iii) and (iv). We remark that for our purposes, inequalities would suffice in part (b) of proposition 2.27 and part (a) of lemma 2.29 (and this is all that is proved in [W3]). We have included the precise formulas for the sake of completeness, since they are not much more difficult to obtain. The only additional observation required is that if ρ is good, then the composite of the natural maps

$$H^1(G_\ell/I_\ell, (\mathrm{ad}^0\rho/\mathrm{ad}^{(-1)}\rho)^{I_\ell}) \to H^1(G_\ell, \mathrm{ad}^0\rho/\mathrm{ad}^{(-1)}\rho) \to H^2(G_\ell, \mathrm{ad}^{(-1)}\rho)$$

is trivial. To prove this, rewrite the composite as

$$H^1(G_\ell/I_\ell, \mathcal{O}/\lambda^n) \to H^1(G_\ell, \mathcal{O}/\lambda^n) \to H^2(G_\ell, \mathrm{ad}^{(-1)}\rho)$$

with the second map given by $\cup c_\rho$ where c_ρ is the class in $H^1(G_\ell, \mathrm{ad}^{(-1)}\rho)$ defining the extension M_ρ and apply lemma 2.25 (c). (In fact, one only needs the easier half of lemma 2.25 (c): if ρ is good then $v(c_\rho) = 0$.)

2.5 The theory of Fontaine and Laffaille

In this section we again assume that ℓ is an odd prime. We mentioned in the last section the importance of understanding good representations of G_ℓ. However the definition of good is somewhat indirect, and this makes computations difficult. The key result we use to address the problem is an equivalence between the category $\mathcal{FF}_\mathcal{O}$ of good finite $\mathcal{O}[G_\ell]$-modules and a category $\mathcal{MF}_\mathcal{O}$ which we define below following Fontaine and Laffaille [FL]. The beauty and utility of the result stems from the elementary algebraic nature of the definition of $\mathcal{MF}_\mathcal{O}$; we can convert questions about good representations into questions in linear algebra.

Remark 2.30 Let $\mathcal{GF}_\mathcal{O}$ denote the category of finite flat commutative group schemes over $\mathbb{Z}_\ell$ with an action of $\mathcal{O}$. By results of Raynaud [Ray], taking $\bar{\mathbb{Q}}_\ell$ points defines an equivalence between the categories $\mathcal{GF}_\mathcal{O}$ and $\mathcal{FF}_\mathcal{O}$. The equivalence between $\mathcal{GF}_\mathcal{O}$ and a category closely related to $\mathcal{MF}_\mathcal{O}$ was first established by Fontaine [Fo2] and [Fo1]. While Fontaine's results would suffice for our purposes here, our formulation will be closer to that in [FL], where an equivalence between $\mathcal{MF}_{\mathcal{O}}$ and $\mathcal{FF}_{\mathcal{O}}$ is defined as part of a more general construction of representations of G_ℓ from linear-algebraic data. We caution however that our formulation is not exactly the same as that of [FL] since we wish to work with covariant functors.

We now turn to the definition of $\mathcal{MF}_{\mathcal{O}}$. The objects are $\mathcal{O}$-modules D of finite cardinality together with a distinguished submodule D^0 and $\mathcal{O}$-linear maps $\phi_{-1} : D \to D$ and $\phi_0 : D^0 \to D$ which satisfy:

- $\phi_{-1}|_{D^0} = \ell\phi_0$,
- $\operatorname{Im}\phi_{-1} + \operatorname{Im}\phi_0 = D$.

The morphisms are $\mathcal{O}$-linear maps compatible with the additional data of the distinguished submodules and maps ϕ.

It is useful to note that if D is an object of $\mathcal{MF}_{\mathcal{O}}$, then there is a surjection

$$\phi_{-1} \oplus \phi_0 : D/D^0 \oplus D^0/\ell D^0 \twoheadrightarrow D/\ell\phi_0(D^0),$$

and on counting orders we see that this is in fact an isomorphism. Thus there is an isomorphism

$$D/(D^0 + \lambda D) \oplus D^0/\lambda D^0 \xrightarrow{\sim} D/\lambda D.$$

It follows that $D^0/\lambda D^0 \to D/\lambda D$ is injective, and hence that D^0 is (non-canonically) a direct summand of D as an $\mathcal{O}$-module. Note also that ϕ_0 is injective, and if $D = D^0 \oplus D'$ as $\mathcal{O}$-modules, then also $D = \phi_0(D^0) \oplus \phi_{-1}(D')$ as $\mathcal{O}$-modules.

It is then straightforward to check that there is a contravariant functor $\cdot^*$ from $\mathcal{MF}_{\mathcal{O}}$ to itself defined by:

- $D^* = \operatorname{Hom}(D, \mathbb{Q}_\ell/\mathbb{Z}_\ell)$;
- $(D^*)^0 = \operatorname{Hom}(D/D^0, \mathbb{Q}_\ell/\mathbb{Z}_\ell)$;
- $\phi_{-1}^*(f)(z) = f(\ell x + y)$, where $z = \phi_{-1}(x) + \phi_0(y)$;
- $\phi_0^*(f)(z) = f(x \bmod D^0)$, where $z \equiv \phi_{-1}(x) \bmod (\phi_0 D^0)$.

Moreover the canonical isomorphism $D \cong (D^*)^*$ of $\mathcal{O}$-modules defines a natural isomorphism in $\mathcal{MF}_{\mathcal{O}}$.

We leave it as an exercise for the reader to use the above observations to define cokernels and then kernels of morphisms in $\mathcal{MF}_{\mathcal{O}}$ and verify that it is an abelian category (or see [FL], sec. 1).

Theorem 2.31 *There is a covariant functor $\mathbb{D} : \mathcal{FF}_{\mathcal{O}} \to \mathcal{MF}_{\mathcal{O}}$ which defines an $\mathcal{O}$-additive equivalence of categories. Moreover if M is an object of $\mathcal{FF}_{\mathcal{O}}$, then we have*

(a) M and $\mathbb{D}(M)$ have the same cardinality;

(b) $\mathbb{D}(M) = \mathbb{D}(M)^0$ if and only if M is unramified.

Remark 2.32 (a) It follows on applying part (a) to $M/\lambda^i M$ for each i that M and $\mathbb{D}(M)$ are in fact (non-canonically) isomorphic as $\mathcal{O}$-modules.

(b) As mentioned above, our formulation differs from that of Fontaine and Laffaille in that we are using covariant functors. To deduce theorem 2.31 from the results in [FL], sec. 9, we define $\mathbb{D}$ as a quasi-inverse of the functor $\underline{U}_S(\cdot^*[1])$, where $\underline{U}_S$ is defined in [FL] and [1] indicates a shift by 1 in filtration degrees. In particular, if $\mathcal{F}$ is a finite flat group scheme over $\mathbb{Z}_\ell$ with an action of $\mathcal{O}$, then the underlying $\mathcal{O}$-module of $\mathbb{D}(\mathcal{F}(\bar{\mathbb{Q}}_\ell))$ can be identified with the covariant Dieudonné module of $\mathcal{F}_{/\mathbb{F}_\ell}$, and ϕ_{-1} with F.

Suppose now that $\rho : G_\ell \to GL_2(\mathcal{O}/\lambda^n)$ is a good continuous representation. Let $D_\rho = \mathbb{D}(M_\rho)$ be the corresponding object of $\mathcal{MF}_\mathcal{O}$. Then $D_\rho \cong (\mathcal{O}/\lambda^n)^2$ as an $\mathcal{O}$-module while $D^0_\rho \cong \mathcal{O}/\lambda^n$.

Lemma 2.33 *The following are isomorphic.*

(a) The group of extensions of M_ρ by itself in the category of good finite $\mathcal{O}/\lambda^n[G_\ell]$-modules.

(b) The group of extensions of D_ρ by itself in the full subcategory of $\mathcal{MF}_\mathcal{O}$ consisting of objects which are $\mathcal{O}/\lambda^n$-modules.

(c) Pairs (α_{-1}, α_0) where $\alpha_{-1} \in \mathrm{Hom}_\mathcal{O}(D_\rho, D_\rho)$, $\alpha_0 \in \mathrm{Hom}_\mathcal{O}(D^0_\rho, D_\rho)$ and $\ell\alpha_0 = \alpha_{-1}|_{D^0_\rho}$, modulo pairs of the form $(a\phi_{\rho,-1} - \phi_{\rho,-1}a, a\phi_{\rho,0} - \phi_{\rho,0}a|_{D^0})$ where $a \in \mathrm{Hom}_\mathcal{O}(D_\rho, D_\rho)$ and $aD^0_\rho \subset D^0_\rho$.

Proof: The first two groups of extensions are isomorphic by the Fontaine-Laffaille theorem. Following Ramakrishna [Ram] we explain how to calculate the second group of extensions. We will write D for D_ρ. If

$$(0) \to D \to E \to D \to (0)$$

is an extension then we have as $\mathcal{O}$-modules that $E \cong D^2$ by an isomorphism such that $E^0 \xrightarrow{\sim} (D^0)^2$. Then E is determined by giving elements $\alpha_{E,-1} \in \mathrm{Hom}(D, D)$ and $\alpha_{E,0} \in \mathrm{Hom}(D^0, D)$ with $\ell\alpha_{E,0} = \alpha_{E,-1}|_{D^0}$. Explicitly

$$\phi_{E,-1} = \begin{pmatrix} \phi_{-1} & \alpha_{E,-1} \\ 0 & \phi_{-1} \end{pmatrix} \quad \text{and} \quad \phi_{E,0} = \begin{pmatrix} \phi_0 & \alpha_{E,0} \\ 0 & \phi_0 \end{pmatrix}.$$

Two such extensions E and E' are isomorphic if there is an element $a \in \mathrm{End}\,(D)$ such that $aD^0 \subset D^0$,

$$\begin{pmatrix} 1 & a \\ 0 & 1 \end{pmatrix} \begin{pmatrix} \phi_{-1} & \alpha_{E,-1} \\ 0 & \phi_{-1} \end{pmatrix} = \begin{pmatrix} \phi_{-1} & \alpha_{E',-1} \\ 0 & \phi_{-1} \end{pmatrix} \begin{pmatrix} 1 & a \\ 0 & 1 \end{pmatrix},$$

and

$$\begin{pmatrix} 1 & a \\ 0 & 1 \end{pmatrix} \begin{pmatrix} \phi_0 & \alpha_{E,0} \\ 0 & \phi_0 \end{pmatrix} = \begin{pmatrix} \phi_0 & \alpha_{E',0} \\ 0 & \phi_0 \end{pmatrix} \begin{pmatrix} 1 & a|_{D^0} \\ 0 & 1 \end{pmatrix};$$

or equivalently if $aD^0 \subset D^0$, $\alpha_{E,-1} - \alpha_{E',-1} = \phi_{-1}a - a\phi_{-1}$ and $\alpha_{E,0} - \alpha_{E',0} = \phi_0 a|_{D^0} - a\phi_0$. The lemma follows. □

Corollary 2.34 *The group, $H^1_\mathrm{f}(G_\ell, \mathrm{ad}\rho)$, of extensions of M_ρ by itself in the category of good, finite $\mathcal{O}/\lambda^n[G_\ell]$-modules is (non-canonically) isomorphic to $(\mathcal{O}/\lambda^n)^2 \oplus H^0(I_\ell, M_\rho)$.*

Proof. Choose generators e_0, e_{-1} of D, such that e_0 is a generator of D^0. With respect to this basis let ϕ_0 have matrix $\binom{x}{z}$ and ϕ_{-1} have matrix $\begin{pmatrix} \ell x & y \\ \ell z & w \end{pmatrix}$. If a has matrix $\begin{pmatrix} a_1 & a_2 \\ 0 & a_3 \end{pmatrix}$ then a direct calculation shows that

$$(a\phi_{-1} - \phi_{-1}a, a\phi_0 - \phi_0 a|_{D^0}) = \left(\begin{pmatrix} \ell a_2 z & (a_1 - a_3)y + a_2(w - \ell x) \\ (a_3 - a_1)\ell z & -\ell z a_2 \end{pmatrix}, \begin{pmatrix} a_2 z \\ (a_3 - a_1)z \end{pmatrix} \right).$$

Thus the Ext group we want is the quotient of the set of pairs of matrices (α_{-1}, α_0) as above by the submodule generated by

$$\left(\begin{pmatrix} \ell z & w - \ell x \\ 0 & -\ell z \end{pmatrix}, \begin{pmatrix} z \\ 0 \end{pmatrix} \right) \quad \text{and} \quad \left(\begin{pmatrix} 0 & -y \\ \ell z & 0 \end{pmatrix}, \begin{pmatrix} 0 \\ z \end{pmatrix} \right).$$

Note that either z or w is a unit in $\mathcal{O}/\lambda$ from which it follows that the Ext group we want is isomorphic to $(\mathcal{O}/\lambda^n)^2 \oplus \mathcal{O}/(z, \lambda^n)$. On the other hand $H^0(I_\ell, M_\rho)$ corresponds to the largest submodule $C \subset D^0$ such that $\phi_0 C = C$, i.e. $C = \{d \in D^0 : zd = 0\} \cong \mathcal{O}/(z, \lambda^n)$. □

Corollary 2.35 *Suppose that $\rho : G_\ell \to GL_2(\mathcal{O}/\lambda^n)$ is a continuous good representation. Then $H^1_{\mathrm{f}}(G_\ell, \mathrm{ad}\rho)$ is isomorphic to $(\mathcal{O}/\lambda^n)^2 \oplus H^0(G_\ell, \mathrm{ad}^0\rho)$.*

Proof. If ρ is not ordinary then $H^0(G_\ell, \mathrm{ad}^0\rho)$ and $H^0(I_\ell, M_\rho)$ are both trivial, so suppose that ρ is ordinary. In this case, let ρ' denote the ordinary representation defined by

$$M_{\rho'} = \mathrm{Hom}(M_\rho^{(0)}, M_\rho) \subset \mathrm{ad}^0(\rho)$$

and let $M_{\rho'}^{(-1)} = \mathrm{ad}^0(\rho)^{(-1)}$. Then

$$H^0(G_\ell, \mathrm{ad}^0\rho) = H^0(G_\ell, M_{\rho'}) \subset H^0(I_\ell, M_{\rho'}) \cong H^0(I_\ell, M_\rho).$$

Since $H^0(I_\ell, M_{\rho'}^{(-1)})$ is trivial, the restriction homomorphism

$$H^1(G_\ell, M_{\rho'}^{(-1)}) \to H^1(I_\ell, M_{\rho'}^{(-1)})$$

is injective. We deduce from the long exact sequences associated to

$$(0) \to M_{\rho'}^{(-1)} \to M_{\rho'} \to \mathcal{O}/\lambda^n \to 0$$

that $H^0(G_\ell, M_{\rho'}) = H^0(I_\ell, M_{\rho'})$. □

2.6 Deformations of representations

In this section we shall review Mazur's theory of deformations of representations of profinite groups (see [Maz3]).

Let $\mathcal{C}_\mathcal{O}$ denote the category whose objects are complete noetherian local $\mathcal{O}$-algebras with residue field k and whose morphisms are $\mathcal{O}$-algebra homomorphisms which are local (i.e. take maximal ideals into maximal ideals). (The structure maps from $\mathcal{O}$ to every object of $\mathcal{C}_\mathcal{O}$ are also assumed to be local.) Let G denote a topologically finitely generated profinite group and let $\bar{\rho}$ denote an absolutely irreducible representation of G into $GL_d(k)$. (In fact all we shall use in the sequel is that k is the centraliser in $M_d(k)$ of the image of $\bar{\rho}$.) Let $\mathcal{D}_0$ denote the category of profinite $\mathcal{O}[G]$-modules with continuous morphisms. We will let $\mathcal{D}$ denote a full subcategory of $\mathcal{D}_0$ which is closed under taking sub-objects, quotients and direct products and which contains $M_{\bar{\rho}}$. Note that if M is an object of $\mathcal{D}_0$ and M_i is a collection of subobjects which have trivial intersection and such that each M/M_i is an object of $\mathcal{D}$, then M is an object of $\mathcal{D}$, since $M \subset \prod_i M/M_i$.

Let $\chi : G \to \mathcal{O}^\times$ be a continuous character such that $\det \bar{\rho} = \chi \bmod \lambda$. By a lifting of $\bar{\rho}$ of type $D = (\mathcal{O}, \chi, \mathcal{D})$ we shall mean an object R of $\mathcal{C}_\mathcal{O}$ and a continuous representation $\rho : G \to GL_d(R)$ such that:

(a) $\rho \bmod \mathfrak{m}_R = \bar{\rho}$,

(b) $\det \rho = \chi$,

(c) M_ρ is an object of $\mathcal{D}$.

Note that if $\phi : R \to R'$ and $\rho : G \to GL_d(R)$ is of type D so is $\phi \circ \rho$.

Theorem 2.36 *There is a lifting $\rho_D^{\mathrm{univ}} : G \to GL_d(R_D)$ of $\bar{\rho}$ of type D such that if $\rho : G \to GL_d(R)$ is any lifting of $\bar{\rho}$ of type D then there is a unique homomorphism of $\mathcal{O}$-algebras $\phi : R_D \to R$ such that ρ is conjugate to $\phi \circ \rho_D^{\mathrm{univ}}$.*

The representation ρ_D^{univ} is referred to as the *universal deformation* of type D. Mazur [Maz3] proved this theorem for $\mathcal{D}_0$ and certain other categories $\mathcal{D}$. Ramakrishna [Ram] observed that the arguments work with any category $\mathcal{D}$ satisfying the above hypotheses. We will sketch a proof which was suggested by Faltings. (We remark that another explicit construction of the deformation ring has been given by de Smit and Lenstra in [dSL].)
Proof of theorem 2.36: Choose a sequence $g_1, ..., g_r$ of topological generators of G and liftings $A_1, ..., A_r$ of $\bar{\rho}(g_1), ..., \bar{\rho}(g_r)$ to $M_d(\mathcal{O})$. Define a mapping $\iota : M_d(\mathcal{O}) \to M_d(\mathcal{O})^r$ which sends x to $(xA_1 - A_1x, ..., xA_r - A_rx)$. Since ι has torsion-free cokernel, so we can decompose

$$M_d(\mathcal{O})^r = \iota(M_d(\mathcal{O})) \oplus V,$$

for some submodule $V \subset M_d(\mathcal{O})^r$. If $\rho : G \to GL_d(R)$ is a lifting of $\bar{\rho}$ of type D set $v_\rho = (\rho(g_1) - A_1, ..., \rho(g_r) - A_r) \in M_d(R)^r$. Note that $v_\rho \equiv 0 \bmod \mathfrak{m}_R$ and that v_ρ completely determines ρ. We call the lifting ρ well-placed if v_ρ belongs to $V \otimes_\mathcal{O} R \subset M_d(R)^r$. The crucial observation is the following result.

Lemma 2.37 *If $\rho : G \to GL_d(R)$ is a lifting of $\bar{\rho}$ of type D then there is a unique conjugate ρ' of ρ which is well-placed.*

The lemma is first proved for algebras R such that $\mathfrak{m}_R^n = (0)$ by induction on n, and then one deduces the general case.

In virtue of the lemma it suffices to find a universal well-placed lifting of type D. Let $e_1, ..., e_s$ be a basis of V as an $\mathcal{O}$-module. If ρ is a well-placed lifting of $\bar{\rho}$ of type D then we can write $v_\rho = \sum_{i=1}^s v_{\rho,i} e_i$ for unique elements $v_{\rho,i} \in \mathfrak{m}_R$ and we can define a homomorphism $\theta_\rho : \mathcal{O}[[T_1, ..., T_s]] \to R$ sending T_i to $v_{\rho,i}$. Note that ρ is completely determined by θ_ρ ($\rho(g_i) = A_i + \sum_{j=1}^s \theta_\rho(T_j) e_{ji}$, where $e_j = (e_{j1}, ..., e_{jr})$). Let I denote the intersection of all ideals J of $\mathcal{O}[[T_1, ..., T_s]]$

such that there is a representation $\rho_J : G \to GL_d(\mathcal{O}[[T_1, ..., T_s]]/J)$ of type D with $\rho_J(g_i) = A_i + \sum_{j=1}^{s} T_j e_{ji}$ for all i. Let R_D denote the quotient of $\mathcal{O}[[T_1, ..., T_s]]$ by I. Then one can check that there is a representation $\rho^{\text{univ}} : G \to GL_d(R_D)$ with $\rho^{\text{univ}}(g_i) = A_i + \sum_{j=1}^{s} T_j e_{ji}$ for all i, and that this is the desired universal representation. □

We will need a few elementary properties of these universal deformations. More precisely we will need to know how these universal rings change when we change the base field, we will need to know how to calculate the equi-characteristic tangent space of these rings and more generally how to calculate $\wp/\wp^2$ for certain prime ideals $\wp$. The first of these lemmas is a remark of Faltings, the second is due to Mazur [Maz3] and the third to Wiles [W3].

Lemma 2.38 *Let K'/K be a finite extension with ring of integers $\mathcal{O}'$ and residue field k'. Let $\mathcal{D}'$ denote the full sub-category of the category of profinite $\mathcal{O}'[G]$-modules such that the underlying object of $\mathcal{D}_0$ is actually an object of $\mathcal{D}$. Let $D' = (\mathcal{O}', \chi, \mathcal{D}')$. Then $R_{D'} = R_D \otimes_{\mathcal{O}} \mathcal{O}'$ and $\rho_{D'}^{\text{univ}} = \rho_D^{\text{univ}} \otimes 1$.*

Proof: Let $\tilde{R}_{D'}$ denote the subring of $R_{D'}$ consisting of elements which reduce modulo the maximal ideal to an element of $k \subset k'$. Similarly let $\tilde{R}_D$ denote the subring of $R_D \otimes_{\mathcal{O}} \mathcal{O}'$ consisting of elements which reduce modulo the maximal ideal to an element of $k \subset k'$. Then $\rho_{D'}^{\text{univ}}$ is in fact valued in $GL_d(\tilde{R}_{D'})$ and $\rho_D^{\text{univ}} \otimes 1$ is in fact valued in $GL_d(\tilde{R}_D)$. The universal properties give natural maps $\alpha : R_D \to \tilde{R}_{D'}$ and $\beta : R_{D'} \to R_D \otimes_{\mathcal{O}} \mathcal{O}'$. Moreover they show that the composite $(\alpha \otimes 1) \circ \beta : R_{D'} \to R_{D'}$ is the identity and that $\beta \circ \alpha : R_D \to \tilde{R}_D$ is the natural embedding. Thus β is an isomorphism. □

We will let $\mathcal{D}^{(n)}$ denote the full subcategory of $\mathcal{D}$ whose objects are killed by λ^n. Suppose that M is an object of $\mathcal{D}_0$ which is finite and free over $\mathcal{O}/\lambda^n$. Recall from section 2.4 that $H^1(G, \text{End}\,(M))$ may be identified with $\text{Ext}^1_{\mathcal{D}_0^{(n)}}(M, M)$. If M is an object of $\mathcal{D}^{(n)}$ which is finite and free over $\mathcal{O}/\lambda^n$, then we have a natural inclusion

$$\text{Ext}^1_{\mathcal{D}^{(n)}}(M, M) \subset \text{Ext}^1_{\mathcal{D}_0^{(n)}}(M, M) \cong H^1(G, \text{End}\,(M)).$$

We define $H^1_{\mathcal{D}}(G, \text{End}\,(M))$ to be the image of $\text{Ext}^1_{\mathcal{D}^{(n)}}(M, M)$ in the group $H^1(G, \text{End}\,(M))$, and $H^1_{\mathcal{D}}(G, \text{End}^0(M))$ to be the intersection

$$H^1(G, \text{End}^0(M)) \cap H^1_{\mathcal{D}}(G, \text{End}\,(M)).$$

Lemma 2.39 *There is a canonical isomorphism of k-vector spaces*

$$\text{Hom}_k(\mathfrak{m}_{R_D}/(\lambda, \mathfrak{m}_{R_D}^2), k) \cong H^1_{\mathcal{D}}(G, \text{ad}^0\bar{\rho}).$$

Proof: There is a natural bijection between $\text{Hom}_k(\mathfrak{m}_{R_D}/(\lambda, \mathfrak{m}_{R_D}^2), k)$ and the set of $\mathcal{O}$-algebra homomorphisms from R_D to the algebra $k[\varepsilon]$ where $\varepsilon^2 = 0$ (the correspondence associates $\phi : R_D \to k[\varepsilon]$ to $\phi|_{\mathfrak{m}_{R_D}}$). Hence there is a bijection to the set of liftings $\rho : G \to GL_d(k[\varepsilon])$ of $\bar{\rho}$ of type D, modulo conjugation by elements of $1 + \varepsilon M_d(k)$. On the other hand, recall from section 2.4 that there is a natural bijection between $\text{Ext}^1_{\mathcal{D}_0^{(1)}}(M_{\bar{\rho}}, M_{\bar{\rho}})$ and the set of of all continuous liftings $\rho : G \to GL_d(k[\varepsilon])$ of $\bar{\rho}$ modulo conjugation by elements of $1 + \varepsilon M_d(k)$. Moreover a lifting ρ is type D if and only if $\det \rho = \det \bar{\rho}$ and the corresponding extension M_ρ is an object of $\mathcal{D}$. The lemma follows on checking linearity. □

Now suppose that $\theta : R_D \to \mathcal{O}$ is an $\mathcal{O}$-algebra homomorphism. Let $\wp$ denote the kernel of θ and let $\rho = \theta \circ \rho_D^{\text{univ}}$. We set

$$H^1_{\mathcal{D}}(G, \mathrm{ad}^0\rho \otimes K/\mathcal{O}) = \lim_{\rightarrow} H^1_{\mathcal{D}}(G, \mathrm{ad}^0\rho \otimes \lambda^{-n}/\mathcal{O}) \subset H^1(G, \mathrm{ad}^0\rho \otimes K/\mathcal{O}).$$

Lemma 2.40 *There is a canonical $\mathcal{O}$-linear isomorphism*

$$\mathrm{Hom}_{\mathcal{O}}(\wp/\wp^2, K/\mathcal{O}) \cong H^1_{\mathcal{D}}(G, \mathrm{ad}^0\rho \otimes K/\mathcal{O}).$$

Proof: One shows in a very similar manner to the proof of lemma 2.39 that for all n there is a natural isomorphism.

$$\mathrm{Hom}_{\mathcal{O}}(\wp/\wp^2, \mathcal{O}/\lambda^n) \cong H^1_{\mathcal{D}}(G, \mathrm{ad}^0\rho \otimes \mathcal{O}/\lambda^n).$$

□

2.7 Deformations of Galois representations

Again in this section we assume that ℓ is an odd prime. Let $\bar{\rho} : G_{\mathbb{Q}} \to GL_2(k)$ denote a continuous absolutely irreducible representation. Suppose moreover that $\det \bar{\rho} = \epsilon$ and that $\bar{\rho}$ is semi-stable in the sense that

- $\bar{\rho}|_{G_\ell}$ is semi-stable,
- and if $p \neq \ell$ then $\#\bar{\rho}(I_p)|\ell$.

Note that if $E/\mathbb{Q}$ is a semistable elliptic curve then $\bar{\rho}_{E,\ell} : G_{\mathbb{Q}} \to GL_2(\mathbb{F}_\ell)$, satisfies these conditions if it is irreducible. By a theorem of Mazur (theorem 2.9) this will be the case if $\ell > 7$.

Let Σ denote a finite set of prime numbers. If R is an object of $\mathcal{C}_{\mathcal{O}}$ then we say that a continuous lifting $\rho : G_{\mathbb{Q}} \to GL_2(R)$ of $\bar{\rho}$ is of type Σ if the following hold.

- $\det \rho = \epsilon$.
- $\rho|_{G_\ell}$ is semi-stable.
- If $\ell \notin \Sigma$ and $\bar{\rho}|_{G_\ell}$ is good then $\rho|_{G_\ell}$ is good.
- If $p \notin \Sigma \cup \{\ell\}$ and $\bar{\rho}$ is unramified at p then ρ is unramified at p.
- If $p \notin \Sigma \cup \{\ell\}$ then $\rho|_{I_p} \sim \begin{pmatrix} 1 & * \\ 0 & 1 \end{pmatrix}$.

Roughly speaking we require that at primes $p \notin \Sigma$, ρ is as unramified as could be hoped for and we require that $\rho|_{G_\ell}$ is semi-stable. Note that if $\Sigma \subset \Sigma'$ and ρ is a lifting of type Σ then it is also a lifting of type Σ'. Note also that if $E/\mathbb{Q}$ is an elliptic curve which is semi-stable at ℓ and for which $\bar{\rho}_{E,\ell}$ is irreducible and semi-stable, then $\rho_{E,\ell} : G_{\mathbb{Q}} \to GL_2(\mathbb{Z}_\ell)$ is a lifting of type Σ if Σ contains all the primes for which E has bad reduction.

Now suppose that $\rho : G_{\mathbb{Q}} \to GL_2(\mathcal{O}/\lambda^n)$ is a lifting of $\bar{\rho}$ of type Σ. We will write $H^1_\Sigma(\mathbb{Q}, \mathrm{ad}^0\rho)$ for $H^1_{\mathcal{L}_\Sigma}(\mathbb{Q}, \mathrm{ad}^0\rho)$, where

- $L_{\Sigma,p} = H^1(G_p/I_p, (\mathrm{ad}^0\rho)^{I_p})$ if $p \notin \Sigma \cup \{\ell\}$;

- $L_{\Sigma,p} = H^1(G_p, (\mathrm{ad}^0 \rho))$ if $p \in \Sigma$ and $p \neq \ell$;
- $L_{\Sigma,\ell} = H^1_{\mathrm{f}}(G_\ell, (\mathrm{ad}^0 \rho))$ if $\ell \notin \Sigma$;
- $L_{\Sigma,\ell} = H^1_{\mathrm{ss}}(G_\ell, (\mathrm{ad}^0 \rho))$ if $\ell \in \Sigma$.

Note that the pairing $\mathrm{ad}^0 \rho \times \mathrm{ad}^0 \rho \to \mathcal{O}/\lambda^n$ given by $(a, b) \mapsto \mathrm{tr}\,(ab)$ is perfect and respects the action of $G_{\mathbb{Q}}$ (i.e. $\mathrm{tr}\,((\mathrm{ad}^0 \rho)(g)a, (\mathrm{ad}^0 \rho)(g)b) = \mathrm{tr}\,(ab)$ for all $g \in G_{\mathbb{Q}}$). Choosing a generator for the $\mathcal{O}$-module $\mathrm{Hom}(\mathcal{O}/\lambda^n, \mathbb{Q}_\ell/\mathbb{Z}_\ell)$, we obtain an isomorphism of $\mathcal{O}[G_{\mathbb{Q}}]$-modules

$$\mathrm{ad}^0 \rho(1) \cong \mathrm{Hom}_{\mathcal{O}}(\mathrm{ad}^0 \rho, \mathcal{O}/\lambda^n)(1) \xrightarrow{\sim} \mathrm{Hom}(\mathrm{ad}^0 \rho, \mathbb{Q}_\ell/\mathbb{Z}_\ell)(1) = (\mathrm{ad}^0 \rho)^*.$$

The $\mathcal{O}$-submodule of $H^1(\mathbb{Q}, \mathrm{ad}^0 \rho(1))$ corresponding to $H^1_{\mathcal{L}^*_\Sigma}(\mathbb{Q}, (\mathrm{ad}^0 \rho)^*)$ is independent of the choice of generator and we denote it $H^1_{\Sigma}(\mathbb{Q}, \mathrm{ad}^0 \rho(1))$. Thus $H^1_{\Sigma}(\mathbb{Q}, \mathrm{ad}^0 \rho(1))$ is defined by the local conditions $\{L^\perp_{\Sigma,v}\}$ where the orthogonality is with respect to the pairing

$$H^1(G_v, \mathrm{ad}^0 \rho) \times H^1(G_v, \mathrm{ad}^0 \rho(1)) \to \mathbb{Q}_\ell/\mathbb{Z}_\ell$$

arising from the above isomorphism. We may equivalently regard the orthogonality as being with respect to the natural perfect $\mathcal{O}$-bilinear pairing defined by the composition

$$\begin{aligned} H^1(G_v, \mathrm{ad}^0 \rho) \times H^1(G_v, \mathrm{ad}^0 \rho(1)) &\to H^2(G_v, \mathrm{ad}^0 \rho \otimes_{\mathcal{O}} \mathrm{ad}^0 \rho(1)) \\ &\to H^2(G_v, \mathcal{O}/\lambda^n(1)) \cong \mathcal{O}/\lambda^n. \end{aligned}$$

Note that if $p \neq \ell$ we have that

- $L^\perp_{\Sigma,p} = H^1(G_p/I_p, (\mathrm{ad}^0 \rho)(1)^{I_p})$ if $p \notin \Sigma$;
- $L^\perp_{\Sigma,p} = (0)$ if $p \in \Sigma$.

If $\rho : G_{\mathbb{Q}} \to GL_2(\mathcal{O})$ is a lifting of $\bar{\rho}$ of type Σ, we will write $H^1_{\Sigma}(\mathbb{Q}, \mathrm{ad}^0 \rho \otimes K/\mathcal{O})$ for the direct limit $\varinjlim H^1_{\Sigma}(\mathbb{Q}, \mathrm{ad}^0 \rho \otimes \lambda^{-n}/\mathcal{O})$, and $H^1_{\Sigma}(\mathbb{Q}, \mathrm{ad}^0 \rho(1) \otimes K/\mathcal{O})$ for the direct limit $\varinjlim H^1_{\Sigma}(\mathbb{Q}, \mathrm{ad}^0 \rho(1) \otimes \lambda^{-n}/\mathcal{O})$.

The main result we need about deformations of $\bar{\rho}$ is the following.

Theorem 2.41 *There is a universal lifting $\rho^{\mathrm{univ}}_{\Sigma} : G_{\mathbb{Q}} \to GL_2(R_\Sigma)$ of $\bar{\rho}$ of type Σ, i.e. $\rho^{\mathrm{univ}}_{\Sigma}$ is a lifting of type Σ and if $\rho : G_{\mathbb{Q}} \to GL_2(R)$ is any lifting of type Σ then there is a unique $\mathcal{O}$-algebra homomorphism $\phi : R_\Sigma \to R$ such that $\rho \sim \phi \circ \rho^{\mathrm{univ}}_{\Sigma}$. Moreover we have the following.*

(a) If K'/K is a finite extension and R'_Σ is the corresponding deformation ring then $R'_\Sigma = R_\Sigma \otimes_{\mathcal{O}} \mathcal{O}'$ and $(\rho^{\mathrm{univ}}_{\Sigma})' = \rho^{\mathrm{univ}}_{\Sigma} \otimes 1$.

(b) R_Σ can be topologically generated as an $\mathcal{O}$-algebra by $\dim_k H^1_{\Sigma}(\mathbb{Q}, \mathrm{ad}^0 \bar{\rho})$ elements.

(c) If $\phi : R_\Sigma \twoheadrightarrow \mathcal{O}$ is a $\mathcal{O}$-algebra homomorphism, if $\rho = \phi \circ \rho_\Sigma$ and if $\wp = \ker \phi$ then $\mathrm{Hom}(\wp/\wp^2, K/\mathcal{O}) \cong H^1_{\Sigma}(\mathbb{Q}, \mathrm{ad}^0 \rho \otimes K/\mathcal{O})$.

Proof: Let L_0 denote the fixed field of $\bar{\rho}$. Let L_n (for $n \in \mathbb{Z}_{>0}$) denote the maximal elementary abelian ℓ-extension of L_{n-1} unramified outside Σ, $\{\ell\}$ and the primes where $\bar{\rho}$ ramifies. Let $L_\infty = \bigcup_n L_n$ and let $G = \mathrm{Gal}\,(L_\infty/\mathbb{Q})$. Note that any lifting of $\bar{\rho}$ of type Σ factors through G. $\mathrm{Gal}\,(L_\infty/L_0)$ is a pro-ℓ-group and its maximal elementary abelian quotient, $\mathrm{Gal}\,(L_1/L_0)$, is finite by theorem 2.2. We deduce from the following lemma that $\mathrm{Gal}\,(L_\infty/L_0)$ and hence G are topologically finitely generated. (See for instance [Koc] Satz 4.10 for a proof of this lemma.)

Lemma 2.42 *Let H be a pro-ℓ-group and $\bar{H}$ its maximal elementary abelian quotient. Suppose $h_1, ..., h_r \in H$ map to a set of generators of $\bar{H}$, then $h_1, ..., h_r$ topologically generate H.*

Let $\mathcal{D}$ denote the category of profinite $\mathcal{O}[G]$-modules M for which

- M is semi-stable as an $\mathcal{O}[G_\ell]$-module,
- if $\ell \notin \Sigma$ and if $\bar{\rho}$ is good then M is good as an $\mathcal{O}[G_\ell]$-module,
- if $p \notin \Sigma \cup \{\ell\}$ and if $\bar{\rho}$ is ramified at p then there is an exact sequence of $\mathcal{O}[I_p]$-modules
$$(0) \to M^{(-1)} \to M \to M^{(0)} \to (0),$$
such that I_p acts trivially on $M^{(-1)}$ and $M^{(0)}$.

Then we see that a lifting $\rho : G_\mathbb{Q} \to GL_2(R)$ of $\bar{\rho}$ is of type Σ if and only if

- ρ factors through G,
- $\det \rho = \epsilon$,
- M_ρ is an object of $\mathcal{D}$.

The existence part of theorem 2.41 now follows from theorem 2.36 and part (a) follows from lemma 2.38. Parts (b) and (c) follow from lemmas 2.39 and 2.40, and the following observation:
If $\rho : G_\mathbb{Q} \to GL_2(\mathcal{O}/\lambda^n)$ is a lifting of type Σ and if $p \notin \Sigma \cup \{l\}$ is a prime where $\bar{\rho}$ ramifies then

$$\ker\,(H^1(G_p, \mathrm{ad}^0\rho) \to H^1(I_p, \mathrm{ad}^0\rho/(\mathrm{ad}^0\rho)^{I_p})) = H^1(G_p/I_p, (\mathrm{ad}^0\rho)^{I_p}). \tag{2.7.1}$$

Equation (2.7.1) follows from the fact that the natural map $H^1(I_p, \mathrm{ad}^0\rho) \to H^1(I_p, \mathrm{ad}^0\rho/(\mathrm{ad}^0\rho)^{I_p})$ is an injection (in fact an isomorphism). It is nothing other than the map $(\mathrm{ad}^0\rho)_{I_p} \to (\mathrm{ad}^0\rho/(\mathrm{ad}^0\rho)^{I_p})_{I_p}$. This completes the proof of theorem 2.41. □

Corollary 2.43 *Suppose that if $\ell = 3$ then $\bar{\rho}|_{G_{\mathbb{Q}(\sqrt{-3})}}$ is absolutely irreducible. Then R_Σ can be topologically generated as an $\mathcal{O}$-algebra by*

$$\dim_k H^1_\Sigma(\mathbb{Q}, \mathrm{ad}^0\bar{\rho}(1)) + d_\ell + \sum_{p\in\Sigma-\{\ell\}} \dim_k H^0(\mathbb{Q}_p, \mathrm{ad}^0\bar{\rho}(1))$$

elements, where $d_\ell = \dim_k H^1_{\mathrm{ss}}(\mathbb{Q}_\ell, \mathrm{ad}^0\bar{\rho}) - \dim_k H^1_{\mathrm{f}}(\mathbb{Q}_\ell, \mathrm{ad}^0\bar{\rho})$ if $\ell \in \Sigma$, while $d_\ell = 0$ if $\ell \notin \Sigma$.

Proof. Note that $H^0(\mathbb{Q}, \mathrm{ad}^0\bar{\rho}(1)) = 0$ unless $\ell = 3$ and $\bar{\rho}|_{G_{\mathbb{Q}(\sqrt{-3})}}$ is not absolutely irreducible. Thus according to theorem 2.18, $\dim_k H^1_\Sigma(\mathbb{Q}, \mathrm{ad}^0\bar{\rho})$ is the sum of the following terms.

- $\dim_k H^1_{\Sigma}(\mathbb{Q}, \mathrm{ad}^0\bar{\rho}(1))$.
- $\dim_k H^1_f(\mathbb{Q}_\ell, \mathrm{ad}^0\bar{\rho}) - \dim_k H^0(\mathbb{Q}_\ell, \mathrm{ad}^0\bar{\rho}) - \dim_k H^0(\mathbb{R}, \mathrm{ad}^0\bar{\rho}) \leq 0$ by proposition 2.27.
- d_ℓ.
- For each $p \in \Sigma - \{l\}$, $\dim_k H^1(\mathbb{Q}_p, \mathrm{ad}^0\bar{\rho}) - \dim_k H^0(\mathbb{Q}_p, \mathrm{ad}^0\bar{\rho}) = \dim_k H^0(\mathbb{Q}_p, \mathrm{ad}^0\bar{\rho}(1))$ by the local Euler characteristic formula (see theorem 2.17).

□

2.8 Special cases

In this section we will restrict attention to the deformation problems associated with sets Q of certain special primes q. These sets Q are chosen so that the associated deformation ring R_Q enjoys a number of special properties which will be crucial in the sequel. In particular we will only consider sets Q of primes q which satisfy the following two properties

- $q \equiv 1 \bmod \ell$,
- $\bar{\rho}$ is unramified at q and $\bar{\rho}(\mathrm{Frob}_q)$ has distinct k-rational eigenvalues.

For such primes q we will let Δ_q denote the maximal quotient of $(\mathbb{Z}/q\mathbb{Z})^\times$ of ℓ-power order. Then Δ_q is naturally a quotient of both G_q and $G_\mathbb{Q}$ via the map

$$\chi_q : G_\mathbb{Q} \twoheadrightarrow \mathrm{Gal}\,(\mathbb{Q}(\zeta_q)/\mathbb{Q}) \cong (\mathbb{Z}/q\mathbb{Z})^\times \twoheadrightarrow \Delta_q.$$

We will let $\Delta_Q = \prod_{q \in Q} \Delta_q$, $\chi_Q = \prod_{q\in Q} \chi_q : G_\mathbb{Q} \twoheadrightarrow \Delta_Q$ and $\mathfrak{a}_Q$ denote the augmentation ideal of $\mathcal{O}[\Delta_Q]$. Note that there is an isomorphism $\mathcal{O}[\Delta_Q] \cong \mathcal{O}[[S_q : q \in Q]]/((1+S_q)^{\#\Delta_q} - 1 : q \in Q)$ under which $\mathfrak{a}_Q$ corresponds to $(S_q : q \in Q)$.

For each $q \in Q$, we choose an eigenvalue α_q of $\bar{\rho}(\mathrm{Frob}_q)$ and denote the other β_q.

Lemma 2.44 *If $q \in Q$ then $\rho_Q^{\mathrm{univ}}|_{G_q}$ is conjugate to $\begin{pmatrix} \xi & 0 \\ 0 & \epsilon\xi^{-1} \end{pmatrix}$ for some character ξ such that $\bar{\xi}(\mathrm{Frob}_q) = \alpha_q$.*

Proof. Choose a lifting $f \in G_q$ of Frob_q. As $\bar{\rho}(f)$ has distinct k-rational eigenvalues it is a simple application of Hensel's lemma to see that we can choose a basis such that $\rho_Q^{\mathrm{univ}}(f) = \begin{pmatrix} \tilde{\alpha}_q & 0 \\ 0 & \tilde{\beta}_q \end{pmatrix}$ where $\tilde{\alpha}_q$ and $\tilde{\beta}_q$ reduce to α_q and β_q in k. It suffices to show that for any $\sigma \in I_q$, $\rho_Q^{\mathrm{univ}}(\sigma)$ is diagonal in this basis. Suppose $\rho_Q^{\mathrm{univ}}(\sigma) = 1_2 + \begin{pmatrix} a & b \\ c & d \end{pmatrix}$ with $a, b, c, d \in \mathfrak{m}_{R_Q}$. Because ρ_Q^{univ} is tamely ramified at q we see that $\rho_Q^{\mathrm{univ}}(f)\rho_Q^{\mathrm{univ}}(\sigma)\rho_Q^{\mathrm{univ}}(f)^{-1} = \rho_Q^{\mathrm{univ}}(\sigma)^q$. Thus

$(\tilde{\alpha}_q/\tilde{\beta}_q - q)b$ and $(\tilde{\beta}_q/\tilde{\alpha}_q - q)c$ lie in $\mathfrak{m}_{R_Q}(b,c)$ so that $(b,c) = \mathfrak{m}_{R_Q}(b,c)$ and (by Nakayama's lemma) $b = c = 0$. □

We let $\xi_{q,Q}$ denote the character $\xi : G_q \to R_Q^\times$ in the conclusion of the lemma. The restriction of $\xi_{q,Q}$ to I_q factors through χ_q. We let ξ_Q denote the character $G_\mathbb{Q} \to R_Q^\times$ which is unramified outside the primes of Q and whose restriction to I_q is $\xi_{q,Q}$ for each q in Q. Thus ξ_Q factors through χ_Q. We wish to regard R_Q as an $\mathcal{O}[\Delta_Q]$-algebra, and it will be most convenient to do so via the map which gives rise to ξ_Q^{-2}. Note the following consequence of the lemma 2.44.

Corollary 2.45 *The natural map $R_Q \to R_\emptyset$ gives rise to an isomorphism $R_Q/\mathfrak{a}_Q R_Q \xrightarrow{\sim} R_\emptyset$.*

Lemma 2.46 *(a) If $q \in Q$ then $H^0(\mathbb{F}_q, \mathrm{ad}^0\bar{\rho}) = H^0(\mathbb{F}_q, \mathrm{ad}^0\bar{\rho}(1)) = k$ and $H^1(\mathbb{F}_q, \mathrm{ad}^0\bar{\rho}) = H^1(\mathbb{F}_q, \mathrm{ad}^0\bar{\rho}(1)) = k$.*

(b) R_Q can be topologically generated as an $\mathcal{O}$-algebra by

$$\#Q + \dim_k H^1_Q(\mathbb{Q}, \mathrm{ad}^0\bar{\rho}(1))$$

elements.

(c) If

$$H^1_\emptyset(\mathbb{Q}, \mathrm{ad}^0\bar{\rho}(1)) \xrightarrow{\sim} \bigoplus_{q\in Q} H^1(\mathbb{F}_q, \mathrm{ad}^0\bar{\rho}(1))$$

then $\#Q = \dim_k H^1_\emptyset(\mathbb{Q}, \mathrm{ad}^0\bar{\rho}(1))$ and R_Q can be topologically generated as an $\mathcal{O}$-algebra by $\#Q$ elements.

Proof: The first part is a direct calculation using the fact that Frob_q acts semi-simply on $\mathrm{ad}^0\bar{\rho}$ with eigenvalues $x, 1, x^{-1}$ for some $x \in \bar{k}\backslash\{0,1\}$. The same is true for $\mathrm{ad}^0\bar{\rho}(1)$ as $q \equiv 1 \bmod \ell$. The second and third parts follow from this and corollary 2.43. □

For the proof of the last theorem of the chapter, we shall need two results on finite groups.

Theorem 2.47 *(a) If H is a finite subgroup of $PGL_2(\mathbb{C})$ then H is isomorphic to one of the following groups: the cyclic group C_n of order n ($n \in \mathbb{Z}_{>0}$), the dihedral group D_{2n} of order $2n$ ($n \in \mathbb{Z}_{>1}$), A_4, S_4 or A_5.*

(b) If H is a finite subgroup of $PGL_2(\bar{\mathbb{F}}_\ell)$ then one of the following holds:

- *H is conjugate to a subgroup of the upper triangular matrices;*
- *H is conjugate to $PSL_2(\mathbb{F}_{\ell^r})$ or $PGL_2(\mathbb{F}_{\ell^r})$ for some $r \in \mathbb{Z}_{>0}$;*
- *H is isomorphic to A_4, S_4, A_5 or the dihedral group D_{2r} of order $2r$ for some $r \in \mathbb{Z}_{>1}$ not divisible by ℓ.*

In fact we shall only need part (b) which is due to Dickson [Dic2], secs. 255, 260 (see also [Hu] II.8.27), but we have included part (a) for later reference.

Lemma 2.48 *Let $\mathbb{F}$ be a finite field of odd characteristic ℓ. If $\#\mathbb{F} \neq 5$, then*

$$H^1(SL_2(\mathbb{F}), \mathrm{End}^0(\mathbb{F}^2)) = 0.$$

Proof. This is a special case of results of [CPS], table 4.5 (assuming $\#\mathbb{F} \neq 3$). In fact we shall only need it in the case $\ell = 3$, but the proof in the general case is no more difficult, and we sketch it here for the reader's convenience.

We let B (resp. U) denote the group of upper-triangular (resp. unipotent) matrices in $G = SL_2(\mathbb{F})$. Since ℓ does not divide the index of B in G, the restriction homomorphism

$$H^1(G, \mathrm{End}^0(\mathbb{F}^2)) \to H^1(B, \mathrm{End}^0(\mathbb{F}^2))$$

is injective, so it suffices to prove the latter group vanishes. Since ℓ does not divide the index of U in B, we have

$$H^i(B, M) \cong H^i(U, M)^{B/U}$$

all integers $i \geq 0$ and $\mathbb{F}[B]$-modules M. If $\#\mathbb{F} = 3$, then one checks directly that for $M = \mathrm{End}^0(\mathbb{F}_3^2)$,

$$H^1(U, M) \cong \ker N/(\sigma - 1)M = 0$$

where σ generates U and $N = 1 + \sigma + \sigma^2$ on M. If $\#\mathbb{F} > 5$, then one proceeds by writing

$$(0) = M_0 \subset M_1 \subset M_2 \subset M_3 = \mathrm{End}^0(\mathbb{F}^2)$$

as $\mathbb{F}[B]$-modules where the successive quotients M_i/M_{i-1}, (for $i = 1, 2, 3$) are one-dimensional over $\mathbb{F}$. The calculation is then straightforward using long exact sequences of cohomology, except in the case $\#\mathbb{F} = 9$ where one must also check that the one-dimensional space of classes in $H^1(U, M_3/M_2)$ fixed by B/U maps injectively to $H^2(U, M_2/M_1)$ via the connecting homomorphism. □

Theorem 2.49 *Keep the assumptions of the last section and suppose moreover that if $L = \mathbb{Q}(\sqrt{(-1)^{(\ell-1)/2}\ell})$ then $\bar{\rho}|_{G_L}$ is absolutely irreducible. Then there exists a non-negative integer r such that for any $n \in \mathbb{Z}_{>0}$ we can find a finite set of primes Q_n with the following properties.*

(a) If $q \in Q_n$ then $q \equiv 1 \bmod \ell^n$.

(b) If $q \in Q_n$ then $\bar{\rho}$ is unramified at q and $\bar{\rho}(\mathrm{Frob}_q)$ has distinct eigenvalues.

(c) $\#Q_n = r$.

(d) R_{Q_n} can be topologically generated by r elements as a $\mathcal{O}$-algebra.

Proof. Take $r = \dim_k H^1_\emptyset(\mathbb{Q}, \mathrm{ad}^0\bar{\rho}(1))$. It suffices to find a set Q_n with the following properties.

(a) If $q \in Q_n$ then $q \equiv 1 \bmod \ell^n$.

(b) If $q \in Q_n$ then $\bar{\rho}$ is unramified at q and $\bar{\rho}(\mathrm{Frob}_q)$ has distinct eigenvalues.

(c) $H^1_\emptyset(\mathbb{Q}, \mathrm{ad}^0\bar{\rho}(1)) \xrightarrow{\sim} \bigoplus_{q \in Q_n} H^1(\mathbb{F}_q, \mathrm{ad}^0\bar{\rho}(1))$.

As each $H^1(\mathbb{F}_q, \mathrm{ad}^0\bar{\rho}(1))$ is one dimensional we may replace the last condition with

$$H^1_\emptyset(\mathbb{Q}, \mathrm{ad}^0\bar{\rho}(1)) \hookrightarrow \bigoplus_{q \in Q_n} H^1(\mathbb{F}_q, \mathrm{ad}^0\bar{\rho}(1)).$$

Thus what we need show is that for each non-zero class $[\psi] \in H^1_\emptyset(\mathbb{Q}, \mathrm{ad}^0\bar{\rho}(1))$ there is a prime q (depending on $[\psi]$) such that

(a) $q \equiv 1 \bmod \ell^n$,

(b) $\bar{\rho}$ is unramified at q and $\bar{\rho}(\mathrm{Frob}_q)$ has distinct eigenvalues,

(c) $\mathrm{res}_q[\psi] \in H^1(\mathbb{F}_q, \mathrm{ad}^0\bar{\rho}(1))$ is nontrivial.

Using the Chebotarev density theorem we see that it will do to find $\sigma \in G_{\mathbb{Q}}$ such that

(a) $\sigma|_{\mathbb{Q}(\zeta_{\ell^n})} = 1$,

(b) $\mathrm{ad}^0\bar{\rho}(\sigma)$ has an eigenvalue other than 1,

(c) $\psi(\sigma) \notin (\sigma - 1)\mathrm{ad}^0\bar{\rho}(1)$.

For $m \geq 0$, let F_m denote the extension of $\mathbb{Q}(\zeta_{\ell^m})$ cut out by $\mathrm{ad}^0\bar{\rho}$; i.e., the field fixed by the kernel of the representation $\mathrm{ad}^0\bar{\rho}$ restricted to $G_{\mathbb{Q}(\zeta_{\ell^m})}$. We will show that $\psi(G_{F_n})$ is non-trivial. For this it suffices to prove that $H^1(\mathrm{Gal}(F_n/\mathbb{Q}), \mathrm{ad}^0\bar{\rho}(1)) = (0)$. Consider the inflation-restriction exact sequence

$$(0) \to H^1(\mathrm{Gal}(F_0/\mathbb{Q}), (\mathrm{ad}^0\bar{\rho}(1))^{G_{F_0}}) \to H^1(\mathrm{Gal}(F_n/\mathbb{Q}), \mathrm{ad}^0\bar{\rho}(1)) \\ \to H^1(\mathrm{Gal}(F_n/F_0), \mathrm{ad}^0\bar{\rho}(1))^{G_{\mathbb{Q}}}.$$

Since F_1/F_0 is an extension of degree prime to ℓ, and since $G_{\mathbb{Q}}$ acts trivially on $\mathrm{Gal}(F_n/F_1)$, we have:

$$H^1(\mathrm{Gal}(F_n/F_0), \mathrm{ad}^0\bar{\rho}(1))^{G_{\mathbb{Q}}} \simeq \mathrm{Hom}(\mathrm{Gal}(F_n/F_1), \mathrm{ad}^0\bar{\rho}(1)^{G_{\mathbb{Q}}}). \tag{2.8.1}$$

Since $\bar{\rho}|_{G_L}$ is absolutely irreducible, the cohomology group in equation (2.8.1) vanishes. On the other hand, G_{F_0} acts trivially on $\mathrm{ad}^0\bar{\rho}$ so the first term vanishes as well unless $\mathrm{Gal}(F_0/\mathbb{Q})$ has order divisible by ℓ and has $\mathrm{Gal}(\mathbb{Q}(\zeta_\ell)/\mathbb{Q})$ as a quotient. Recall that $\mathrm{Gal}(F_0/\mathbb{Q})$ is isomorphic to the projective image of $\bar{\rho}$, so by theorem 2.47 we are reduced to the case $\ell = 3$ and the map

$$\mathrm{Gal}(F_0/\mathbb{Q}(\zeta_3)) \to PSL_2(\bar{k})$$

has image conjugate to $PSL_2(\mathbb{F}_{3^r})$ for some $r \geq 1$. It suffices to prove that in this case

$$H^1(\mathrm{Gal}(F_0/\mathbb{Q}(\zeta_3)), \mathrm{ad}^0\bar{\rho} \otimes_k \bar{k})) = (0),$$

but this follows directly from lemma 2.48.

Now note that since $\bar{\rho}|_{G_L}$ is absolutely irreducible, so is $\bar{\rho}|_{G_{\mathbb{Q}(\zeta_{\ell^n})}}$. Regarding $\psi(G_{F_n})$ as a $\mathrm{Gal}(F_n/\mathbb{Q}(\zeta_{\ell^n}))$-submodule of $\mathrm{ad}^0(\bar{\rho})$, we find that some $g \in \mathrm{Gal}(F_n/\mathbb{Q}(\zeta_{\ell^n}))$ of order not dividing ℓ fixes a non-zero element of $\psi(G_{F_n})$. Let σ_0 be a lifting of g to $G_{\mathbb{Q}(\zeta_{\ell^n})}$. We will look for $\sigma = \tau\sigma_0$ with $\tau \in G_{F_n}$. We only need choose τ so that $\psi(\tau\sigma_0) = \psi(\tau) + \psi(\sigma_0) \notin (\sigma_0 - 1)\mathrm{ad}^0\bar{\rho}(1)$. This is possible because $\psi(G_F) \not\subset (g-1)\mathrm{ad}^0\bar{\rho}(1)$. □

3 Modular forms and Galois representations

3.1 From modular forms to Galois representations

We suppose in this section that $f = \sum a_n(f)q^n$ is a newform of weight two and level N_f (see definition 1.21). Let K_f denote the number field in $\mathbb{C}$ generated by the Fourier coefficients $a_n(f)$. Let ψ_f denote the character of f, i.e., the homomorphism $(\mathbb{Z}/N_f\mathbb{Z})^\times \to K_f^\times$ defined by mapping d to the eigenvalue of $\langle d \rangle$ on f.

Recall that a construction of Shimura (section 1.7) associates to f an abelian variety A_f of dimension $[K_f : \mathbb{Q}]$. This abelian variety is a certain quotient of $J_1(N_f)$, and the action of the Hecke algebra on $J_1(N_f)$ provides an embedding

$$K_f \hookrightarrow \operatorname{End}_{\mathbb{Q}}(A_f) \otimes \mathbb{Q}.$$

We saw also that for each prime ℓ the Tate module $T_\ell(A_f) \otimes_{\mathbb{Z}_\ell} \mathbb{Q}_\ell$ becomes a free module of rank two over $K_f \otimes \mathbb{Q}_\ell$ (lemma 1.48). The action of the Galois group $G_{\mathbb{Q}}$ on the Tate module commutes with that of K_f, so that a choice of basis for the Tate module provides a representation

$$G_{\mathbb{Q}} \to GL_2(K_f \otimes \mathbb{Q}_\ell). \tag{3.1.1}$$

As $K_f \otimes \mathbb{Q}_\ell$ can be identified with the product of the completions of K_f at its primes over ℓ, we obtain from f certain 2-dimensional ℓ-adic representations of $G_{\mathbb{Q}}$.

ℓ-adic representations: In this discussion, we fix a prime ℓ and a finite extension K of $\mathbb{Q}_\ell$. We let $\mathcal{O}$ denote the ring of integers of K, λ the maximal ideal and k the residue field. We shall consider ℓ-adic representations with coefficients in finite extensions of our fixed field K. We regard K as a subfield of $\bar{\mathbb{Q}}_\ell$ and fix embeddings $\bar{\mathbb{Q}} \hookrightarrow \bar{\mathbb{Q}}_\ell$ and $\bar{\mathbb{Q}} \hookrightarrow \bar{\mathbb{C}}$. If K' is a finite extension of K with ring of integers $\mathcal{O}'$, then we say that an ℓ-adic representation $G_\ell \to GL_2(K')$ is good (respectively, ordinary, semistable) if it is conjugate over K' to a representation $G_\ell \to GL_2(\mathcal{O}')$ which is good (respectively, ordinary, semistable) in the sense of section 2.4.

Let K'_f denote the K-algebra in $\bar{\mathbb{Q}}_\ell$ generated by the Fourier coefficients of f. Thus K'_f is a finite extension of K, and it contains the completion of K_f at the prime over ℓ determined by our choice of embeddings. We let $\mathcal{O}'_f$ denote the ring of integers of K'_f and write k_f for its residue field. We define

$$\rho_f : G_{\mathbb{Q}} \to GL_2(K'_f)$$

as the pushforward of (3.1.1) by the natural map $K_f \otimes \mathbb{Q}_\ell \to K'_f$. We assume the basis is chosen so that ρ_f factors through $GL_2(\mathcal{O}'_f)$. We also let ψ'_f denote the finite order ℓ-adic character

$$G_{\mathbb{Q}} \twoheadrightarrow \operatorname{Gal}(\mathbb{Q}(\zeta_{N_f})/\mathbb{Q}) \to (K'_f)^\times$$

obtained from ψ_f.

The following theorem lists several fundamental properties of the ℓ-adic representations ρ_f obtained from Shimura's construction. The result is a combination of the work of many mathematicians. We discuss some of the proofs and provide references below. In the statement we fix f as above and write simply N, a_n, ρ, ψ, ψ' and K' for N_f, $a_n(f)$, ρ_f, ψ_f, ψ'_f and K'_f respectively.

Theorem 3.1 *The ℓ-adic representation*

$$\rho : G_{\mathbf{Q}} \to GL_2(K')$$

has the following properties.

(a) If $p \nmid N\ell$ then ρ is unramified at p and $\rho(\mathrm{Frob}_p)$ has characteristic polynomial

$$X^2 - a_pX + p\psi(p).$$

(b) $\det(\rho)$ is the product of ψ' with the ℓ-adic cyclotomic character ϵ, and $\rho(c)$ is conjugate to $\begin{pmatrix} 1 & 0 \\ 0 & -1 \end{pmatrix}$.

(c) ρ is absolutely irreducible.

(d) The conductor $N(\rho)$ is the prime-to-ℓ-part of N.

(e) Suppose that $p \neq \ell$ and $p||N$. Let χ denote the unramified character $G_p \to (K')^\times$ satisfying $\chi(\mathrm{Frob}_p) = a_p$. If p does not divide the conductor of ψ, then $\rho|_{G_p}$ is of the form

$$\begin{pmatrix} \chi\epsilon & * \\ 0 & \chi \end{pmatrix}.$$

If p divides the conductor of ψ, then $\rho|_{G_p}$ is of the form

$$\chi^{-1}\epsilon\psi'|_{G_p} \oplus \chi.$$

(f) If $\ell \nmid 2N$, then $\rho|_{G_\ell}$ is good. Moreover $\rho|_{G_\ell}$ is ordinary if and only if a_ℓ is a unit in the ring of integers of K', in which case $\rho_{I_\ell}(\mathrm{Frob}_\ell)$ is the unit root of the polynomial $X^2 - a_\ell X + \ell\psi(\ell)$.

(g) If ℓ is odd and $\ell||N$, but the conductor of ψ is not divisible by ℓ, then $\rho|_{G_\ell}$ is ordinary and $\rho_{I_\ell}(\mathrm{Frob}_\ell) = a_\ell$.

Proof: Part (a) was established by Shimura ([Shi2], [Shi3]). The key ingredient is the Eichler-Shimura congruence relation, theorem 1.29. Recall that $J_1(N)$ has good reduction at primes p not dividing N. So the action of G_p on $T_\ell(A_f)\otimes_{\mathbf{Z}_\ell}\mathbf{Q}_\ell$ is unramified and is in fact described by the action of $\mathrm{Frob}_p \in G_{\mathbf{F}_p}$ on the Tate module of the reduction. But this is given by the Frobenius endomorphism F whose characteristic polynomial is computed in corollary 1.41.

The first assertion of (b) follows from (a) on applying the Chebotarev density theorem. The second assertion then follows on noting that $\psi(-1) = 1$.

Part (c) was proved by Ribet (see section 2 of [R3]). Assuming reducibility of the representation, he applies algebraicity results of Lang and Serre to obtain a contradiction to the estimate on the Fourier coefficients stated in theorem 1.24.

Parts (d) and (e) follow from a deep result of Carayol [Ca1], Thm. (A), building on the work of Langlands [Ll1], Deligne and others. In fact, this result and the local Langlands correspondence characterize $\rho|_{G_p}$ in terms of $\psi|_{G_p}$ and the L- and ϵ-factors at p of twists of f. The descriptions in the case of $p||N$ are based on the analysis of Deligne-Rapoport of the reduction mod p of $J_1(N)$ (see [DR], [Ll1]).

The first assertion of (f) follows from the fact that A_f has good reduction at ℓ if ℓ does not divide N. The second assertion of (f) (respectively, all of (g)) follows from the Eichler-Shimura congruence relation (respectively, the work of Deligne-Rapoport), and general results on ℓ-divisible groups and the reduction of abelian varieties; see thm. 2.2 of [W2], lemma 2.1.5 of [W1], §12 of [Gro] and thms. 2.5 and 2.6 of [Edi]. The restriction to odd ℓ in (f) and (g) is made primarily out of lack of suitable definitions. □

Mod ℓ representations: We maintain the notation used in the discussion of ℓ-adic representations. Define

$$\bar{\rho}_f : G_{\mathbf{Q}} \to GL_2(k_f)$$

to be the *semi-simplification* of the reduction of ρ_f. (See the discussion following proposition 2.6.) Assertions analogous to those in theorem 3.1 hold for $\bar{\rho} = \bar{\rho}_f$, except that

- The representation need not be absolutely irreducible (as in (c)). However if ℓ is odd, one checks using (b) that $\bar{\rho}$ is irreducible if and only if it is absolutely irreducible.
- In (d) one only has divisibility of the prime-to-ℓ part of N_f by $N(\bar{\rho})$ in general.

The various possibilities for $m_p(\bar{\rho})$ to be strictly less than the exponent of p in N (where $p \neq \ell$) were classified independently by Carayol [Ca2] and Livné [Liv]. We record the following consequence of their results (cf. the introduction of [DT1]):

Proposition 3.2 *Suppose that p is a prime such that $p|N_f$, $p \not\equiv 1 \bmod \ell$ and $\bar{\rho}_f$ is unramified at p. Then $\operatorname{tr}(\bar{\rho}_f(\operatorname{Frob}_p))^2 = (p+1)^2$ in k_f.*

Artin representations: The theory of Hecke operators and newforms (see section 1.3) extends to modular forms on $\Gamma_1(N)$ of arbitrary weight. The construction of ℓ-adic representations associated to newforms was generalized to weight greater than 1 by Deligne [De] using étale cohomology. There are also Galois representations associated to newforms of weight 1 by Deligne and Serre [DS], but an essential difference is that these are Artin representations.

Theorem 3.3 *If $g = \sum a_n(g)q^n$ is a newform of weight one, level N_g and character ψ_g, then there is an irreducible Artin representation*

$$\rho_g : G_{\mathbf{Q}} \longrightarrow GL_2(\mathbf{C})$$

of conductor N_g with the following property: If $p \nmid N_g$, then the characteristic polynomial of $\rho_g(\operatorname{Frob}_p)$ is

$$X^2 - a_p(g)X + \psi_g(p).$$

Remark 3.4 Note that $\det(\rho_g)$ is the character of $G_{\mathbf{Q}}$ corresponding to ψ and $\rho_g(c)$ is conjugate to $\begin{pmatrix} 1 & 0 \\ 0 & -1 \end{pmatrix}$ (see theorem 3.1).

Remark 3.5 A basis can be chosen so that the representation ρ_g takes values in $GL_2(K_g)$ (where K_g is the number field generated by the $a_n(g)$). Moreover suppose that K is a finite extension of $\mathbb{Q}_\ell$ in $\bar{\mathbb{Q}}_\ell$ and we have fixed embeddings of $\bar{\mathbb{Q}}$ in $\mathbb{C}$ and $\bar{\mathbb{Q}}_\ell$). If K_g is contained in K, then we can view ρ_g as giving rise to an ℓ-adic representation $G_\mathbb{Q} \to GL_2(K)$ and hence a mod ℓ representation $G_\mathbb{Q} \to GL_2(k)$.

Remark 3.6 A key idea in the construction of ρ_g is to first construct the mod ℓ representations using those already associated to newforms of higher weight. More precisely, suppose that $K_g \hookrightarrow K$ as in remark 3.5. One can show that for some newform f of weight 2 and level N_f dividing $N\ell$ we have

$$a_p(g) \equiv a_p(f), \qquad \psi_g(p)) \equiv p\psi_f(p)$$

for all $p \nmid N\ell$, the congruence being modulo the maximal ideal of the ring of integers of K'_f. Thus $\bar{\rho}_f$ is the semisimplification of the desired mod ℓ representation (with scalars extended to k_f).

3.2 From Galois representations to modular forms

It is conjectured that certain types of two-dimensional representations of $G_\mathbb{Q}$ always arise from the constructions described in section 3.1. We now state some of the conjectures and the results known prior to [W3] and [TW].

Artin representations:

Conjecture 3.7 *Let $\rho : G_\mathbb{Q} \to GL_2(\mathbb{C})$ be a continuous irreducible representation with* $\det(\rho(c)) = -1$. *Then ρ is equivalent to ρ_g for some newform g of weight one.*

Recall that ρ_g is the Artin representation associated to g by the Deligne-Serre construction (theorem 3.3).

Remark 3.8 Conjecture 3.7 is equivalent to the statement that the Artin L-functions attached to ρ and to all its twists by one-dimensional characters are entire. (The Artin conjecture predicts that the Artin L-function $L(s, \rho)$ is entire, for an arbitrary irreducible, non-trivial Artin representation $\rho : G_\mathbb{Q} \longrightarrow GL_d(\mathbb{C})$.)

A large part of conjecture 3.7 was proved by Langlands in [Ll2], and the results were extended by Tunnell [Tu].

Theorem 3.9 *Let $\rho : G_\mathbb{Q} \to GL_2(\mathbb{C})$ be a continuous irreducible representation such that $\rho(G_\mathbb{Q})$ is solvable and* $\det(\rho(c)) = -1$. *Then ρ is equivalent to ρ_g for some newform g of weight one.*

Remark 3.10 Note that by theorem 2.47, part (a), the solvability hypothesis excludes only the case where the projective image of ρ is isomorphic to A_5.

Remark 3.11 If the projective image of ρ is dihedral, then ρ is induced from a character of a quadratic extension of $\mathbb{Q}$. In this case the result can already be deduced from work of Hecke.

Mod ℓ representations: We fix notation as in the discussion of ℓ-adic and mod ℓ representations in section 3.1.

Definition 3.12 We say that a representation $\bar{\rho} : G_{\mathbb{Q}} \to GL_2(k)$ is *modular (of level N)* if for some newform f of weight 2 (and level N), $\bar{\rho}$ is equivalent over k_f to $\bar{\rho}_f$.

By proposition 1.32 the notion is independent of the choices in section 3.1 of embeddings $K \hookrightarrow \bar{\mathbb{Q}}_\ell$, $\bar{\mathbb{Q}} \hookrightarrow \bar{\mathbb{Q}}_\ell$ and $\bar{\mathbb{Q}} \hookrightarrow \mathbb{C}$. Moreover if K' is a finite extension of K with residue field k', then $\bar{\rho}$ is modular if and only if $\bar{\rho} \otimes_k k'$ is modular.

The following was conjectured by Serre [Se7], (3.2.3). (See also [Dal] for further discussion and references.)

Conjecture 3.13 *Let $\bar{\rho} : G_{\mathbb{Q}} \to GL_2(k)$ be a continuous absolutely irreducible representation with* $\det(\bar{\rho}(c)) = -1$. *Then $\bar{\rho}$ is modular.*

Some cases of Serre's conjecture can be deduced from theorem 3.9.

Theorem 3.14 *Let $\bar{\rho} : G_{\mathbb{Q}} \to GL_2(k)$ be a continuous absolutely irreducible representation with* $\det(\bar{\rho}(c)) = -1$. *Suppose that one of the following holds:*

(a) $k = \mathbb{F}_3$;

(b) the projective image of $\bar{\rho}$ is dihedral.

Then $\bar{\rho}$ is modular.

Sketch of proof: For case (a), we consider the surjection

$$GL_2(\mathbb{Z}[\sqrt{-2}]) \to GL_2(\mathbb{F}_3)$$

defined by reduction mod $(1+\sqrt{-2})$. One checks that there is a section $s : GL_2(\mathbb{F}_3) \to GL_2(\mathbb{Z}[\sqrt{-2}])$ and applies theorem 3.9 to $s \circ \bar{\rho}$. The resulting representation arises from a weight one newform, and hence its reduction $\bar{\rho}$ is equivalent to $\bar{\rho}_f$ for some f (see remark 3.6).

In case (b), $\bar{\rho}$ is equivalent to a representation of the form $\operatorname{Ind}_{G_F}^{G_{\mathbb{Q}}} \bar{\xi}$ where F is a quadratic extension of $\mathbb{Q}$ and $\bar{\xi}$ is a character $G_F \to k^\times$. (We have here enlarged K if necessary.) Let n be the order of $\bar{\xi}$, choose an embedding $\mathbb{Q}(e^{2\pi i/n}) \hookrightarrow K$ and lift $\bar{\xi}$ to a character $\xi : G_F \to \mathbb{Z}[e^{2\pi i/n}]^\times$. We may always choose ξ so that the Artin representation $\rho = \operatorname{Ind}_{G_F}^{G_{\mathbb{Q}}} \xi$ is odd, i.e., $\det(\rho(c)) = -1$. (In the case $\ell = 2$ and F real quadratic, we may have to multiply ξ by a suitable quadratic character of G_F.) We then apply theorem 3.9 to ρ and deduce as in case (a) that $\bar{\rho}$ is modular. □

Serre also proposed a refinement ([Se7], (3.2.4)) of the conjecture which predicts that $\bar{\rho}$ is associated to a newform of specified weight, level and character. Through work of Mazur, Ribet [R5], Carayol [Ca2], Gross [Gro] and others, this refinement is now known to be equivalent to conjecture 3.13 if ℓ is odd. (One also needs to impose a mild restriction in the case $\ell = 3$.) See [R6] and [Di1] for statements of the results and further references; here we give a variant which applies to newforms of weight two. Before doing so, we assume ℓ is odd and define an integer $\delta(\bar{\rho})$ as follows:

- $\delta(\bar{\rho}) = 0$ if $\bar{\rho}|_{G_\ell}$ is good;

- $\delta(\bar{\rho}) = 1$ if $\bar{\rho}|_{G_\ell}$ is not good and $\bar{\rho}|_{I_\ell} \otimes_k \bar{k}$ is of the form

$$\begin{pmatrix} \epsilon^a & * \\ 0 & 1 \end{pmatrix}, \quad \begin{pmatrix} \epsilon & * \\ 0 & \epsilon^a \end{pmatrix} \quad \text{or} \quad \begin{pmatrix} \psi^a & 0 \\ 0 & \psi^{\ell a} \end{pmatrix}$$

  for some positive integer $a < \ell$. (Recall that ϵ is the cyclotomic character and ψ is the character of I_ℓ defined after lemma 2.22.)

- $\delta(\bar{\rho}) = 2$ otherwise.

Theorem 3.15 *Suppose that ℓ is odd and $\bar{\rho}$ is absolutely irreducible and modular. If $\ell = 3$, then suppose also that $\bar{\rho}|_{G_{\mathbf{Q}(\sqrt{-3})}}$ is absolutely irreducible. Then there exists a newform f of weight two such that*

- *$\bar{\rho}$ is equivalent over k_f to $\bar{\rho}_f$;*

- *$N_f = N(\bar{\rho})\ell^{\delta(\bar{\rho})}$;*

- *the order of ψ_f is not divisible by ℓ.*

Proof: The existence of such an f follows from [Dil] thm. 1.1, thm. 5.1 and lemma 2.1, but with N_f dividing $N(\bar{\rho})\ell^{\delta(\bar{\rho})}$. By lemma 2.7 above, we see that N_f is divisible by $N(\bar{\rho})$. The divisibility of N_f by $\delta(\bar{\rho})$ follows from results in sec. 8 of [Gro] and sec. 2.4 of [Edi]. □

ℓ-adic representations: We again use the notation of section 3.1. Let $\rho : G_{\mathbf{Q}} \to GL_2(K)$ be an ℓ-adic representation.

Definition 3.16 We say that ρ is *modular* if for some weight 2 newform f, ρ is equivalent over K'_f to ρ_f.

The notion is independent of the choices of embeddings and well-behaved under extension of scalars by proposition 1.32 (cf. definition 3.12).

The following is a special case of a conjecture of Fontaine and Mazur [FM].

Conjecture 3.17 *If $\rho : G_{\mathbf{Q}} \to GL_2(K)$ is an absolutely irreducible ℓ-adic representation and $\rho|_{G_\ell}$ is semistable (in the sense of section 2.4), then ρ is modular.*

(Recall that for us ℓ-adic representations are defined to be unramified at all but finitely many primes. Recall also that if $\rho|_{G_\ell}$ is semistable, then by definition $\det \rho|_{I_\ell}$ is the cyclotomic character ϵ.)

Remark 3.18 Relatively little was known about this conjecture before Wiles' work [W3]. Wiles proves that under suitable hypotheses, the modularity of $\bar{\rho}$ implies that of ρ.

Remark 3.19 The conjecture stated in [FM] is stronger than the one here; in particular, the semistability hypothesis could be replaced with a suitable notion of potential semistability. On the other hand, one expects that if $\rho|_{G_\ell}$ is semistable, then it is equivalent to ρ_f (over K'_f) for some f on $\Gamma_1(N(\rho)) \cap \Gamma_0(\ell)$ (and on $\Gamma_1(N(\rho))$ if $\rho|_{G_\ell}$ is good).

The Shimura-Taniyama conjecture: Conjecture 1.54 can be viewed in the framework of the problem of associating modular forms to Galois representations. Let E be an elliptic curve defined over $\mathbb{Q}$. For each prime ℓ, we let $\rho_{E,\ell}$ denote the ℓ-adic representation $G_{\mathbb{Q}} \to GL_2(\mathbb{Q}_\ell)$ defined by the action of $G_{\mathbb{Q}}$ on the Tate module of E (see section 2.2).

Proposition 3.20 *The following are equivalent:*

(a) E is modular.

(b) $\rho_{E,\ell}$ is modular for all primes ℓ.

(c) $\rho_{E,\ell}$ is modular for some prime ℓ.

Proof: If E is modular, then E is isogenous to A_f for some weight two newform f with $K_f = \mathbb{Q}$ (see section 1.8). It follows that for each prime ℓ, $\rho_{E,\ell}$ is equivalent to the ℓ-adic representation ρ_f. Hence (a) $\Rightarrow$ (b) $\Rightarrow$ (c).

To show (c) $\Rightarrow$ (b), suppose that for some ℓ and some f, the representations $\rho_{E,\ell}$ and ρ_f are equivalent. First observe that for all but finitely primes p, we have

$$\mathrm{tr}\,(\rho_f(\mathrm{Frob}_p)) = \mathrm{tr}\,(\rho_{E,\ell}(\mathrm{Frob}_p)).$$

We deduce from proposition 2.11 and theorem 3.1, part (a) that for all but finitely many primes p

$$a_p(f) = p + 1 - \#\bar{E}_p(\mathbb{F}_p) \in \mathbb{Z}. \tag{3.2.1}$$

Applying proposition 2.6, we find that for each prime ℓ, $\rho_{E,\ell}$ is equivalent to ρ_f and is therefore modular.

We finally show that (b) $\Rightarrow$ (a). The equality (3.2.1) holds for all primes p not dividing N_f, which by theorem 3.1, part (d), is the conductor of E. Since $\det(\rho_f) = \det(\rho_{E,\ell}) = \epsilon$, we see by 3.1, Part (b) that ψ_f is trivial. By theorem 1.27 parts (b) and (d) (or [AL] thm. 3), a_p is in $\{0, \pm 1\}$ for primes p dividing N_f. Thus $K_f = \mathbb{Q}$ and A_f is an elliptic curve. Faltings' isogeny theorem (see [CS], sec. II.5) now tells us that E and A_f are isogenous and we conclude that E is modular. □

Remark 3.21 Note that the equivalence (b) $\Leftrightarrow$ (c) does not require Faltings' isogeny theorem.

Proposition 3.22 *If the Fontaine-Mazur conjecture (conjecture 3.17) holds for some prime ℓ, then the Shimura-Taniyama conjecture (conjecture 1.54) holds. If Serre's conjecture (conjecture 3.13) holds for infinitely many ℓ, then conjecture 1.54 holds.*

Proof: The first assertion is immediate from proposition 3.20 and the irreducibility of $\rho_{E,\ell}$. See [Se7], sec. 4.6 for a proof of the second. (We have implicitly chosen the field K to be $\mathbb{Q}_\ell$ in the statements of conjectures 3.17 and 3.13, but it may be replaced by a finite extension.) □

Remark 3.23 Note that to prove a given elliptic curve E is modular, it suffices to prove that conjecture 3.17 holds for a single ℓ at which E has semistable reduction. Wiles' approach is to show that certain cases of conjecture 3.13 imply cases of conjecture 3.17 and hence cases of conjecture 1.54.

3.3 Hecke algebras

In this section fix the following notation. Let ℓ be an odd prime, let K be a finite extension of $\mathbb{Q}_\ell$, let $\mathcal{O}$ denote the ring of integers of K, let λ denote its maximal ideal and k its residue field. Fix embeddings $K \hookrightarrow \bar{\mathbb{Q}}_\ell$, $\bar{\mathbb{Q}} \hookrightarrow \bar{\mathbb{Q}}_\ell$ and $\bar{\mathbb{Q}} \hookrightarrow \mathbb{C}$. Let $\bar{\rho} : G_\mathbb{Q} \to GL_2(k)$ denote a continuous representation with the following properties

(a) $\bar{\rho}$ is irreducible,

(b) $\bar{\rho}$ is modular,

(c) $\det \bar{\rho} = \epsilon$,

(d) $\bar{\rho}|_{G_\ell}$ is semi-stable,

(e) and if $p \neq \ell$ then $\#\bar{\rho}(I_p)|\ell$.

Let us first record the following lemma.

Lemma 3.24 *The representation $\bar{\rho}|_{G_L}$ is absolutely irreducible where $L = \mathbb{Q}(\sqrt{(-1)^{(\ell-1)/2}\ell})$.*

Proof: If it were not then we see that $\ell \not| \#\bar{\rho}(G_\mathbb{Q})$ and so $\bar{\rho}$ is unramified at all $p \neq \ell$. Moreover we can check that $\bar{\rho}|_{I_\ell} \sim \begin{pmatrix} \epsilon & 0 \\ 0 & 1 \end{pmatrix}$. If $\ell > 3$ we can use theorem 3.15 to deduce that $\bar{\rho}$ is modular of weight 2 and level 1 and hence obtain a contradiction. If $\ell = 3$ we see that the splitting field of $\bar{\rho}$ is everywhere unramified over $\mathbb{Q}(\sqrt{-3})$ and hence must equal $\mathbb{Q}(\sqrt{-3})$, a contradiction. □

Let Σ denote a finite set of primes. For the application to modularity of semistable elliptic curves, it suffices to consider sets Σ contained in $\Sigma_{\bar{\rho}}$ where $\Sigma_{\bar{\rho}}$ is defined as follows

Definition 3.25 For a representation $\bar{\rho}$ as above, we let $\Sigma_{\bar{\rho}}$ denote the set of primes p satisfying

- $p = \ell$ and $\bar{\rho}|_{G_\ell}$ is good and ordinary; or
- $p \neq \ell$ and $\bar{\rho}$ is unramified at p.

We shall sometimes assume that $\Sigma \subset \Sigma_{\bar{\rho}}$ in order to simplify statements and proofs. Let $\mathcal{N}_\Sigma$ denote the set of newforms f such that

$$\rho_f : G_\mathbb{Q} \to GL_2(\mathcal{O}'_f)$$

is equivalent to a lifting of $\bar{\rho} \otimes_k k_f$ of type Σ and $\ell^2 \not| N_f$ (the last condition is presumably not necessary, cf. remark 3.19). From theorem 3.1 and lemma 2.7 one deduces the following description of $\mathcal{N}_\Sigma$.

Lemma 3.26 *The set $\mathcal{N}_\Sigma$ consists of newforms f such that*

- $\bar{\rho}_f \cong \bar{\rho} \otimes_k k_f$,
- *ψ_f is trivial,*

- *N_f divides $\ell^\delta N(\bar\rho) \prod_{p\in\Sigma-\{\ell\}} p^{\dim \bar\rho^{I_p}}$, where $\delta = 0$ if $\bar\rho$ is good and $\ell \notin \Sigma$ and $\delta = 1$ otherwise.*

As $\bar\rho$ is modular it follows from theorem 3.15 that for all Σ, $\mathcal{N}_\Sigma \neq \emptyset$. Set $\tilde{\mathbb{T}}_\Sigma = \prod_{f\in\mathcal{N}_\Sigma} \mathcal{O}'_f$. If p is a prime not in Σ and not dividing $\ell N(\bar\rho)$, we let T_p denote the element $(a_p(f))_f$ in $\tilde{\mathbb{T}}_\Sigma$. Then define $\mathbb{T}_\Sigma$ to be the $\mathcal{O}$-subalgebra of $\tilde{\mathbb{T}}_\Sigma$ generated by the elements T_p for such primes p. Then $\mathbb{T}_\Sigma$ is a complete noetherian local $\mathcal{O}$-algebra with residue field k. Moreover it is reduced and it is a finitely generated free $\mathcal{O}$-module.

Lemma 3.27 *There is a continuous representation*

$$\rho_\Sigma^{\mathrm{mod}} : G_\mathbb{Q} \to GL_2(\mathbb{T}_\Sigma)$$

such that if $p \nmid \ell N(\bar\rho)$ and $p \notin \Sigma$ then $\rho_\Sigma^{\mathrm{mod}}$ is unramified at p and we have $\mathrm{tr}\,\rho_\Sigma^{\mathrm{mod}}(\mathrm{Frob}_p) = T_p$. *Moreover we have the following.*

(a) $\rho_\Sigma^{\mathrm{mod}}$ is a lift of $\bar\rho$ of type Σ and there is a unique surjection $\phi_\Sigma : R_\Sigma \twoheadrightarrow \mathbb{T}_\Sigma$ such that $\rho_\Sigma^{\mathrm{mod}} \sim \phi_\Sigma \circ \rho_\Sigma^{\mathrm{univ}}$.

(b) If $\Sigma' \supset \Sigma$ then there is a unique surjection $\mathbb{T}_{\Sigma'} \twoheadrightarrow \mathbb{T}_\Sigma$ such that $\rho_{\Sigma'}^{\mathrm{mod}}$ pushes forward to $\rho_\Sigma^{\mathrm{mod}}$ and T_p maps to T_p for $p \nmid \ell N(\bar\rho)$ and $p \notin \Sigma'$.

(c) If K' is a finite extension of K and $\mathbb{T}'_\Sigma$ is constructed in the same way as $\mathbb{T}_\Sigma$ but with K' replacing K then $\mathbb{T}'_\Sigma \cong \mathbb{T}_\Sigma \otimes_\mathcal{O} \mathcal{O}_{K'}$.

Proof: Consider $\tilde\rho_\Sigma^{\mathrm{mod}} = \prod \rho_f : G_\mathbb{Q} \to GL_2(\tilde{\mathbb{T}}_\Sigma)$. Choose some complex conjugation, c, and conjugate $\bar\rho$ so that $\bar\rho(c) = \begin{pmatrix}1 & 0\\ 0 & -1\end{pmatrix}$. By conjugating $\tilde\rho_\Sigma^{\mathrm{mod}}$ we may assume that modulo every maximal ideal it reduces to $\bar\rho$ and that $\tilde\rho_\Sigma^{\mathrm{mod}}(c) = \begin{pmatrix}1 & 0\\ 0 & -1\end{pmatrix}$. Call the corresponding basis $\{e_+, e_-\}$. Note that by continuity and the Chebotarev density theorem $\mathrm{tr}\,\tilde\rho_\Sigma^{\mathrm{mod}}$ is valued in $\mathbb{T}_\Sigma$. Thus for any $g \in G_\mathbb{Q}$ the diagonal entries of $\tilde\rho_\Sigma^{\mathrm{mod}}(g)$ lie in $\mathbb{T}_\Sigma$, because $\mathrm{tr}\,\tilde\rho_\Sigma^{\mathrm{mod}}(g)$ and $\mathrm{tr}\,\tilde\rho_\Sigma^{\mathrm{mod}}(cg)$ both do.

By irreducibility of $\bar\rho$ we can find some $\sigma \in G_\mathbb{Q}$ such that $\bar\rho(\sigma) = \begin{pmatrix}* & b\\ * & *\end{pmatrix}$ where $b \neq 0$. Rescaling e_+ we may assume that $\tilde\rho_\Sigma^{\mathrm{mod}}(\sigma) = \begin{pmatrix}a & 1\\ c & d\end{pmatrix}$. Then for all $g \in G_\mathbb{Q}$ the lower left entry of $\tilde\rho_\Sigma^{\mathrm{mod}}(g)$ lies in $\mathbb{T}_\Sigma$ (look at the upper left entry of $\tilde\rho_\Sigma^{\mathrm{mod}}(\sigma g)$). Again using the irreducibility of $\bar\rho$ we can find $\tau \in G_\mathbb{Q}$ such that $\tilde\rho_\Sigma^{\mathrm{mod}}(\tau) = \begin{pmatrix}* & *\\ e & *\end{pmatrix}$ where $e \in \mathbb{T}_\Sigma^\times$. Looking at the lower right entry of $\tilde\rho_\Sigma^{\mathrm{mod}}(\tau g)$ we see that for any $g \in G_\mathbb{Q}$ the upper right entry of $\tilde\rho_\Sigma^{\mathrm{mod}}(g)$ lies in $\mathbb{T}_\Sigma$. Thus $\tilde\rho_\Sigma^{\mathrm{mod}}$ is now in fact valued in $GL_2(\mathbb{T}_\Sigma)$ and will be our candidate for $\rho_\Sigma^{\mathrm{mod}}$. We leave the verification of the other properties of $\rho_\Sigma^{\mathrm{mod}}$ as an exercise. □

Example 3.28 Let $\bar\rho = \bar\rho_{f_{57B},3}$ where f_{57B} is the newform of level 57 discussed in the example of section 1.6. As f_{57B} is not congruent modulo 3 to any form of level 19 or 3 we see by theorem 3.15 that $\bar\rho$ is ramified at 19 and $\bar\rho|_{G_3}$ is not good. On the other hand $\bar\rho$ is semi-stable. The facts that $\bar\rho(\mathrm{Frob}_2)$ has order 8 (see the table below) and $3|\#\bar\rho(I_{19})$ (from the discussion above) imply that $\bar\rho : G_\mathbb{Q} \twoheadrightarrow GL_2(\mathbb{F}_3)$. The table below also shows that $\mathbb{T}_\emptyset \cong \{(x,y) \in \mathbb{Z}_3^2 : x \equiv y \bmod 3\}$. There is a unique continuous representation $\rho_\emptyset^{\mathrm{mod}} : G_\mathbb{Q} \to GL_2(\mathbb{T}_\emptyset)$ such that if $p \not| 57$ then $\rho_\emptyset^{\mathrm{mod}}$ is unramified at p and $\mathrm{tr}\,\rho_\emptyset^{\mathrm{mod}}(\mathrm{Frob}_p) = T_p$. We have an isomorphism $\mathbb{T}_\emptyset/3\mathbb{T}_\emptyset \xrightarrow{\sim} \mathbb{F}_3[\varepsilon]$ (where $\varepsilon^2 = 0$) given by $(x, y) \mapsto x + \frac{y-x}{3}\varepsilon$. Thus we get a representation $\rho : G_{\mathbb{Q}} \to GL_2(\mathbb{F}_3[\varepsilon])$. The following table lists the traces of Frobenius elements for the first few unramified primes under these representations:

p	2	5	7	11	13	17	23	29
$\bar\rho$	1	1	0	0	0	0	1	-1
$\rho_\emptyset^{\mathrm{mod}}$	$(1,-2)$	$(-2,1)$	$(0,3)$	$(0,-3)$	$(6,-6)$	$(-6,3)$	$(4,4)$	$(2,-10)$
ρ	$1-\varepsilon$	$1+\varepsilon$	ε	$-\varepsilon$	$-\varepsilon$	0	1	$-1-\varepsilon$

Exercise 3.29 Show that there are three algebra homomorphisms $\mathbb{T}_\emptyset \to \mathbb{Z}/9\mathbb{Z}$ and hence show that there are at least three liftings of $\bar\rho$ of type $\emptyset$ to $\mathbb{Z}/9\mathbb{Z}$.

Exercise 3.30 What is the image of ρ?

We will need two deeper properties of the Hecke algebras $\mathbb{T}_\Sigma$ (theorem 3.31 and theorem 3.36 below). These results will be proved in the next chapter (sections 4.3 and 4.4).

Theorem 3.31 *Let Q be a finite set of primes as in section 2.8. Then $\mathbb{T}_Q$ is a free $\mathcal{O}[\Delta_Q]$-module, where $\mathbb{T}_Q$ inherits the structure of an $\mathcal{O}[\Delta_Q]$-module from R_Q via the homomorphism $\phi_Q : R_Q \twoheadrightarrow \mathbb{T}_Q$.*

Corollary 3.32 $\mathbb{T}_\emptyset = \mathbb{T}_Q/\mathfrak{a}_Q$.

Proof: From the definitions and corollary 2.45 we have that

$$\mathbb{T}_\emptyset \otimes_{\mathcal{O}} K = (\mathbb{T}_Q \otimes_{\mathcal{O}} K)/\mathfrak{a}_Q,$$

and from the theorem we have that $\mathbb{T}_Q/\mathfrak{a}_Q$ is torsion free. The corollary follows. □

For the second of these results we will need some additional notation and we restrict our attention to sets of primes contained in $\Sigma_{\bar\rho}$. Suppose that $\Sigma \subset \Sigma_{\bar\rho}$ and that f is an element of $\mathcal{N}_\Sigma$ whose Fourier coefficients are in $\mathcal{O}$. Then $\mathcal{O}'_f = \mathcal{O}$ and projection to the component corresponding to f gives rise to an $\mathcal{O}$-algebra homomorphism

$$\pi = \pi_f : \mathbb{T}_\Sigma \to \mathcal{O}.$$

The pushforward of $\rho_\Sigma^{\mathrm{mod}}$ by π is equivalent to

$$\rho_f : G_{\mathbb{Q}} \to GL_2(\mathcal{O}).$$

Remark 3.33 Most of the objects considered in the rest of the section will depend on the choice of newform f. We also remark that to give an $\mathcal{O}$-algebra homomorphism $\mathbb{T}_\Sigma \to \mathcal{O}$ is equivalent to giving a newform f in $\mathcal{N}_\Sigma$ with coefficients in $\mathcal{O}$. Indeed given such a homomorphism, there exists a newform f in $\mathcal{N}_\Sigma$ such that the homomorphism is defined by $T_p \mapsto a_p(f)$ for all $p \notin \Sigma$ with p not dividing $\ell N(\bar\rho)$. The fact that $a_p(f) \in \mathcal{O}$ for such p implies (using for example parts (b) and (d) of theorem 1.27 and lemma 4.1 below) that all the Fourier coefficients of f are in $\mathcal{O}$. The uniqueness of f is a consequence of the theory of newforms, theorem 1.22.

For Σ' satisfying

$$\Sigma \subset \Sigma' \subset \Sigma_{\bar{\rho}},$$

let $\pi_{\Sigma'}$ denote the composite $\mathbb{T}_{\Sigma'} \to \mathbb{T}_{\Sigma} \to \mathcal{O}$. Let

$$\eta_{\Sigma'} = \pi_{\Sigma'}(\operatorname{Ann}_{\mathbb{T}_{\Sigma'}}(\ker \pi_{\Sigma'})). \tag{3.3.1}$$

Note that because $\mathbb{T}_{\Sigma'}$ is reduced, $\eta_{\Sigma'} \neq (0)$. Also let $\wp_{\Sigma'}$ denote the kernel of $\phi_{\Sigma'} \circ \pi_{\Sigma'}$. Recall that

$$\#\wp_{\Sigma'}/\wp_{\Sigma'}^2 = \#H^1_{\Sigma'}(\mathbb{Q}, \operatorname{ad}^0 \rho_f \otimes_{\mathcal{O}} K/\mathcal{O}).$$

Remark 3.34 We have not yet shown that these groups are finite, but if either is finite, then so is the other and their cardinality is the same.

Note that if ℓ is in $\Sigma_{\bar{\rho}} - \Sigma$ then $T_\ell := (a_\ell(g))_g \in \tilde{\mathbb{T}}_\Sigma$ is actually in $\mathbb{T}_\Sigma$. This follows from theorem 3.1(f), which shows that $T_\ell = \alpha_\ell + \ell\alpha_\ell^{-1}$ where α_ℓ is the eigenvalue of Frob_ℓ on the free rank one $\mathbb{T}_\Sigma$-module $M^{(0)}$ where M is the module underlying $\rho_\Sigma^{\mathrm{mod}}$ (see lemma 2.20). Now that we have shown that $T_\ell \in \mathbb{T}_\Sigma$, we may characterize α_ℓ as the unit root in $\mathbb{T}_\Sigma$ of

$$X^2 - T_\ell X + \ell = 0. \tag{3.3.2}$$

For primes p in $\Sigma_{\bar{\rho}} - \Sigma$ we define an element $c_p \in \mathbb{T}_\Sigma$ by

$$c_p = (p-1)(T_p^2 - (p+1)^2).$$

Note that $\pi(c_p) = (p-1)(a_p(f)^2 - (p+1)^2)$, which is non-zero by theorem 1.27(a). A calculation using theorem 2.17 and theorem 3.1(a) shows that if p is in $\Sigma_{\bar{\rho}} - (\Sigma \cup \ell)$ then

$$\begin{aligned}&\#H^1(G_p, \operatorname{ad}^0\rho_f \otimes_{\mathcal{O}} K/\mathcal{O})/H^1(G_p/I_p, \operatorname{ad}^0\rho_f \otimes_{\mathcal{O}} K/\mathcal{O}) \\ &\quad = \#H^0(G_p, \operatorname{ad}^0\rho_f \otimes_{\mathcal{O}} K/\mathcal{O}(1)) = \#\mathcal{O}/\pi(c_p).\end{aligned}$$

If ℓ is in $\Sigma_{\bar{\rho}} - \Sigma$ then we see from proposition 2.27 and theorem 3.1(f) that

$$\#H^1_{\mathrm{ss}}(G_\ell, \operatorname{ad}^0\rho_f \otimes_{\mathcal{O}} K/\mathcal{O})/H^1_{\mathrm{f}}(G_\ell, \operatorname{ad}^0\rho_f \otimes_{\mathcal{O}} K/\mathcal{O}) = \#\mathcal{O}/\pi(c_\ell).$$

For the next proposition define groups H_p for $p \in \Sigma_{\bar{\rho}} - \Sigma$ by

- $H_p = H^1(G_p, \operatorname{ad}^0\rho_f \otimes_{\mathcal{O}} K/\mathcal{O})/H^1(G_p/I_p, \operatorname{ad}^0\rho_f \otimes_{\mathcal{O}} K/\mathcal{O})$ if $p \neq \ell$
- and $H_\ell = H^1_{\mathrm{ss}}(G_\ell, \operatorname{ad}^0\rho_f \otimes_{\mathcal{O}} K/\mathcal{O})/H^1_{\mathrm{f}}(G_\ell, \operatorname{ad}^0\rho_f \otimes_{\mathcal{O}} K/\mathcal{O})$.

Then we have the following result.

Proposition 3.35 *If $\Sigma \subset \Sigma' \subset \Sigma_{\bar{\rho}}$, then*

$$\#H^1_{\Sigma'}(\mathbb{Q}, \operatorname{ad}^0\rho_f \otimes_{\mathcal{O}} K/\mathcal{O})/H^1_{\Sigma}(\mathbb{Q}, \operatorname{ad}^0\rho_f \otimes_{\mathcal{O}} K/\mathcal{O}) \leq \#(\mathcal{O}/\pi(\prod_{p \in \Sigma' - \Sigma} c_p)).$$

Moreover if we have equality then the sequence

$$(0) \to H^1_{\Sigma}(\mathbb{Q}, \operatorname{ad}^0\rho_f \otimes_{\mathcal{O}} K/\mathcal{O}) \to H^1_{\Sigma'}(\mathbb{Q}, \operatorname{ad}^0\rho_f \otimes_{\mathcal{O}} K/\mathcal{O}) \to \bigoplus_{p \in \Sigma' - \Sigma} H_p \to (0)$$

is exact.

Finally we state the second theorem on Hecke algebras which shall be proved in section 4.4.

Theorem 3.36 *If $\Sigma \subset \Sigma' \subset \Sigma_{\bar{\rho}}$ and f is a newform in $\mathcal{N}_\Sigma$ with coefficients in $\mathcal{O}$, then we have*

$$\eta_{\Sigma'} \subset \pi(\prod_{p \in \Sigma' - \Sigma} c_p)\eta_\Sigma.$$

Corollary 3.37 *With the above notation*

$$\#H^1_{\Sigma'}(\mathbb{Q}, \mathrm{ad}^0 \rho_f \otimes_{\mathcal{O}} K/\mathcal{O})/H^1_\Sigma(\mathbb{Q}, \mathrm{ad}^0 \rho_f \otimes_{\mathcal{O}} K/\mathcal{O}) \leq \#(\eta_\Sigma/\eta'_\Sigma).$$

3.4 Isomorphism criteria

The main thrust of Wiles' approach is to prove that in many circumstances the map $\phi_\Sigma : R_\Sigma \to \mathbb{T}_\Sigma$ is an isomorphism. For this we will need two criteria from commutative algebra. The first was found by Wiles [W3] (but is presented here in a slightly stronger form due to Lenstra [Len]); the second was developed by Faltings, from the original arguments of [TW]. In both criteria the notion of complete intersection plays a vital part.

Proofs of all the results in this section, together with some background, references and examples, is given in chapter 5.

Definition 3.38 Suppose that A is an object of $\mathcal{C}_{\mathcal{O}}$ which is finitely generated and free as an $\mathcal{O}$-module. Then we call A a *complete intersection* if and only if for some $r \in \mathbb{Z}_{\geq 0}$ and some $f_1, ..., f_r \in \mathcal{O}[[X_1, ..., X_r]]$ we have

$$A \cong \mathcal{O}[[X_1, ..., X_r]]/(f_1, ..., f_r)$$

(i.e. there are the same number of generators as relations).

We first record a lemma about complete intersections.

Lemma 3.39 *Suppose that K'/K is a finite extension with ring of integers $\mathcal{O}'$ and that A is an object of $\mathcal{C}_{\mathcal{O}}$ which is finitely generated and free as an $\mathcal{O}$-module. Then A is a complete intersection if and only if $A \otimes_{\mathcal{O}} \mathcal{O}'$ is.*

For the proof, see chapter 5, lemma 5.30.

Now fix objects R and T of $\mathcal{C}_{\mathcal{O}}$ and a surjection of $\mathcal{O}$-algebras $\phi : R \twoheadrightarrow T$. Also assume that T *is a finitely generated, free $\mathcal{O}$-module.* The first criterion is as follows.

Theorem 3.40 *Suppose that $\pi : T \twoheadrightarrow \mathcal{O}$. Let $\wp = \ker(\pi \circ \phi) \lhd R$ and let $\eta = \pi(\mathrm{Ann}_T(\ker \pi)) \lhd \mathcal{O}$. Suppose also that $\eta \neq (0)$. Then the following are equivalent.*

(a) The inequality $\#\wp/\wp^2 \leq \#\mathcal{O}/\eta$ is satisfied.

(b) The equality $\#\wp/\wp^2 = \#\mathcal{O}/\eta$ is satisfied.

(c) The rings R and T are complete intersections, and the map $\phi : R \longrightarrow T$ is an isomorphism.

The proof is explained in chapter 5, sections 5.1 to 5.8. (See theorem 5.3.)

For the second criterion let us also fix a non-negative integer r. If $J \lhd \mathcal{O}[[S_1, ..., S_r]]$ is an ideal contained in $(S_1, ..., S_r)$, then by a J-structure we mean a commutative diagram in $\mathcal{C}_\mathcal{O}$

$$\begin{array}{ccccc} & & \mathcal{O}[[S_1, ..., S_r]] & & \\ & & \downarrow & & \\ \mathcal{O}[[X_1, ..., X_r]] & \twoheadrightarrow & R' & \twoheadrightarrow & T' \\ & & \downarrow & & \downarrow \\ & & R & \twoheadrightarrow & T, \end{array}$$

such that

(a) $T'/(S_1, ..., S_r) \stackrel{\sim}{\to} T$ and $R'/(S_1, ..., S_r) \twoheadrightarrow R$,

(b) for each ideal $I \supset J$, $I = \ker(\mathcal{O}[[S_1, ..., S_r]] \to T'/I)$.

Theorem 3.41 *Suppose there exist a sequence of ideals $J_n \lhd \mathcal{O}[[S_1, ..., S_r]]$ such that $J_0 = (S_1, ..., S_r)$, $J_n \supset J_{n+1}$, $\bigcap_n J_n = (0)$ and for each n there exists a J_n-structure. Then the map $R \longrightarrow T$ is an isomorphism and these rings are complete intersections.*

The proof of this statement is explained in sections 5.9 and 5.10.

3.5 The main theorem

We are now in a position to deduce the main theorems. We will keep the notation and assumptions from the start of section 3.3.

Theorem 3.42 *Keep the notation and assumptions of section 3.3. Then, for all finite sets $\Sigma \subset \Sigma_{\bar{\rho}}$, $\phi_\Sigma : R_\Sigma \longrightarrow \mathbb{T}_\Sigma$ is an isomorphism and these rings are complete intersections.*

Remark 3.43 There seems to be a deep link between the fact that ϕ_Σ is an isomorphism and the fact that $\mathbb{T}_\Sigma$ is a complete intersection. The proof of the theorem divides into two parts. One first proves it in the minimal case where $\Sigma = \emptyset$. One then deduces the full theorem from this special case by a different argument. In both these steps the facts that ϕ_Σ is an isomorphism and that $\mathbb{T}_\Sigma$ is a complete intersection are proved simultaneously.

Proof of theorem 3.42: Note that to prove the theorem we may extend scalars if necessary (by lemma 3.39) and hence assume that both of the following hold:

- The eigenvalues of all elements of the image of ρ are rational over k.
- There is a newform f in $\mathcal{N}_\emptyset$ with coefficients in $\mathcal{O}$, hence an $\mathcal{O}$-algebra homomorphism $\mathbb{T}_\emptyset \to \mathcal{O}$.

We first prove that $\phi_\emptyset : R_\emptyset \longrightarrow \mathbb{T}_\emptyset$ is an isomorphism and that the rings $R_\emptyset$ and $\mathbb{T}_\emptyset$ are complete intersections.

Note that according to theorem 2.49 and lemma 3.24 we can find an integer $r \geq 0$ and for each $n \in \mathbb{Z}_{>0}$ we can find a set Q_n of r primes such that

- if $q \in Q_n$ then $q \equiv 1 \bmod \ell^n$;

- if $q \in Q_n$ then $\bar{\rho}$ is unramified at q and $\bar{\rho}(\mathrm{Frob}_q)$ has distinct eigenvalues;
- R_{Q_n} can be topologically generated by r elements as a $\mathcal{O}$-algebra.

Let $J'_n = ((S_i + 1)^{\ell^{n_i}} - 1 : i = 1, \dots, r) \lhd \mathcal{O}[[S_1, \dots, S_r]]$, where the n_i are chosen such that $\mathcal{O}[[S_1, \dots, S_r]]/J_n \cong \mathcal{O}[\Delta_{Q_n}]$. Then for each n we have a diagram

$$\begin{array}{cccc} & \mathcal{O}[[S_1, \dots, S_r]] & & \\ & \downarrow & & \\ \mathcal{O}[[X_1, \dots, X_r]] \twoheadrightarrow & R_{Q_n} & \twoheadrightarrow & \mathbb{T}_{Q_n} \\ & \downarrow & & \downarrow \\ & R_\emptyset & \twoheadrightarrow & \mathbb{T}_\emptyset, \end{array}$$

where

(a) $R_{Q_n}/(S_1, \dots, S_r) \tilde{\to} R_\emptyset$ (see corollary 2.45),

(b) $\mathbb{T}_{Q_n}$ is finite and free over $\mathcal{O}[[S_1, \dots, S_r]]/J'_n$ (see theorem 3.31),

(c) $\mathbb{T}_{Q_n}/(S_1, \dots, S_r) \tilde{\to} \mathbb{T}_\emptyset$ (see corollary 3.32).

Let $J_n = ((S_i + 1)^{\ell^n} - 1 : i = 1, \dots, r)$. Replacing $\mathbb{T}_{Q_n}$ and R_{Q_n} by $\mathbb{T}_{Q_n}/J_n$ and R_{Q_n}/J_n we see that we have a J_n-structure for every n. Theorem 3.42 for $\Sigma = \emptyset$ now follows from the criterion of theorem 3.41.

We now turn to the proof of theorem 3.42 in the general case. By theorem 2.41 and theorem 3.40 we see that

$$\# H^1_\emptyset(\mathbb{Q}, \mathrm{ad}^0 \rho_f \otimes_{\mathcal{O}} K/\mathcal{O}) = \# \mathcal{O}/\eta_\emptyset, \tag{3.5.1}$$

and so applying corollary 3.37 we see that for any $\Sigma \subset \Sigma_{\bar{\rho}}$

$$\# H^1_\Sigma(\mathbb{Q}, \mathrm{ad}^0 \rho_f \otimes_{\mathcal{O}} K/\mathcal{O}) \le \# \mathcal{O}/\eta_\Sigma. \tag{3.5.2}$$

A second application of theorems 2.41 and 3.40 allows us to deduce theorem 3.42. □

Remark 3.44 In certain cases where $\eta_\emptyset = (1)$, the bound (3.5.1) on the order of the Selmer group $H^1_\emptyset(\mathbb{Q}, \mathrm{ad}^0 \rho_f \otimes_{\mathcal{O}} K/\mathcal{O})$ also follows from the previous work of Flach, by a different method. See [Fl1] for details.

Corollary 3.45 *Keep the notation of theorem 3.42 and suppose that f is a newform in $\mathcal{N}_\Sigma$ with coefficients in $\mathcal{O}$.*

(a) We have

$$\# H^1_\Sigma(\mathbb{Q}, \mathrm{ad}^0 \rho_f \otimes_{\mathcal{O}} K/\mathcal{O}) = \# \mathcal{O}/\eta_\Sigma < \infty,$$

where η_Σ was defined in equation (3.3.1) after remark 3.33.

(b) If $\Sigma \subset \Sigma' \subset \Sigma_{\bar{\rho}}$ then

$$\begin{aligned}(0) \to H^1_\Sigma(\mathbb{Q}, \mathrm{ad}^0 \rho_f \otimes_{\mathcal{O}} K/\mathcal{O}) \to H^1_{\Sigma'}(\mathbb{Q}, \mathrm{ad}^0 \rho_f \otimes_{\mathcal{O}} K/\mathcal{O}) \\ \to \textstyle\bigoplus_{p \in \Sigma' - \Sigma} H_p \to (0)\end{aligned}$$

is exact, where the groups H_p and H_ℓ were defined before proposition 3.35.

Proof: The first part now follows as a direct consequence of theorem 3.42 and another application of theorems 2.41 and 3.40. The second part follows from the first, together with proposition 3.35 and theorem 3.36. □

Corollary 3.46 *Suppose $\rho : G_{\mathbb{Q}} \to GL_2(K)$ is a continuous representation and let $\bar{\rho}$ denote its reduction. Suppose also that*

(a) $\bar{\rho}$ is irreducible and modular,

(b) if $p \neq \ell$ then $\rho|_{I_p} \sim \begin{pmatrix} 1 & * \\ 0 & 1 \end{pmatrix}$,

(c) $\rho|_{G_\ell}$ is semi-stable,

(d) $\det \rho = \epsilon$.

Then ρ is modular.

Proof: We let Σ denote the set of primes in $\Sigma_{\bar{\rho}}$ at which ρ is ramified. Then $\rho : G_{\mathbb{Q}} \to GL_2(\mathcal{O})$ is a deformation of $\bar{\rho}$ of type Σ, so there is an $\mathcal{O}$-algebra homomorphism $R_\Sigma \to \mathcal{O}$ such that $\rho = \rho_\Sigma^{\mathrm{univ}} \otimes_{R_\Sigma} \mathcal{O}$. Since $\phi_\Sigma : R_\Sigma \to \mathbb{T}_\Sigma$ is an isomorphism by theorem 3.42, it follows that there is a homomorphism $\mathbb{T}_\Sigma \to \mathcal{O}$ sending T_p to $\mathrm{tr}\,(\rho(\mathrm{Frob}\,_p))$. Since such a homomorphism is necessarily of the form $T_p \mapsto a_p(f)$ for some newform f, it follows that ρ is equivalent to ρ_f and hence is modular. □

Corollary 3.47 *Suppose that $E/\mathbb{Q}$ is a semi-stable elliptic curve such that $\bar{\rho}_{E,3}$ is irreducible. Then E is modular.*

Proof: One need only apply the last corollary with $\ell = 3$ and theorem 3.14. □

3.6 Applications

The Shimura-Taniyama conjecture for semi-stable elliptic curves:

Theorem 3.48 *If $E/\mathbb{Q}$ is a semistable elliptic curve, then E is modular.*

Proof: By corollary 3.47, it is enough to show that E is modular when its associated mod 3 representation $\bar{\rho}_{E,3}$ is *reducible*, i.e., when E has a subgroup of order 3 defined over $\mathbb{Q}$. Consider the group $E[5]$ of 5-division points of E. The mod 5 Galois representation $\bar{\rho}_{E,5}$ associated to $E[5]$ is irreducible: for otherwise, E would have a subgroup of order 15 defined over $\mathbb{Q}$, and would give rise to a (non-cuspidal) rational point on the modular curve $X_0(15)$. This curve is of genus one, and is known to have only 4 non-cuspidal rational points, which do not correspond to semi-stable elliptic curves (and, at any rate, are known to correspond to modular elliptic curves). Hence we know that $\bar{\rho}_{E,5}$ satisfies all the assumptions of corollary 3.46, except the (crucial!) modularity property. To show that $\bar{\rho}_{E,5}$ is modular, one starts with

Lemma 3.49 *There is an auxiliary (semi-stable) elliptic curve $A/\mathbb{Q}$ which satisfies*

(a) $A[5] \simeq E[5]$ as $G_{\mathbb{Q}}$-modules.

(b) $A[3]$ is an irreducible $G_{\mathbb{Q}}$-module.

Proof: Let $Y'(5)$ be the curve over $\mathbb{Q}$ which classifies elliptic curves A together with an isomorphism $E[5] \simeq A[5]$ compatible with Weil pairings. Elliptic curves over $\mathbb{Q}$ satisfying (a) correspond to rational points in $Y'(5)(\mathbb{Q})$. Adjoining a finite set of points to $Y'(5)$ yields its compactification $X'(5)$ which is a twist of the modular curve $X(5)$ with full level 5 structure. (I.e., it becomes isomorphic to this curve, over $\bar{\mathbb{Q}}$.) As was shown by Klein, the modular curve $X(5)$ over $\mathbb{C}$ has genus 0. Since $X'(5)$ has a point x_0 defined over $\mathbb{Q}$ corresponding to E, it is isomorphic over $\mathbb{Q}$ to $\mathbb{P}^1$. The rational points of $Y'(5)$ therefore give a plentiful supply of elliptic curves satisfying condition (a). Now consider the curve $Y'(5,3)$ classifying elliptic curves A with an isomorphism $E[5] \simeq A[5]$ (respecting Weil pairings) and a subgroup of A of order 3. One checks that the compactification of $Y'(5,3)$ has genus greater than 1, hence has only finitely many rational points by Faltings' theorem (the Mordell conjecture). It follows that only finitely many points $Y'(5)(\mathbb{Q})$ are in the image of $Y'(5,3)(\mathbb{Q})$ under the natural map $Y'(5,3) \to Y'(5)$. Hence for all but finitely many points x in $Y'(5)(\mathbb{Q})$, the corresponding elliptic curve A satisfies (b) since it has no rational subgroup of order 3. Choosing x arbitrarily close in the 5-adic topology to x_0, we find that the elliptic curve A associated to x is semistable and satisfies the two conditions in the lemma. □

We can now finish the proof of theorem 3.48. Applying corollary 3.47 to the curve A, we find that A is modular. Hence so is the mod 5 representation $\bar{\rho}_{A,5} \simeq \bar{\rho}_{E,5}$. Now applying corollary 3.46 with $\ell = 5$ and ρ the representation of $G_{\mathbb{Q}}$ acting on the 5-adic Tate module of E, we find that $\rho_{E,5}$ is modular, and hence, so is E, as was to be shown.

Remark 3.50 Wiles' original argument uses Hilbert's irreducibility theorem where we have used Faltings' theorem. The alternative presented here is based on a remark of Karl Rubin.

Remark 3.51 The results of [W3] and [TW] actually apply to a larger class of elliptic curves than those which are semistable. In [Di2], their methods are further strengthened to prove that all elliptic curves which have semi-stable reduction at 3 and 5 are modular.

Remark 3.52 Rubin and Silverberg observed that an elliptic curve of the form $y^2 = x(x-a)(x+b)$ has a twist with semi-stable reduction at all odd primes, hence is modular by [Di2]. In fact it is shown in [DK] that their observation together with the general results of [W3] and [TW] already imply modularity.

Fermat's Last Theorem: As was already mentioned in the introduction, the Shimura-Taniyama conjecture for semi-stable elliptic curves (and, more precisely, for the elliptic curves that arise in Frey's construction explained in section 2.2) implies Fermat's Last Theorem.

More precisely, suppose that there is a non-trivial solution to the Fermat equation $x^\ell + y^\ell = z^\ell$, with $\ell > 3$. By theorem 2.15 the Frey curve constructed from this solution (cf. section 2.2) is a semistable elliptic curve $E/\mathbb{Q}$ whose associated mod ℓ representation $\bar{\rho}_{E,\ell}$ is irreducible, unramified outside 2ℓ and is good at ℓ. Serre's conjecture predicts that $\bar{\rho}_{E,\ell}$ arises from a newform of weight

2 and level 2; the "lowering the level" result of Ribet [R5] (cf. theorem 3.15) actually proves this, once we know that E is modular, i.e., $\bar{\rho}_{E,\ell}$ arises from a modular form of weight 2 and *some* level. But this is a contradiction, since there are no modular forms of weight two and level two: such forms would correspond to holomorphic differentials on the modular curve $X_0(2)$ which is of genus 0. This contradiction completes the proof.

Values of L-functions: Also mentioned in the introduction was the relationship between the calculation of the Selmer group (3.5.1) and certain cases of a conjecture of Bloch-Kato [BK], called the Tamagawa number conjecture. It was in this context that partial results were obtained by Flach in [Fl1] (cf. remark 3.44).

If f is a newform of weight 2, then one can associate to f a certain "symmetric square" L-function $L(\mathrm{Symm}^2 f, s)$. We shall recall the definition in section 4.4 and explain how a method of Hida establishes a relationship between $L(\mathrm{Symm}^2 f, 2)$ and $\mathcal{O}/\eta_\Sigma$ in the setting of corollary 3.45. We may therefore regard part (a) of that corollary as a relationship between $L(\mathrm{Symm}^2 f, 2)$ and the size of a Selmer group. While the result is in the spirit of the Tamagawa number conjecture of [BK], we have not verified that the relevant cases of the conjecture can be deduced from it. We shall however state a partial result in the context of semistable elliptic curves. The reader can consult [Fl1] and [Fl2] for a discussion of the relation to the Tamagawa number conjecture.

Suppose that E is a semistable elliptic curve over $\mathbb{Q}$ of conductor N_E and (minimal) discriminant

$$\Delta_E = \prod_{p|N_E} p^{d_p}.$$

The symmetric square L-function associated to E is defined by

$$L(\mathrm{Symm}^2 E, s) = \prod_p L_p(\mathrm{Symm}^2 E, s)$$

where the Euler factors $L_p(\mathrm{Symm}^2 E, s)$ are defined as follows:

- If $p \nmid N_E$, then

  $$L_p(\mathrm{Symm}^2 E, s) = ((1 - \alpha_p^2 p^{-s})(1 - p^{1-s})(1 - \beta_p^2 p^{-s}))^{-1}$$

  where α_p and β_p are the roots roots of $X^2 - a_p X + p$ with $a_p = p + 1 - N_p$ as in section 1.1).

- If $p|N_E$, then $L_p(\mathrm{Symm}^2 E, s) = (1 - p^{-s})^{-1}$.

Let

$$\Omega_E = \int_{E(\mathbb{C})} \omega_E^{\mathrm{Neron}} \wedge \bar{\omega}_E^{\mathrm{Neron}}$$

where $\omega_E^{\mathrm{Neron}}$ is the Néron differential defined in section 1.1. Since E is modular by theorem 3.48, a method of Shimura (see [Shi4] and the introduction of [St]) establishes the analytic continuation of $L(\mathrm{Symm}^2 E, s)$ to an entire function and shows that $L(\mathrm{Symm}^2 E, 2)$ is a non-zero rational multiple of $i\pi\Omega_E$. We now explain how to deduce the following theorem from Wiles' results and a formula of Hida, corollary 4.21.

Theorem 3.53 *Suppose that E is a semistable elliptic curve and ℓ is a prime such that*

- *$\bar\rho_{E,\ell}$ is irreducible, and*
- *ℓ does not divide $2\prod_{p|N_E} d_p$.*

Then the ℓ-part of

$$\frac{N_E L(\mathrm{Symm}^2 E, 2)}{\pi i \Omega_E}$$

is the order of

$$H^1_\emptyset(\mathbb{Q}, \mathrm{ad}^0\rho_{E,\ell} \otimes_{\mathbb{Z}_\ell} \mathbb{Q}_\ell/\mathbb{Z}_\ell).$$

Sketch of proof: Since E is semistable, it is modular by theorem 3.48. Letting f denote the associated newform, we have $N_f = N_E$ and $L(\mathrm{Symm}^2 E, s) = L(\mathrm{Symm}^2 f, s)$. Furthermore the representations $\rho_{E,\ell}$ and

$$\rho_f : G_\mathbb{Q} \to GL_2(\mathbb{Z}_\ell)$$

are equivalent, since they are equivalent over $\mathbb{Q}_\ell$ and $\bar\rho = \bar\rho_{E,\ell}$ is irreducible. The conditions on E and ℓ ensure that $\bar\rho$ and f satisfy the hypotheses of corollary 3.45 with $\Sigma = \emptyset$, so that

$$\#H^1_\emptyset(\mathbb{Q}, \mathrm{ad}^0\rho_{E,\ell} \otimes_{\mathbb{Z}_\ell} \mathbb{Q}_\ell/\mathbb{Z}_\ell) = \#(\mathbb{Z}_\ell/\eta_\emptyset).$$

To apply Hida's formula, corollary 4.21, it remains to relate $\det A$ and Ω_E where the matrix $A \in GL_2(\mathbb{C})$ is defined in section 4.4. Using that E is semistable, ℓ is odd and $\bar\rho_{E,\ell}$ is irreducible, one obtains a modular parametrization $\pi : X_0(N_E) \to E$ such that

- $\pi^* : H^1(E, \mathbb{Z}_\ell) \to H^1(X_0(N_E), \mathbb{Z}_\ell)$ has torsion-free cokernel;
- the Manin constant for π is not divisible by ℓ ([Maz2], sec. 4(a)).

One can then verify that $\Omega_E^{-1} \det A$ is an ℓ-adic unit and theorem 3.53 follows from corollary 4.21. □

4 Hecke algebras

4.1 Full Hecke algebras

Suppose that K is a finite extension of $\mathbb{Q}_\ell$ for some prime ℓ. Let $\mathcal{O}$ denote its ring of integers and let $k = \mathcal{O}/\lambda$ where λ is the maximal ideal of $\mathcal{O}$. Fix embeddings $K \hookrightarrow \bar{\mathbb{Q}}_\ell$, $\bar{\mathbb{Q}} \hookrightarrow \bar{\mathbb{Q}}_\ell$ and $\bar{\mathbb{Q}} \hookrightarrow \mathbb{C}$. Recall that we defined "Hecke algebras" over $\mathcal{O}$ in two different contexts:

- In section 1.6 as an algebra $\mathbb{T}_\mathcal{O}$ generated by the full set of Hecke operators acting on a space of modular forms;
- In section 3.3 as a certain subring $\mathbb{T}_\Sigma$ of a product of fields of Fourier coefficients of newforms giving rise to the same mod ℓ representation.

The first of these provides a concrete geometric description useful for establishing properties of the fine structure of the algebra; the second yields a reduced ring which is more easily interpreted as the coefficient ring of a Galois representation. In the next section we shall relate the two notions by identifying the "reduced Hecke algebras" of the form $\mathbb{T}_\Sigma$ as localizations of the "full Hecke algebras" of the form $\mathbb{T}_\mathcal{O}$. Before doing so we need to recall some fundamental properties of the algebras $\mathbb{T}_K$, $\mathbb{T}_\mathcal{O}$ and $\mathbb{T}_k$.

Let $\Gamma = \Gamma_H(N)$ for some positive integer N and subgroup H of $(\mathbb{Z}/N\mathbb{Z})^\times$ (see section 1.2). We let $\mathbb{T}_\mathbb{Z}$ denote the subring of $\mathrm{End}\,(S_2(\Gamma))$ generated by the operators T_n for all positive integers n and $\langle d \rangle$ for all $d \in (\mathbb{Z}/N\mathbb{Z})^\times$. If R is a ring, then $\mathbb{T}_R$ denotes the R-algebra $\mathbb{T}_\mathbb{Z} \otimes R$. Recall that $\mathbb{T}_R$ acts faithfully on $S_2(\Gamma, R)$ and is finitely generated and free as an R-module. (This holds for $R = \mathbb{Z}$, hence for arbitrary R.)

We first record the following lemma:

Lemma 4.1 *(a) $\mathbb{T}_R$ is generated as an R-algebra by either of the following sets of elements:*

- *T_n for all positive integers n.*
- *T_p for all primes p and $\langle d \rangle$ for all d in $(\mathbb{Z}/N\mathbb{Z})^\times$.*

(b) Suppose that D is a positive integer relatively prime to N. If either D is odd or 2 is invertible in R, then $\mathbb{T}_R$ is generated as an R-algebra by either of the following sets of elements:

- *T_n for all positive integers n relatively prime to D.*
- *T_p for all primes p not dividing D, and $\langle d \rangle$ for all d in $(\mathbb{Z}/N\mathbb{Z})^\times$.*

For a proof of (a), see [DI], prop. 3.5.1; for (b), see p. 491 of [W3].

The spectrum of $\mathbb{T}_\mathcal{O}$: First note that $\mathbb{T}_K$ and $\mathbb{T}_k$ are Artinian, hence have only a finite number of prime ideals, all of which are maximal. Since $\mathbb{T}_\mathcal{O}$ is finitely generated and free as an $\mathcal{O}$-module, its maximal (resp. minimal) prime ideals are those lying over the prime λ (resp. (0)) of $\mathcal{O}$. (This follows from the going-up and going-down theorems, [Mat] thms. 9.4 and 9.5, for example.) It follows that the natural maps

$$\mathbb{T}_\mathcal{O} \hookrightarrow \mathbb{T}_\mathcal{O} \otimes_\mathcal{O} K \cong \mathbb{T}_K; \quad \text{and} \quad \mathbb{T}_\mathcal{O} \twoheadrightarrow \mathbb{T}_\mathcal{O} \otimes_\mathcal{O} k \cong \mathbb{T}_k$$

induce bijections

$$\{\text{maximal ideals of } \mathbb{T}_K\} \leftrightarrow \{\text{minimal primes of } \mathbb{T}_\mathcal{O}\} \quad \text{and}$$
$$\{\text{maximal ideals of } \mathbb{T}_k\} \leftrightarrow \{\text{maximal primes of } \mathbb{T}_\mathcal{O}\}.$$

Moreover, since $\mathcal{O}$ is complete we have (by [Mat] thms. 8.7 and 8.15, for example) that the natural map

$$\mathbb{T}_\mathcal{O} \to \prod_\mathfrak{m} \mathbb{T}_\mathfrak{m}$$

is an isomorphism, where the product is over the finite set of maximal ideals $\mathfrak{m}$ of $\mathbb{T}_\mathcal{O}$ and $\mathbb{T}_\mathfrak{m}$ denotes the localization of $\mathbb{T}_\mathcal{O}$ at $\mathfrak{m}$. Furthermore each $\mathbb{T}_\mathfrak{m}$ is a complete local $\mathcal{O}$-algebra which is finitely generated and free as an $\mathcal{O}$-module, and each minimal prime $\mathcal{P}$ of $\mathbb{T}_\mathcal{O}$ is contained in a unique $\mathfrak{m}$.

Now suppose that $f = \sum a_n q^n$ is a normalized eigenform in $S_2(\Gamma, \bar{K})$ for the operators T_n for all $n \geq 1$. Then $T_n \mapsto a_n$ defines a map $\mathbb{T}_\mathbb{Z} \to \bar{K}$ and induces a K-algebra homomorphism $\theta_f : \mathbb{T}_K \to \bar{K}$. The image is the finite extension of K generated by the a_n, and the kernel is a maximal ideal of $\mathbb{T}_K$ which depends only on the G_K-conjugacy class of f. Similarly a G_k-conjugacy class of normalized eigenforms in $S_2(\Gamma, \bar{k})$ gives rise to a maximal ideal of $\mathbb{T}_k$.

Recall also that a normalized eigenform f in $S_2(\Gamma, \bar{K})$ has coefficients in $\mathcal{O}_{\bar{K}}$, hence gives rise by reduction to a normalized eigenform $\bar{f}$ in $S_2(\Gamma, \bar{k})$. Furthermore if f and g are G_K-conjugate, then $\bar{f}$ and $\bar{g}$ are G_k-conjugate.

We have thus constructed a diagram of maps of finite sets whose commutativity is easily verified.

$$\begin{array}{ccc}
\left\{\begin{matrix}\text{normalized eigenforms in} \\ S_2(\Gamma, \bar{K}) \text{ modulo } G_K\text{-conjugacy}\end{matrix}\right\} & \to & \left\{\begin{matrix}\text{normalized eigenforms in} \\ S_2(\Gamma, \bar{k}) \text{ modulo } G_k\text{-conjugacy}\end{matrix}\right\} \\
\downarrow & & \downarrow \\
\{\text{maximal ideals of } \mathbb{T}_K\} & & \{\text{maximal ideals of } \mathbb{T}_k\} \\
\updownarrow & & \updownarrow \\
\{\text{minimal primes of } \mathbb{T}_\mathcal{O}\} & \twoheadrightarrow & \{\text{maximal primes of } \mathbb{T}_\mathcal{O}\}.
\end{array} \tag{4.1.1}$$

Proposition 4.2 *The vertical maps are bijective, and the horizontal maps are surjective.*

Proof. For the injectivity of the upper-left vertical arrow, note that if $\mathfrak{p}$ is a maximal ideal of $\mathbb{T}_K$, then all K-algebra homomorphisms $\mathbb{T}_K/\mathfrak{p} \hookrightarrow \bar{K}$ are obtained from a single one by composing with an element of G_K. For the surjectivity, let $K' = \mathbb{T}_K/\mathfrak{p}$ and $\mathfrak{p}'$ denote the kernel of the natural K'-algebra homomorphism $\mathbb{T}_{K'} \twoheadrightarrow K'$. Since $\mathbb{T}_{K'}$ acts faithfully on $S_2(\Gamma, K')$, the localization $S_2(\Gamma, K')_\mathfrak{p}$ is non-zero, hence so is $S_2(\Gamma, K')[\mathfrak{p}']$. (For an R-module M and an ideal I of R, we write $M[I]$ for the intersection over the elements r in I of the kernels of $r : M \to M$.) It follows that there is a normalized eigenform f in $S_2(\Gamma, K')$ so that $\mathfrak{p}'$ is the kernel of the $\theta'_f : \mathbb{T}_{K'} \to K'$, and therefore $\mathfrak{p}$ is the kernel of θ_f. To prove that the upper-right vertical arrow is bijective, note that the above arguments carry over with K replaced by k. □

Remark 4.3 The surjectivity of the top arrow is called the Deligne-Serre lifting lemma ([DS], lemma 6.11).

Suppose that $\mathfrak{m}$ is a maximal ideal of $\mathbb{T}_{\mathcal{O}}$. Note that the maximal ideals of $\mathbb{T}_K$ mapping to $\mathfrak{m}$ are precisely those $\mathfrak{p}$ for which $\mathfrak{p} \cap \mathbb{T}_{\mathcal{O}}$ is contained in $\mathfrak{m}$. Note also that the natural map

$$\alpha_{\mathfrak{m}} : \mathbb{T}_{\mathfrak{m}} \otimes_{\mathcal{O}} K \to \prod_{\mathfrak{p}} \mathbb{T}_{\mathfrak{p}}$$

is an isomorphism, where the product is over such $\mathfrak{p}$.

It is straightforward to check that the above constructions are well-behaved with respect to replacing the field K by an extension K'. More precisely, for each set $\mathcal{S}$ in the above diagram, there is a natural surjective map from $\varpi : \mathcal{S}' \to \mathcal{S}$ where $\mathcal{S}'$ is defined by replacing K with K', and these maps are compatible with the maps in the diagram. Furthermore the maps

$$\mathbb{T}_{\mathfrak{p}} \otimes_K K' \to \prod_{\mathfrak{p}' \in \varpi^{-1}(\mathfrak{p})} \mathbb{T}_{\mathfrak{p}'} \quad \text{and} \quad \mathbb{T}_{\mathfrak{m}} \otimes_{\mathcal{O}} \mathcal{O}' \to \prod_{\mathfrak{m}' \in \varpi^{-1}(\mathfrak{m})} \mathbb{T}_{\mathfrak{m}'}$$

are isomorphisms by which $\alpha_{\mathfrak{m}} \otimes_K K'$ can be identified with $\prod_{\mathfrak{m}' \in \varpi^{-1}(\mathfrak{m})} \alpha_{\mathfrak{m}'}$.

Associated Galois representations: Suppose that $\mathfrak{p}$ is a maximal ideal of $\mathbb{T}_K$ and $\mathfrak{m}$ is the associated maximal ideal of $\mathbb{T}_{\mathcal{O}}$. By lemma 1.39,

$$\mathcal{T}_\ell(J_H(N)) \otimes_{\mathbb{Z}_\ell} K$$

is free of rank two over $\mathbb{T}_K$, so reduction mod $\mathfrak{p}$ yields a two-dimensional vector space over the field $\mathbb{T}_K/\mathfrak{p}$ endowed with an action of $G_{\mathbb{Q}}$. The resulting Galois representation

$$\rho_{\mathfrak{p}} : G_{\mathbb{Q}} \to GL_2(\mathbb{T}_K/\mathfrak{p})$$

is unramified at all primes p not dividing $N\ell$, and for such p the characteristic polynomial of $\rho_{\mathfrak{p}}(\mathrm{Frob}_p)$ is

$$X^2 - T_p X + p\langle p \rangle \mod \mathfrak{p}.$$

If ℓ is odd, then $\bar{\rho}_{\mathfrak{p}}$ is defined over $\mathbb{T}_{\mathcal{O}}/\mathfrak{m}$ and we write $\rho_{\mathfrak{m}}$ for its semisimplification

$$\rho_{\mathfrak{m}} : G_{\mathbb{Q}} \to GL_2(\mathbb{T}_{\mathcal{O}}/\mathfrak{m}).$$

Thus $\rho_{\mathfrak{m}}$ is unramified at primes p not dividing $N\ell$ and the characteristic polynomial of Frob_p is

$$X^2 - T_p X + p\langle p \rangle \mod \mathfrak{m}.$$

Suppose that g is in the G_K-conjugacy class of eigenforms in $S_2(\Gamma)$ corresponding to $\mathfrak{p}$. If g is a newform then $\mathbb{T}_K/\mathfrak{p}$ is isomorphic to the field denoted K_g' in section 3.1 and $\rho_{\mathfrak{p}}$ can be identified with the representation

$$\rho_g : G_{\mathbb{Q}} \to GL_2(K_g')$$

considered in theorem 3.1. If also ℓ is odd then $\bar{\rho}_g$ is obtained from $\rho_{\mathfrak{m}}$ by extending scalars. More generally suppose that g is not necessarily a newform and consider the associated newform f. Let D denote the product of the primes which divide N but not N_f. Let $\mathbb{T}_K^{(D)}$ denote the K-subalgebra of $\mathbb{T}_K$ generated by the operators T_n for n relatively prime to D and $\langle d\rangle$ for d in $(\mathbb{Z}/N\mathbb{Z})^\times$. Let $\Gamma' = \Gamma_{H'}(N_f)$ where H' is the image of H in $(\mathbb{Z}/N_f\mathbb{Z})^\times$, and let $\mathbb{T}'_K = \mathbb{T}'_\mathbb{Z} \otimes K$ where $\mathbb{T}'_\mathbb{Z}$ is the Hecke algebra acting on $S_2(\Gamma')$. Restriction of operators defines a natural homomorphism $\mathbb{T}_K^{(D)} \to \mathbb{T}'_K$ which is surjective by lemma 4.1. The composite with $\mathbb{T}'_K \twoheadrightarrow K'_f$ factors through the field $\mathbb{T}_K^{(D)}/(\mathfrak{p}\cap\mathbb{T}_K^{(D)})$, so we may identify K'_f with a subfield of $\mathbb{T}_K/\mathfrak{p}$ and $\rho_\mathfrak{p}$ is then equivalent to the extension of scalars of ρ_f. If ℓ is odd, then $\rho_\mathfrak{m}$ and $\bar\rho_f$ are defined and equivalent over a common subfield of $\mathbb{T}_\mathcal{O}/\mathfrak{m}$ and k_f.

The structure of $\mathbb{T}_K$: We now give an explicit description of $\mathbb{T}_K$ in the case that K contains the coefficients of all eigenforms of level dividing N. Let $\mathcal{N}_\Gamma$ denote the set of newforms in $S_2(\Gamma)$, i.e., the set of newforms f of level N_f dividing N such that H is contained in the kernel of the character

$$(\mathbb{Z}/N\mathbb{Z})^\times \to (\mathbb{Z}/N_f\mathbb{Z})^\times \xrightarrow{\psi_f} K^\times.$$

By theorem 1.22

$$S_2(\Gamma, K) = \bigoplus_{f\in\mathcal{N}_\Gamma} S_{K,f},$$

where $S_{K,f}$ is spanned by the linear independent elements

$$\{\, f(a\tau) \mid a \text{ divides } N/N_f \,\}.$$

For each $f = \sum a_n(f)q^n$ in $\mathcal{N}_\Gamma$, let

- $\mathbb{T}_{K,f}$ denote the image of $\mathbb{T}_K$ in $\operatorname{End}_K S_{K,f}$;
- $A_{K,f}$ denote the polynomial ring over K in the variables $u_{f,p}$ indexed by the prime divisors of N/N_f;
- $I_{K,f}$ denote the ideal in $A_{K,f}$ generated by the polynomials

  $$P_{f,p}(u_{f,p}) = u_{f,p}^{v_p(N/N_f)-1}(u_{f,p}^2 - a_p(f)u_{f,p} + \psi_f(p)p),$$

  for primes p dividing N/N_f (setting $\psi_f(p) = 0$ if p divides N_f).

Consider the K-algebra homomorphism $A_{K,f} \twoheadrightarrow \mathbb{T}_{K,f}$ defined by mapping $u_{f,p}$ to the operator T_p. Since $P_{f,p}$ is the characteristic polynomial of T_p on the span of $\{\, f(ap^i\tau) \mid i = 1,\dots, v_p(N/N_f)\,\}$ for each a dividing $N/(N_f p^{v_p(N/N_f)})$, we see that $I_{K,f}$ is contained in the kernel of $A_{K,f} \twoheadrightarrow \mathbb{T}_{K,f}$. Taking the product over f in $\mathcal{N}_\Gamma$, we have a surjective K-algebra homomorphism

$$\prod_f A_{K,f}/I_{K,f} \twoheadrightarrow \prod_f \mathbb{T}_{K,f}.$$

Since the natural map $\mathbb{T}_K \to \prod_f \mathbb{T}_{K,f}$ is injective and

$$\dim_K \mathbb{T}_K = \dim_\mathbb{C} S_2(\Gamma, \mathbb{C}) = \sum_f \sigma_0(N/N_f) = \sum_f \dim_K A_{K,f}/I_{K,f},$$

we conclude

Lemma 4.4 *There is an isomorphism of K-algebras:*

$$\alpha : \mathbb{T}_K \to \prod_{f \in \mathcal{N}_\Gamma} A_{K,f}/I_{K,f}$$

defined by

- $\alpha(T_p)_f = a_p(f)$ *if p is a prime not dividing N/N_f;*
- $\alpha(T_p)_f = u_{f,p} \bmod I_{K,f}$ *if p is a prime dividing N/N_f;*
- $\alpha(\langle d \rangle)_f = \psi_f(d)$ *if d is relatively prime to N.*

Remark 4.5 It follows that the algebra $\mathbb{T}_K$ is a "complete intersection" over K in the sense that it is a finite-dimensional K-algebra of the form

$$K[X_1, \dots, X_r]/(P_1, \dots, P_r)$$

for some r.

4.2 Reduced Hecke algebras

As in section 4.1, we suppose K is a finite extension of $\mathbb{Q}_\ell$ with ring of integers $\mathcal{O}$ and residue field k, and fix embeddings $K \hookrightarrow \bar{\mathbb{Q}}_\ell$, $\bar{\mathbb{Q}} \hookrightarrow \bar{\mathbb{Q}}_\ell$ and $\bar{\mathbb{Q}} \hookrightarrow \mathbb{C}$. We assume in this section that ℓ is odd and we fix a representation

$$\bar{\rho} : G_{\mathbb{Q}} \to GL_2(k)$$

which is modular (definition 3.12). Thus $\bar{\rho}$ is equivalent to $\rho_{\mathfrak{m}}$ (over $\mathbb{T}_{\mathcal{O}}/\mathfrak{m}$) for some $\Gamma = \Gamma_H(N)$ and maximal ideal $\mathfrak{m}$ of $\mathbb{T}_{\mathcal{O}}$.

We suppose also that $\bar{\rho}$ has the properties listed at the beginning of section 3.3 and that Σ is a finite set of primes contained in $\Sigma_{\bar{\rho}}$ (definition 3.25). We shall show that Γ and $\mathfrak{m}$ can be chosen so that the reduced Hecke algebra $\mathbb{T}_\Sigma$ can be identified with a localization $\mathbb{T}_{\mathfrak{m}}$ of the full Hecke algebra $\mathbb{T}_{\mathcal{O}}$. The main result is due to Wiles ([W3], prop. 2.15), but we also explain an important variant ([TW], lemma 1) which arises when considering special sets of primes Q as in section 2.8 above.

$\mathbb{T}_\Sigma$ and the full Hecke algebra: Let $\Gamma = \Gamma_0(N_\Sigma)$ where

$$N_\Sigma = \ell^\delta N(\bar{\rho}) \prod_{p \in \Sigma - \{\ell\}} p^{\dim \bar{\rho}^{I_p}} = \ell^\delta \prod_{p | N(\bar{\rho})} p \prod_{p \in \Sigma - \{\ell\}} p^2 \qquad (4.2.1)$$

with $\delta = 0$ if $\bar{\rho}$ is good and $\ell \notin \Sigma$ and $\delta = 1$ otherwise.

Suppose that f is a newform in $\mathcal{N}_\Sigma$. Recall that $\mathcal{N}_\Sigma$, defined in section 3.3, is the set of newforms f in $S_2(\Gamma)$ such that $\bar{\rho}$ is equivalent to $\bar{\rho}_f$ over k_f (lemma 3.26). Note that these representations are equivalent if and only if

$$a_p(f) \bmod \lambda' = \mathrm{tr}\,(\bar{\rho}(\mathrm{Frob}_p)) \qquad \text{for all } p \nmid N_\Sigma \ell,$$

where λ' is maximal ideal of the ring of integers of K'_f. There is then a normalized eigenform g in $S_2(\Gamma, K'_f)$ characterized by

- $a_p(g) = a_p(f)$ if p does not divide N_Σ/N_f;

- $a_p(g) = 0$ if $p \neq \ell$ and p divides N_Σ/N_f;
- $a_\ell(g)$ is the unit root of $X^2 - a_\ell(f)X + \ell$ if ℓ divides N_Σ/N_f.

Note that if ℓ divides N_Σ/N_f, then $\delta = 1$ and $\bar\rho$ is flat, hence $\bar\rho$ is ordinary and $a_\ell(f)$ is a unit by theorem 3.1 (f).

Lemma 4.6 *For f and g as above, the reduction $\bar g$ is the normalized eigenform in $S_2(\Gamma, k')$ characterized by:*

- $a_p(\bar g) = \operatorname{tr} \bar\rho_{I_p}(\operatorname{Frob}_p)$ *if $p = \ell$ or p is not in Σ (recall that $\bar\rho_{I_p}$ denotes the representation on I_p-coinvariants);*
- $a_p(\bar g) = 0$ *otherwise.*

Furthermore, if K is as in lemma 4.4, then g is the unique normalized eigenform in $S_{K,f}$ with this reduction.

Proof. If p is not in $\Sigma \cup \{\ell\}$ then the formula for $a_p(\bar g) = a_p(\bar f)$ follows from theorem 3.1, parts (a) and (e). If $p = \ell$, then we use parts (f) and (g) of theorem 3.1. Otherwise, we have $a_p(\bar g) = 0$ by theorem 1.27(d) (if $p^2 | N_f$ and f has trivial character, then $a_p(f) = 0$). The uniqueness follows from the fact that if $p || N_f$, then $a_p(f) = \pm 1$ (by theorem 1.27(b)). □

Note in particular that $\bar g$ has coefficients in k and is independent of the choice of f in $\mathcal{N}_\Sigma$. We let $\mathfrak{m}$ denote the corresponding maximal ideal of $\mathbb{T}_\mathcal{O}$.

Proposition 4.7 *There is an isomorphism $\mathbb{T}_\Sigma \xrightarrow{\sim} \mathbb{T}_\mathfrak{m}$ of $\mathcal{O}$-algebras such that $T_p \mapsto T_p$ for all primes p not dividing $N_\Sigma \ell$.*

Proof. Replacing K by a larger field K', we have $\mathbb{T}_\Sigma \otimes_\mathcal{O} \mathcal{O}' \xrightarrow{\sim} \mathbb{T}'_\Sigma$ (lemma 3.27) and $\mathbb{T}_\mathfrak{m} \otimes_\mathcal{O} \mathcal{O}' \xrightarrow{\sim} \mathbb{T}'_{\mathfrak{m}'}$ (since there is a unique maximal ideal $\mathfrak{m}'$ of $\mathbb{T}_{\mathcal{O}'}$ over $\mathfrak{m}$). We are therefore reduced to the case where K is as in lemma 4.4. Note that $\mathcal{N}_\Sigma$ is the set of newforms f in $\mathcal{N}_\Gamma$ such that

$$a_p(\bar f) = \operatorname{tr}(\bar\rho(\operatorname{Frob}_p)) \qquad \text{for all } p \nmid N_\Sigma \ell.$$

(We have $K = K'_f$ and write $\bar f$ for $f \bmod \lambda$.)

We now define an isomorphism of K-algebras

$$\kappa : \mathbb{T}_\mathfrak{m} \otimes K \xrightarrow{\sim} \prod_{f \in \mathcal{N}_\Sigma} K,$$

such that for each $f \in \mathcal{N}_\Sigma$:

- $\kappa(T_p)_f = a_p(f)$ if p is not in Σ;
- $\kappa(T_p)_f = 0$ if p is in $\Sigma - \{\ell\}$;
- $\kappa(T_\ell)_f$ is the unit root of $X^2 - a_\ell(f)X + \psi_f(\ell)\ell$ if ℓ is in Σ.

Recall that $\mathbb{T}_\mathfrak{m} \otimes K \cong \prod_\mathfrak{p} \mathbb{T}_\mathfrak{p}$ where $\mathfrak{p}$ runs over the primes of $\mathbb{T}_K$ whose preimage in $\mathbb{T}_\mathcal{O}$ is contained in $\mathfrak{m}$. Thus according to lemma 4.4 we have

$$\mathbb{T}_\mathfrak{m} \otimes K \xrightarrow{\sim} \prod_{f \in \mathcal{N}_\Gamma} \prod_{\mathfrak{p} \in \mathcal{M}_f} (A_{K,f}/I_{K,f})_\mathfrak{p},$$

where $\mathcal{M}_f$ is the set of prime ideals in $A_{K,f}/I_{K,f}$ whose preimage in $\mathbb{T}_\mathcal{O}$ is contained in $\mathfrak{m}$. If f is not in $\mathcal{N}_\Sigma$, then $\mathcal{M}_f$ is empty. If f is in $\mathcal{N}_\Sigma$, then $\mathcal{M}_f$ consists only of the kernel $\mathfrak{p}_f$ of the map defined by $u_{p,f} \mapsto a_p(g)$ where g is the eigenform of lemma 4.6. Furthermore the maps $u_{p,f} \mapsto a_p(g)$ induce isomorphisms $(A_{K,f}/I_{K,f})_{\mathfrak{p}_f} \xrightarrow{\sim} K$, and taking the product over $f \in \mathcal{N}_\Sigma$, we obtain the desired isomorphism κ.

Identifying $\mathbb{T}_\Sigma$ with the $\mathcal{O}$-subalgebra of $\prod_{f\in\mathcal{N}_\Sigma} K$ generated by the elements $T_p = (a_p(f))_f$ for all p not dividing $N_\Sigma\ell$, it suffices to prove that $\mathbb{T}_\Sigma$ contains $\kappa(T_p)$ for all p dividing $N(\bar\rho)\ell$. Now observe that for each f in $\mathcal{N}_\Sigma$ the representation ρ_f is isomorphic to

$$G_\mathbb{Q} \to GL_2(\mathbb{T}_\Sigma) \to GL_2(K)$$

obtained by composing $\rho_\Sigma^{\mathrm{mod}}$ with the projection $\mathbb{T}_\Sigma \to K$. It follows that for p dividing $N(\bar\rho)\ell^\delta$, the image of $(\rho_\Sigma^{\mathrm{mod}})_{I_p}(\mathrm{Frob}_p)$ in K is $(\rho_f)_{I_p}(\mathrm{Frob}_p) = a_p(f)_f$ and therefore

$$(\rho_\Sigma^{\mathrm{mod}})_{I_p}(\mathrm{Frob}_p) = \kappa(T_p).$$

Finally if $\delta = 0$, then we conclude from lemma 4.1 that $\mathbb{T}_\Sigma$ contains $\kappa(T_\ell)$. □

A twisted variant: Suppose now that Σ is a finite set of primes contained in $\Sigma_{\bar\rho} - \{\ell\}$ such that if p is in Σ then $\bar\rho(\mathrm{Frob}_p)$ has eigenvalues α_p and β_p in k satisfying

$$\alpha_p/\beta_p \neq p^{\pm 1}. \tag{4.2.2}$$

Note that (4.2.2) is equivalent to

$$\mathrm{tr}\,(\bar\rho(\mathrm{Frob}_p))^2 \neq (p+1)^2. \tag{4.2.3}$$

Let Q denote the set of primes in Σ such that $q \equiv 1 \bmod \ell$. The set of primes Q is therefore as in section 2.8. Choosing an eigenvalue α_q of $\bar\rho(\mathrm{Frob}_q)$ for each $q \in Q$ as in section 2.8, we regard R_Q and hence $\mathbb{T}_Q$ as an $\mathcal{O}[\Delta_Q]$-algebra. (Recall that Δ_Q is the the maximal quotient of $(\mathbb{Z}/(\prod_{q\in Q} q)\mathbb{Z})^\times$ of ℓ-power order.)

Instead of working with $\Gamma_0(N_\Sigma)$ as in proposition 4.7, we shall now work with the group

$$\Gamma = \Gamma_0(N_\emptyset) \cap \Gamma_1(M), \tag{4.2.4}$$

where

$$M = \prod_{p\in\Sigma-Q} p^2 \prod_{q\in Q} q.$$

Remark 4.8 We are about to relate $\mathbb{T}_Q$ to a localization of $\mathbb{T}_\mathcal{O}$, where $\mathbb{T}_\mathcal{O}$ is now defined using the Hecke operators on $S_2(\Gamma)$, with Γ defined by (4.2.4). Recall that $\mathbb{T}_Q$ is defined using modular forms with trivial character, but note that the modular forms involved in the definition of $\mathbb{T}_\mathcal{O}$ may have non-trivial character. The purpose of establishing this relationship is to give a concrete realization of the image of Δ_Q in $\mathbb{T}_Q$ for the purpose of proving theorem 3.31.

Suppose that f is a newform pair in $\mathcal{N}_\emptyset$. For each q in Q, let $\tilde\alpha_q(f)$ be the root of $X^2 - a_q(f)X + q = 0$ (in K'_f) whose image in k_f is α_q. We let g denote the unique normalized eigenform in $S_2(\Gamma, K'_f)$ of trivial character such that

- $a_p(g) = a_p(f)$ if p is not in Σ;
- $a_p(g) = \tilde{\alpha}_p(f)$ if p is in Q;
- $a_p(g) = 0$ otherwise.

The reduction $\bar{g}$ is the unique normalized eigenform in $S_2(\Gamma, k')$ of trivial character such that

- $a_p(\bar{g}) = \operatorname{tr} \bar{\rho}_{I_p}(\operatorname{Frob}_p)$ if p is not in Σ;
- $a_p(\bar{g}) = \alpha_p$ if p is in Q;
- $a_p(\bar{g}) = 0$ otherwise.

Thus $\bar{g}$ has coefficients in k and is independent of the choice of f in $\mathcal{N}_\emptyset$. We let $\mathfrak{m}$ denote the corresponding maximal ideal of $\mathbb{T}_\mathcal{O}$.

Suppose for the moment that K contains the coefficients of all eigenforms of level dividing N_Σ. Let $\mathcal{N}'$ denote the set of newforms g in $\mathcal{N}_\Gamma$ such that

- $\bar{\rho}_g \sim \bar{\rho}$;
- $a_p(\bar{g}) = \alpha_p$ for all p dividing $N_g/N_\emptyset$.

Suppose we are given a newform $g \in \mathcal{N}'$. Let ψ_g denote its character and write Q_g for the conductor of ψ_g. Note that ψ_g has trivial reduction and hence ℓ-power order, and that Q_g divides Q. By proposition 3.2 we see that if $p \in \Sigma - Q$, then ρ_g is unramified at p and hence N_g is not divisible by p. Furthermore, by theorem 3.1 (e) and (4.2.2), $N_g = N_\emptyset Q_g$. Let ξ_g denote the character of $(\mathbb{Z}/Q_g\mathbb{Z})^\times$ of ℓ-power order such that ξ_g^{-2} is the primitive character associated to ψ_g. Then

$$g \otimes \xi_g = \sum \xi_g(n) a_n(g) q^n$$

is in $\mathcal{N}_{Q_g} \subset \mathcal{N}_Q$.

Lemma 4.9 *The map $g \mapsto g \otimes \xi_g$ defines a bijection between $\mathcal{N}'$ and $\mathcal{N}_Q$.*

Proof: Suppose we are given a newform f in $\mathcal{N}_Q$; i.e., f is in $\mathcal{N}_{\Gamma_0(Q)}$ and $\bar{\rho}_f \sim \bar{\rho}$. For $q \in Q$, we have (by lemma 2.44 for example) that

$$\rho_f|_{G_q} \sim \begin{pmatrix} \xi_{f,q} & 0 \\ 0 & \epsilon\xi_{f,q}^{-1} \end{pmatrix},$$

where $\xi_{f,q} : G_q \to K^\times$ is a character whose reduction is the unramified character sending Frob_q to α_q. Note that the characters $\xi_{f,q}|_{I_q}$ have ℓ-power order. In particular $N_f/N_\emptyset$ is Q_f^2 where Q_f is the product of the primes $q \in Q$ such that $\xi_{f,q}$ is ramified (theorem 3.1 (d)). Note also that there is a unique character

$$\xi_f : (\mathbb{Z}/Q_f\mathbb{Z})^\times \to K^\times$$

such that

$$I_q \to G_\mathbb{Q} \twoheadrightarrow \operatorname{Gal}(\mathbb{Q}(\zeta_{Q_f})/\mathbb{Q}) \xrightarrow{\xi_f} K^\times$$

coincides with the restriction of $\xi_{f,q}$ to I_q for each $q|Q_f$. (We have written ξ_f for the character of $G_\mathbb{Q}$ as well as the corresponding Dirichlet character. We shall also write ξ_f for the corresponding character of Δ_Q.) Let g denote the newform associated to the eigenform

$$\sum \xi_f^{-1}(n)a_n(f)q^n \in S_2(\Gamma_0(N_\emptyset) \cap \Gamma_1(Q_f^2), K).$$

By proposition 2.6 and theorem 3.1 we have

- $\rho_g \sim \rho_f \otimes \xi_f^{-1}$,
- $\psi_g = \xi_f^{-2}$,
- $N_g = N_\emptyset Q_f$ and
- $a_q(g) = (\xi_f|_{G_q}^{-1}\xi_{f,q})(\mathrm{Frob}_q)$ has reduction $\bar{\xi}_{f,q}(\mathrm{Frob}_q) = \alpha_q$ for all q dividing Q_f.

Therefore g is in $\mathcal{N}'$ and $f = g \otimes \xi_g$. In particular $g \mapsto g \otimes \xi_g$ is surjective and $\xi_g = \xi_{g\otimes\xi_g}$. Injectivity follows as well on noting that if $g \otimes \xi_g = g' \otimes \xi_{g'}$ then $\xi_g = \xi_{g'}$, hence $g = g'$. □

Recall that $\mathbb{T}_\mathcal{O}$ contains the operators $\langle d \rangle$ for d in $(\mathbb{Z}/M\mathbb{Z})^\times$, and note that $\langle d \rangle - 1 \in \mathfrak{m}$ for all d. Let H' denote the kernel of

$$(\mathbb{Z}/M\mathbb{Z})^\times \twoheadrightarrow (\mathbb{Z}/(\prod_{q\in Q} q)\mathbb{Z})^\times \twoheadrightarrow \Delta_Q.$$

Since the order of H' is not divisible by ℓ, we find that $\sum_{d\in H'} \langle d \rangle \notin \mathfrak{m}$. It follows that if $d \in H'$ then $\langle d \rangle = 1$ in $\mathbb{T}_\mathfrak{m}$. We may therefore regard $\mathbb{T}_\mathfrak{m}$ as an $\mathcal{O}[\Delta_Q]$-algebra via the map $d \mapsto \langle d \rangle$. Recall that $\mathbb{T}_Q$ is considered an $\mathcal{O}[\Delta_Q]$-algebra via the map

$$\mathcal{O}[\Delta_Q] \to R_Q \to \mathbb{T}_Q.$$

If p is a prime not in Q, let x_p denote the unique element of Δ_Q such that $x_p^{-2} = \bar{p}$.

Proposition 4.10 *There is an isomorphism $\mathbb{T}_Q \xrightarrow{\sim} \mathbb{T}_\mathfrak{m}$ of $\mathcal{O}[\Delta_Q]$-algebras such that $T_p \mapsto x_p T_p$ for all primes $p \notin \Sigma$ with $p \nmid \ell N(\bar{\rho})$.*

Proof: We may enlarge K so that we are in the setting of lemma 4.9. We then define an isomorphism of K-algebras

$$\kappa : \mathbb{T}_\mathfrak{m} \otimes K \xrightarrow{\sim} \prod_{g\in\mathcal{N}'} K$$

such that for each $g \in \mathcal{N}'$:

- $\kappa(T_p)_g = a_p(g)$ if p is not in Σ or if p divides Q_g,
- $\kappa(T_p)_g = 0$ if p is in $\Sigma - Q$,
- $\kappa(T_p)_g$ is the root of $X^2 - a_p(g) + \psi_g(p)p$ with reduction α_p if p is in Q but does not divide Q_g, and
- $\kappa(d)_g = \psi_g(d)$ for all $d \in \Delta_Q$.

The existence of such an isomorphism follows from lemma 4.4 which gives

$$\mathbb{T}_{\mathfrak{m}} \otimes K \xrightarrow{\sim} \prod_{g \in \mathcal{N}_\Gamma} \prod_{\mathfrak{p} \in \mathcal{M}_g} (A_{K,g}/I_{K,g})_{\mathfrak{p}},$$

where $\mathcal{M}_g$ is the set of prime ideals in $A_{K,g}/I_{K,g}$ whose preimage in $\mathbb{T}_{\mathcal{O}}$ is contained in $\mathfrak{m}$. If g is not in $\mathcal{N}'$, then $\mathcal{M}_g$ is empty. If g is in $\mathcal{N}'$, then $\mathcal{M}_g$ consists of the prime ideal corresponding to the eigenform whose eigenvalues are prescribed as above, and one checks that $(A_{K,g}/I_{K,g})_{\mathfrak{p}} = K$.

Viewing $\mathbb{T}_Q$ as a subalgebra of $\prod_{f \in \mathcal{N}_Q} K$ and matching indices via the bijection $f = g \otimes \xi_g \leftrightarrow g$, we obtain an injective homomorphism of $\mathcal{O}[\Delta_Q]$-algebras

$$\kappa' : \mathbb{T}_Q \longrightarrow \prod_{g \in \mathcal{N}'} K$$

such that

$$\kappa'(T_p) = (\xi_g(p) a_p(g))_g = \kappa(x_p T_p)$$

for primes $p \notin \Sigma$ such that $p \nmid \ell N_\emptyset$. Since $\mathbb{T}_Q$ is generated over $\mathcal{O}$ by the set of such T_p, we see that $\kappa'(\mathbb{T}_Q)$ is contained in $\kappa(\mathbb{T}_{\mathfrak{m}})$. On the other hand, the image of κ' contains the image of Δ_Q, hence contains $\kappa(\langle d \rangle)$ for d in $(\mathbb{Z}/M\mathbb{Z})^\times$ as well as $\kappa(T_p)$ for $p \notin \Sigma$ with $p \nmid N_\emptyset \ell$.

It remains to prove that $\kappa(T_p)$ is in the image of $\kappa'(\mathbb{T}_Q)$ for p dividing $N_\emptyset \ell \prod_{q \in Q} q$. Consider the composite

$$R_Q \xrightarrow{\phi_Q} \mathbb{T}_Q \xrightarrow{\kappa'} \prod_{g \in \mathcal{N}'} K.$$

The pushforward of ρ_Q^{univ} is equivalent to $\prod \rho_g \otimes \xi_g$ and therefore $\prod \rho_g$ is equivalent to the pushforward of

$$\rho_Q^{\text{univ}} \otimes \xi_Q^{-1}$$

(where ξ_Q was defined in section 2.8). For primes p dividing $N_\emptyset$, we recover $\kappa(T_p)$ as the image of Frob_p on the I_p-coinvariants, hence $\kappa(T_p)$ is in $\kappa'(\mathbb{T}_Q)$. For $q \in Q$ and $g \in \mathcal{N}'$, the pushforward of $\xi_Q|_{G_q}^{-1} \cdot \xi_{q,Q}$ to the g-component is an unramified summand of $\rho_g|_{G_q}$ whose reduction sends Frob_q to α_q. It follows that this character maps Frob_q to $\kappa(T_q)$ and we conclude that $\kappa(T_q) \in \kappa'(\mathbb{T}_Q)$. Finally in the case that ℓ does not divide $N_\emptyset$, we appeal to lemma 4.1 to conclude that $\kappa(T_\ell) \in \kappa'(\mathbb{T}_Q)$. □

Auxiliary primes: For Σ as in proposition 4.10, i.e. a set of primes satisfying (4.2.2), a somewhat simpler argument provides a similar a description of $\mathbb{T}_\emptyset$ with Γ replaced by

$$\Gamma' = \Gamma_0(N_\emptyset) \cap \Gamma_H(M) \tag{4.2.5}$$

where H is the ℓ-Sylow subgroup of $(\mathbb{Z}/M\mathbb{Z})^\times$. While we shall make no direct use of this, the group Γ' will play a role in the proof of theorem 3.31 and we shall need to choose Σ so that Γ' has no elliptic elements; i.e., non-trivial elements of finite order. This is the case for example if Σ contains a prime $p \not\equiv 1 \bmod \ell$

with $p > 3$. The group Γ' is then contained in $\Gamma_1(N)$ for some integer $N > 3$, and therefore has no elliptic elements. Indeed if $\left(\begin{smallmatrix} a & b \\ c & d \end{smallmatrix}\right) \in \Gamma$ has finite order then the roots of $X^2 - (a+d)X + 1$ are roots of unity and we deduce that this matrix is the identity.

In order to show that Σ can be so chosen, we appeal to the following lemma (cf. [DT2], lemma 3).

Lemma 4.11 *Suppose that G is a finite group, $\chi : G \to \bar{k}^\times$ is a character of order d and $\bar{\rho} : G \to GL_2(\bar{k})$ is a representation. Suppose that for all $g \notin \ker \chi$ we have*

$$(\operatorname{tr} \bar{\rho}(g))^2 / \det \bar{\rho}(g) = (1 + \chi(g))^2 / \chi(g). \tag{4.2.6}$$

- *If $d > 3$, then $\bar{\rho}$ is reducible.*
- *If $d = 3$, then $\bar{\rho}$ is reducible or has projective image isomorphic to A_4.*
- *If $d = 2$, then $\bar{\rho}|_{\ker \chi}$ is reducible.*

Proof: Note that if $\bar{\rho}(g)$ is a scalar then $\chi(G) = 1$. Hence χ induces induces a surjective homomorphism $\chi' : G' \to C_d$ where G' is the projective image of $\bar{\rho}$ and C_d is cyclic of order d. Furthermore if $d = 2$, then every element of $G' - \ker \chi'$ has order 2. The lemma then follows from theorem 2.47(b). □

4.3 Proof of theorem 3.31

We shall give a proof of theorem 3.31 which is based on the q-expansion principle rather than the method of de Shalit [dS] employed in [TW]. It will be more convenient to consider the action of the Hecke operators on the full space of modular forms $M_2(\Gamma)$. The Riemann-Roch theorem shows that the dimension of $M_2(\Gamma)$ is $g + s - 1$ where g is the genus of the modular curve X_Γ associated to Γ and s is the number of cusps on X_Γ (see for example [Shi2] thm. 2.23 or [DI] (12.1.5)).

Eisenstein maximal ideals: One can give an explicit description of a space of Eisenstein series $G_2(\Gamma)$ so that

$$M_2(\Gamma) = S_2(\Gamma) \oplus G_2(\Gamma).$$

(See for example [Hi3], lemma 5.2.)

There is a natural action on $M_2(\Gamma)$ by the Hecke operators $\langle d \rangle$ for all $d \in (\mathbb{Z}/N\mathbb{Z})^\times$ and T_p for all primes p. We shall only need to use the operators $\langle d \rangle$ and T_p for p not dividing N, and we let $\tilde{\mathbb{T}}_{\mathbb{Z}}$ denote the subring of $\operatorname{End}(M_2(\Gamma))$ these generate. The ring $\tilde{\mathbb{T}}_{\mathbb{Z}}$ is commutative and is finitely generated and free as a $\mathbb{Z}$-module. That $\tilde{\mathbb{T}}_{\mathbb{Z}}$ is finitely generated is proved for example by showing that $M_2(\Gamma, \mathbb{Z})$, the set of forms with integer Fourier coefficients at ∞, is stable under the Hecke operators and contains a basis for $M_2(\Gamma)$ (see [DI], cor. 12.3.12 and prop. 12.4.1). Alternatively, one can show that $\tilde{\mathbb{T}}_{\mathbb{Z}}$ acts faithfully on the cohomology of the non-compact modular curve Y_Γ (cf. section 1.3).

For any ring A, we write $M_2(\Gamma, A)$ for $M_2(\Gamma, \mathbb{Z}) \otimes A$ and regard this as a module for $\tilde{\mathbb{T}}_A := \tilde{\mathbb{T}}_{\mathbb{Z}} \otimes A$. If $\mathfrak{n}$ is a maximal ideal of $\tilde{\mathbb{T}}_{\mathcal{O}}$ we write $\tilde{\mathbb{T}}_{\mathfrak{n}}$ for the localization. If $\mathfrak{m}$ is a maximal ideal of $\mathbb{T}_{\mathcal{O}}$ then we let $\tilde{\mathfrak{m}}$ denote its preimage in $\tilde{\mathbb{T}}_{\mathcal{O}}$.

We say that a maximal ideal $\mathfrak{n}$ of $\tilde{\mathbb{T}}_{\mathcal{O}}$ is *Eisenstein* if

$$T_p \equiv p+1 \bmod \mathfrak{n} \quad \text{for all } p \equiv 1 \bmod N.$$

One sees from the explicit description of the Eisenstein series that $T_p = p+1$ on $G_2(\Gamma)$ if $p \equiv 1 \bmod N$. We shall be interested in the *non-Eisenstein* maximal ideals because of the following lemma (see [R5], thm. 5.2(c)).

Lemma 4.12 *Suppose that ℓ is odd. The representation*

$$\rho_{\mathfrak{m}} : G_{\mathbb{Q}} \to GL_2(\mathbb{T}_{\mathcal{O}}/\mathfrak{m})$$

is absolutely irreducible if and only if $\tilde{\mathfrak{m}}$ is not Eisenstein.

Proof. If $\rho_{\mathfrak{m}}$ is not absolutely irreducible then $\rho_{\mathfrak{m}} \sim \chi_1 \oplus \chi_2$ with N divisible by the product of the conductors of χ_1 and χ_2. Furthermore if ℓ does not divide N, then one of the characters is unramified at ℓ while the other coincides with the cyclotomic character ϵ on I_ℓ. Therefore $\tilde{\mathfrak{m}}$ is Eisenstein. Conversely if $\tilde{\mathfrak{m}}$ is Eisenstein then proposition 2.6 implies that $\rho_{\mathfrak{m}}$ restricted to $G_{\mathbb{Q}(\zeta_{N\ell})}$ has trivial semisimplification from which it follows that $\rho_{\mathfrak{m}}$ is also reducible. □

Differentials: Suppose now that Σ is a finite set of primes satisfying (4.2.2) Moreover we assume that Σ contains a prime $p > 3$ with $p \not\equiv 1 \bmod \ell$. Let Γ and Γ' be defined as in (4.2.4) and (4.2.5); i.e.

$$\begin{aligned} \Gamma &= \Gamma_0(N_\emptyset) \cap \Gamma_1(M) \\ \Gamma' &= \Gamma_0(N_\emptyset) \cap \Gamma_H(M) \end{aligned}$$

where H is the ℓ-Sylow subgroup of $(\mathbb{Z}/M\mathbb{Z})^\times$. Let X and X' denote the modular curves associated to Γ and Γ'.

We first consider the case where $\ell \nmid N_\emptyset$. The curve X has a smooth proper model $\mathcal{X}$ over $\mathbb{Z}[1/N_\emptyset M]$ such that the complement of the cusps parametrizes cyclic isogenies $(E_1, i_1) \to (E_2, i_2)$ of degree $N_\emptyset$ where E_j is an elliptic curve and i_j is an embedding $\mu_M \hookrightarrow E_j$. (See [DR] or [Kat] for example.) The action of H on X extends to $\mathcal{X}$ and the quotient $\mathcal{X}' = \mathcal{X}/H$ is a smooth model over $\mathbb{Z}[1/N_\emptyset M]$ for X'. The natural projection $\mathcal{X} \to \mathcal{X}'$ is étale. (The fact that it is étale on the complement of the cusps follows from the natural moduli-theoretic description of $\mathcal{X}'$ using the fact that $\Gamma_H(M) \subset \Gamma_1(p)$ for some $p > 3$. One then need only verify that $X \to X'$ is unramified at the cusps.)

For a $\mathbb{Z}[1/N_\emptyset M]$-algebra A, we define

- $\Omega_A = H^0(\mathcal{X}_A, \Omega^1_{\mathcal{X}_A/A})$,
- $\Omega'_A = H^0(\mathcal{X}'_A, \Omega^1_{\mathcal{X}'_A/A})$,
- $\tilde{\Omega}_A = H^0(\mathcal{X}_A, \Omega^1_{\mathcal{X}_A/A}(\mathcal{D}))$,
- $\tilde{\Omega}'_A = H^0(\mathcal{X}'_A, \Omega^1_{\mathcal{X}'_A/A}(\mathcal{D}'))$,

where $\mathcal{D}$ (resp. $\mathcal{D}'$) denotes the reduced divisor defined by the cusps of $\mathcal{X}$ (resp. $\mathcal{X}'$). The q-expansion principle and standard base-change arguments allow us to identify these with $S_2(\Gamma, A)$, $S_2(\Gamma', A)$, $M_2(\Gamma, A)$ and $M_2(\Gamma', A)$ (see [Kat], sec. 1.6, 1.7). We may therefore regard the first two of these as modules for $\mathbb{T}_A$ and all of them as modules for $\tilde{\mathbb{T}}_A$.

Lemma 4.13 *Suppose that $A = k$ or K.*

(a) $\tilde{\Omega}_{\mathcal{O}} \otimes_{\mathcal{O}} A \cong \tilde{\Omega}_A$ and $\tilde{\Omega}'_{\mathcal{O}} \otimes_{\mathcal{O}} A \cong \tilde{\Omega}'_A$.

(b) The natural map $\mathcal{X} \to \mathcal{X}'$ induces an isomorphism

$$\tilde{\Omega}'_A \xrightarrow{\sim} \tilde{\Omega}^H_A.$$

(c) If $\mathfrak{n}$ is a non-Eisenstein maximal ideal of $\tilde{\mathbb{T}}_{\mathcal{O}}$, then the localization at $\mathfrak{n}$ of $\Omega_{\mathcal{O}} \to \tilde{\Omega}_{\mathcal{O}}$ is an isomorphism.

(d) If $\mathfrak{m}$ is a maximal ideal of $\mathbb{T}_A$ then $\Omega_A[\mathfrak{m}]$ is one-dimensional over $\mathbb{T}_A/\mathfrak{m}$.

Sketch of proof:

(a) In the case $A = K$ this follows from the fact that K is flat over $\mathcal{O}$, so suppose $A = k$. Identify $\tilde{\Omega}_{\mathcal{O}}/\lambda\tilde{\Omega}_{\mathcal{O}}$ with the direct image of $\tilde{\Omega}_k$ under $\mathcal{X}_k \to \mathcal{X}_{\mathcal{O}}$ and use that $H^1(\mathcal{X}_k, \Omega^1_{\mathcal{X}_k/k}(\mathcal{D}))$ vanishes by Serre duality. The case of $\tilde{\Omega}'$ is similar. (See [Maz1] sec. II.3.)

(b) Using the fact that $\mathcal{X} \to \mathcal{X}'$ is étale one identifies the pull-back of $\Omega^1_{\mathcal{X}'_A/A}$ with $\Omega^1_{\mathcal{X}_A/A}$ and that of $\mathcal{D}'$ with $\mathcal{D}$.

(c) The map is injective and one proves that its cokernel is free and annihilated by the operators $T_p - (p+1)$ for all $p \equiv 1 \bmod N_\emptyset M$. To prove the latter assertion, first observe that it holds with $\mathcal{O}$ replaced by $\mathbb{C}$, then by $\mathbb{Z}[1/N_\emptyset M]$.

(d) This follows from

$$\Omega_A[\mathfrak{m}] \cong S_2(\Gamma, A)[\mathfrak{m}] \cong \mathrm{Hom}_A(\mathbb{T}_A, A)[\mathfrak{m}] \cong \mathrm{Hom}_A(\mathbb{T}_A/\mathfrak{m}, A),$$

where the middle isomorphism is that of proposition 1.34, but let us reformulate the argument in a way more easily generalized to the case of ℓ dividing $N_\emptyset$ discussed below. Note that $\Omega_A[\mathfrak{m}]$ is non-zero since $\mathbb{T}_{\mathfrak{m}}$ acts faithfully on $\Omega_{\mathfrak{m}}$. To prove that the dimension is at most one we can enlarge the field A and assume $\mathbb{T}_A/\mathfrak{m} \cong A$. One then shows that an eigenform f in Ω_A (for all the Hecke operators T_p and $\langle d\rangle$) is determined by its eigenvalues and the first coefficient of its q-expansion. This follows from the fact that f is determined by its q-expansion and for all n, $a_n(f) = a_1(T_n f)$ and T_n can be expressed in terms of the T_p and $\langle d\rangle$. (See [Maz1], sec. II.9.) □

We now explain how the situation changes if ℓ divides $N_\emptyset$. In that case X has a regular model $\mathcal{X}$ over $\mathbb{Z}[\ell/N_\emptyset M]$ with the same moduli-theoretic description as above. This model is smooth over $\mathbb{Z}[1/N_\emptyset M]$, but $\mathcal{X}_{\mathbb{F}_\ell}$ has two smooth irreducible components crossing at ordinary double points as in [DR] sec. V.1.

The quotient $\mathcal{X}' = \mathcal{X}/H$ is a regular model for X' over $\mathbb{Z}[\ell/N_\emptyset M]$ and $\mathcal{X} \to \mathcal{X}'$ is étale. For a $\mathbb{Z}[\ell/N_\emptyset M]$-algebra A we define Ω_A, Ω'_A, $\tilde{\Omega}_A$ and $\tilde{\Omega}'_A$ as before, except that Ω^1 is replaced by the sheaf of regular differentials (see [DR] sec. I.2, [Maz1] sec. II.6 or [MRi] sec. 7). Formation of these modules again commutes with change of the base A, but we can no longer identify them with $S_2(\Gamma, A)$, $S_2(\Gamma', A)$, $M_2(\Gamma, A)$ and $M_2(\Gamma', A)$ if ℓ is not invertible in A. For every A, there is a natural action of $\mathbb{T}_A$ (resp. $\tilde{\mathbb{T}}_A$) on Ω_A and Ω'_A (resp. $\tilde{\Omega}_A$ and $\tilde{\Omega}'_A$). In the case of $\mathbb{T}_A$ this is proved by identifying Ω_A (resp. Ω'_A) with the cotangent space at the origin for the Néron model over A for the Jacobian of X (resp. X'). In the case of $\tilde{\mathbb{T}}_A$ the action is defined using Grothendieck-Serre duality (as in [Maz1] sec. II.6 or [MRi] sec. 7).

Lemma 4.14 *If ℓ divides $N_\emptyset$ then lemma 4.13 carries over with the above notation and the additional hypothesis $T_\ell \notin \mathfrak{m}$ for part (d).*

The proof is essentially the same except that part (d) is more delicate in the case $A = k$. For the proof in that case we refer the reader to [W3], lemma 2.2 (from which the result stated here is immediate). We remark however that we shall only use lemma 4.14 if $\rho_\mathfrak{m}$ is not good. In that case one can also deduce the statement here from the argument used in the proof of [MRi] prop. 20, which shows instead that $\dim_k \Omega_k[\mathfrak{m}'] = 1$ where $\mathfrak{m}'$ is a certain maximal ideal corresponding to $\mathfrak{m}$, but contained in an algebra defined using the operators $\langle d \rangle$, T_p for $p \neq \ell$ and w_ℓ.

Remark 4.15 The approach in [TW] to proving theorem 3.31 is based on the result that under certain hypotheses $(\mathcal{T}_\ell(J) \otimes_{\mathbb{Z}_\ell} \mathcal{O})_\mathfrak{m}$ is free of rank two over $\mathbb{T}_\mathfrak{m}$ (see [W3], thm. 2.1 and its corollaries). This result generalizes work of Mazur Ribet, and Edixhoven (see secs. 14 and 15 of ch. II of [Maz1], thm. 5.2 of [R5], [MRi] and sec. 9 of [Edi]), and the key to its proof is lemmas 4.13(d) and 4.14. We shall instead give a proof of theorem 3.31 which is based directly on lemma 4.13, (d).

The fact that $(\mathcal{T}_\ell(J) \otimes_{\mathbb{Z}_\ell} \mathcal{O})_\mathfrak{m}$ is free of rank two over $\mathbb{T}_\mathfrak{m}$ actually underlies Wiles' approach to many intermediate results along the way to proving the Shimura-Taniyama conjecture for semistable elliptic curves. We shall appeal to a special case in the course of proving theorem 3.36 below.

Proof of the theorem: Suppose we are given a representation $\bar{\rho}$ as in section 3.3 and a set of primes Q as in the statement of theorem 3.31. We apply lemma 4.11 with $G = G_\mathbb{Q}$ and $\chi = \epsilon$ to choose an auxiliary prime p not dividing $6N_\emptyset \ell \prod_{q \in Q} q$ such that

- $p \not\equiv 1 \bmod \ell$
- $\operatorname{tr}(\bar{\rho}(\operatorname{Frob}_p))^2 \not\equiv (p+1)^2$.

To prove theorem 3.31 we may replace K by a larger field and assume that k contains the eigenvalues of $\bar{\rho}(\operatorname{Frob}_p)$.

Let $\Sigma = Q \cup \{p\}$. Thus Σ is a set of primes as in proposition 4.10 and Σ contains a prime $p > 3$ such that $p \not\equiv 1 \bmod \ell$. We let Γ be as in (4.2.4); thus

$$\Gamma = \Gamma_0(N_\emptyset) \cap \Gamma_1(M)$$

where $M = p^2 \prod_{q \in Q} q$, and we choose the maximal ideal $\mathfrak{m}$ of $\mathbb{T}_{\mathcal{O}}$ as in proposition 4.10. Let H denote the ℓ-Sylow subgroup of $(\mathbb{Z}/M\mathbb{Z})^\times$ and regard $\mathbb{T}_{\mathcal{O}}$ and hence $\mathbb{T}_{\mathfrak{m}}$ as an $\mathcal{O}[H]$-algebra via $d \mapsto \langle d \rangle$. In view of proposition 4.10, theorem 3.31 is equivalent to the following:

Theorem 4.16 *$\mathbb{T}_{\mathfrak{m}}$ is free over $\mathcal{O}[H]$.*

Proof: Consider the $\tilde{\mathbb{T}}_{\mathcal{O}}$-module

$$\tilde{L} = \mathrm{Hom}_{\mathcal{O}}(\tilde{\Omega}_{\mathcal{O}}, \mathcal{O})$$

and the $\mathbb{T}_{\mathcal{O}}$-module

$$L = \mathrm{Hom}_{\mathcal{O}}(\Omega_{\mathcal{O}}, \mathcal{O}).$$

Since $X \to X'$ is unramified, the Riemann-Hurwitz formula implies that

$$\dim M_2(\Gamma) = g + s - 1 = \#H(g' + s' - 1) = \#H \dim M_2(\Gamma'),$$

where g (resp. g') is the genus and s (resp. s') is the number of cusps of X (resp. X'). It follows from lemma 4.13 (a) (and lemma 4.14) that

$$\dim_K(\tilde{L} \otimes_{\mathcal{O}} K) = \#H \dim_k \tilde{\Omega}'_k \qquad (4.3.1)$$

On the other hand, by (b) we have that

$$\tilde{L}/(\mathfrak{a}, \lambda)\tilde{L} \cong \mathrm{Hom}_k(\tilde{\Omega}_k^H, k) \cong \mathrm{Hom}_k(\tilde{\Omega}'_k, k) \qquad (4.3.2)$$

where $\mathfrak{a} = \ker(\mathcal{O}[H] \to \mathcal{O})$ is the augmentation ideal. By Nakayama's lemma and (4.3.2), $\tilde{L}$ is generated as an $\mathcal{O}[H]$-module by d elements where $d = \dim_k \tilde{\Omega}'_k$. By (4.3.1) any surjective homomorphism

$$\mathcal{O}[H]^d \to \tilde{L}$$

is in fact an isomorphism. Hence $\tilde{L}$ is free over $\mathcal{O}[H]$, as is

$$L_{\mathfrak{n}} = \mathrm{Hom}_{\mathcal{O}}(\Omega_{\mathfrak{n}}, \mathcal{O}) \cong \mathrm{Hom}_{\mathcal{O}}(\tilde{\Omega}_{\mathfrak{n}}, \mathcal{O})$$

for each non-Eisenstein maximal ideal $\mathfrak{n}$ of $\tilde{\mathbb{T}}_{\mathcal{O}}$ (by part (c) of lemmas 4.13 and 4.14).

Since $\rho_{\mathfrak{m}} \sim \bar{\rho}$ is absolutely irreducible, $\tilde{\mathfrak{m}}$ is not Eisenstein (by lemma 4.12), and it follows that $L_{\mathfrak{m}}$ is free over $\mathcal{O}[H]$.

Since $L \otimes_{\mathcal{O}} K$ is free of rank one over $\mathbb{T}_K$ (lemma 1.34), we have that $L_{\mathfrak{m}} \otimes_{\mathcal{O}} K$ is free of rank one over $\mathbb{T}_{\mathfrak{m}} \otimes_{\mathcal{O}} K$. If $\ell \nmid N_\emptyset$ then lemma 4.13 implies that

$$L/\mathfrak{m}L \cong \mathrm{Hom}_k(\Omega_k[\mathfrak{m}], k)$$

is one-dimensional over k. The same assertion holds if $\ell | N_\emptyset$ by lemma 4.14 (from the definition of $\mathfrak{m}$ in this case, we see that it does not contain T_ℓ). It follows from Nakayama's lemma that $L_{\mathfrak{m}}$ is free of rank one over $\mathbb{T}_{\mathfrak{m}}$. Therefore $\mathbb{T}_{\mathfrak{m}}$ is free over $\mathcal{O}[H]$. □

4.4 Proof of theorem 3.36

Our proof of theorem 3.36 is based on Wiles' arguments in ch. 2 of [W3], which in turn are based on a method of Ribet [R4]. We shall reformulate the proof somewhat to underscore the relationship observed by Doi and Hida [Hi1] between the size of $\mathcal{O}/\eta$ and the value of an L-function, but after doing so we shall also sketch the more direct argument used by Wiles. Two important ingredients appear in both versions of the argument, and we shall discuss their proofs in section 4.5. These ingredients are

- the generalization of a result of Ihara [Ih] used in Ribet's argument;
- the generalization of a result of Mazur [Maz1] on the structure of the Tate module $\mathcal{T}_\ell(J_0(N))$ as a module for $\mathbb{T}_{\mathbb{Z}_\ell}$.

Assume now that we are in the setting of theorem 3.36. In particular, $\bar{\rho}$ is a representation as in section 3.3, Σ is a finite set of primes contained in $\Sigma_{\bar{\rho}}$ and π is an $\mathcal{O}$-algebra homomorphism $\mathbb{T}_\Sigma \to \mathcal{O}$ arising from a newform f in $\mathcal{N}_\Sigma$ with coefficients in $\mathcal{O}$.

The symmetric square L-function: For each prime p let $\alpha_p(f)$ and $\beta_p(f)$ denote the roots of the polynomial $X^2 - a_p(f)X + \delta_p p = 0$, where $\delta_p = 0$ or 1 according to whether p divides N_f. Thus we have

$$L_p(f,s) = (1-\alpha_p(f)p^{-s})^{-1}(1-\beta_p(f)p^{-s})^{-1}.$$

If $p = \ell$ divides N/N_f, then we require that $\alpha_\ell(f)$ be the root which is a unit in $\mathcal{O}$, i.e. $a_\ell(g)$. We define the symmetric square L-function associated to f as

$$L(\mathrm{Symm}^2 f, s) = \prod_p L_p(\mathrm{Symm}^2 f, s),$$

where

$$L_p(\mathrm{Symm}^2 f, s) = (1-\alpha_p^2(f)p^{-s})^{-1}(1-\alpha_p(f)\beta_p(f)p^{-s})^{-1}(1-\beta_p^2(f)p^{-s})^{-1}.$$

(We caution the reader we have defined the Euler factors rather naively at primes p such that p^2 divides N_f.) The product converges absolutely for real part of $s > 2$ and can be analytically continued to an entire function.

The calculation of η_Σ: We write $\wp$ for the kernel of π and I for the annihilator of $\wp$ in $\mathbb{T}_\Sigma$. Since $\mathbb{T}_\Sigma$ is reduced, we have $\wp \cap I = 0$ and $\wp \oplus I$ has finite index in $\mathbb{T}_\Sigma$. Note that

$$\mathcal{O}/\eta_\Sigma \cong \mathbb{T}_\Sigma/(\wp \oplus I).$$

Suppose that we are given a $\mathbb{T}_\Sigma$-module L which is finitely generated and free over $\mathcal{O}$, and such that $L \otimes K$ is free of rank d over $\mathbb{T}_\Sigma \otimes_{\mathcal{O}} K$. Suppose also that L is endowed with a perfect $\mathcal{O}$-bilinear pairing

$$\begin{array}{ccc} L \times L & \to & \mathcal{O} \\ (x,y) & \mapsto & \langle x,y\rangle \end{array}$$

such that $\langle Tx, y\rangle = \langle x, Ty\rangle$ for all $x, y \in L$ and $T \in \mathbb{T}_\Sigma$. Note that to give such a pairing is equivalent to giving an isomorphism

$$\begin{array}{ccc} L & \xrightarrow{\sim} & \mathrm{Hom}_{\mathcal{O}}(L,\mathcal{O}) \\ x & \mapsto & \langle x,\cdot\rangle \end{array}$$

of $\mathbb{T}_\Sigma$-modules. We shall refer to the module L together with the pairing $\langle\ ,\ \rangle$ as a *self-dual* $\mathbb{T}_\Sigma$-module of *rank d*. We then can give a lower bound for the size of $\mathcal{O}/\eta_\Sigma$ in terms of a basis $\{x_1, x_2, \ldots, x_d\}$ for the free $\mathcal{O}$-module $L[\wp]$:

Lemma 4.17 *We have*

$$\eta_\Sigma^d \subset \mathcal{O}\det(\langle x_i, x_j\rangle)_{i,j},$$

and equality holds if L is free over $\mathbb{T}_\Sigma$.

Proof: The modules $L[\wp]$ and $L/L[I]$ are free of rank d over $\mathcal{O}$ and the pairing $\langle\ ,\ \rangle$ induces an isomorphism

$$L/L[I] \xrightarrow{\sim} \mathrm{Hom}_\mathcal{O}(L[\wp], \mathcal{O}). \tag{4.4.1}$$

The $\mathcal{O}$-module $M = L/(L[\wp] \oplus L[I])$ is annihilated by η_Σ and is generated by d elements. Furthermore M is isomorphic to $(\mathcal{O}/\eta_\Sigma)^d$ if L is free over $\mathbb{T}_\Sigma$. It follows that

$$\#(\mathcal{O}/\eta_\Sigma)^d \geq \#M,$$

with equality if L is free. The cardinality of M is that of the cokernel of the map

$$L[\wp] \hookrightarrow \mathrm{Hom}_\mathcal{O}(L[\wp], \mathcal{O})$$

arising from the pairing $\langle\ ,\ \rangle$, and this is precisely

$$\mathcal{O}/\det(\langle x_i, x_j\rangle)_{i,j}.$$

□

Recall that proposition 4.7 establishes an isomorphism between $\mathbb{T}_\Sigma$ and $\mathbb{T}_\mathfrak{m}$, the localization at a certain maximal ideal $\mathfrak{m}$ of $\mathbb{T}_\mathcal{O} = \mathbb{T}_\mathbf{Z} \otimes \mathcal{O}$, where

$$\mathbb{T}_\mathbf{Z} \subset \mathrm{End}\,(S_2(\Gamma_0(N))),$$

and where $N = N_\Sigma$ is defined in (4.2.1). Write X for $X_0(N)$ and J for $J_0(N)$. Recall that $H_1(X, \mathbb{Z})$ is endowed with the structure of a $\mathbb{T}_\mathbf{Z}$-module and

$$\mathcal{T}_\ell(J) \otimes_{\mathbf{Z}_\ell} \mathcal{O} \cong H_1(X, \mathbb{Z}) \otimes \mathcal{O}$$

with that of a $\mathbb{T}_\mathcal{O}$-module. We also regard

$$H^1(X, \mathcal{O}) \cong \mathrm{Hom}(H_1(X, \mathbb{Z}), \mathcal{O})$$

as a $\mathbb{T}_\mathcal{O}$-module; as such it is naturally isomorphic to

$$\mathrm{Hom}_\mathcal{O}(\mathcal{T}_\ell(J) \otimes_{\mathbf{Z}_\ell} \mathcal{O}, \mathcal{O}).$$

The Weil pairing defines a perfect pairing

$$\mathcal{T}_\ell(J) \times \mathcal{T}_\ell(J) \to \mathbb{Z}_\ell$$

(with $(e^{2\pi i/\ell^n})_n$ as the chosen generator for $\varprojlim \mu_{\ell^n}(\mathbb{C})$). The composite

$$\mathcal{T}_\ell(J) \overset{w}{\rightrightarrows} \mathcal{T}_\ell(J) \to \mathrm{Hom}_{\mathbf{Z}_\ell}(\mathcal{T}_\ell(J), \mathbb{Z}_\ell) \tag{4.4.2}$$

is an isomorphism of $\mathbb{T}_{\mathbb{Z}_\ell}$-modules where w_* is induced by the involution $w = w_N$ of X (see section 1.4). Tensoring with $\mathcal{O}$ and localizing at $\mathfrak{m}$, we regard

$$L_{\mathcal{T}} = (\mathcal{T}_\ell(J) \otimes_{\mathbb{Z}_\ell} \mathcal{O})_{\mathfrak{m}}$$

as a self-dual $\mathbb{T}_\Sigma$-module of rank 2 via the isomorphism $\mathbb{T}_\Sigma \cong \mathbb{T}_{\mathfrak{m}}$ of proposition 4.7. We shall write $\langle , \rangle_{\mathcal{T}}$ for the pairing obtained from (4.4.2). Similarly, using w^*, the cup product and Poincaré duality, we regard

$$L_H = H^1(X, \mathcal{O})_{\mathfrak{m}}$$

as a self-dual $\mathbb{T}_\Sigma$-module of rank 2, with $\langle x, y\rangle_H$ defined by the image of $x \cup w^*y$ under the canonical isomorphism of $H^2(X, \mathcal{O})$ with $\mathcal{O}$. Note that L_H is canonically isomorphic as a $\mathbb{T}_\Sigma$-module to $\mathrm{Hom}_{\mathcal{O}}(L_{\mathcal{T}}, \mathcal{O})$. In the next section we shall discuss the following generalization of a result of Mazur [Maz1].

Theorem 4.18 *The $\mathbb{T}_\Sigma$-modules $L_{\mathcal{T}}$ and L_H are free.*

Corollary 4.19 *If $\{x, y\}$ is a basis for $L_{\mathcal{T}}[\wp]$ (resp. $L_H[\wp]$), then η_Σ is generated by $\langle x, y\rangle_{\mathcal{T}}$ (resp. $\langle x, y\rangle_H$).*

This follows from lemma 4.17, theorem 4.18 and skew-symmetry of the pairings.

Hida's formula: We now explain how the value of $\langle x, y\rangle_H$ in corollary 4.19 is related to $L(\mathrm{Symm}^2 f, 2)$ by a formula of Hida (see [Hi1] sec. 5, [W3] sec. 4.1).

Recall that the homomorphism

$$\mathbb{T}_{\mathcal{O}} \to \mathbb{T}_{\mathfrak{m}} \cong \mathbb{T}_\Sigma \xrightarrow{\pi} \mathcal{O}$$

arises as $T_n \mapsto a_n(g)$ for a normalized eigenform g in $S_2(\Gamma_0(N), K)$ whose associated newform is f. We shall write $\mathcal{P}$ for the kernel of this homomorphism; thus $\mathcal{P}$ is the preimage in $\mathbb{T}_{\mathcal{O}}$ of the ideal $\wp$ of $\mathbb{T}_\Sigma$. Note that $\mathcal{P}_{\mathfrak{m}}$ corresponds to $\wp$ under the isomorphism $\mathbb{T}_{\mathfrak{m}} \cong \mathbb{T}_\Sigma$.

Choose a number field K_0 containing the Fourier coefficients of g and let $\mathcal{O}_0$ denote the valuation ring $\mathcal{O} \cap K_0$. We choose the basis $\{x, y\}$ for $L_H[\wp]$ in the image of

$$H^1(X, \mathcal{O}_0)[\mathcal{P}_0] \subset H^1(X, \mathcal{O})[\mathcal{P}] \xrightarrow{\sim} H^1(X, \mathcal{O})_{\mathfrak{m}}[\mathcal{P}] = L_H[\wp],$$

where $\mathcal{P}_0 = \mathcal{P} \cap \mathbb{T}_{\mathcal{O}_0}$. Let $\mathcal{P}_{\mathbb{C}}$ denote the kernel of the homomorphism

$$\theta_g : \mathbb{T}_{\mathbb{C}} \to \mathbb{C}$$

associated to the eigenform g. The two-dimensional complex vector space

$$(H^1(X, \mathcal{O}_0)[\mathcal{P}_0]) \otimes_{\mathcal{O}_0} \mathbb{C} = H^1(X, \mathbb{C})[\mathcal{P}_{\mathbb{C}}]$$

has a canonical basis

$$\{\omega_g, \bar{\omega}_{g^c}\},$$

where we view these differential forms as cohomology classes and the superscript c indicates complex conjugation applied to the Fourier coefficients. (Recall that ω_g is the holomorphic differential on X defined by $\sum a_n(g)q^{n-1}dq$, so $\bar{\omega}_{g^c}$ is the antiholomorphic differential defined by $\sum a_n(g)\bar{q}^{n-1}d\bar{q}$ where $q = e^{2\pi i\tau}$.) Let A denote the matrix in $GL_2(\mathbb{C})$ such that

$$(\omega_g, \bar{\omega}_{g^c}) = (x, y)A.$$

Theorem 4.20 *With the above notation we have (up to sign)*

$$\langle x, y\rangle = (i\pi \det A)^{-1} N_f L(\mathrm{Symm}^2 f, 2) \prod_{p|N/N_f} c_p(f)),$$

where

$$c_p(f) = \begin{cases} p^{\delta_p} L_p(\mathrm{Symm}^2 f, 2)^{-1} & \text{if } p \neq \ell \\ (\alpha_\ell(f) - \beta_\ell(f))(1 - \alpha_\ell^{-2}(f)) & \text{if } p = \ell. \end{cases}$$

Proof: We have

$$\langle x, y\rangle_H \det A = \int_X \omega_g \wedge \bar{\omega}_{wg^c} = 8\pi^2 i \langle g, (wg)^c\rangle$$

where the last $\langle , \rangle$ denotes the Petersson inner product and we have used that $wg^c = (wg)^c$ for forms on $\Gamma_0(N)$. We then appeal to a formula of Shimura (see [Shi5] (2.5) and [Hi1] (5.13)) to obtain

$$\langle g, (wg)^c\rangle = (48\pi)^{-1}[SL_2(\mathbb{Z}) : \Gamma_0(N)]\mathrm{res}_{s=2} D(g, wg, s).$$

By [Shi5] lemma 1, the Dirichlet series $D(g, h, s)$ is defined by $\sum a_n(g) a_n(h) n^{-s}$ and if g and h are normalized eigenforms then this has an Euler product expression in which the factors are

$$(1 - \alpha_p(g)\beta_p(g)\alpha_p(h)\beta_p(h)p^{-2s})(1 - \alpha_p(g)\alpha_p(h)p^{-s})^{-1} \cdot$$
$$(1 - \alpha_p(g)\beta_p(h)p^{-s})^{-1}(1 - \beta_p(g)\alpha_p(h)p^{-s})^{-1}(1 - \beta_p(g)\beta_p(h)p^{-s})^{-1},$$

(where the α_p's and β_p's are defined as they were for f). Using the recipe for obtaining g from f (see lemma 4.6) we find that if ℓ is not in Σ then

$$D(g, wg, s) = \pm D_\Sigma(f, f, s) \prod_{p|N/N_f} p^{-1},$$

(where the subscript Σ indicates that the primes in Σ are removed from the defining Euler product). If ℓ is in Σ, then we must also multiply by the factor

- $(1 - \ell^{-s})^{-1}$ if ℓ divides N_f;
- $(1 - \ell^{1-s})^{-1}(1 - \alpha_\ell^2(f)\ell^{-s})^{-1}(\beta_\ell(f) - \alpha_\ell^{2-s}(f))$ otherwise.

To obtain the formula in this last case, consider the eigenform

$$f_\alpha = f - \beta_\ell(f) f(\ell\tau)$$

on $\Gamma_0(N_f \ell)$. Since

$$w_{N_f \ell} f_\alpha = \pm(\ell f(\ell\tau) - \alpha_\ell^{-1}(f) f)$$

and

$$D(f_\alpha, \ell f(\ell\tau), s) = \alpha_\ell(f)\ell^{1-s} D(f_\alpha, f, s),$$

we get

$$D(f_\alpha, w_{N_f\ell} f_\alpha, s) = \pm\ell^{-1}(\beta_\ell(f) - \alpha_\ell(f)\ell^{2-s}) D(f_\alpha, f, s)$$

and use the Euler product. The Euler product expression also gives

$$D_\Sigma(f, f, s) = \frac{\zeta_N(s-1) L_\Sigma(\mathrm{Symm}^2 f, s)}{\zeta_N(2s-2)},$$

where $\zeta_N(s)$ is the Riemann zeta function with the Euler factors removed at primes dividing N. Thus

$$\mathrm{res}_{s=2}D(g,wg,s) = \pm\frac{6}{\pi^2}\frac{NL_\Sigma(\mathrm{Symm}^2 f,2)\prod_{p|N/N_f} p}{[SL_2(\mathbb{Z}):\Gamma_0(N)]},$$

with the extra factor of $(1-\ell^{-2})^{-1}$ or

$$(\beta_\ell(f)-\alpha_\ell(f))(1-\beta_\ell^{-2}(f))^{-1}(1-\ell^{-1})^{-1}$$

if ℓ is in Σ. The theorem follows. □

Corollary 4.21 *With the above notation (in particular, $\Sigma \subset \Sigma_{\bar\rho}$), we have*

$$\frac{\ell^m N_f L_\Sigma(\mathrm{Symm}^2 f,2)}{i\pi \det A} \in \mathcal{O}_0$$

is a generator of η_Σ, where

$$m = \begin{cases} 0, & \text{if } \ell \notin \Sigma, \\ 2, & \text{if } \ell \in \Sigma \text{ and } \ell | N_f, \\ 3, & \text{if } \ell \in \Sigma \text{ and } \ell \nmid N_f. \end{cases}$$

Remark 4.22 Without theorem 4.18 we obtain instead that η_Σ is contained in the ideal generated by the expression in the corollary.

Comparing η_Σ's: Suppose that $\Sigma' \subset \Sigma_{\bar\rho}$ is a finite set of primes containing Σ. In view of corollary 4.21, theorem 3.36 reduces to a comparison of $\det A$ and $\det A'$ where A' is defined using Σ' instead of Σ. We shall make the desired comparison using Wiles' generalization of a result of Ihara.

Applying proposition 4.7 to Σ', we have an isomorphism between $\mathbb{T}_{\Sigma'}$ and $\mathbb{T}'_{\mathfrak{m}'}$ where $\mathfrak{m}'$ is a certain maximal ideal of $\mathbb{T}'_{\mathcal{O}} = \mathbb{T}'_{\mathbb{Z}} \otimes \mathcal{O}$ with $\mathbb{T}'_{\mathbb{Z}}$ acting on $S_2(\Gamma_0(N_{\Sigma'}))$. We write $\mathcal{P}'$ for the preimage in $\mathbb{T}'_{\mathcal{O}}$ of $\wp'$. We let $N' = N_{\Sigma'}$, $X' = X_0(N')$, $J' = J_0(N')$ and define self-dual $\mathbb{T}'_{\mathfrak{m}'}$-modules $L'_{\mathcal{T}}$ and L'_H as above.

We now define a homomorphism

$$\phi_{\mathcal{T}} : L'_{\mathcal{T}} \to L_{\mathcal{T}}$$

of $\mathbb{T}'_{\mathfrak{m}'}$-modules in the case that $\Sigma' = \Sigma \cup \{p\}$ for some prime $p \in \Sigma_{\bar\rho} - \Sigma$. We first suppose that $p \neq \ell$, in which case $N' = Np^2$ and the surjective homomorphism of $\mathcal{O}$-algebras $\mathbb{T}'_{\mathfrak{m}'} \to \mathbb{T}_{\mathfrak{m}}$ (corresponding to $\mathbb{T}_{\Sigma'} \mapsto \mathbb{T}_\Sigma$) is defined by $T_r \mapsto T_r$ if $r \neq p$ and $T_p \mapsto 0$. The maps $\tau \mapsto \tau$, $\tau \mapsto p\tau$ and $\tau \mapsto p^2\tau$ on the upper half-plane induce maps $X' \to X$ which we denote α, β and γ. These give rise by functoriality to a map

$$\begin{array}{rcl} J' & \to & J \times J \times J \\ x & \mapsto & (\alpha_* x, \beta_* x, \gamma_* x) \end{array}$$

which induces an $\mathcal{O}$-linear homomorphism

$$\mathcal{T}_\ell(J') \otimes_{\mathbb{Z}_\ell} \mathcal{O} \to (\mathcal{T}_\ell(J) \otimes_{\mathbb{Z}_\ell} \mathcal{O})^3.$$

We compose with the projection to

$$L_T^3 = (\mathcal{T}_\ell(J) \otimes_{\mathbf{Z}_\ell} \mathcal{O})_{\mathfrak{m}}^3$$

followed by the map $L_T^3 \to L_T$ defined by the matrix $(1, -p^{-1}T_p, p^{-1})$. The reason for this choice of matrix is that

$$g' = (\alpha^* - p^{-1}\beta^* T_p + p^{-1}\gamma^*)g \tag{4.4.3}$$

is the eigenform corresponding to $\mathcal{P}'$.

Lemma 4.23 *The composite*

$$\mathcal{T}_\ell(J') \otimes_{\mathbf{Z}_\ell} \mathcal{O} \to L_T \tag{4.4.4}$$

is $\mathbb{T}'_{\mathcal{O}}$-linear where L_T is regarded as a $\mathbb{T}'_{\mathcal{O}}$-module via the surjection

$$\mathbb{T}'_{\mathcal{O}} \to \mathbb{T}'_{\mathfrak{m}'} \to \mathbb{T}_{\mathfrak{m}}.$$

We leave the proof to the reader after pointing out that the map on Hecke algebras can be rewritten as

$$\begin{array}{ccccccc} \mathbb{T}'_{\mathcal{O}} & \to & \mathbb{T}_{\mathcal{O}}[u]/(f(u)) & \to & \mathbb{T}_{\mathfrak{m}}[u]/(f(u)) & \to & \mathbb{T}_{\mathfrak{m}} \\ T_p & \mapsto & u & & u & \mapsto & 0, \end{array}$$

where $f(u) = u^3 - T_p u^2 + pu$.

It follows that (4.4.4) factors through L'_T and we thus obtain the desired homomorphism

$$\phi_T : L'_T \to L_T$$

of $\mathbb{T}'_{\mathfrak{m}'}$-modules. Note that ϕ_T is also the composite of the canonical splitting

$$L'_T \hookrightarrow \mathcal{T}_\ell(J') \otimes_{\mathbf{Z}_\ell} \mathcal{O}$$

with (4.4.4).

The construction of ϕ_T is similar for $p = \ell$ except that we have only two copies of J and the map $L_T^2 \to L_T$ is given by the matrix $(1, -\alpha_\ell^{-1})$ where α_ℓ is the unit root of (3.3.2). For arbitrary Σ' satisfying $\Sigma \subset \Sigma' \subset \Sigma_{\bar{\rho}}$ we define ϕ_T as a composite of the maps defined above. It is independent of the choice of ordering of $\Sigma' - \Sigma$. Recall that ϕ'_T is used to denote the dual map $L_T \to L'_T$. We define $\phi_H : L'_H \to L_H$ as the adjoint of the map ϕ'_H which renders the diagram

$$\begin{array}{cccc} \phi'_H: & H^1(X, \mathcal{O})_{\mathfrak{m}} & \to & H^1(X', \mathcal{O})_{\mathfrak{m}'} \\ & \downarrow & & \downarrow \\ & \mathrm{Hom}_{\mathcal{O}}(L_T, \mathcal{O}) & \to & \mathrm{Hom}_{\mathcal{O}}(L'_T, \mathcal{O}) \end{array}$$

commutative (where the vertical maps are the natural isomorphisms and the bottom one is dual to ϕ_T).

The second crucial result whose discussion we postpone until the next section is the following generalization of a lemma of Ihara.

Lemma 4.24 *ϕ_T and ϕ_H are surjective.*

Suppose again we are in the case $\Sigma' = \Sigma \cup \{p\}$ with $p \neq \ell$. One need only unravel the definition of ϕ'_H to see that the diagram

$$\begin{array}{ccc} H^1(X, \mathcal{O}_0)[\mathcal{P}_0] & \to & H^1(X', \mathcal{O}_0)[\mathcal{P}'_0] \\ \downarrow & & \downarrow \\ L_H & \overset{\phi'_H}{\to} & L'_H, \end{array}$$

commutes, where the top arrow is defined as

$$\alpha^* - p^{-1}\pi(T_p)\beta^* + p^{-1}\gamma^*.$$

Extending scalars to $\mathbb{C}$, this map sends ω_g to $\omega_{g'}$ and $\bar{\omega}_{g^c}$ to $\bar{\omega}_{g'^c}$ where g' is defined by (4.4.3). In the case $\Sigma' = \Sigma \cup \{\ell\}$, the same assertion holds if $\mathcal{O}_0$ is chosen so that it contains $\pi(\alpha_\ell) = a_\ell(g')$ and the top arrow is defined as

$$\alpha^* - \pi(\alpha_\ell)^{-1}\beta^*.$$

We conclude that if $\{x, y\}$ is a basis for $H^1(X, \mathcal{O}_0)[\mathcal{P}_0]$ and the matrices A and A' are defined using the bases $\{x, y\}$ for $H^1(X, \mathcal{O}_0)[\mathcal{P}_0]$ and $\{\phi'_H x, \phi'_H y\}$ for $H^1(X', \mathcal{O}_0)[\mathcal{P}'_0]$, then $A' = A$. Theorem 3.36 is now immediate from corollary 4.21. In fact, we have proved that

$$\eta_{\Sigma'} = \pi(\prod_{p \in \Sigma' - \Sigma} c_p)\eta_\Sigma.$$

□

Remark 4.25 Note that to obtain the inclusion stated in theorem 3.36 it suffices to apply theorem 4.18 for Σ, rather than both Σ and Σ' (cf. remark 4.22).

Wiles' argument: The method of [W3] sec. 2.2, like that of Ribet in [R4], is more direct than the one given above but does not explicitly illustrate the relation with values of L-functions. The approach is simply to compute the composite $\phi_T \phi'_T$. It suffices to consider the case $\Sigma' = \Sigma \cup \{p\}$ and we suppose first that $p \neq \ell$. Let Δ denote the endomorphism of J^3 defined by the matrix

$$\Delta = \begin{pmatrix} \alpha_* \\ \beta_* \\ \gamma_* \end{pmatrix} w'_*(\alpha^*, \beta^*, \gamma^*)w_*.$$

Using the relations $\alpha w' = w\gamma$ and $\beta w' = w\beta$, we find that

$$\Delta = \begin{pmatrix} \alpha_*\gamma^* & \alpha_*\beta^* & \alpha_*\alpha^* \\ \beta_*\gamma^* & \beta_*\beta^* & \beta_*\alpha^* \\ \gamma_*\gamma^* & \gamma_*\beta^* & \gamma_*\alpha^* \end{pmatrix},$$

which can be computed for example by considering its effect on the cotangent space $S_2(\Gamma_0(N))^3$. The result is that

$$\Delta = \begin{pmatrix} T_p^2 - (p+1) & pT_p & p(p+1) \\ pT_p & p(p+1) & pT_p \\ p(p+1) & pT_p & T_p^2 - (p+1) \end{pmatrix},$$

which we note commutes with the action of $\mathbb{T}_{\mathbf{Z}}$ on J^3. One then checks that

$$\begin{aligned}\phi_{\mathcal{T}}\phi'_{\mathcal{T}} &= (1, -p^{-1}T_p, p^{-1})\ (\mathcal{T}_\ell(\Delta) \otimes_{\mathbf{Z}_\ell} \mathcal{O})_{\mathfrak{m}} \begin{pmatrix} 1 \\ -p^{-1}T_p \\ p^{-1} \end{pmatrix} \\ &= -p^{-2}(p-1)(T_p^2 - (p+1)^2).\end{aligned}$$

The case of $p = \ell$ is similar but simpler. One uses

$$\Delta = \begin{pmatrix} \alpha_* \\ \beta_* \end{pmatrix} w'_*(\alpha^*, \beta^*)w_* = \begin{pmatrix} T_\ell & \ell+1 \\ \ell+1 & T_\ell \end{pmatrix}$$

and gets

$$\phi_{\mathcal{T}}\phi'_{\mathcal{T}} = -(\alpha_\ell - \ell\alpha_\ell^{-1})(1 - \alpha_\ell^{-2}).$$

Thus in either case we find that $\phi_{\mathcal{T}}\phi'_{\mathcal{T}}$ acts on $L'_{\mathcal{T}}[\mathcal{P}]$ by an element of $\mathcal{O} = \mathbb{T}_{\mathfrak{m}}/\mathcal{P}$ which is a unit times $\pi(c_p)$. Theorem 3.36 then follows from corollary 4.19 and lemma 4.24.

4.5 Homological results

In this section we sketch the proofs of theorem 4.18 and lemma 4.24, but shall often refer the reader to ch. 2 of [W3] for more details.

Multiplicities: We first consider theorem 4.18, generalizing a result proved by Mazur in sections II.14 and II.15 of [Maz1]. Recall that $L_{\mathcal{T}}$ (resp. L_H) is defined as $(\mathcal{T}_\ell(J_0(N))\otimes_{\mathcal{O}})_{\mathfrak{m}}$ (resp. $H^1(X_0(N), \mathcal{O})_{\mathfrak{m}}$) where $\mathfrak{m}$ is a maximal of the ideal of the Hecke algebra $\mathbb{T}_{\mathcal{O}}$ generated by the Hecke operators on $S_2(\Gamma_0(N))$. We are assuming moreover that $\rho_{\mathfrak{m}}$ is irreducible and that one of the following holds:

- ℓ does not divide N;
- ℓ^2 does not divide N and $T_\ell \notin \mathfrak{m}$.

We wish to prove that $L_{\mathcal{T}}$ and L_H are free over $\mathbb{T}_{\mathfrak{m}}$.

Since L_H and $L_{\mathcal{T}}$ are isomorphic as $\mathbb{T}_{\mathfrak{m}}$-modules, it suffices to prove that $L_{\mathcal{T}}$ is free. Furthermore, to prove that $L_{\mathcal{T}}$ is free, it suffices to prove that the localization of $\mathcal{T}_\ell(J_0(N))$ at $\mathfrak{m} \cap \mathbb{T}_{\mathbf{Z}_\ell}$ is free; i.e., we may replace $\mathcal{O}$ by $\mathbf{Z}_\ell$ and $\mathfrak{m}$ by its intersection with $\mathbb{T}_{\mathbf{Z}_\ell}$ and consider

$$\mathcal{T}_\ell(J_0(N))_{\mathfrak{m}}.$$

Since $\mathcal{T}_\ell(J_0(N)) \otimes_{\mathbf{Z}_\ell} \mathbf{Q}_\ell$ is free is free of rank two over $\mathbb{T}_{\mathbf{Q}_\ell}$, it suffices to prove that

$$\dim_{\mathbb{T}/\mathfrak{m}}(L_{\mathcal{T}}/\mathfrak{m}L_{\mathcal{T}}) = 2.$$

Since

$$\mathcal{T}_\ell(J_0(N))/\ell\mathcal{T}_\ell(J_0(N)) \cong J_0(N)[\ell] \cong \mathrm{Hom}_{\mathbf{F}_\ell}(J_0(N)[\ell], \mathbf{F}_\ell)$$

as $\mathbb{T}_{\mathbf{F}_\ell}$-modules, it suffices to prove the following cases of [W3] thm. 2.1.

Theorem 4.26 *Let $\mathfrak{m}$ be a maximal ideal of $\mathbb{T}_{\mathbf{Z}_\ell}$. Suppose that $\rho_{\mathfrak{m}}$ is irreducible and that either*

(a) ℓ does not divide N, or

(b) $\ell^2 \nmid N$ *and* $T_\ell \notin \mathfrak{m}$.

Then

$$\dim_{\mathbf{T}_{\mathbb{Z}_\ell}/\mathfrak{m}} J_0(N)[\mathfrak{m}] = \dim_{\mathbf{T}_{\mathbb{Z}_\ell}/\mathfrak{m}}(J_0(N)[\ell]/\mathfrak{m}) = 2.$$

The proof of theorem 4.26 in case (b) requires more of the theory of group schemes and Néron models than we wish to delve into here. We shall therefore only explain the proof in the case (a) (which is [R5] thm. 5.2(b), see Ribet's paper for more details) and refer to [W3] sec. 2.1 for the general case. We remark however that the proof of theorem 3.42 appeals to theorem 3.36 only in the case $\Sigma = \emptyset$ (cf. remarks 4.22 and remark 4.25). Recall that ℓ divides $N_\emptyset$ only if $\bar{\rho}$ is not good. So for the purposes of proving the Shimura-Taniyama conjecture for semistable elliptic curves, (b) can be replaced by the stronger hypothesis

(b') $\ell^2 \nmid N$ and $\rho_\mathfrak{m}|_{G_\ell}$ is not good.

The result in this case is due to Mazur and Ribet (the main result of [MRi]) and the proof is slightly easier than in the case of (b).

Returning to case (a), we appeal to a general property of the functor $\mathbb{D}$ of theorem 2.31 which follows from cor. 5.11 of [Oda].

Theorem 4.27 *Suppose that* A *is an abelian variety over* $\mathbb{Z}_\ell$ *(i.e., the Néron model of an abelian variety over* $\mathbb{Q}_\ell$ *with good reduction). There is a canonical and functorial isomorphism of vector spaces over* $\mathbb{F}_\ell$

$$\mathbb{D}(A(\bar{\mathbb{Q}}_\ell)[\ell]^*)^0 \cong \mathrm{Cot}_0(A/\mathbb{F}_\ell),$$

where Cot_0 *denotes the cotangent space at the origin.*

In the case that A is the Jacobian of a smooth proper curve X over $\mathbb{Z}_\ell$, we have also a canonical isomorphism

$$\mathrm{Cot}_0(A/\mathbb{F}_\ell) \cong H^0(X, \Omega^1_{X/\mathbb{F}_\ell})$$

and in the case $X = X_0(N)$ with $\ell \nmid N$, this provides an isomorphism of $\mathbf{T}_{\mathbb{F}_\ell}$-modules

$$\mathbb{D}(J_0(N)(\bar{\mathbb{Q}}_\ell)[\ell]^*)^0 \cong S_2(\Gamma_0(N), \mathbb{F}_\ell). \tag{4.5.1}$$

Consider now the action of $G_\mathbb{Q}$ on the points of

$$V = J_0(N)[\ell]^*[\mathfrak{m}] \cong (J_0(N)[\ell]/\mathfrak{m})^*.$$

By the argument in [Maz1] prop. II.14.2 or by the main result of [BLR], we know that every Jordan-Hölder constituent of the representation of $G_\mathbb{Q}$ on this $\mathbf{T}/\mathfrak{m}$-vector space is isomorphic to $\rho_\mathfrak{m}^* \cong \rho_\mathfrak{m}$. It follows that

$$\dim_{\mathbf{T}/\mathfrak{m}} \mathbb{D}(V) = 2 \dim_{\mathbf{T}/\mathfrak{m}} \mathbb{D}(V)^0$$

where we now regard V as a good G_ℓ-module. On the other hand (4.5.1) implies that

$$\mathbb{D}(V)^0 \cong S_2(\Gamma_0(N), \mathbb{F}_\ell)[\mathfrak{m}],$$

which is one-dimensional over $\mathbb{T}/\mathfrak{m}$ by the q-expansion principle (lemma 1.34 for example). We have now shown that $\mathbb{D}(V)$ is two-dimensional over $\mathbb{T}/\mathfrak{m}$, and it follows that so is V.

Ihara's lemma: We now sketch the proof of lemma 4.24, which proceeds by analyzing the behavior of the homology of modular curves under certain degeneracy maps. Note that it suffices to prove the lemma for $\phi_{\mathcal{T}}$.

If N and M are positive integers, then we let $\Gamma_1(N,M) = \Gamma_1(N) \cap \Gamma_0(M)$ and write $Y_1(N,M)$ (resp. $X_1(N,M)$) for the associated non-compactified (resp. compactified) modular curve. The key intermediate result is the following:

Lemma 4.28 *Suppose that N is a positive integer and p is a prime not dividing N.*

(a) The map

$$\begin{array}{ccc} H_1(X_1(N,p),\mathbb{Z}) & \to & H_1(X_1(N),\mathbb{Z})^2 \\ x & \mapsto & (\alpha_* x, \beta_* x) \end{array}$$

is surjective, where α is defined by $\tau \mapsto \tau$ and β by $\tau \mapsto p\tau$.

(b) If $N > 3$ and ℓ is an odd prime different from p, then the sequence

$$\begin{array}{ccccc} H_1(Y_1(Np,p^2),\mathbb{Z}_\ell) & \to & H_1(Y_1(Np),\mathbb{Z}_\ell)^2 & \to & H_1(Y_1(N),\mathbb{Z}_\ell) \\ x & \mapsto & (\alpha_{1*}x, \beta_{1*}x); \quad (y,z) & \mapsto & \beta_{2*}y - \alpha_{2*}z \end{array}$$

is exact, where the maps α_1 and α_2 are defined by $\tau \mapsto \tau$ and β_1 and β_2 by $\tau \mapsto p\tau$.

To prove the lemma, one first translates the statement into one about the homology (or cohomology) of the congruence subgroups involved. Part (a) of the lemma is due to Ihara [Ih], but see also the proof of thm. 4.1 of [R4]. Part (b) due to Wiles is more elementary and is established in the course of proving lemma 2.5 of [W3] (see the sequence (2.13)). The result stated there is in terms of cohomology rather than homology, but the one given here is immediate from it.

Remark 4.29 A method of Khare [Kh], sec. 2 using modular symbols yields an alternate proof of part (b) (after localization at a non-Eisenstein maximal ideal).

To deduce lemma 4.24 from lemma 4.28 we also need the following result:

Lemma 4.30 *Suppose that $\mathcal{O}$ is the ring of integers of a finite extension of $\mathbb{Q}_\ell$. Let M be a positive integer and H a subgroup of $(\mathbb{Z}/M\mathbb{Z})^\times$. Let $\tilde{\mathbb{T}}_{\mathcal{O}} = \tilde{\mathbb{T}}_{\mathbb{Z}} \otimes \mathcal{O}$ where $\tilde{\mathbb{T}}_{\mathbb{Z}}$ is the subring of endomorphisms of $M_2(\Gamma_1(M))$ generated by the operators $\langle d \rangle$ and T_r for primes $r \nmid M$. If $\mathfrak{n}$ is a non-Eisenstein maximal ideal then the localization at $\mathfrak{n}$ of the natural map*

(a) $H_1(Y_1(M),\mathcal{O}) \to H_1(X_1(M),\mathcal{O})$ is an isomorphism;

(b) $H_1(X_1(M),\mathcal{O}) \to H_1(X_H(M),\mathcal{O})$ is surjective.

To prove part (a), one checks that

$$H_1(Y_1(M),\mathbb{Z}) \to H_1(X_1(M),\mathbb{Z})$$

is surjective and that $T_r = r+1$ on the kernel for primes $r \equiv 1 \bmod M$ (cf. proposition 4.13, part (c)).

Part (b) is most easily proved by showing that $G_{\mathbb{Q}}$ acts trivially on the cokernel of the natural map

$$\mathcal{T}_\ell(J_1(M)) \to \mathcal{T}_\ell(J_H(M))$$

(using [LO], prop. 6 for example) and then applying lemma 4.12.

Finally we explain how to use lemmas 4.28 and 4.30 to prove lemma 4.24. It suffices to consider the case $\Sigma' = \Sigma \cup \{p\}$. We discuss only the more difficult case of $p \neq \ell$. We define $N = N_\Sigma$, $N' = Np^2$, $\mathfrak{m}$ and $\mathfrak{m}'$ as in the preceding section. Thus $\mathfrak{m}'$ is a maximal ideal of $\mathbb{T}'_{\mathcal{O}}$ where $\mathbb{T}'_{\mathbb{Z}}$ acts on $S_2(\Gamma_0(N'))$. All the maps we consider in the argument below will respect the action of the Hecke operators in $\tilde{\mathbb{T}}''_{\mathbb{Z}}$ where $\tilde{\mathbb{T}}''_{\mathbb{Z}}$ is the subring of endomorphisms of $M_2(\Gamma_1(Np,p^2))$ generated by the operators $\langle d\rangle$ and T_r for primes r not dividing Np. We let $\mathfrak{n}$ denote the preimage of $\mathfrak{m}'$ in $\tilde{\mathbb{T}}''_{\mathcal{O}} = \tilde{\mathbb{T}}''_{\mathbb{Z}} \otimes \mathcal{O}$. We first apply lemma 4.28(b) to show that

$$H_1(Y_1(Np,p^2),\mathcal{O})_{\mathfrak{n}} \to H_1(Y_1(Np),\mathcal{O})^2_{\mathfrak{n}} \to H_1(Y_1(N),\mathcal{O})_{\mathfrak{n}}$$

is exact. Next we apply lemma 4.30(a) to obtain the exactness of

$$H_1(X_1(Np,p^2),\mathcal{O})_{\mathfrak{n}} \to H_1(X_1(Np),\mathcal{O})^2_{\mathfrak{n}} \to H_1(X_1(N),\mathcal{O})_{\mathfrak{n}}.$$

Applying lemma 4.30(b) with $M = Np$, we find that

$$H_1(X_1(N,p^2),\mathcal{O})_{\mathfrak{n}} \to H_1(X_1(N,p),\mathcal{O})^2_{\mathfrak{n}} \to H_1(X_1(N),\mathcal{O})_{\mathfrak{n}}$$

is exact. We then apply lemma 4.28(a) to conclude that the map

$$\begin{array}{ccc} H_1(X_1(N,p^2),\mathcal{O})_{\mathfrak{n}} & \to & H_1(X_1(N),\mathcal{O})^3_{\mathfrak{n}} \\ x & \mapsto & (\alpha_* x, \beta_* x, \gamma_* x) \end{array}$$

is surjective where α, β and γ are defined respectively by $\tau \mapsto \tau$, $\tau \mapsto p\tau$ and $\tau \mapsto p^2\tau$. Finally, we use lemma 4.30(b) again (now with $M = N$) to get the surjectivity of

$$H_1(X_0(Np^2),\mathcal{O})_{\mathfrak{n}} \to H_1(X_0(N),\mathcal{O})^3_{\mathfrak{n}},$$

hence that of (4.4.4) upon localizing at $\mathfrak{m}'$.

5 Commutative algebra

In this section we collect some basic facts of commutative algebra that are used in the proof. We recall that $\mathcal{O}$ is the ring of integers in a finite extension K of $\mathbb{Q}_\ell$, and that $\mathcal{O}$ has residue field k. Let $\mathcal{C}_\mathcal{O}$ denote as in section 2.6 the category of complete noetherian local $\mathcal{O}$-algebras with residue field k.

5.1 Wiles' numerical criterion

In this section we state a numerical criterion discovered by Wiles for a map between two rings in $\mathcal{C}_\mathcal{O}$ to be an isomorphism.

Complete intersections: We say that a ring A in $\mathcal{C}_\mathcal{O}$ is *finite flat* if it is finitely generated and torsion free as an $\mathcal{O}$-module. A key ingredient in Wiles' isomorphism criterion is played by the concept of *complete intersection*, for which we give the following naive definition.

Definition 5.1 An object A in $\mathcal{C}_\mathcal{O}$ which is finite flat is called a *complete intersection* if it can be expressed as a quotient

$$A \simeq \mathcal{O}[[X_1, \dots, X_n]]/(f_1, \dots, f_n),$$

where there are as many relations as there are variables.

Remark 5.2 It is also true that if an object $\mathcal{O}[[Y_1, \dots, Y_r]]/J$ in $\mathcal{C}_\mathcal{O}$ which is finite flat is a complete intersection, then necessarily J can be generated by r elements. See for example [Mat], thm. 21.2 (and lemma 5.11 below). We will not use this fact in this chapter, although a special case is proved in the course of establishing lemma 5.30.

The category $\mathcal{C}_\mathcal{O}^\bullet$: The numerical criterion of Wiles is stated more naturally in terms of rings A in the category $\mathcal{C}_\mathcal{O}$ which are endowed with some *extra structure*: namely, a surjective $\mathcal{O}$-algebra homomorphism $\pi_A : A \longrightarrow \mathcal{O}$. Let $\mathcal{C}_\mathcal{O}^\bullet$ be the category whose objects are pairs (A, π_A), where A is an object of $\mathcal{C}_\mathcal{O}$ and $\pi_A : A \longrightarrow \mathcal{O}$ is a surjective $\mathcal{O}$-algebra homomorphism, also called the *augmentation map* attached to A. Morphisms in $\mathcal{C}_\mathcal{O}^\bullet$ are local ring homomorphisms which are compatible in the obvious way with the augmentation maps. By abuse of notation one often omits mentioning the augmentation map π_A when talking of objects in $\mathcal{C}_\mathcal{O}^\bullet$, and simply uses A to denote (A, π_A), when this causes no confusion. Objects of $\mathcal{C}_\mathcal{O}^\bullet$ will also be called *augmented rings.*

The invariants Φ_A and η_A: One associates to an augmented ring (A, π_A) two basic invariants:

$$\begin{aligned} \Phi_A &= (\ker \pi_A)/(\ker \pi_A)^2; \\ \eta_A &= \pi_A(\mathrm{Ann}_A \ker \pi_A). \end{aligned}$$

Here $\mathrm{Ann}_A(I)$ denotes the annihilator ideal of the ideal I in A.

The invariant Φ_A can be thought of as a tangent space for the object A. (More precisely, it is the cotangent space of the scheme spec(A) at the point $\ker \pi_A$.) It is a finitely generated $\mathcal{O}$-module.

The invariant η_A seems less familiar at first sight. It is called the *congruence ideal.* (The reason for this terminology should become clearer shortly.) We are now ready to state Wiles' numerical criterion:

Theorem 5.3 *Let $\phi : R \longrightarrow T$ be a surjective morphism of augmented rings. Assume that T is finitely generated and torsion-free as an $\mathcal{O}$-module, and that $\eta_T \neq (0)$. (And hence, in particular, $\#(\mathcal{O}/\eta_T) < \infty$.) Then the following are equivalent:*

(a) The inequality $\#\Phi_R \leq \#(\mathcal{O}/\eta_T)$ is satisfied.

(b) The equality $\#\Phi_R = \#(\mathcal{O}/\eta_T)$ is satisfied.

(c) The rings R and T are complete intersections, and the map $\phi : R \longrightarrow T$ is an isomorphism.

Remark 5.4 The above theorem is slightly different from the one that appears in [W3], where it is assumed from the outset that the ring T is Gorenstein. In the form in which we state it above, the theorem is due to H. Lenstra [Len]. Our presentation follows Lenstra's very closely. For the original (and slightly different) point of view, the reader should consult the appendix of [W3].

Some examples: Before going further, it may be good to pause and consider some examples of objects of $\mathcal{C}_{\mathcal{O}}^{\bullet}$ and the invariants associated to them. While logically independent of the proof, the examples should help the reader develop some intuition. (For a systematic way to compute the tangent spaces Φ_A, see the paragraph at the end of section 5.2.)

Example 1:

$$\begin{aligned} A &= \{(a,b) \in \mathcal{O} \times \mathcal{O}, \quad a \equiv b \pmod{\lambda^n}\} \\ &\simeq \mathcal{O}[[T]]/(T(T-\lambda^n)), \qquad \text{with } \pi_A(a,b) := a. \end{aligned}$$

$$\Phi_A \simeq \mathcal{O}/\lambda^n\mathcal{O}, \qquad \eta_A = (\lambda^n).$$

Example 2:

$$\begin{aligned} A &= \{(a,b,c) \in \mathcal{O} \times \mathcal{O} \times \mathcal{O}, a \equiv b \equiv c \pmod{\lambda}\} \\ &\simeq \mathcal{O}[[X,Y]]/(X(X-\lambda), Y(Y-\lambda), XY), \qquad \text{with } \pi_A(a,b,c) := a. \end{aligned}$$

$$\Phi_A \simeq \mathcal{O}/\lambda\mathcal{O} \times \mathcal{O}/\lambda\mathcal{O}, \qquad \eta_A = (\lambda).$$

Example 3:

$$A = \mathcal{O}[[X]]/(X^2), \qquad \text{with } \pi_A(f) := f(0).$$

$$\Phi_A \simeq \mathcal{O}, \qquad \eta_A = 0.$$

Example 4:

$$\begin{aligned} A &= \left\{ (a,b,c,d) \in \mathcal{O} \times \cdots \times \mathcal{O}, \begin{matrix} a \equiv b \equiv c \equiv d \pmod{\lambda}, \\ a + d \equiv b + c \pmod{\lambda^2} \end{matrix} \right\} \\ &\simeq \mathcal{O}[[X,Y]]/(X(X-\lambda), Y(Y-\lambda)), \qquad \text{with } \pi_A(a,b,c,d) := a. \end{aligned}$$

$$\Phi_A \simeq \mathcal{O}/\lambda\mathcal{O} \times \mathcal{O}/\lambda\mathcal{O}, \qquad \eta_A = (\lambda^2).$$

Example 5:

$$A = \mathcal{O}[[X_1, \dots, X_n]], \qquad \text{with } \pi_A(f) := f(0).$$

$$\Phi_A \simeq \mathcal{O}^n, \qquad \eta_A = (0).$$

Example 6: $(\lambda = \ell \equiv -1 \pmod 4), \quad \mathcal{O} = \mathbb{Z}_\ell)$.

$$\begin{aligned} A &= \{(a, b+ci) \in \mathbb{Z}_\ell \times \mathbb{Z}_\ell[i], \quad a \equiv b \pmod{\ell^2}, \quad c \equiv 0 \pmod{\ell}\} \\ &\simeq \mathbb{Z}_\ell[[X]]/(X(X^2+\ell^2)), \qquad \text{with } \pi_A(a, b+ci) := a. \end{aligned}$$

$$\Phi_A \simeq \mathbb{Z}/\ell^2\mathbb{Z}, \qquad \eta_A = (\ell^2).$$

Example 7:

$$A = \mathcal{O}[[T]]/(\lambda T) \simeq \mathcal{O} \oplus kT \oplus kT^2 \oplus \cdots, \qquad \text{with } \pi_A(f) = f(0).$$

$$\Phi_A \simeq k, \qquad \eta_A = (\lambda).$$

5.2 Basic properties of Φ_A and η_A

In this section we collect some of the basic properties of the invariants Φ_A and η_A, and prove the equivalence of (a) and (b) in theorem 5.3.

Behaviour of Φ_A under morphisms: The assignment $A \mapsto \Phi_A$ is a functor from the category $\mathcal{C}_\mathcal{O}^\bullet$ to the category of $\mathcal{O}$-modules; a morphism $A \longrightarrow B$ in $\mathcal{C}_\mathcal{O}^\bullet$ induces a homomorphism $\Phi_A \longrightarrow \Phi_B$ of $\mathcal{O}$-modules. Moreover, if $A \longrightarrow B$ is surjective, then so is the induced map on the tangent spaces. Therefore, when A maps surjectively onto B we have

$$\#\Phi_A \geq \#\Phi_B. \tag{5.2.1}$$

There is also a converse to this, which will be useful later:

Lemma 5.5 *If the homomorphism $\Phi_A \longrightarrow \Phi_B$ is surjective, then $A \longrightarrow B$ is also surjective.*

Proof: This follows from Nakayama's lemma. (Cf. [Ha], ch. II, sec. 7.4 and [Mat], th. 8.4.) □

Behaviour of η_A under (surjective) morphisms: Unlike the assignment $A \mapsto \Phi_A$, the assignment $A \longrightarrow \eta_A$ is not functorial, but it does have a nice behaviour under surjective morphisms: namely, if $\phi : A \longrightarrow B$ is surjective, then

$$\eta_A \subset \eta_B, \quad \text{i.e.,} \quad \#(\mathcal{O}/\eta_A) \geq \#(\mathcal{O}/\eta_B). \tag{5.2.2}$$

This is simply because in that case ϕ induces a map

$$\mathrm{Ann}_A \ker \pi_A \longrightarrow \mathrm{Ann}_B \ker \pi_B.$$

Relation between the invariants Φ_A and η_A: In general, we have the following inequality:

$$\#\Phi_A \geq \#(\mathcal{O}/\eta_A). \tag{5.2.3}$$

The key behind proving this identity is to interpret $\#\Phi_A$ in terms of Fitting ideals.

Digression on Fitting ideals: If R is a ring (in $\mathcal{C}_{\mathcal{O}}$, say) and M is a finitely generated R-module, we express M as a quotient of R^n for some n:

$$0 \longrightarrow M' \longrightarrow R^n \longrightarrow M \longrightarrow 0. \tag{5.2.4}$$

The *Fitting ideal* of M, denoted $\mathrm{Fitt}_R(M)$, is the ideal of R generated by the determinants $\det(v_1, \ldots, v_n)$, where the vectors $v_i \in R^n$ range over all possible choices of elements of $M' \subset R^n$. One checks that this ideal does not depend on the choice of exact sequence (5.2.4), and hence is an invariant of the R-module M. For example, if M is a finitely generated $\mathcal{O}$-module, we may write

$$M = \mathcal{O}^r \oplus \mathcal{O}/(\lambda^{n_1}) \oplus \mathcal{O}/(\lambda^{n_2}) \oplus \cdots \oplus \mathcal{O}/(\lambda^{n_k}),$$

with $n_1 \geq n_2 \geq \cdots \geq n_k$, and the sequence n_i is completely determined by M. The Fitting ideal $\mathrm{Fitt}_{\mathcal{O}}(M)$ is then the ideal of $\mathcal{O}$ generated by $\lambda^{n_1+\cdots+n_k}$, if $r = 0$, and is the 0-ideal if $r > 0$. Note in particular that, if M is a finite $\mathcal{O}$-module, then

$$\#M = \#(\mathcal{O}/\mathrm{Fitt}_{\mathcal{O}}(M)). \tag{5.2.5}$$

Furthermore, if M is any R-module, it follows directly from the definition that

$$\mathrm{Fitt}_R(M) \subset \mathrm{Ann}_R(M). \tag{5.2.6}$$

Fitting ideals behave well under tensor products: in particular, if M is a finitely generated A-module, where A is an object in $\mathcal{C}^{\bullet}_{\mathcal{O}}$, then:

$$\pi_A(\mathrm{Fitt}_A(M)) = \mathrm{Fitt}_{\mathcal{O}}(M \otimes_A \mathcal{O}), \tag{5.2.7}$$

where the tensor product is taken with respect to the augmentation map π_A. For more details and references on the Fitting ideal, see [Len] for example.

Now we are ready to prove equation (5.2.3). For, noting that $\Phi_A = \ker \pi_A \otimes_A \mathcal{O}$, where the tensor product is taken with respect to π_A, and applying equation (5.2.7) with $M = \ker \pi_A$, we have:

$$\mathrm{Fitt}_{\mathcal{O}}(\Phi_A) = \pi_A(\mathrm{Fitt}_A(\ker \pi_A)) \subset \pi_A(\mathrm{Ann}_A \ker \pi_A) = \eta_A, \tag{5.2.8}$$

where the containment follows from equation (5.2.6). Now the inequality (5.2.3) follows by combining (5.2.5) and (5.2.8).

As a consequence, we have

Corollary 5.6 *The statements* (a) *and* (b) *in theorem 5.3 are equivalent.*

Proof: If $R \longrightarrow T$ is a surjective map of augmented rings, then $\#\Phi_R \geq \#\Phi_T$, by equation (5.2.1). But equation (5.2.3) gives the inequality $\#\Phi_T \geq \#(\mathcal{O}/\eta_T)$. Hence the inequality $\#\Phi_R \geq \#(\mathcal{O}/\eta_T)$ always holds, so that (a) implies (b) in theorem 5.3. The reverse implication is clear. □

Remark 5.7 (Computing the tangent spaces Φ_A): Any object (A, π_A) in $\mathcal{C}_{\mathcal{O}}^{\bullet}$ can be expressed as a quotient of the object $U = \mathcal{O}[[X_1, \ldots, X_n]]$ of example 5 with augmentation map given by $\pi_U(f) = f(0)$. Indeed, one can take $a_1, \ldots, a_n$ to be A-module generators of the finitely generated A-module $\ker \pi_A$, and obtain the desired quotient map by sending X_i to a_i.

The tangent space Φ_U of U is a free $\mathcal{O}$-module of rank n which can be written down canonically as

$$\mathcal{O}X_1 \oplus \mathcal{O}X_2 \oplus \cdots \oplus \mathcal{O}X_n,$$

the natural map from $(\ker \pi_U)$ being simply the map which sends a power series $f \in U$ with no constant term to its degree 1 term, which we will denote by $\bar{f}$.

If A is expressed as a quotient $U/(f_1, \ldots, f_r)$, then one has

$$\Phi_A = \Phi_U/(\bar{f}_1, \ldots, \bar{f}_r). \tag{5.2.9}$$

5.3 Complete intersections and the Gorenstein condition

In this section we show that complete intersections satisfy a Gorenstein condition, and that (c) implies (a) and (b) in the statement of Wiles isomorphism criterion (theorem 5.3).

Definition 5.8 An object A in $\mathcal{C}_{\mathcal{O}}$ which is finite flat is said to be *Gorenstein* if

$$\mathrm{Hom}_{\mathcal{O}}(A, \mathcal{O}) \simeq A \quad \text{as } A\text{-modules}.$$

Proposition 5.9 *Suppose A in $\mathcal{C}_{\mathcal{O}}$ is finite flat. If A is a complete intersection, then A is Gorenstein.*

The remainder of this section will be devoted to proving proposition 5.9. Since the proof is a bit long and involved, and the concepts it uses are not used elsewhere, the reader is advised on a first reading to take it on faith and skip to the next section. A more direct proof of proposition 5.9 which bypasses the arguments of this section is explained in [Len].

We let A be a ring which is finite flat, and is a complete intersection. (Hence, A can be written as $\mathcal{O}[[X_1, \ldots, X_n]]/(f_1, \ldots, f_n)$.) We assume that the augmentation map for A is induced from the map on $\mathcal{O}[[X_1, \ldots, X_n]]$ sending f to $f(0)$. This implies that $f_i(0) = 0$, i.e., the f_i have no constant term.

We recall some definitions from commutative algebra that we will need. An ideal I of a local ring R is said to be *primary* if $I \neq R$ and every zero divisor in R/I is nilpotent. If $(x_1, \ldots, x_n)$ generates a primary ideal of R, and $n = \dim R$, then $(x_1, \ldots, x_n)$ is called a *system of parameters* for R.

Lemma 5.10 *The sequence $(f_1, \ldots, f_n, \lambda)$ is a system of parameters for $U = \mathcal{O}[[X_1, \ldots, X_n]]$.*

Proof: The quotient ring $U/(f_1, \ldots, f_n, \lambda)$ is local and is finitely generated as a k-vector space; therefore every element in its maximal ideal is nilpotent. □

A sequence $(x_1, \ldots, x_n)$ in a ring R is said to be a *regular sequence* if x_i is not a zero-divisor in $R/(x_1, \ldots, x_{i-1})$ for $i = 1, \ldots, n$.

Lemma 5.11 *The sequence $(f_1, \ldots, f_n)$ is a regular sequence for U.*

Proof: The ring U is Cohen Macaulay, since $\lambda, X_1, \ldots, X_n$ is a system of parameters of U which is also a regular sequence. Hence, by theorem 17.4 (iii) of [Mat], $(f_1, \ldots, f_n, \lambda)$ is a regular sequence in U. A fortiori, the sequence $(f_1, \ldots, f_n)$ is also a regular sequence. □

The proofs of lemma 5.10 and 5.11 use only the fact that A is finitely generated as an $\mathcal{O}$-module, and not that A is flat. As a corollary of this proof, we therefore have:

Corollary 5.12 *If R is an object of $\mathcal{C}_{\mathcal{O}}^{\bullet}$ which is finitely generated as an $\mathcal{O}$-module, and $R \simeq \mathcal{O}[[X_1, \ldots, X_n]]/(f_1, \ldots, f_n)$, then R is also flat, and hence is a complete intersection.*

To go further, we will introduce the Koszul complex

$$K(\underline{x}, R) := \oplus_{p=0}^{n} K_p(\underline{x}, R)$$

associated to a local ring R and a sequence $\underline{x} = (x_1, \ldots, x_n)$ of elements in its maximal ideal. For more details on the Koszul complex and its relation to regular sequences, the reader can consult [Mat], especially §16, or [Bour] X, or [Se3]. This complex is defined to be the free graded differential algebra generated by symbols $u_1, \ldots, u_n$:

$$K_p(\underline{x}, R) := \oplus_{i_1 < i_2 < \cdots < i_p} R \cdot u_{i_1} \wedge \cdots \wedge u_{i_p},$$

with differential $d : K_p \longrightarrow K_{p-1}$ defined by

$$d(u_{i_1} \wedge \cdots \wedge u_{i_p}) = \sum_{t=1}^{p} (-1)^t x_t \cdot u_{i_1} \wedge \cdots \wedge u_{i_{t-1}} \wedge u_{i_{t+1}} \wedge \cdots \wedge u_{i_p}.$$

We denote by $H_p(\underline{x}; R)$ the homology groups of this complex. We record here the main properties of this complex that we will use.

Proposition 5.13 *(a) $H_0(\underline{x}; R) = R/(\underline{x})$.*

(b) There is a long exact homology sequence

$$\cdots \longrightarrow H_p(\underline{x}; R) \longrightarrow H_p(\underline{x}, x_{n+1}; R) \longrightarrow H_{p-1}(\underline{x}; R) \xrightarrow{\pm x_{n+1}}$$

$$H_{p-1}(\underline{x}; R) \longrightarrow H_{p-1}(\underline{x}, x_{n+1}; R) \longrightarrow H_{p-2}(\underline{x}; R) \longrightarrow \cdots$$

(c) $H_p(\underline{x}; R)$ is annihilated by the ideal $(\underline{x})$, i.e., it has a natural $R/(\underline{x})$-module structure.

(d) If $\underline{x}$ is a regular sequence, then $H_p(\underline{x}; R) = 0$ for all $p > 0$ (i.e., the complex $K_p(\underline{x}; R)$ is a free resolution of $R/(\underline{x})$.)

Proof: The first assertion follows directly from the definition. For (*b*) and (*c*), see [Mat], th. 16.4. The assertion (*d*) can be proved by a direct induction argument on n, using the long exact homology sequence: For $p > 1$, this sequence becomes

$$0 \longrightarrow H_p(\underline{x}, x_{n+1}; R) \longrightarrow 0,$$

and for $p = 1$, it is

$$0 \longrightarrow H_1(\underline{x}, x_{n+1}; R) \longrightarrow H_0(\underline{x}; R) \xrightarrow{x_{n+1}} H_0(\underline{x}; R).$$

But the assumption that $\underline{x}, x_{n+1}$ is a regular sequence means that multiplication by x_{n+1} is injective on $H_0(\underline{x}; R) = R/(\underline{x})$. Hence, the assertion (*d*) follows. □

Now, we turn to the proof of proposition 5.9, following a method of Tate which is explained in the appendix of [MRo]. For any ring R, write $R[[\underline{X}]] := R[[X_1, \dots, X_n]]$. Let $a_1, \dots, a_n$ be the images in A of $X_1, \dots, X_n$ by the natural map

$$\alpha : \mathcal{O}[[\underline{X}]] \longrightarrow A = \mathcal{O}[[\underline{X}]]/(f_1, \dots, f_n),$$

and let

$$\beta : A[[\underline{X}]] \longrightarrow A$$

be the natural map which sends X_i to a_i. The sequence $(g_i) = (X_i - a_i)$ generates the kernel of β. Since the f_i, viewed as polynomials in $A[[\underline{X}]]$, also belong to $\ker \beta$, we have:

$$(f_1, \dots, f_n) = (g_1, \dots, g_n)M, \tag{5.3.1}$$

where M is an $n \times n$ matrix with coefficients in $A[[\underline{X}]]$. Let

$$D = \det(M) \in A[[\underline{X}]].$$

Our goal is to construct an A-module isomorphism

$$\mathrm{Hom}_{\mathcal{O}}(A, \mathcal{O}) \longrightarrow A.$$

We begin by constructing a surjective ($\mathcal{O}$-linear) map

$$\mathrm{Hom}_{\mathcal{O}[[\underline{X}]]}(A[[\underline{X}]], \mathcal{O}[[\underline{X}]]) \longrightarrow A.$$

Lemma 5.14 *(Tate): The function $\Phi(f) = \alpha(f(D))$ induces an isomorphism of $\mathcal{O}[[\underline{X}]]$-modules*

$$\mathrm{Hom}_{\mathcal{O}[[\underline{X}]]}(A[[\underline{X}]], \mathcal{O}[[\underline{X}]])/(g_1, \dots, g_n) \longrightarrow A.$$

Proof: By lemma 5.11, the sequence $(\underline{f}) = (f_1, \dots, f_n)$ is a regular $\mathcal{O}[[\underline{X}]]$-sequence. One can see directly from the definition that the sequence $(\underline{g}) = (g_i) = (X_i - a_i)$ is a regular $A[[\underline{X}]]$-sequence. Let $K(\underline{f})$ and $K(\underline{g})$ be the Koszul complexes associated to these two sequences. It follows from proposition 5.13 that the Koszul complex $K(\underline{f})$ is a resolution of A by free $\mathcal{O}[[\underline{X}]]$-modules, and the Koszul complex $K(\underline{g})$ is a resolution of A by free $A[[\underline{X}]]$-modules, and hence a fortiori, by free $\mathcal{O}[[\underline{X}]]$-modules. We define a map $\Phi : K(\underline{f}) \longrightarrow K(\underline{g})$ of complexes by letting

$$\Phi_0 : K_0(\underline{f}) \longrightarrow K_0(\underline{g})$$

be the natural inclusion of $\mathcal{O}[[\underline{X}]]$ into $A[[\underline{X}]]$, and letting

$$\Phi_1 : K_1(\underline{f}) \longrightarrow K_1(\underline{g})$$

be the map defined by

$$(\Phi_1(u_1), \dots, \Phi_1(u_n)) = (v_1, \dots, v_n)M,$$

and extending it by skew-linearity to a map of exterior algebras. One can check that the resulting map Φ is a morphism of complexes which induces the identity map $A \longrightarrow A$, and satisfies

$$\Phi_n(u_1 \wedge \cdots \wedge u_n) = D \cdot v_1 \wedge \cdots \wedge v_n.$$

Applying the functor $\mathrm{Hom}_{\mathcal{O}[[\underline{X}]]}(-, \mathcal{O}[[\underline{X}]])$ to these two free resolutions, and taking the homology of the resulting complexes, we find that since Φ is a homotopy equivalence of complexes it induces an isomorphism on the cohomology, and in particular, on the nth cohomology:

$$\Phi_n : \mathrm{Hom}_{\mathcal{O}[[\underline{X}]]}(A[[\underline{X}]], \mathcal{O}[[\underline{X}]])/(g_1, \ldots, g_n) \xrightarrow{\simeq}$$

$$\mathrm{Hom}_{\mathcal{O}[[\underline{X}]]}(\mathcal{O}[[\underline{X}]], \mathcal{O}[[\underline{X}]])/(f_1, \ldots, f_n) \simeq A,$$

which is given explicitly by the formula:

$$\Phi_n(f) = \alpha(f(D)).$$

□

We finally come to the proof of proposition 5.9, which we can state in a more precise form.

Lemma 5.15 *The map* $\Psi : \mathrm{Hom}_{\mathcal{O}}(A, \mathcal{O}) \longrightarrow A$ *defined by* $\Psi(f) = \alpha(\tilde{f}(D))$, *where* $\tilde{f} : A[[\underline{X}]] \longrightarrow \mathcal{O}[[\underline{X}]]$ *is the base change of* f, *is an* A*-module isomorphism, and hence,* A *is Gorenstein.*

Proof: The key point is to show that Ψ is A-linear. By definition, if

$$a = \alpha(a') \in A, \text{ with } a' \in \mathcal{O}[[\underline{X}]],$$

then

$$\Psi(af) = \alpha(\tilde{f}(aD)) = \alpha(\tilde{f}((a - a')D)) + \alpha(\tilde{f}(a'D)).$$

Since $a - a' \in \ker \beta$, it can be written as an $A[[\underline{X}]]$-linear combination of the g_i. By multiplying the relation

$$(f_1, \ldots, f_n) = (g_1, \ldots, g_n)M$$

by the matrix $D \cdot M^{-1} \in M_n(A[[\underline{X}]])$, one sees that the Dg_i can be written as $A[[\underline{X}]]$-linear combinations of the f_i's. Hence, so can the expression $(a - a')D$. By the $\mathcal{O}[[\underline{X}]]$-linearity of $\tilde{f}$, and the fact that each f_i belongs to $\ker \alpha$, it follows that

$$\alpha(\tilde{f}((a - a')D)) = 0.$$

Therefore,

$$\Psi(af) = \alpha(a'\tilde{f}(D)) = a\Psi(f).$$

This shows that Ψ is A-linear. To show that Ψ is surjective, observe that if $f_1, \ldots, f_r$ is a $\mathcal{O}$-basis of $\mathrm{Hom}_{\mathcal{O}}(A, \mathcal{O})$, then $\tilde{f}_1, \ldots, \tilde{f}_r$ is a $\mathcal{O}[[\underline{X}]]$-basis for $\mathrm{Hom}_{\mathcal{O}[[\underline{X}]]}(A[[\underline{X}]], \mathcal{O}[[\underline{X}]])$. Hence, for all $a \in A$, there exist $p_1, \ldots, p_r$ such that

$$\Phi_n(p_1\tilde{f}_1 + \cdots + p_r\tilde{f}_r) = a.$$

But this means that

$$\Psi(\alpha(p_1)f_1 + \cdots + \alpha(p_r)f_r) = a,$$

so that Ψ is surjective. Finally, since $\mathrm{Hom}_{\mathcal{O}}(A, \mathcal{O})$ and A are free $\mathcal{O}$-modules of the same finite rank, and Ψ is surjective, it must also be injective. Hence Ψ is an isomorphism, as was to be shown. □

Example 5.16 Let $A \subset \mathcal{O} \times \mathcal{O}$ be the ring of example 1 in section 5.1. Then, $f = T^2 - \lambda^n T$, and $g = T - (0, \lambda^n)$. Hence, $f = (T - (\lambda^n, 0))g$, so that $D = T - (\lambda^n, 0)$. It follows that

$$\Phi(h) = \alpha(\tilde{h}(T - (\lambda^n, 0))) = \alpha(Th(1,1) - h(\lambda^n, 0)) = (0, \lambda^n)h(1,1) - h(\lambda^n, 0).$$

The reader can check directly that Φ is indeed an A-linear isomorphism from the A-module $\mathrm{Hom}_{\mathcal{O}}(A, \mathcal{O})$ to A.

Example 5.17 Let $A = \mathcal{O}[\epsilon]/(\epsilon^2)$ be the ring of example 3 in section 5.1. Then, $f = T^2$, and $g = T - \epsilon$. Hence, $f = (T + \epsilon)g$, so that $D = T + \epsilon$. It follows that

$$\Phi(f) = \alpha(\tilde{f}(T + \epsilon)) = \alpha(Tf(1) + f(\epsilon)) = \epsilon f(1) + f(\epsilon).$$

Exercise 5.18 Write down explicitly an isomorphism $A \xrightarrow{\sim} \mathrm{Hom}_{\mathcal{O}}(A, \mathcal{O})$ for the complete intersection $A = \mathcal{O}[[X, Y]]/(X(X - \lambda), Y(Y - \lambda))$ of example 4 of section 5.1.

5.4 The Congruence ideal for complete intersections

Let A be an object of $\mathcal{C}_{\mathcal{O}}^{\bullet}$, which is finite flat and is a complete intersection as in the previous section, so that $A \simeq \mathcal{O}[[X_1, \ldots, X_n]]/(f_1, \ldots, f_n)$.

Using the result of the previous section, we will give an explicit formula for computing η_A in this case, and prove that (*c*) implies (*b*) in theorem 5.3.

Let $A^\vee := \mathrm{Hom}_{\mathcal{O}}(A, \mathcal{O})$, and let $\pi_A^\vee : \mathcal{O}^\vee \longrightarrow A^\vee$ be the dual map. From the Gorenstein property of A, we may identify $A^\vee$ with A (as A-modules). Fix any identification $\Psi : A^\vee \xrightarrow{\simeq} A$. (Any two such differ by a unit in A.) One checks that

$$\Psi\pi_A^\vee(\mathcal{O}^\vee) = \mathrm{Ann}_A \ker \pi_A.$$

Hence, η_A is the image of the map $\pi_A \Psi \pi_A^\vee$. By using the explicit construction of Ψ given in lemma 5.15, we find:

$$\pi_A \Psi \pi_A^\vee(\mathcal{O}^\vee) = \pi\alpha\tilde{\pi}(D),$$

where D is the determinant defined in section 5.3. By a direct calculation using equation (5.3.1), one sees that the right hand side is equal to $\det(\partial f_i / \partial X_j(0))$. Hence we have shown:

Proposition 5.19 *Suppose that A is an object of $\mathcal{C}_{\mathcal{O}}^{\bullet}$ which is finite flat. If $A = \mathcal{O}[[X_1, \ldots, X_n]]/(f_1, \ldots, f_n)$ is a complete intersection, then*

$$\eta_A = (\det(\partial f_i / \partial X_j(0))).$$

This proposition implies:

Corollary 5.20 *Let A in $\mathcal{C}_{\mathcal{O}}^{\bullet}$ be finite flat. If A is a complete intersection, then $\#\Phi_A = \#(\mathcal{O}/\eta_A)$. Hence, the statement (c) implies statement (b) in theorem 5.3.*

Proof: The equation (5.2.9) of remark 5.7 implies that

$$\#\Phi_A = \#\mathcal{O}/(\det(\partial f_i/\partial X_j(0))).$$

The result follows. □

5.5 Isomorphism theorems

The usefulness of the notion of complete intersections comes from the following two (vaguely stated) principles:
1. Isomorphisms to complete intersections can often be recognized by looking at their effects on the tangent spaces.
2. Isomorphisms from complete intersections can often be recognized by looking at their effects on the invariants η.

These vague principles are made precise in theorems 5.21 and 5.24 respectively.

Theorem 5.21 *Let $\phi : A \longrightarrow B$ be a surjective morphism of augmented rings, with B a (finite, flat) complete intersection. If ϕ induces an isomorphism from Φ_A to Φ_B, and these modules are finite, then ϕ is an isomorphism.*

Remark 5.22 Let $A = \mathcal{O}[[X,Y]]/(X(X-\lambda), Y(Y-\lambda))$ be the ring of example 4, let $B = \mathcal{O}[[X,Y]]/(X(X-\lambda), Y(Y-\lambda), XY)$ be the ring of example 2, and let $\phi : A \longrightarrow B$ be the natural projection. The map ϕ induces an isomorphism $\Phi_A \longrightarrow \Phi_B$, even though ϕ is not an isomorphism. The assumption that B is a complete intersection is crucial for concluding that ϕ is an isomorphism.

Remark 5.23 The natural map

$$\mathcal{O}[[X]]/(X^3) \longrightarrow \mathcal{O}[[X]]/(X^2)$$

is a surjective morphism inducing an isomorphism on tangent spaces, and the target ring is a complete intersection. Yet this map is *not* an isomorphism. This shows that the assumption on the finiteness of the tangent spaces cannot be dispensed with.

Proof of theorem 5.21: Recall that $U = \mathcal{O}[[X_1, \dots, X_n]]$ is the augmented ring of example 5 of section 5.1. Let

$$\nu_B : U \longrightarrow B$$

be a surjective morphism of augmented rings with $\ker \nu_B = (f_1, \dots, f_n)$. Let $b_1, \dots, b_n \in \ker \pi_B$ denote the images of $X_1, \dots, X_n$ by ν_B, and let $a_1, \dots, a_n \in \ker \pi_A$ denote inverse images of $b_1, \dots, b_n$ by ϕ. Since ϕ is an isomorphism on tangent spaces, the elements a_i generate $(\ker \pi_A)/(\ker \pi_A)^2$. Hence the morphism

$$\nu_A : \mathcal{O}[[X_1, \dots, X_n]] \longrightarrow A$$

defined by $\nu_A(X_i) = a_i$ induces a surjection $\Phi_U \longrightarrow \Phi_A$, and so it is surjective, by lemma 5.5.

We claim that $\ker \nu_B$ is contained in $\ker \nu_A$ (and hence, $\ker \nu_B = \ker \nu_A$). For, let $g_1, \ldots, g_n$ be elements of $\ker \nu_A$ whose linear terms $\bar{g}_1, \ldots, \bar{g}_n$ generate the kernel of

$$\bar{\nu}_A : \Phi_U \longrightarrow \Phi_A.$$

Since $\ker \nu_A \subset \ker \nu_B$, it follows that there exists an $n \times n$ matrix $M \in M_n(U)$ with entries in U such that

$$(g_1, \ldots, g_n) = (f_1, \ldots, f_n)M.$$

Let $\bar{M}$ be the matrix of constant terms of the matrix M. Then we have

$$(\bar{g}_1, \ldots, \bar{g}_n) = (\bar{f}_1, \ldots, \bar{f}_n)\bar{M}.$$

Since $(\bar{g}_1, \ldots, \bar{g}_n)$ and $(\bar{f}_1, \ldots, \bar{f}_n)$ generate the same submodules of rank n and finite index in Φ_U, it follows that $\det \bar{M}$ is a unit in $\mathcal{O}$. Hence, M is invertible, and therefore the f_i can be expressed as a U-linear combination of the g_j. This implies that $\ker \nu_B \subset \ker \nu_A$. Now we see that $\nu_A \nu_B^{-1}$ gives a well-defined inverse to ϕ, so that ϕ is an isomorphism. □

Theorem 5.24 *Let $\phi : A \longrightarrow B$ be a surjective morphism of augmented rings. Suppose that A and B are finite flat, and that A a complete intersection. If $\eta_A = \eta_B \neq 0$, then ϕ is an isomorphism.*

Remark 5.25 The torsion-freeness assumption on B is essential: if n is large enough, then $B = A/(\ker \pi_A)^n$ satisfies $\eta_A = \eta_B$, although the natural map $A \longrightarrow B$ is not injective when $A \neq \mathcal{O}$.

Proof: By proposition 5.9, A is Gorenstein, i.e., it satisfies

$$A^\vee := \mathrm{Hom}_{\mathcal{O}}(A, \mathcal{O}) \simeq A \text{ as } A\text{-modules.}$$

Now, we observe that

$$\ker \pi_A \cap \mathrm{Ann}_A \ker \pi_A = 0, \tag{5.5.1}$$

and likewise for B. For, let x be a non-zero element of η_A, and let $x' \in \mathrm{Ann}_A \ker \pi_A$ satisfy $\pi_A(x') = x$. For all $a \in \ker \pi_A \cap \mathrm{Ann}_A \ker \pi_A$, we have

$$0 = a(x - x') = ax,$$

the first equality because a belongs to $\mathrm{Ann}_A \ker \pi_A$ and $(x - x')$ belongs to $\ker \pi_A$, the second equality because a belongs to $\ker \pi_A$ and x' belongs to $\mathrm{Ann}_A \ker \pi_A$. Hence a belongs to the $\mathcal{O}$-torsion submodule of A, and therefore is 0.

It follows from (5.5.1) that the homomorphism π_A (resp. π_B) induces an isomorphism from $\mathrm{Ann}_A \ker \pi_A$ (resp. $\mathrm{Ann}_B \ker \pi_B$) to η_A (resp. η_B). Since η_A is isomorphic to η_B, it follows that ϕ induces an isomorphism from $\mathrm{Ann}_A \ker \pi_A$ to $\mathrm{Ann}_B \ker \pi_B$, i.e.,

$$\phi \mathrm{Ann}_A \ker \pi_A = \mathrm{Ann}_B \ker \pi_B.$$

From (5.5.1) it also follows a fortiori that

$$\ker\phi \cap \mathrm{Ann}_A \ker \pi_A = 0,$$

hence there is an exact sequence of A-modules:

$$0 \longrightarrow \ker\phi \oplus \mathrm{Ann}_A \ker \pi_A \longrightarrow A. \tag{5.5.2}$$

The cokernel of the last map is

$$A/(\ker\phi \oplus \mathrm{Ann}_A \ker \pi_A) \simeq B/(\phi \mathrm{Ann}_A \ker \pi_A) \simeq B/(\mathrm{Ann}_B \ker \pi_B),$$

which is torsion-free, since there is a natural injection

$$B/(\mathrm{Ann}_B \ker \pi_B) \hookrightarrow \mathrm{End}_{\mathcal{O}}(\ker \pi_B).$$

Hence, the exact sequence (5.5.2) splits over $\mathcal{O}$. Taking $\mathcal{O}$ duals in (5.5.2) and using the Gorenstein condition for A, we thus get an exact sequence of A-modules:

$$A \longrightarrow (\ker\phi)^\vee \oplus (\mathrm{Ann}_A \ker \pi_A)^\vee \longrightarrow 0.$$

Applying the functor $- \otimes_A k$ (relative to the map $A \longrightarrow k$), we find

$$1 = \dim_k(A \otimes_A k) \geq \dim_k((\ker\phi)^\vee \otimes_A k) + \dim_k((\mathrm{Ann}_A \ker \pi_A)^\vee \otimes_A k).$$

Since $\eta_A \neq 0$, it follows that $(\mathrm{Ann}_A \ker \pi_A)^\vee \otimes_A k \neq 0$, and hence we must have

$$(\ker\phi)^\vee \otimes_A k = 0.$$

Therefore by Nakayama's lemma and duality, $\ker\phi = 0$, which proves the theorem. □

5.6 A resolution lemma

It turns out that objects in $\mathcal{C}^\bullet_{\mathcal{O}}$ can be "resolved" (in a weak sense) by a complete intersection, namely,

Theorem 5.26 *Let A be an augmented ring which is finite flat over $\mathcal{O}$. Then there is a morphism $\tilde{A} \longrightarrow A$ in the category $\mathcal{C}^\bullet_{\mathcal{O}}$ such that:*

(a) the ring $\tilde{A}$ is finite flat over $\mathcal{O}$ and is a complete intersection;

(b) the map $\tilde{A} \longrightarrow A$ induces an isomorphism $\Phi_{\tilde{A}} \longrightarrow \Phi_A$.

Proof: Write A as a quotient of $U = \mathcal{O}[[X_1, \dots, X_n]]$ (with $\pi_U : U \longrightarrow \mathcal{O}$ the map which sends each X_i to 0.) Let $f_1, \dots, f_n$ be elements in the kernel of the natural map $U \longrightarrow A$, such that $\bar{f}_1, \dots, \bar{f}_n$ generate the kernel of $\Phi_U \longrightarrow \Phi_A$. Now letting

$$\tilde{A} = U/(f_1, \dots, f_n)$$

would give the desired ring $\tilde{A}$, provided $\tilde{A}$ is finitely generated: indeed, the flatness of $\tilde{A}$ would follow from corollary 5.12.

Thus we need to show that the f_i can be chosen so that $\tilde{A}$ is finitely generated. Let $a_1, \dots, a_n$ be $\mathcal{O}$-module generators of the finitely generated module $\ker \pi_A$, and define a homomorphism ϕ from the polynomial ring

$$V = \mathcal{O}[X_1, \dots, X_n]$$

to A by sending X_j to a_j. Clearly ϕ is surjective. Let $f_1, \ldots, f_n$ be elements of $\ker\phi$ chosen as above, and let m denote their maximal degree. Since the elements a_i^2 belong to $\ker \pi_A$, we may write

$$a_i^2 = h_i(a_1, \ldots, a_n),$$

where $h_i(X_1, \ldots, X_n)$ is a linear polynomial. Now, replacing the relations f_i by the relations

$$f_i + X_i^m h_i - X_i^{m+2},$$

we find that the ring $V/(f_1, \ldots, f_n)$ is a finitely generated $\mathcal{O}$-module: it can be generated by the images of the monomials of degree $\leq n(m+1)$, since the relations allow us to rewrite any monomial of higher degree in terms of ones of lower degree. Completing at the ideal $(\lambda, X_1, \ldots, X_n)$, we find that the ring

$$\tilde{A} = U/(f_1, \ldots, f_n)$$

has the desired properties: the natural homomorphism from $\tilde{A}$ to A induces an isomorphism on the tangent spaces, since the linear terms of the f_i generate the kernel of the induced map $\Phi_U \longrightarrow \Phi_A$ on the tangent spaces, and $\tilde{A}$ is a finitely generated $\mathcal{O}$-module, since $V/(f_1, \ldots, f_n)$ is. □

5.7 A criterion for complete intersections

The results we have accumulated so far allow us to give an important criterion for an object A to be a complete intersection:

Theorem 5.27 *Let A be an augmented ring which is a finitely generated torsion-free $\mathcal{O}$-module. If $\#\Phi_A \leq \#(\mathcal{O}/\eta_A) < \infty$, then A is a complete intersection.*

Proof: Let $\phi : \tilde{A} \longrightarrow A$ be the surjective morphism given by the resolution theorem (theorem 5.26). Then we have

$$\#(\mathcal{O}/\eta_A) \geq (\#\Phi_A) = (\#\Phi_{\tilde{A}}) \geq \#(\mathcal{O}/\eta_{\tilde{A}}),$$

where the first inequality is by assumption, the second by the choice of $\tilde{A}$, and the third is by the equation (5.2.3). On the other hand, by equation (5.2.2), we have

$$\#(\mathcal{O}/\eta_{\tilde{A}}) \geq \#(\mathcal{O}/\eta_A).$$

It follows that

$$\eta_A = \eta_{\tilde{A}},$$

so that ϕ is an isomorphism by theorem 5.24. It follows that A is a complete intersection. □

5.8 Proof of Wiles' numerical criterion

Theorem 5.28 *Let R and T be augmented rings such that T is a finitely generated torsion-free $\mathcal{O}$-module, and let $\phi : R \longrightarrow T$ be a surjective morphism. If*

$$\#\Phi_R \leq \#(\mathcal{O}/\eta_T) < \infty,$$

then R and T are complete intersections, and ϕ is an isomorphism.

Proof: We have:

$$\#(\mathcal{O}/\eta_T) \le \#\Phi_T \le \#\Phi_R \le \#(\mathcal{O}/\eta_T),$$

where the first inequality is by equation (5.2.3), the second follows from the surjectivity of ϕ, and the third is by hypothesis. Therefore,

$$\#\Phi_T = \#(\mathcal{O}/\eta_T),$$

and hence T is a complete intersection by theorem 5.27. Since the orders of Φ_R and Φ_T are the same, ϕ induces an isomorphism between them. Hence ϕ is an isomorphism $R \longrightarrow T$, by theorem 5.21. This completes the proof. □

Theorem 5.28 shows that the statements (a) and (b) in theorem 5.3 imply the statement (c). Combining corollaries 5.6, 5.20, and theorem 5.28 completes the proof of theorem 5.3.

5.9 A reduction to characteristic ℓ

Let $\mathcal{C}_k$ be the category of complete local noetherian k-algebras with residue field k. Again all morphisms are assumed to be local. There is a natural functor $A \mapsto \bar{A}$ from $\mathcal{C}_{\mathcal{O}}$ to $\mathcal{C}_k$ which send A to $\bar{A} := A/\lambda$.

We say that an object A of $\mathcal{C}_k$ which is finite dimensional as a k-vector space is a complete intersection if it is isomorphic to a quotient

$$A = k[[X_1, \dots, X_r]]/(f_1, \dots, f_r).$$

Note that if an object A of $\mathcal{C}_{\mathcal{O}}$ is a complete intersection, then $\bar{A}$ is a complete intersection in $\mathcal{C}_k$. As a partial converse, we have:

Lemma 5.29 *Suppose that $R \mapsto T$ is a map in the category $\mathcal{C}_{\mathcal{O}}$, and that T is finitely generated and free as an $\mathcal{O}$-module. Then $R \longrightarrow T$ is an isomorphism of complete intersections, if and only if $\bar{R} \longrightarrow \bar{T}$ is.*

Proof: This is an exercise and is left to the reader. □

We now come to the proof of lemma 3.39 of section 3.4:

Lemma 5.30 *Suppose that $K \subset K'$ are local fields with rings of integers $\mathcal{O} \subset \mathcal{O}'$ and that A is an object of $\mathcal{C}_{\mathcal{O}}$ which is finitely generated and free as an $\mathcal{O}$-module. Then A is a complete intersection if and only if $A \otimes_{\mathcal{O}} \mathcal{O}'$ is.*

Proof: One implication is clear. Let k and k' be the residue fields of $\mathcal{O}$ and $\mathcal{O}'$ respectively. By lemma 5.29 it is enough to prove that, if R is an object of $\mathcal{C}_k$ which is finite dimensional as a k-vector space, then

$$R' = R \otimes_k k' \text{ is a complete intersection} \quad \Rightarrow \quad R \text{ is a complete intersection.}$$

Let $\mathfrak{m}$ and $\mathfrak{m}'$ denote the maximal ideals of R and R' respectively.

By assumption, we have $R' = k'[[Y_1, \dots, Y_r]]/J$, where the ideal J can be generated by r elements. We can assume without loss of generality (by adding extra variables and relations if necessary) that the images of $Y_1, \dots Y_r$ generate $\mathfrak{m}'$ as a k'-vector space. Now, let

$$\phi : k[[X_1, \dots, X_r]] \longrightarrow R$$

be a ring homomorphism, such that the images of $X_1, \ldots, X_r$ generate $\mathfrak{m}$ as a k-vector space, and let

$$\phi' : k'[[X_1, \ldots, X_r]] \longrightarrow R'$$

be the extension of scalars. Let I and $I' = I \otimes_k k'$ be the kernels of these two maps. We claim that I' can be generated by r elements. (In fact, this is also true without the assumption that the images of $X_1, \ldots, X_r$ generate $\mathfrak{m}'$ as a k' vector space, although we use this assumption in the proof below. Cf. remark 5.2.) To see this, choose an isomorphism of k'-vector spaces $k'X_1 \oplus \cdots \oplus k'X_r \longrightarrow k'Y_1 \oplus \cdots \oplus k'Y_r$ such that

$$\begin{array}{ccc} k'X_1 \oplus \cdots \oplus k'X_r & \longrightarrow & k'Y_1 \oplus \cdots \oplus k'Y_r \\ \downarrow & & \downarrow \\ \mathfrak{m}' & = & \mathfrak{m}' \end{array}$$

commutes. Extending this map to an isomorphism of k'-algebras,

$$\nu : k'[[X_1, \ldots, X_r]] \longrightarrow k'[[Y_1, \ldots, Y_r]],$$

one sees that $I' = \nu^{-1}(J)$, and hence can be generated by r elements, as claimed.

In particular we have $\dim_{k'}(I'/\mathfrak{m}'I') \leq r$, and

$$\dim_k(I/\mathfrak{m}I) = \dim_{k'}((I/\mathfrak{m}I) \otimes_k k') \leq \dim_{k'}(I'/\mathfrak{m}'I') \leq r.$$

Nakayama's lemma now implies that I can be generated by r elements, and hence R is a complete intersection. This proves the lemma. □

5.10 J-structures

We now turn to the proof of theorem 3.41. In view of the last section we will work in characteristic ℓ. Thus let $\pi : R \twoheadrightarrow T$ be a surjective morphism in the category $\mathcal{C}_k$, where R and T are finite dimensional as k-vector spaces. Let r be a non-negative integer. If $J \lhd k[[S_1, ..., S_r]]$ and $J \subset (S_1, ..., S_r)$ then by a strong J-structure we shall mean a commutative diagram in $\mathcal{C}_k$

$$\begin{array}{cccc} & & k[[S_1, ..., S_r]] & \\ & & \downarrow & \\ k[[X_1, ..., X_r]] & \twoheadrightarrow & R' & \twoheadrightarrow T' \\ & & \downarrow & \downarrow \\ & & R & \twoheadrightarrow T, \end{array}$$

such that

(a) $T'/(S_1, ..., S_r)T' \xrightarrow{\sim} T$ and $R'/(S_1, ..., S_r)R' \twoheadrightarrow R$,

(b) for each ideal $I \supset J$, $I = \ker(k[[S_1, ..., S_r]] \to T'/I)$,

(c) $J = \ker(k[[S_1, ..., S_r]] \to T')$ and $R' \hookrightarrow T' \oplus R$.

We refer to these as *strong* J-structures because we have slightly altered the conditions from section 3.4 for technical convenience in the rest of this section.

We will prove the following result.

Theorem 5.31 *Suppose there exist a sequence of ideals $J_n \lhd k[[S_1, \ldots, S_r]]$ such that $J_0 = (S_1, \ldots, S_r)$, $J_n \supset J_{n+1}$, $\bigcap_n J_n = (0)$ and for each n there exists a strong J_n-structure. Then $R \xrightarrow{\sim} T$ and these rings are complete intersections.*

Before proving theorem 5.31 we shall explain how to deduce theorem 3.41 from it. If J is an ideal of $\mathcal{O}[[S_1, \ldots, S_r]]$ we will use $\bar{J}$ to denote its image in $k[[S_1, \ldots, S_r]]$. If $\mathcal{S}$ given by

$$\begin{array}{ccccc} & & \mathcal{O}[[S_1, \ldots, S_r]] & & \\ & & \downarrow & & \\ \mathcal{O}[[X_1, \ldots, X_r]] & \twoheadrightarrow & R' & \twoheadrightarrow & T' \\ & & \downarrow & & \downarrow \\ & & R & \twoheadrightarrow & T, \end{array}$$

is a J-structure for $R \to T$, let $\mathfrak{a} = \ker(R \longrightarrow T)$, and let R'' denote the ring $\mathrm{Im}\,(R' \to (R/\mathfrak{m}_R\mathfrak{a} \oplus (T'/J)) \otimes_{\mathcal{O}} k)$. Then $\bar{\mathcal{S}}$ given by

$$\begin{array}{ccccc} & & k[[S_1, \ldots, S_r]] & & \\ & & \downarrow & & \\ k[[X_1, \ldots, X_r]] & \twoheadrightarrow & R'' & \twoheadrightarrow & (T'/J) \otimes_{\mathcal{O}} k \\ & & \downarrow & & \downarrow \\ & & (R/\mathfrak{m}_R\mathfrak{a}) \otimes_{\mathcal{O}} k & \twoheadrightarrow & T \otimes_{\mathcal{O}} k, \end{array}$$

is a strong $\bar{J}$-structure for $(R/\mathfrak{m}_R\mathfrak{a}) \otimes_{\mathcal{O}} k \to T \otimes_{\mathcal{O}} k$. If J_n is a sequence of ideals as in theorem 3.41 then $\bar{J}_n$ is a sequence of nested ideals with $\bar{J}_0 = (S_1, \ldots, S_r)$ and $\bigcap_n \bar{J}_n = (0)$. Moreover if for each n there is a J_n-structure $\mathcal{S}_n$ for $R \to T$ then for each n there is a strong $\bar{J}_n$-structure $\bar{\mathcal{S}}_n$ for

$$(R/\mathfrak{m}_R\mathfrak{a}) \otimes_{\mathcal{O}} k \to T \otimes_{\mathcal{O}} k.$$

Then by theorem 5.31 we see that this map is an isomorphism of complete intersections. Lemma 5.29 shows that theorem 3.41 follows. □

Before returning to the proof of theorem 5.31, let us first make some remarks about strong J-structures.

- The set of strong J structures for all ideals $J \subset (S_1, \ldots, S_r) \subset k[[S_1, \ldots, S_r]]$ forms a category, with the obvious notion of morphism.

- If $\mathcal{S}$ is a strong J-structure and if $(S_1, \ldots, S_r) \supset J' \supset J$ then there is a natural J'-structure $\mathcal{S}$ mod J' obtained by replacing T' by T'/J' and R' by the image of $R' \to R \oplus (T'/J')$.

- If R is a finite dimensional k-vector space and if J has finite index in $k[[S_1, \ldots, S_r]]$ then there are only finitely many isomorphism classes of strong J-structure. (This follows because we can bound the order of R' in any J-structure. Explicitly we must have $\#R' \leq (\#R)(\#k[[S_1, \ldots, S_r]]/J)^{\dim_k T}$.)

Lemma 5.32 *Suppose that R is a finite dimensional k vector space. Suppose also that $\{J_n\}$ is a nested (decreasing) sequence of ideals and that $J = \bigcap_n J_n$. If for each n a strong J_n structure exists then a strong J structure exists.*

Proof: We may suppose that each J_n has finite index in $k[[S_1, \ldots, S_r]]$. Let $\mathcal{S}_n$ denote a strong J_n-structure. Let $\mathcal{S}_{n,m} = \mathcal{S}_n$ mod J_m if $m \leq n$. Because there are only finitely many isomorphism classes of strong J_m structure, we may recursively choose integers $n(m)$ such that

- $\mathcal{S}_{n(m),m} \cong \mathcal{S}_{n,m}$ for infinitely many n,
- if $m > 1$ then $\mathcal{S}_{n(m),m-1} \cong \mathcal{S}_{n(m-1),m-1}$.

Let $\mathcal{S}'_m = \mathcal{S}_{n(m),m}$. Then $\mathcal{S}'_m$ is a strong J_m structure and if $m \geq m_1$ then $\mathcal{S}'_m \bmod J_{m_1} \cong \mathcal{S}'_{m_1}$. One checks that $\mathcal{S} = \varprojlim \mathcal{S}'_m$ is the desired strong J-structure. □

Lemma 5.33 *Suppose that a strong* (0) *structure exists. Then the map* $R \longrightarrow T$ *is an isomorphism, and these rings are complete intersections.*

Proof: Because $k[[S_1, ..., S_r]] \hookrightarrow T'$ (and T' is a finitely generated $k[[S_1, ..., S_r]]$-module by Nakayama's lemma, cf. [Mat], thm. 8.4) we see that the Krull dimension of T' is at least r. On the other hand $k[[X_1, ..., X_r]] \twoheadrightarrow T'$ and so by Krull's principal ideal theorem this map must be an isomorphism. Thus

$$k[[X_1, ..., X_r]] \xrightarrow{\sim} R' \xrightarrow{\sim} T'.$$

Hence we have that

$$k[[X_1, ..., X_r]]/(S_1, ..., S_r) \xrightarrow{\sim} R'/(S_1, ..., S_r) \xrightarrow{\sim} T,$$

and the lemma follows. □

Theorem 5.31 follows at once from these two lemmas. □

References

[Ant2] W. Kuyk, P. Deligne, eds., Modular Functions of One Variable II, Lecture Notes in Math. **349**, Springer-Verlag, New York, Berlin, Heidelberg, 1973.

[Ant3] W. Kuyk, J.-P. Serre, eds., Modular Functions of One Variable III, Lecture Notes in Math. **350**, Springer-Verlag, New York, Berlin, Heidelberg, 1973.

[Ant4] B. Birch, W. Kuyk, eds., Modular Functions of One Variable IV, Lecture Notes in Math. **476**, Springer-Verlag, New York, Berlin, Heidelberg, 1975.

[Ant5] J.-P. Serre, D. Zagier, eds., Modular Functions of One Variable V, Lecture Notes in Math. **601**, Springer-Verlag, New York, Berlin, Heidelberg, 1977.

[AL] A.O.L. Atkin and J. Lehner, *Hecke operators on* $\Gamma_0(m)$, Math. Ann. **185** (1970), 134–160.

[BCEM] J. Buhler, R. Crandall, R. Ernvall, T. Metsänkylä, *Irregular primes and cyclotomic invariants to four million*, Math. Comp. **61** (1993), 151–153.

[BK] S. Bloch, K. Kato, *L-functions and Tamagawa numbers of motives*, in: The Grothendieck Festschrift I, Progress in Math. **86**, Birkhäuser, Boston, Basel, Berlin, 1990, pp. 333–400.

[BLR] N. Boston, H. Lenstra, K. Ribet, *Quotients of group rings arising from two-dimensional representations*, C. R. Acad. Sc. **t312** (1991), 323–328.

[BM] N. Boston, B. Mazur, *Explicit universal deformations of Galois representations*, Algebraic Number Theory, 1–21, Adv. Stud. Pure Math. **17**, Acad. Press, Boston MA, 1989.

[Bour] N. Bourbaki, Algèbre (Éléments de Mathématiques, Fasc. XXIII), Hermann, Paris, 1958.

[Ca1] H. Carayol, *Sur les représentations p-adiques associées aux formes modulaires de Hilbert*, Ann. Sci. Ec. Norm. Super. **19** (1986), 409–468.

[Ca2] H. Carayol, *Sur les représentations galoisiennes modulo ℓ attachées aux formes modulaires*, Duke Math. J. **59** (1989), 785–801.

[Ca3] H. Carayol, *Formes modulaires et représentations galoisiennes à valeurs dans un anneau local complet*, in: p-adic Monodromy and the Birch-Swinnerton-Dyer Conjecture (B. Mazur and G. Stevens, eds.), Contemporary Math. **165**, 1994, pp. 213–237.

[Co] J. Coates, S.T. Yau, eds., Elliptic Curves, Modular Forms and Fermat's Last Theorem, International Press, Cambridge, 1995.

[CPS] E. Cline, B. Parshall, L. Scott, *Cohomology of finite groups of Lie type I*, Publ. Math. IHES **45** (1975), 169–191.

[Cr] J. Cremona, Algorithms for Modular Elliptic Curves, Cambridge Univ. Press, Cambridge, 1992.

[CR] C.W. Curtis, I. Reiner, Representation Theory of Finite Groups and Associative Algebras, Wiley Interscience, New York, 1962.

[CS] G. Cornell and J.H. Silverman, eds., Arithmetic Geometry, Springer-Verlag, New York, Berlin, Heidelberg, 1986.

[CW] J. Coates, A. Wiles, *On the conjecture of Birch and Swinnerton-Dyer*, Inv. Math. **39** (1977), 223–251.

[Da1] H. Darmon, *Serre's conjectures*, in [Mu2], pp. 135–153.

[Da2] H. Darmon, *The Shimura-Taniyama Conjecture, (d'après Wiles)*, (in Russian) Uspekhi Mat. Nauk 50 (1995), no. 3(303), 33–82. (English version to appear in Russian Math Surveys)

[De] P. Deligne, *Formes modulaires et représentations ℓ-adiques*, in: Lecture Notes in Math. **179**, Springer-Verlag, New York, Berlin, Heidelberg, 1971, pp. 139–172.

[DR] P. Deligne, M. Rapoport, *Les schémas de modules de courbes elliptiques*, in [Ant2], pp. 143–316.

[dS] E. de Shalit, *On certain Galois representations related to the modular curve $X_1(p)$*, to appear in Compositio Math.

[dSL] B. de Smit, H.W. Lenstra, *Explicit construction of universal deformation rings*, to appear in the Proceedings of the Conference on Fermat's Last Theorem, Boston University, August 9–18, 1995.

[DS] P. Deligne, J.-P. Serre, *Formes modulaires de poids* 1, Ann. Sci. Ec. Norm. Sup. **7** (1974), 507–530.

[Di1] F. Diamond, *The refined conjecture of Serre*, in [Co], pp. 22–37.

[Di2] F. Diamond, *On deformation rings and Hecke rings*, Annals of Math. **144** (1996), 137–166.

[Dic1] L.E. Dickson, History of the Theory of Numbers, Vol. II, Chelsea Publ. Co., New York, 1971.

[Dic2] L.E. Dickson, Linear Groups with an Exposition of the Galois Field Theory, Teubner, Leipzig, 1901.

[Dir] P.G.L. Dirichlet, *Mémoire sur l'impossibilité de quelques équations indéterminées du cinquième degré*, Jour. für Math. (Crelle) **3** (1828), 354–375.

[DI] F. Diamond and J. Im, *Modular forms and modular curves*, in [Mu2], pp. 39–133.

[DK] F. Diamond and K. Kramer, *Modularity of a family of elliptic curves* Math. Research Letters **2** (1995), 299–305.

[DO] K. Doi, M. Ohta, *On some congruences between cusp forms on* $\Gamma_0(N)$, in [Ant5], pp. 91–106.

[Dr] V.G. Drinfeld, *Elliptic modules*, (Russian) Math Sbornik **94** (1974), 594–627. (English translation: Math. USSR, Sbornik **23** (1973).)

[DT1] F. Diamond and R. Taylor, *Non-optimal levels of mod ℓ modular representations*, Invent. Math. **115** (1994) 435–462.

[DT2] F. Diamond and R. Taylor, *Lifting modular mod ℓ representations*, Duke Math. J. **74** (1994) 253–269.

[Edi] B. Edixhoven, *The weight in Serre's conjectures on modular forms*, Invent. Math. **109** (1992), 563–594.

[Edw] H.M. Edwards, Fermat's Last Theorem: A genetic introduction to algebraic number theory, Graduate Texts in Math. **50**, Springer-Verlag, New York, Berlin, Heidelberg, 1977.

[Ei] M. Eichler, *Quaternäre quadratische Formen und die Riemannsche Vermutung für die Kongruenzzetafunktion*, Arch. Math. **5** (1954) 355–366.

[Fa] G. Faltings, *Endlichkeitssätze für abelsche Varietäten über Zahlkörpern*, Inv. Math. **73** (1983), 349–366. (English translation in [CS].)

[Fl1] M. Flach, *A finiteness theorem for the symmetric square of an elliptic curve*, Inv. Math. **109** (1992), 307–327.

[Fl2] M. Flach, *On the degree of modular parametrizations*, Séminaire de théorie des nombres, Paris, 1991-92, 23-36; Prog. in Math. **116**, Birkhäuser, Boston, MA 1993.

[FL] J.-M. Fontaine and G. Laffaille, *Construction de représentations p-adiques*, Ann. Sci. Ec. Norm. Super. **15** (1982), 547–608.

[FM] J.-M. Fontaine and B. Mazur, *Geometric Galois representations*, in [Co], 41–78.

[Fo1] J.-M. Fontaine, *Groupes finis commutatifs sur les vecteurs de Witt*, C. R. Acad. Sc. **280** (1975), 1423–1425.

[Fo2] J.-M. Fontaine, *Groupes p-divisibles sur les corps locaux*, Astérisque **47–48** (1977).

[For] O. Forster, Lecures on Riemann Surfaces, Springer-Verlag, New York, Berlin, Heidelberg, 1981.

[Fr] G. Frey, *Links between solutions of $A - B = C$ and elliptic curves*, in: Number Theory, Ulm 1987, Proceedings, Lecture Notes in Math. **1380**, Springer-Verlag, New York, Berlin, Heidelberg, 1989, pp. 31–62.

[Gra] A. Granville, *The set of exponents for which Fermat's Last Theorem is true, has density one*, Comptes Rendus de l'Académie des Sciences du Canada **7** (1985), 55-60.

[Gre] R. Greenberg, *Iwasawa theory for p-adic representations*, in: Algebraic Number Theory, in honor of K. Iwasawa, Adv. Studies in Pure Math. **17**, Academic Press, San Diego, 1989, pp. 97–137.

[Gro] B.H. Gross, *A tameness criterion for Galois representations associated to modular forms* mod$p$, Duke Math. J. **61** (1990), 445–517.

[Ha] R. Hartshorne, Algebraic Geometry, Graduate Texts in Math. **52**, Springer-Verlag, New York, Berlin, Heidelberg, 1977.

[HB] D.R. Heath-Brown, *Fermat's Last Theorem for "almost all" exponents*, Bull. London. Math. Soc. 17 (1985), 15-16

[He] Y. Hellegouarch, *Points d'ordre $2p^h$ sur les courbes elliptiques*, Acta Arith. **26** (1974/75) 253–263.

[Hi1] H. Hida, *Congruences of cusp forms and special values of their zeta functions*, Inv. Math. **63** (1981), 225–261.

[Hi2] H. Hida, *A p-adic measure attached to the zeta functions associated with two elliptic modular forms. I*, Inv. Math. **79** (1985), 159–195.

[Hi3] H. Hida, *Galois representations into $GL_2(\mathbb{Z}_p[[X]])$ attached to ordinary cusp forms,* Inv. Math. **85** (1986), 545–613.

[Hu] B. Huppert, Endliche Gruppen I, Grundlehren Math. Wiss. **134**, Springer-Verlag, New York, Berlin, Heidelberg, 1983.

[Ig] J. Igusa, *Kroneckerian model of fields of elliptic modular functions*, Amer. J. Math. **81**, 561–577 (1959).

[Ih] Y. Ihara, *On modular curves over finite fields*, in: Proc. Intl. Coll. on Discrete Subgroups of Lie Groups and Applications to Moduli, Bombay, 1973, pp. 161–202.

[Kam] S. Kamienny, *Torsion points on elliptic curves over fields of higher degree*, Inter. Math. Res. Not. **6**, 1992, 129–133.

[Kat] N.M. Katz, *p-adic properties of modular schemes and modular forms*, in [Ant3], pp. 70–189.

[Kh] C. Khare, *Congruences between cusp forms: the (p, p) case*, to appear in Duke Math. J.

[Ki] F. Kirwan, Complex Algebraic Curves, LMS Student Texts **23**, Cambridge Univ. Press, Cambridge, 1993.

[KL] D.S. Kubert, S. Lang, Modular Units. Springer-Verlag, 1981.

[KM] N.M. Katz, and B. Mazur, Arithmetic Moduli of Elliptic Curves, Annals of Math. Studies **108**, Princeton Univ. Press, Princeton, 1985.

[Kn] A.W. Knapp, Elliptic Curves, Princeton Univ. Press, Princeton, 1992.

[Koc] H. Koch, Galoissche Theorie der p-Erweiterungen, Springer-Verlag, New York, Berlin, Heidelberg, 1970.

[Kol] V.A. Kolyvagin, *Finiteness of $E(\mathbb{Q})$ and sha$(E/\mathbb{Q})$ for a subclass of Weil curves*, Izv. Akad. Nauk. SSSR Ser. Mat. **52** (3) (1988), 522–540. (English translation: Math. USSR, Izvestiya **32** (1989).)

[Ku] E. Kunz, Introduction to Commutative Algebra and Algebraic Geometry, Birkhäuser, Boston, Basel, Berlin, 1985.

[La1] S. Lang, Algebraic Number Theory, Addison-Wesley, Reading, MA 1970.

[La2] S. Lang, Introduction to Modular Forms, Grundlehren Math. Wiss. **222**, Springer-Verlag, New York, Berlin, Heidelberg, 1976.

[Leg] A.M. Legendre, *Sur quelques objets d'analyse indéterminée et particulièrement sur le théorème de Fermat*, Mém. Acad. R. Sc. de l'Institut de France **6**, Paris, 1827.

[Len] H.W. Lenstra, *Complete intersections and Gorenstein rings*, in [Co], pp. 99–109.

[Liv] R. Livné, *On the conductors of mod ℓ representations coming from modular forms*, J. Number Theory **31** (1989), 133–141.

[Ll1] R.P. Langlands, *Modular forms and ℓ-adic representations*, in [Ant2], pp. 361–500.

[Ll2] R.P. Langlands, Base Change for $GL(2)$, Annals of Math. Studies **96**, Princeton Univ. Press, Princeton, 1980.

[LO] S. Ling, J. Oesterlé, *The Shimura subgroup of $J_0(N)$*, Astérisque **196–197** (1991), 171–203.

[Mat] H. Matsumura, Commutative Ring Theory, Cambridge Studies in Adv. Math. **8**, Cambridge Univ. Press, Cambridge, 1986.

[Maz1] B. Mazur, *Modular curves and the Eisenstein ideal*, Publ. Math. IHES **47** (1977), 33–186.

[Maz2] B. Mazur, *Rational isogenies of prime degree*, Inv. Math. **44** (1978), 129–162.

[Maz3] B. Mazur, *Deforming Galois representations*, in: Galois groups over $\mathbb{Q}$, Y. Ihara, K. Ribet, J-P. Serre, eds., MSRI Publ. **16**, Springer-Verlag, New York, Berlin, Heidelberg, 1989, pp. 385–437.

[Mc] W. McCallum, *On the method of Coleman and Chabauty*, Math. Ann. **299** (1994), 565–596.

[Mer] L. Merel, *Bornes pour la torsion des courbes elliptiques sur les corps de nombres*, Inventiones Math. **124** (1996), 437–449.

[Mes] J-F. Mestre, *Constructions polynomiales et théorie de Galois*, Proceedings of the ICM, Zurich, August 1994.

[Mi] J.S. Milne, Arithmetic Duality Theorems, Perspectives in Math., Academic Press, San Diego, 1986.

[MRi] B. Mazur, K. Ribet, *Two-dimensional representations in the arithmetic of modular curves*, Astérisque **196–197** (1991), 215–255.

[MRo] B. Mazur, L. Roberts, *Local Euler characteristics*, Inv. Math. **9** (1970), 201–234.

[MT] B. Mazur, J. Tilouine, *Représentations galoisiennes, différentielles de Kähler et "conjectures principales"*, Publ. Math. IHES **71** (1990), 65–103.

[Mu1] V.K. Murty, Introduction to Abelian Varieties, CRM Monograph Series **3**, AMS Publ., Providence, 1993.

[Mu2] V.K. Murty, ed., Seminar on Fermat's Last Theorem, CMS Conference Proceedings **17**, AMS Publ., Providence, 1995.

[MW] B. Mazur, A. Wiles, *Class fields of abelian extensions of* $\mathbb{Q}$, Inv. Math. **76** (1984), 179–330.

[Na] K. Nagao, *An example of an elliptic curve over* $\mathbb{Q}(T)$ *with rank* ≥ 13, Proc. of The Japan Acad., **70**, 1994, 152-153.

[Oda] T. Oda, *The first De Rham cohomology group and Dieudonné modules*, Ann. Sci. Ec. Norm. Super. **2** (1969), 63–135.

[Oe] J. Oesterlé, *Nouvelles aproches du "théorème" de Fermat"*, Séminaire Bourbaki no. 694 (1987-88), Astérisque **161–162** (1988), 165–186.

[Ogg] A. Ogg, Modular Forms and Dirichlet Series, Benjamin, New York, 1969.

[Oo] F. Oort, Commutative Group Schemes, Lecture Notes in Math. **15**, Springer-Verlag, New York, Berlin, Heidelberg, 1966.

[Ram] R. Ramakrishna, *On a variation of Mazur's deformation functor*, Compositio Math. **87** (1993), 269–286.

[Ray] M. Raynaud, *Schémas en groupes de type* $(p, p, ..., p)$, Bull. Soc. Math. France **102** (1974), 241–280.

[R1] K. Ribet, *A modular construction of unramified p-extensions of* $\mathbb{Q}(\mu_p)$, Inv. Math. **34**, 151-162.

[R2] K. Ribet, *Twists of modular forms and endomorphisms of abelian varieties*, Math. Ann. **253** (1980), 43-62.

[R3] K.A. Ribet, *The ℓ-adic representations attached to an eigenform with Nebentypus: a survey*, in [Ant5], pp. 17–52.

[R4] K.A. Ribet, *Congruence relations between modular forms*, Proc. ICM, 1983, 503–514.

[R5] K.A. Ribet, *On modular representations of* $\mathrm{Gal}(\bar{\mathbb{Q}}/\mathbb{Q})$ *arising from modular forms*, Inv. Math. **100** (1990), 431–476.

[R6] K.A. Ribet, *Report on mod ℓ representations of* $\mathrm{Gal}(\bar{\mathbb{Q}}/\mathbb{Q})$, in: Motives, Proc. Symp. in Pure Math. **55** (2), 1994, pp. 639–676.

[R7] K.A. Ribet, *Abelian varieties over $\mathbb{Q}$ and modular forms*, 1992 Proceedings of KAIST Mathematics Workshop, Korea Advanced Institute of Science and Technology, Taejon, 1992, pp. 53–79.

[R8] K.A. Ribet, *Galois representations and modular forms*, Bull. AMS. **32** (1995), 375–402.

[Ri] P. Ribenboim, 13 Lectures on Fermat's Last Theorem, Springer-Verlag, New York, Berlin, Heidelberg, 1979.

[Sa] T. Saito, *Conductor, discriminant and the Noether formula for arithmetic surfaces*, Duke Math. J. **57** (1988), 151–173.

[Se1] J.-P. Serre, Cohomologie Galoisienne, Lecture Notes in Math. **5**, Springer-Verlag, New York, Berlin, Heidelberg, 1964.

[Se2] J.-P. Serre, Corps Locaux, Hermann, Paris, 1962.

[Se3] J.-P. Serre, Algèbre Locale, Multiplicités, Cours au Collège de France, 1957-58, Lecture Notes in Math. **11**, Springer-Verlag, New York, Berlin, Heidelberg, 1965.

[Se4] J.-P. Serre, A Course in Arithmetic, Graduate Texts in Math. **7**, Springer-Verlag, New York, Berlin, Heidelberg, 1973.

[Se5] J.-P. Serre, Abelian ℓ-adic Representations and Elliptic Curves, Benjamin, New York, 1968.

[Se6] J.-P. Serre, *Propriétés galoisiennes des points d'ordre fini des courbes elliptiques*, Inv. Math. **15** (1972), 259–331.

[Se7] J.-P. Serre, *Sur les représentations modulaires de degré* 2 *de* $\mathrm{Gal}(\bar{\mathbb{Q}}/\mathbb{Q})$, Duke Math. J. **54** (1987), 179–230.

[Sha] S. Shatz, *Group schemes, formal groups, and p-divisible groups*, in [CS], pp. 29–78.

[Shi1] G. Shimura, *Correspondances modulaires et les fonctions ζ de courbes algébriques*, J. Math. Soc. Japan **10** (1958), 1–28.

[Shi2] G. Shimura, Introduction to the Arithmetic Theory of Automorphic Functions, Princeton Univ. Press, Princeton, 1971.

[Shi3] G. Shimura, *On the factors of the jacobian variety of a modular function field*, J. Math. Soc. Japan **25** (1973), 523–544.

[Shi4] G. Shimura, *On the holomorphy of certain Dirichlet series*, Proc. LMS (3) **31** (1975), 79–98.

[Shi5] G. Shimura, *The special values of the zeta functions associated with cusp forms*, Comm. Pure Appl. Math. **29** (1976), 783–804.

[Si1] J. Silverman, The Arithmetic of Elliptic Curves, Graduate Texts in Math. **106**, Springer-Verlag, New York, Berlin, Heidelberg, 1986.

[Si2] J. Silverman, Advanced Topics in the Arithmetic of Elliptic Curves, Graduate Texts in Math. **151**, Springer-Verlag, New York, Berlin, Heidelberg, 1994.

[St] J. Sturm, *Special values of zeta functions, and Eisenstein series of half integral weight*, Amer. J. Math. **102** (1980), 219–240.

[ST] J.-P. Serre, J.T. Tate, *Good reduction of abelian varieties*, Annals of Math. **88** (1968), 492–517.

[Su] J. Suzuki, *On the generalized Wieferich criterion*, Proc. Japan Acad. **70** (1994), 230-234.

[Ta] J.T. Tate, *Algorithm for determining the type of a singular fiber in an elliptic pencil*, in [Ant4], pp. 33–52.

[Te] G. Terjanian, *Sur l'équation $x^{2p} + y^{2p} = z^{2p}$*, C.R. Acad. Sci. Paris, **285** (1977) 973-975.

[Th] F. Thaine, *On the ideal class groups of real abelian number fields*, Annals of Math. **128** (1988), 1–18.

[Tu] J. Tunnell, *Artin's conjecture for representations of octahedral type*, Bull. AMS **5** (1981), 173–175.

[TW] R. Taylor, A. Wiles, *Ring theoretic properties of certain Hecke algebras*, Annals of Math. **141** (1995), 553–572.

[Wa] L. Washington, Introduction to Cyclotomic Fields, Graduate Texts in Math. **83**, Springer-Verlag, New York, Berlin, Heidelberg, 1982.

[We1] A. Weil, Variétés Abéliennes et Courbes Algébriques, Hermann, Paris, 1948.

[We2] A. Weil, *Über die Bestimmung Dirichletscher Reihen durch Funktionalgleichungen*, Math. Ann. **168** (1967), 165–172.

[W1] A. Wiles, *On p-adic representations for totally real fields*, Annals of Math. **123** (1986), 407–456.

[W2] A. Wiles, *On ordinary λ-adic representations associated to modular forms*, Inv. Math. **94** (1988), 529–573.

[W3] A. Wiles, *Modular elliptic curves and Fermat's Last Theorem*, Annals of Math. **141** (1995), 443–551.

Conference on Elliptic Curves and Modular Forms
Hong Kong, December 18-21, 1993

Characteristic distributions for non-torsion Λ-modules.

P.N. Balister
Dept. of Pure Mathematics and Mathematical Statistics
University of Cambridge
16 Mill Lane, Cambridge CB2 1SB England

1. Introduction. Let G be a profinite Abelian group which is isomorphic to $\mathbb{Z}_p^d$ for some d and has topological generators γ_i, $1 \leq i \leq d$, and let G_k denote the subgroup $p^k G$ of G. Let K be a finite extension of $\mathbb{Q}_p$ with ring of integers $\mathcal{O}_K$. Denote by Λ the completed group ring $\mathcal{O}_K[[G]]$. It is known that Λ is Noetherian and an integral domain (in fact a unique factorization domain), and there is an isomorphism of topological rings

$$\Lambda \cong \mathcal{O}_K[[T_1, \ldots, T_d]]$$

obtained by sending topological generators γ_i of G to $1 + T_i$. There is also a structure theorem for Λ-modules which states that any finitely generated Λ-module Y can be placed in an exact sequence

$$0 \to C \to Y \to R \oplus \Lambda/(f_1) \oplus \ldots \oplus \Lambda/(f_n) \to D \to 0$$

where C and D are annihilated be height two ideals of Λ and R is Λ-torsion free. If Y is a torsion Λ-module, (so $R = 0$), we can define a *characteristic element* (or *characteristic power series*) $\operatorname{char}(Y)$ of Y as any generator of the ideal $(f_1 \ldots f_n)$. This is uniquely determined up to multiplication by a unit in Λ. This characteristic element has the following crucial property: for any exact sequence of finitely generated torsion Λ-modules

$$0 \to A \to B \to C \to 0$$

we have $\operatorname{char}(B) = \operatorname{char}(A)\operatorname{char}(C)$ (up to multiplication by a unit in Λ). Unfortunately, this property does not extend to non-torsion modules, even if we define the characteristic element to be that of the torsion submodule. For example, consider the sequence

$$0 \longrightarrow \Lambda \xrightarrow{\times f} \Lambda \longrightarrow \Lambda/(f) \longrightarrow 0$$

where $\operatorname{char}(\Lambda_{\mathrm{tors}}) = 1$ and $\operatorname{char}(\Lambda/(f)) = f$. We can however recover the multiplicative property if we can quotient out by suitable "parametrising modules", e.g.,

$$\begin{array}{ccccccccl}
0 \to & \Lambda & \xrightarrow{\times 1} & \Lambda & \longrightarrow & 0 & \to 0 & & \text{(parametrising modules)} \\
& \downarrow{\scriptstyle \times g} & & \downarrow{\scriptstyle \times fg} & & \downarrow & & & \\
0 \to & \Lambda & \xrightarrow{\times f} & \Lambda & \longrightarrow & \Lambda/(f) & \to 0 & & \text{(original sequence)} \\
& \downarrow & & \downarrow & & \downarrow & & & \\
0 \to & \Lambda/(g) & \longrightarrow & \Lambda/(fg) & \longrightarrow & \Lambda/(f) & \to 0 & & \text{(quotient sequence)}
\end{array}$$

This results in torsion modules for which the characteristic elements behave well. The problem is therefore to find some canonical submodules which we can quotient out by. In the case of interest, the modules are duals of Selmer groups of abelian varieties with non-ordinary reduction at p. In this case there is in general no obvious choice for such submodules, however, we can often find modules with canonical maps at each finite level, i.e., a "nice" module X and a collection of maps $f_k : X_k \to Y_k$ (where $X_k = X_{G_k}$ etc. are the coinvariants). X and f_k are constructed from the formal group of the variety defined locally at primes dividing p. See **[PR]** and **[K]** for details of this construction. The maps f_k are not in general compatible with the projection maps and so do not give a map $f : X \to Y$, but they are related by certain more complicated compatibility relations of the form

$$\sum_{j=0}^{n} a_j \mathrm{pr}^Y_{k+j,k} f_{k+j} \mathrm{pr}^X_{k+j}(x) = 0 \qquad \forall x \in X$$

where pr denotes the projection maps. In this paper, I will show how one can use such a collection of maps to recover a definition of characteristic element with the good properties that we want.

2. Characteristic element of a Λ-morphism. Instead of attaching a characteristic element to a module, we will attach it to maps between modules. Define for any Λ-morphism f,

$$\mathrm{char}(f) = \begin{cases} \mathrm{char}(\mathrm{Coker}\, f)/\mathrm{char}(\mathrm{Ker}\, f), & \text{if } f \text{ has torsion kernel and cokernel;} \\ 0, & \text{if only the kernel is torsion;} \\ \text{undefined}, & \text{if the kernel is not torsion.} \end{cases}$$

char(f) is an element of the field of fractions L of Λ and is defined only up to multiplication by a unit of Λ. We now have the following results.

Lemma 2.1. *If we have a commutative diagram of finitely generated Λ-modules with exact rows as follows*

$$\begin{array}{ccccccccc} 0 & \longrightarrow & A' & \longrightarrow & B' & \longrightarrow & C' & \longrightarrow & 0 \\ & & \downarrow{\scriptstyle\alpha} & & \downarrow{\scriptstyle\beta} & & \downarrow{\scriptstyle\gamma} & & \\ 0 & \longrightarrow & A & \longrightarrow & B & \longrightarrow & C & \longrightarrow & 0 \end{array}$$

then $\mathrm{char}(\beta) = \mathrm{char}(\alpha)\mathrm{char}(\gamma)$ *(when defined, and up to a unit of Λ).*

Proof. Apply the Snake Lemma to the above diagram. If all kernels and cokernels are torsion, the result follows from the fact that char is multiplicative in exact sequences of torsion modules. If all the kernels are torsion but some of the cokernels are not, then the equation reduces to $0 = 0$. If some of the kernels are non-torsion, then the right hand side of the equation is undefined. □

Note that if $A' = B' = C' = 0$ and A, B and C are torsion, then we recover the multiplicativity of characteristic elements of torsion modules.

Lemma 2.2. *If $A \xrightarrow{f} B \xrightarrow{g} C$ are maps between finitely generated Λ-modules, then* $\operatorname{char}(gf) = \operatorname{char}(g)\operatorname{char}(f)$ *(when defined, and up to a unit of Λ).*

Proof. Apply Lemma 2.1 to the following diagram,

$$\begin{array}{ccccccccc} 0 & \longrightarrow & A & \xrightarrow{1\oplus f} & A \oplus B & \xrightarrow{f-1} & B & \longrightarrow & 0 \\ & & \downarrow{\scriptstyle f} & & \downarrow{\scriptstyle gf\oplus 1} & & \downarrow{\scriptstyle g} & & \\ 0 & \longrightarrow & B & \xrightarrow{g\oplus 1} & C \oplus B & \xrightarrow{1-g} & C & \longrightarrow & 0 \end{array}$$

and note that $\operatorname{char}(gf \oplus 1)$ is the same as $\operatorname{char}(gf)$. □

Lemma 2.3. *If $f : \Lambda^r \longrightarrow \Lambda^s$ is injective then* $\operatorname{char}(f) = \det(f)$ *(up to a unit).*

Proof. If $r < s$, define $\det(f) = 0$ so that the equation holds in this case. The only other case is $r = s$. Let (A_{ij}) be the matrix corresponding to f. The result clearly holds when (A_{ij}) is diagonal since the cokernel of f is then just $\Lambda/(A_{11}) \oplus \ldots \oplus \Lambda/(A_{rr})$. The result also holds when f (and so (A_{ij})) is invertible, since both sides are then units. Also both sides of the equation are multiplicative under composition of maps. Without loss of generality we can assume $A_{11} \neq 0$ (permutation of rows and columns corresponds to composition with invertible maps, also f is assumed injective so some A_{ij} must be non-zero). Multiply all the rows $i \neq 1$ by A_{11} (equivalent to composition with a diagonal matrix) and subtract suitable multiples of the first row from the others to obtain $A_{i1} = 0$ for all $i \neq 1$ (equivalent to composition by an invertible map). Repeat with rows 2 to r to get a diagonal matrix for which the result is known. Now use the fact that Λ is an integral domain and the fact that all the maps we composed with were injective, so had non-zero determinant. □

3. Semi-morphisms.

We will be considering families $\{f_k\}$ of maps $f_k : X_k \to Y_k$, where X and Y are any finitely generated Λ-modules. They will be subject to relations of the form

$$\sum_{j \geq k} r_j \operatorname{pr}^Y_{jk}(f_j(\operatorname{pr}^X_j(x))) = 0 \qquad \forall x \in X$$

where $j, k \in \mathbb{N}$, $r_j \in \mathcal{O}_K$ and only finitely many r_j are non-zero.

$$\operatorname{pr}^Y_{jk} : Y_j \to Y_k$$

$$\operatorname{pr}^X_j : X \to X_j$$

are projection maps. We will write such relations as $r = ((r_j)_{j \geq k}, k)$. Denote by $\mathcal{R}$ the set of all such relations imposed.

The main example of a relation set is $\mathcal{R} = \{((a_{j-k})_{j \geq k}, k) : k \in \mathbb{N}\}$, where $R(T) = \sum_{i=0}^{h} a_i T^i$ is a monic polynomial in $\mathcal{O}_K[T]$ with non-zero constant term.

Now let F be any extension of K. Define the *scalar solution space* V $(= V_F)$ *of* $\mathcal{R}$ *over* F by

$$V = \{(\alpha_j)_{j\in\mathbb{N}} \mid \forall r \in \mathcal{R} : \sum r_j\alpha_j = 0\} \subseteq \prod_{j\in\mathbb{N}} F = F^{\mathbb{N}}$$

I.e., V is the set of sequences in F satisfying the relations of $\mathcal{R}$. For $i \in \mathbb{N}$ define

$$V^{\geq i} = \{(\alpha_j)_{j\geq i} \mid \forall r \in \mathcal{R} \text{ for which } k \geq i \ : \sum r_j\alpha_j = 0\}$$

We will write $\mathcal{O}^{\geq k}$ to denote the set

$$\mathcal{O}^{\geq k} = \{(\alpha_j)_{j\in\mathbb{N}} \in V \mid \alpha_j \in \mathcal{O}_F \text{ for all } j \geq k\} \subseteq V$$

We will call $\mathcal{R}$ an *admissible relation set* if it satisfies the following two conditions;

(A) V is a finite dimensional F-vector space,

(B) $\forall i \in \mathbb{N}$, the projection map $V \to V^{\geq i}$ is an isomorphism.

We will call $\mathcal{R}$ *strongly admissible* if it also satisfies the condition;

(C) If pr_j is the projection onto the jth component of V, then for any j, $p^k\mathrm{pr}_j(\mathcal{O}^{\geq k}) \to 0$ in F as $k \to \infty$. Equivalently, $p^k\mathcal{O}^{\geq k} \to 0$ in V under the standard topology of V as a finite dimensional F-vector space.

Note that since $V_F = V_K \otimes_K F$, these conditions do not depend on F.

Lemma 3.1.

(a) *If $\mathcal{R}$ is derived as above from a monic polynomial $R(T) \in \mathcal{O}_K[T]$ with non-zero constant term, then it is admissible.*

(b) *If all the roots of $R(T)$ (in some algebraic closure of K) satisfy $|\,p\,|_p < |\,\alpha\,|_p \leq 1$ then $\mathcal{R}$ is strongly admissible.*

Proof.

(a) Condition (A) follows from the fact that any recurrence sequence is uniquely determined by its first $\deg R(T)$ terms which can be chosen arbitrarily, so $\dim V = \deg R(T)$ is finite. For (B) we have $\dim V = \dim V^{\geq i}$ by symmetry, and the map $V \to V^{\geq i}$ given by "forgetting" coordinates $j < i$, is injective since the α_j for $j \geq i$ determine those for $j < i$ (here we need $R(0) \neq 0$).

(b) Write $R(T) = \sum a_iT^i = \prod_\alpha (T-\alpha)$, and let v_p denote the valuation written additively and normalised so that $v_pp = 1$. We now need to prove (C) under the extra condition that all the roots satisfy $v_p\alpha < 1$. Now a_0 is the product of all n roots, and a_i is a sum of products of $n-i$ of the roots, so $v_pa_0 \leq v_pa_i + i\lambda$, where $\lambda = \max v_p\alpha < 1$. If $(\alpha_j) \in \mathcal{O}^{\geq k}$ we have

$$a_0\alpha_j = -a_1\alpha_{j+1} - a_2\alpha_{j+2} - \cdots \quad \text{with } \alpha_j \in \mathcal{O}_F \text{ for } j \geq k$$

so we can prove by induction on decreasing j that

$$v_p\alpha_j \geq (j-k)\lambda \quad \text{for } j \leq k$$

$$v_p(p^k\alpha_j) \geq j\lambda + k(1-\lambda).$$

Fixing j, letting $k \to \infty$ and using $\lambda < 1$ gives the result. □

If a family of Λ-morphisms $f_k : X_k \to Y_k$ satisfies all the relations of $\mathcal{R}$, we will call (f_k) a *semi-morphism from X to Y with respect to $\mathcal{R}$* and write

$$f = ((f_k)_{k \in \mathbb{N}}, \mathcal{R}) : X \Longrightarrow Y.$$

If $\mathcal{R}$ is derived from a polynomial $R(T)$ as above, then we will also call f a *semi-morphism from X to Y with respect to $R(T)$.*

If $f : X \to Y$ is a Λ-morphism, then the induced maps $f_k : X_k \to Y_k$ form a semi-morphism $X \Longrightarrow Y$ with respect to $R(T) = T - 1$, so the concept of a semi-morphism includes that of normal Λ-morphisms.

4. Distributions. Let Step denote the set of locally constant functions (step functions) from G to $\mathcal{O}_K$ and let A denote any normed $\mathcal{O}_K$-module (in what follows A will be $\mathcal{O}_K$, F with the standard norm extending that on $\mathcal{O}_K$ or V with any norm compatible with the one on F). We normalise the norm on $\mathcal{O}_K$ by $|p|_p = p^{-1}$.

Let $\mathcal{D}(G, A) = \varprojlim(\Lambda_k \otimes_{\mathcal{O}_K} A)$ where the inverse limit is taken with respect to the natural projection maps. Then $\mathcal{D}(G, A)$ can be considered as the set of (unbounded) A-valued distributions μ on G, that is to say, the set of $\mathcal{O}_K$-linear maps from Step to A. The correspondence is given by defining $\int 1_{gG_k} d\mu$ to be the coefficient of gG_k in the projection of μ to $A[G/G_k] = \Lambda_k \otimes_{\mathcal{O}_K} A$, and then extending by linearity to Step. 1_{gG_k} is the function that is 1 on gG_k and 0 elsewhere. Since G is compact, all locally constant functions are a finite linear combination of functions of this type. If A is a ring, then as a ring, $\mathcal{D}(G, A)$ is just $A[[G]]$, the completed group ring of G with coefficients in A. Unfortunately, when A is p-divisible it has some rather unpleasant properties, in particular it is not an integral domain (unlike $\Lambda = \mathcal{D}(G, \mathcal{O}_K)$, which is). To prove this, let $\Gamma \cong \mathbb{Z}_p$ be a closed subgroup of G with a topological generator γ. Denote by $\mathrm{Haar}_\Gamma = (\frac{1}{p^k} \sum_{i=0}^{p^k-1} \gamma^i G_k)_{k \in \mathbb{N}} \in \varprojlim(\Lambda_k \otimes_{\mathcal{O}_K} A)$, the Haar distribution supported on Γ. Then it is easy to check that

$$(\gamma - 1)\mathrm{Haar}_\Gamma = 0 \quad \text{in } \mathcal{D}(G, A)$$

but both Haar_Γ and $(\gamma - 1)$ are non-zero. The problem comes from the fact that our distributions are unbounded.

Let $\mathcal{D}_a(G, A) = \{\mu : \max_{gG_k \in G/G_k} |p^k \int 1_{gG_k} d\mu|_p \to 0 \text{ as } k \to \infty\}$. Define a norm on $\mathcal{D}_a(G, A)$ by $|\mu|_p = \sup_{g,k} |p^k \int 1_{gG_k} d\mu|_p < \infty$. For functions $f : G \to \mathcal{O}_K$ define the norm $|f|_p = \sup_g |f(g)|_p$.

Lemma 4.1. *If $\mathcal{R}$ is strongly admissible, and let μ be an element of $\mathcal{D}(G, V) = \varprojlim(\Lambda_k \otimes V)$ whose image under each projection map is contained in $\Lambda_k \otimes \mathcal{O}^{\geq k}$. Then $\mu \in \mathcal{D}_a(G, V)$.*

Proof. The definition of strongly admissible implies that $|p^k \mathcal{O}^{\geq k}|_p \to 0$ and the condition in the statement implies that $\int 1_{gG_k} d\mu \in \mathcal{O}^{\geq k}$ for all $g \in G$, so $\max_g |p^k \int 1_{gG_k} d\mu|_p \leq |p^k \mathcal{O}^{\geq k}|_p \to 0$ as $k \to \infty$. □

Lemma 4.2. *If $\mu \in \mathcal{D}_a(G, A)$ then μ can be extended to a linear functional on differentiable functions $f : G \to \mathcal{O}_K$ and we have for any differentiable f, $|\int f\, d\mu|_p \leq |\mu|_p \max(pC, |f|_p)$ where $C = \sup_{g,h} |f(g) - f(h)|_p / |g-h|_p < \infty$.*

Proof. Let $f : G \to \mathcal{O}_K$ be differentiable (considering G as $\mathbb{Z}_p^d$ and f as a function in d distinct $\mathbb{Z}_p$-valued variables). Then for each $g \in G$ there is a neighbourhood of g in which

$$|f(h) - f(g)|_p \leq C_g |h - g|_p \qquad \text{for some } C_g.$$

(The norm on G is given by the maximum norm of its components in $\mathbb{Z}_p^d$). f is bounded, so this equation also holds (with a different constant) when h is bounded away from g. Therefore

$$\exists C_g' : \forall h \in G : |f(h) - f(g)|_p \leq C_g' |h - g|_p .$$

Now using the triangle inequality and compactness of G we get that

$$\exists C'' : \forall g, h \in G : |f(h) - f(g)|_p \leq C'' |h - g|_p .$$

The C in the statement is therefore finite and bounded by C''. Now define $\int f\, d\mu$ to be the limit as $k \to \infty$ of $\int \sum f(g) 1_{gG_k} d\mu$ where the sum is over any set of representatives for G/G_k. Now

$$\begin{aligned}
&\left| \int \sum_{g'} f(g') 1_{g'G_{k+1}} d\mu - \int \sum_{g} f(g) 1_{gG_k} d\mu \right|_p \\
&= \left| \int \sum (f(g') - f(g)) 1_{g'G_{k+1}} d\mu \right|_p \\
&\leq \max_{g'} |f(g') - f(g)|_p \left| \int 1_{g'G_{k+1}} d\mu \right|_p \\
&\leq Cp^{-k} \max \left| \int 1_{g'G_{k+1}} d\mu \right|_p \\
&\to 0 \text{ as } k \to \infty
\end{aligned}$$

The sums are over representatives g' of G/G_{k+1} and representatives g of G/G_k, where in the right hand side we have grouped the sum so that $gG_k = g'G_k$ in each term. The limit therefore converges to a value independent of the choices of representatives. Also each term above is bounded by $pC|\mu|_p$ and the initial approximation $\int f(g) 1_G$ is bounded by $|f|_p |\mu|_p$. This gives the bound in the statement of the Lemma. □

Let Poly denote the set of polynomials in d variables with coefficients in $\mathcal{O}_K$. Elements of Poly can be considered as maps from G to $\mathcal{O}_K$.

Let $\mathcal{D}_p(G, A) = \mathrm{Hom}_{\mathcal{O}_K}(\mathrm{Poly}, A)$, the set of linear functionals on polynomials. Recall we have an isomorphism $\Lambda \cong \mathcal{O}_K[[T_1, \ldots, T_d]]$ given by sending topological generators γ_i of G to $1 + T_i$. Using this isomorphism, $\mathcal{D}_p(G, A)$ can be considered as the set of power series $A[[T_1, \ldots, T_d]]$. The correspondence is given by mapping μ to $\sum \int \binom{x_1}{n_1} \ldots \binom{x_d}{n_d} d\mu T_1^{n_1} \ldots T_d^{n_d}$. It is clear that functions of the form $\binom{x_1}{n_1} \ldots \binom{x_d}{n_d}$ form a basis for Poly so this map is an isomorphism. Also $\mu = \gamma_1^{m_1} \ldots \gamma_d^{m_d}$ is mapped to $\sum \binom{m_1}{n_1} \ldots \binom{m_d}{n_d} T_1^{n_1} \ldots T_d^{n_d} = \prod_i (1 + T_i)^{m_i}$.

Lemma 4.3. *$\mathcal{D}_a(G, A)$ can be considered a subset of $\mathcal{D}_p(G, A) = A[[T_1, ..., T_d]]$.*

Proof. Polynomial functions are differentiable so we get a map $\mathcal{D}_a(G, A) \to \mathcal{D}_p(G, A)$. We now need to show that if $\mu \neq 0$ then $\int f \, d\mu \neq 0$ for some polynomial f. Pick any g and k such that $\int 1_{gGk} d\mu \neq 0$. 1_{gGk} is continuous, so can be uniformly approximated with a polynomial function on the compact set G, i.e., there exists polynomial P with $| P - 1_{gGk} |_p \leq p^{-1}$. Write $P_i = P^{p^i}$ and $f_i = P_i - 1_{gG_k}$, then $| f_i |_p \to 0$. If h and h' are in different cosets of G/G_k then $| h' - h |_p \geq p^{-k}$ and $| f_i(h') - f_i(h) |_p / | h' - h |_p \leq p^k | f_i |_p \to 0$. If h and h' are in the same coset then $| f_i(h') - f_i(h) |_p / | h' - h |_p = c | f_{i-1}(h') - f_{i-1}(h) |_p / | h' - h |_p$ where $c = |(f_i(h') - f_i(h))/(f_{i-1}(h') - f_{i-1}(h)) |_p = | \sum P_{i-1}(h)^j P_{i-1}(h')^{p-j} |_p$. Since $P_{i-1}(h) \equiv P_{i-1}(h') \mod p$ and the sum above contains p terms, $c \leq p^{-1}$. Therefore both the C for f_i and $| f_i |_p$ in Lemma 4.2 tend to 0 and $\int P_i d\mu \to \int 1_{gG_k} d\mu \neq 0$ as $i \to \infty$, so μ is non-trivial on some polynomial. □

Lemma 4.4. *The image of any $\mu \in \mathcal{D}_a(G, A)$ in $A[[T_1, \ldots, T_d]]$ is a power series that converges in the ball $\{(T_1, \ldots, T_d) : \max_i | T_i |_p < 1\}$.*

Proof. We need to show that the coefficients $\int \binom{x_1}{n_1} \ldots \binom{x_d}{n_d} d\mu$ do not grow too rapidly, and this will follow from Lemma 4.2 provided we can show that $|(\binom{x}{n} - \binom{y}{n})/(x - y) |_p$ is bounded independently of x and y by a function of n which grows subexponentially in n. Now $((1 + T)^x - (1 + T)^y)/(x - y) = (1 + T)^y((1 + T)^{x-y} - 1)/(x - y)$ and the derivative of $((1 + T)^{x-y} - 1)/(x - y)$ with respect to T is an integral power series in T. Therefore the n'th term of $((1 + T)^x - (1 + T)^y)/(x - y)$ is bounded $O(n)$. This gives the result. □

Note that if A is an integral domain, then so is $\mathcal{D}_p(G, A)$. However, $\mathcal{D}_a(G, A)$ is not necessarily even a ring since it is not closed under multiplication (convolution). The above Lemma allows us to embed $\mathcal{D}_a(G, A)$ into a ring. Note also that in some sense the conditions defining $\mathcal{D}_a(G, F)$ are the weakest necessary conditions to make it embeddable in an integral domain. In the example of the Haar distribution, $p^k \int 1_{gG_k} d\mathrm{Haar}_\Gamma = 1$ if $g \in \Gamma$.

5. Characteristic distribution of a semi-morphism.

Given a semi-morphism $X \Longrightarrow Y$ we will construct a characteristic distribution with good properties.

First we will simplify the situation. Let $\alpha : \Lambda^r \to X$ and $\beta : Y \to \Lambda^s$ be any two maps with torsion kernel and cokernel (so $r = \mathrm{rank} X$ and $s = \mathrm{rank} Y$). The semi-morphism $f : X \Longrightarrow Y$ clearly gives a semi-morphism $g : \Lambda^r \Longrightarrow \Lambda^s$ with the same relations $\mathcal{R}$ by composing f_k with the induced maps $\alpha_k : \Lambda_k^r \to X_k$ and $\beta_k : Y_k \to \Lambda_k^s$.

Now let $x \in \Lambda^r$. The sequence of points $(\mathrm{pr}_{jk}(g_j(\mathrm{pr}_j(x))))_{j\geq k}$ satisfy all the relations in $\mathcal{R}$ when defined. Since Λ_k^s is a finitely generated $\mathcal{O}_K$-module, $\Lambda_k^s \otimes_{\mathcal{O}_K} F$ is a finite dimensional F-vector space. The points $(\mathrm{pr}_{jk}(g_j(\mathrm{pr}_j(x)))\otimes 1)_{j\geq k}$ form a sequence in $\Lambda_k^s \otimes F$ satisfying relations in $\mathcal{R}$, so by considering components with respect to any basis of $\Lambda_k^s \otimes F$, we see that the sequence defines an element of $(\Lambda_k^s \otimes_{\mathcal{O}_K} F) \otimes_F V^{\geq k}$, which is canonically isomorphic to $\Lambda_k^s \otimes_{\mathcal{O}_K} V$ by condition (B). The sequence lies in the subspace $\Lambda_k^s \otimes \mathcal{O}^{\geq k}$, and the maps $\Lambda^r \to \Lambda_k^s \otimes V$ are compatible with the projection maps $\Lambda_j^s \to \Lambda_k^s$ for $j \geq k$. Now let N denote any ring containing $\mathcal{D}(G,F) \otimes_\Lambda L$ in the general case, or $\mathcal{D}_a(G,F) \otimes_\Lambda L$ in the case of a strongly admissible relation set. Then we get a map $\bar{g} : \Lambda^r \to N^s \otimes_F V$.

Now let Π denote any projection from V to F. Composing with $\bar{g}$ gives a map $\bar{g}_\Pi : \Lambda^r \to N^s$. Define

$$\mathrm{char}_\Pi(g) = \begin{cases} \det(\bar{g}_\Pi), & \text{if } \bar{g}_\Pi \text{ is injective;} \\ \text{undefined}, & \text{otherwise.} \end{cases}$$

The determinant is taken with respect to the natural bases of Λ^r and N^s and gives an element of N. (We define the determinant to be zero if $r < s$).

Finally define

$$\mathrm{char}_\Pi(f) = \mathrm{char}_\Pi(g)\mathrm{char}(\alpha)^{-1}\mathrm{char}(\beta)^{-1} \in N$$

where $\mathrm{char}(\alpha)$ and $\mathrm{char}(\beta)$ are defined as in section 2.

Lemma 5.1.

(a) *If α and α' are two maps $\Lambda^r \to X$ with torsion kernel and cokernel, then there exists i and i' injections $\Lambda^r \to \Lambda^r$ such that $\alpha i = \alpha' i'$ and this map has torsion kernel and cokernel.*

(b) *If β and β' are two maps $Y \to \Lambda^s$ with torsion kernel and cokernel, then there exist j and j' injections $\Lambda^s \to \Lambda^s$ such that $j\beta = j'\beta'$ and this map has torsion kernel and cokernel.*

Proof. Both results are clear if we tensored the modules with the field of fractions L of Λ, since then each module would be a finite dimensional L-vector space of the same dimension and all the maps would be isomorphisms. Define a *lattice* of an L-vector space L^n to be a finitely generated Λ submodule which generates the whole space when tensored with L. Every lattice in L^n contains a sublattice of the form $x\Lambda^n$ and is contained in a superlattice of the form $\frac{1}{y}\Lambda^n$ which are both of course isomorphic to Λ^n. Now for (a), restrict i and i' to a sublattice of $i^{-1}(\Lambda^r) \cap i'^{-1}(\Lambda^r) \subseteq L^r$ which is isomorphic to Λ^r. This gives us i and i' as injective maps from Λ^r to Λ^r with $\alpha i = \alpha' i'$ up to torsion. Now multiply both maps i and i' by any annihilator in Λ of this torsion. For (b), consider j and j' as maps to a superlattice of $j(\Lambda^s) \cup j'(\Lambda^s) \subseteq L^s$ isomorphic to Λ^s, then $j\beta = j'\beta'$. All the maps have torsion kernel and cokernel because the kernel and cokernels vanish on tensoring with L (tensoring with L is exact on finitely generated Λ-modules). □

Theorem 5.2. *The definition of* $\mathrm{char}_\Pi(f)$ *is independent of the maps* α *and* β, *and coincides with the earlier definition of* $\mathrm{char}(f)$ *when* f *is a* Λ*-morphism (semi-morphism with respect to* $R(T) = T - 1$*).*

Proof. We will prove the second statement first. If f is actually a Λ-morphism, then $V = \{(\alpha_i) : \text{ all } \alpha_i \text{ equal}\} \cong F$. We take Π to be the projection onto any component $\Pi((\alpha_i)) = \alpha_1$. The map $\bar{f'}_\Pi : \Lambda^r \to N^s$ is then just the same as $f' \otimes 1$, and $\det(\bar{f'}_\Pi) = \det(f') = \mathrm{char}(f')$ by Lemma 2.3. $\mathrm{char}_\Pi(f) = \mathrm{char}(f)$ then follows from the definition of $\mathrm{char}_\Pi(f)$ and Lemma 2.2.
For the first part. Multiplicativity of char and determinants under composition means that the definition is unaltered if we replace α by αi and β by $j\beta$ where $i : \Lambda^r \to \Lambda^r$ and $j : \Lambda^s \to \Lambda^s$ are any maps with torsion kernal and cokernal. Now Lemma 5.1 gives the result. □

6. Properties of char_Π.

Theorem 6.1. *If we have a commutative diagram of finitely generated modules with exact rows as follows*

$$\begin{array}{ccccccccc} 0 & \longrightarrow & A' & \longrightarrow & B' & \longrightarrow & C' & \longrightarrow & 0 \\ & & \Downarrow{\scriptstyle\alpha} & & \Downarrow{\scriptstyle\beta} & & \Downarrow{\scriptstyle\gamma} & & \\ 0 & \longrightarrow & A & \longrightarrow & B & \longrightarrow & C & \longrightarrow & 0 \end{array}$$

(by commutative we mean that

$$\begin{array}{ccccccc} A'_k & \longrightarrow & B'_k & & B'_k & \longrightarrow & C'_k \\ \downarrow{\scriptstyle\alpha_k} & & \downarrow{\scriptstyle\beta_k} & & \downarrow{\scriptstyle\beta_k} & & \downarrow{\scriptstyle\gamma_k} \\ A_k & \longrightarrow & B_k & & B_k & \longrightarrow & C_k \end{array}$$

all commute). Then $\mathrm{char}_\Pi(\beta) = \mathrm{char}_\Pi(\alpha)\mathrm{char}_\Pi(\gamma)$ *(when defined, and up to a unit of* Λ*).*

Proof. First we reduce to the case of free modules. Tensoring each row with L we get rows that are isomorphic to split exact sequences of the form

$$0 \longrightarrow L^r \xrightarrow{(1,0)} L^r \oplus L^s \xrightarrow{(0,1)} L^s \longrightarrow 0.$$

Arguing as in Lemma 5.1 we can take sublattices and superlattices of the images of the modules in these vector spaces and construct maps so that we have the following diagram,

$$\begin{array}{ccccccccc} 0 & \longrightarrow & \Lambda^{r'} & \xrightarrow{(1,0)} & \Lambda^{r'} \oplus \Lambda^{s'} & \xrightarrow{(0,1)} & \Lambda^{s'} & \longrightarrow & 0 \\ & & \downarrow & & \downarrow & & \downarrow & & \\ 0 & \longrightarrow & A' & \longrightarrow & B' & \longrightarrow & C' & \longrightarrow & 0 \\ & & \Downarrow{\scriptstyle\alpha} & & \Downarrow{\scriptstyle\beta} & & \Downarrow{\scriptstyle\gamma} & & \\ 0 & \longrightarrow & A & \longrightarrow & B & \longrightarrow & C & \longrightarrow & 0 \\ & & \downarrow & & \downarrow & & \downarrow & & \\ 0 & \longrightarrow & \Lambda^{r} & \xrightarrow{(1,0)} & \Lambda^{r} \oplus \Lambda^{s} & \xrightarrow{(0,1)} & \Lambda^{s} & \longrightarrow & 0 \end{array}$$

with all vertical maps having torsion kernel and cokernel. By the definition of char_Π, we need now only consider free modules. The definition in this case gives the char_Π's as the determinants of the vertical maps in the commutative diagram

$$\begin{array}{ccccccccc}
0 \longrightarrow & \Lambda^{r'} & \stackrel{(1,0)}{\longrightarrow} & \Lambda^{r'} \oplus \Lambda^{s'} & \stackrel{(0,1)}{\longrightarrow} & \Lambda^{s'} & \longrightarrow 0 \\
& \big\downarrow \bar{\alpha} & & \big\downarrow \bar{\beta} & & \big\downarrow \bar{\gamma} & \\
0 \longrightarrow & N^r \otimes V & \stackrel{(1,0)}{\longrightarrow} & (N^r \oplus N^s) \otimes V & \stackrel{(0,1)}{\longrightarrow} & N^s \otimes V & \longrightarrow 0 \\
& \big\downarrow 1\otimes\Pi & & \big\downarrow 1\otimes\Pi & & \big\downarrow 1\otimes\Pi & \\
0 \longrightarrow & N^r & \stackrel{(1,0)}{\longrightarrow} & N^r \oplus N^s & \stackrel{(0,1)}{\longrightarrow} & N^s & \longrightarrow 0.
\end{array}$$

The result follows from elementary linear algebra.

Theorem 6.2.

(a) *If $f : X \Longrightarrow Y$ and $g : Y \Longrightarrow Z$ are semi-morphisms with respect to two relation sets $\mathcal{R}$ and $\mathcal{R}'$, then the composite $gf : X \Longrightarrow Z$ is a semi-morphism with respect to another relation set $\mathcal{R}''$*

(b) *The scalar solution space of $\mathcal{R}''$ is the tensor product of the scalar solution spaces for $\mathcal{R}$ and $\mathcal{R}'$.*

(c) *If f is a Λ-morphism we can take $\mathcal{R}'' = \mathcal{R}'$, If g is a Λ-morphism we can take $\mathcal{R}'' = \mathcal{R}$.*

(d) $\mathrm{char}_{\Pi'\otimes\Pi}(gf) = \mathrm{char}_{\Pi'}(g)\mathrm{char}_\Pi(f)$.

Proof.
Pick an $x \in X$. Let $g_m f_n(x)$ denote the element $\mathrm{pr}^Z_{m,k} g_m \mathrm{pr}^Y_{n,m} f_n \mathrm{pr}^X_n(x) \in Z_k$, where $n \geq m \geq k$. As m and n vary, these elements satisfy the various relations in $\mathcal{R}$ and $\mathcal{R}'$. We wish to find enough relationships between the elements with $m = n$. Tensoring Z_k with F gives a finite dimensional vector space, and elements $g_m f_n(x) \otimes 1$ satisfying relations $\mathcal{R}$ and $\mathcal{R}'$. By decomposing into one dimensional subspaces we are left with the problem of constucting relationships between $a_{nn} \in F$ when we know $\{a_{mn} \in F : n \geq m \geq k\}$ satisfy relations $\mathcal{R}$ in the index n and $\mathcal{R}'$ in the index m when defined. Fixing m, the sequence $(a_{mn})_{n\geq m}$ gives an element of $V^{\geq m}$. but this is isomorphic to V, so we can extend the definition of a_{mn} to all n. Similarly, fixing n we can now vary m over all $m \geq k$, and they must satisfy the conditions of $\mathcal{R}'$ (each relation in $\mathcal{R}'$ involves only finitely many m, and the a_{mn} must satisfy these relations for sufficiently large n. Since a_{mn} for large n determine those for small n in a linear manner, the linear relations must hold for all n.). It is now clear that the elements a_{mn} determine an element of $V \otimes V'$ which is a finite dimensional F-vector space. Any set of more than $\dim(V \otimes V')$ terms of the form a_{nn} must be linearly dependent by a relation of the form $\sum r_n a_{nn} = 0$ with $r_n \in \mathcal{O}_K$. Now $\sum r_n g_n f_n(X)$ must be p-torsion and finitely generated, so by multiplying r_n by a power of p we can assume $\sum r_n g_n f_n(x) = \sum r_n (gf)_n = 0$ for all X. Constuct $\mathcal{R}''$ out of all the relations of this form. $\mathcal{R}''$ clearly contains enough relations to determine $V \otimes V'$ and is an admissible relation set with scalar solution space $V \otimes V'$. gf is a semi-morphism with respect to $\mathcal{R}''$ by construction. This proves parts (a) and (b). Part (c) is clear. It remains to prove part (d). First note that without loss of generality we can assume that X and Z are free. I will show that

we can also assume Y is free. Let $Y \xrightarrow{\delta} \Lambda^r$ be any map with torsion kernel and cokernel, then by an argument similar to those in Lemma 5.1 we can constuct a map $\Lambda^r \xrightarrow{\eta} Y$ with $\eta\delta$ acting on Y as multiplication by some non-zero $x \in \Lambda$. By replacing f by δf, g by $g\eta$ and gf by xgf, we may assume Y is free. The result for free modules will follow from the definition of char provided we can show that $\overline{gf}_{\Pi' \otimes \Pi} = \bar{g}_{\Pi'} \bar{f}_{\Pi}$, however this is clear from the construction of these maps and the identification of $V \otimes V'$ as doubly indexed sequences above.

References

[C] P. Colmez, *Theorie d'Iwasawa des Representations de de Rham d'un Corps Local*, to be published.

[K] H. Knospe, *Iwasawa-theory of abelian varieties at primes of non-ordinary reduction*, Manuscripta math., **87** (1995), 225–258.

[PR] B. Perrin-Riou, *Theorie d'Iwasawa p-adique locale et globale*, Invent. Math., **99(2)** (1990), 247–292.

Conference on Elliptic Curves and Modular Forms
Hong Kong, December 18-21, 1993

On the symmetric square of a modular elliptic curve

J. COATES
A. SYDENHAM
DEPT. OF PURE MATHEMATICS AND MATHEMATICAL STATISTICS
UNIVERSITY OF CAMBRIDGE
16 MILL LANE, CAMBRIDGE CB2 1SB ENGLAND

1. Introduction. Let E be a modular elliptic curve over $\mathbb{Q}$, and write $\mathrm{Sym}^2(E)$ for the 3-dimensional motive whose l-adic realisations are the Galois representations $\mathrm{Sym}^2(H_l^1(E))$, where

$$H_l^1(E) = \mathrm{Hom}(V_l(E), \mathbb{Q}_l),$$

and, as usual, $V_l(E)$ is the tensor product with $\mathbb{Q}_l$ over $\mathbb{Z}_l$ of the Tate module $T_l(E)$ (see §2). Apart from the Tate modules of roots of unity, this is perhaps the simplest of all genuine motives. Interest in it has been heightened by the key role it plays in Wiles' [20] work on Taniyama-Weil conjecture. The work of Bloch and Kato [1] enables us to define two analogues for $\mathrm{Sym}^2(E)$ of the classical Selmer group for E; these Selmer groups of $\mathrm{Sym}^2(E)$ are certain subgroups of

$$H^1(\mathbb{Q}, \mathrm{Sym}^2(E_{p^\infty})(n)) \qquad (n = -1, 0)$$

where $\mathrm{Sym}^2(E_{p^\infty}) = \varinjlim \mathrm{Sym}^2(E_{p^k})$, and E_{p^k} denotes the Galois module of p^k-division points on E (see §3 for the precise definition). It can easily be shown (see Proposition 3.5) that the finiteness of one of these two Selmer groups is equivalent to the finiteness of the other one. In fact, it has been conjectured for about ten years that both of these Selmer groups are finite for all primes p. This conjecture was proven for all but a finite number of primes p for each modular elliptic curve E by Flach [7] and Wiles [20]. Their results have subsequently been extended by Diamond [5] and Langer [11]. The present note does not throw any new light on the finiteness of these Selmer groups, but rather investigates some of the cohomological consequences of this finiteness. Our results are parallel to those established in [4] for the elliptic curve E itself when the L-function of E has a zero at $s = 1$ of order ≤ 1. Our arguments are well known to the experts, but we thought that there might be some interest in giving them in some detail. We also derive some simple consequences of the finiteness of the Selmer groups over $\mathbb{Q}$ for the Iwasawa modules given by the analogues of these Selmer groups over the cyclotomic $\mathbb{Z}_p$-extension of $\mathbb{Q}$. The present article is a slightly modified, and hopefully improved, version of the one given in the first edition of this volume.

2. The critical twists of $\mathrm{Sym}^2(E)$ and their local behaviour. Let E be an elliptic curve defined over $\mathbb{Q}$. We begin by briefly recalling the definition of the motive $\mathrm{Sym}^2(E)$ and its two Tate twists which are critical. We then derive a number of basic local properties of the two critical twists.

Let l be a prime number. For each $n \geq 1$, denote the group of l^n-division points by E_{l^n} and set

$$T_l(E) = \varprojlim E_{l^n}, \qquad V_l(E) = T_l(E) \otimes \mathbb{Q}_l.$$

These spaces are endowed with a canonical $G_{\mathbb{Q}}$-action, and the Weil pairing gives a canonical isomorphism of $G_{\mathbb{Q}}$-modules

$$V_l(E) \xrightarrow{\sim} \mathrm{Hom}(V_l(E), \mathbb{Q}_l(1)). \tag{2.1}$$

here $\mathbb{Q}_l(1) = (\varprojlim \mu_{l^n}) \otimes \mathbb{Q}_l$, where μ_{l^n} denotes the group of l^n-th roots of unity.

Let $\mathrm{Sym}^2(E)$ denote the motive whose l-adic realisations are given by $\mathrm{Sym}^2(H_l^1(E))$, where $H_l^1(E) = \mathrm{Hom}(V_l(E), \mathbb{Q}_l)$. It has just the two critical Tate twists

$$A = (\mathrm{Sym}^2(E))(1), \qquad B = (\mathrm{Sym}^2(E))(2). \tag{2.2}$$

By (2.1) it can easily be seen that the l-adic realisations of these motives are given by

$$H_l(A) = (\mathrm{Sym}^2(V_l(E)))(-1), \qquad H_l(B) = (\mathrm{Sym}^2(V_l(E))), \tag{2.3}$$

and that $\check{A} = A$, where $\check{A}$ denotes the dual motive of A. Hence $B = \check{A}(1)$ and so A and B correspond under the functional equation. From the fact that $d^+(V_l(E)) = d^-(V_l(E)) = 1$, we have that

$$d^+(H_l(A)) = d^-(H_l(B)) = 1, \qquad d^-(H_l(A)) = d^+(H_l(B)) = 2, \tag{2.4}$$

where $d^{\pm}(V)$ denote the dimensions of the subspaces of the vector space V on which complex conjugation acts like ± 1. The vector spaces $H_l(A)$ and $H_l(B)$ contain the $G_{\mathbb{Q}}$-invariant lattices

$$T_l(A) = (\mathrm{Sym}^2(T_l(E)))(-1), \qquad T_l(B) = (\mathrm{Sym}^2(T_l(E))), \tag{2.5}$$

and so we can form the discrete, divisible, torsion $G_{\mathbb{Q}}$-modules

$$\mathfrak{A}_{l^\infty} = H_l(A)/T_l(A), \qquad \mathfrak{B}_{l^\infty} = H_l(B)/T_l(B). \tag{2.6}$$

The Weil pairing (2.1) now gives rise to the canonical isomorphism of $G_{\mathbb{Q}}$-modules

$$\mathfrak{A}_{l^n} \xrightarrow{\sim} \mathrm{Hom}(\mathfrak{B}_{l^n}, \mu_{l^n}), \tag{2.7}$$

where $\mathfrak{A}_{l^n}$ (resp. $\mathfrak{B}_{l^n}$) denotes the kernel of multiplication by l^n in $\mathfrak{A}_{l^\infty}$ (resp. $\mathfrak{B}_{l^\infty}$).

Now for each finite prime v of $\mathbb{Q}$, let $L_v(\mathrm{Sym}^2(E), s)$ be the Euler factor at v of the L-function of the motive $\mathrm{Sym}^2(E)$. We refer the reader to **[2]**, where $L_v(\mathrm{Sym}^2(E), s)$ is calculated explicitly for all v. Of course, the Euler factors for A and B are then given by

$$L_v(A, s) = L_v(\mathrm{Sym}^2(E), s+1), \qquad L_v(B, s) = L_v(\mathrm{Sym}^2(E), s+2),$$

and the next lemma is then plain from the explicit formulae of **[2]**.

Lemma 2.1.
(i) *$L_v(A,s)$ has a pole at $s=0$ for all but finitely many v, and this occurs, in particular, when $\mathrm{Sym}^2(E)$ has good reduction at v.*
(ii) *If $L_v(A,s)$ does have a pole at $s=0$, then it is simple.*
(iii) *$L_v(B,s)$ is holomorphic at $s=0$ for all finite primes v.*

We now fix a prime number p. Let v be any finite prime of $\mathbb{Q}$ different from p. We write I_v for the inertial subgroup of $G_v = \mathrm{Gal}(\overline{\mathbb{Q}}_v/\mathbb{Q}_v)$, and Frob_v for the Frobenius automorphism in G_v/I_v. Let M denote either the motive A or the motive B. Following Bloch-Kato, we then define

$$H^1_f(\mathbb{Q}_v, H_p(M)) = H^1(G_v/I_v, H_p(M)^{I_v}) \qquad (v \neq p). \tag{2.8}$$

Lemma 2.2. *Suppose $v \neq p$. Then*
(i) $H^1_f(\mathbb{Q}_v, H_p(B)) = 0$,
(ii) $\dim_{\mathbb{Q}_p} H^1_f(\mathbb{Q}_v, H_p(A)) = 1$ *(resp. 0) if the Euler factor $L_v(A,s)$ has (resp. does not have) a pole at $s=0$.*

Proof. Since G_v/I_v is topologically generated by Frob_v^{-1}, we have

$$H^1(G_v/I_v, H_p(M)^{I_v}) = H_p(M)^{I_v}/(1-\mathrm{Frob}_v^{-1})H_p(M)^{I_v}.$$

On the other hand, as $v \neq p$, we have

$$L_v(M,s) = \det(1-\mathrm{Frob}_v^{-1}.v^{-s} \mid H_p(M)^{I_v})^{-1},$$

and so the required result follows immediately from Lemma 2.1. □

We next consider the case when $v = p$. We recall that, in this case, Bloch and Kato define

$$H^1_f(\mathbb{Q}_p, H_p(M)) = \mathrm{Ker}(H^1(\mathbb{Q}_p, H_p(M)) \longrightarrow H^1(\mathbb{Q}_p, H_p(M) \otimes_{\mathbb{Q}_p} B_{crys})),$$

where B_{crys} is the ring defined by Fontaine (see [**1**]). Prior to calculating the dimension of $H^1_f(\mathbb{Q}_p, H_p(M))$, we need the following basic lemma.

Lemma 2.3.
(i) $H^0(\mathbb{Q}_p, H_p(B)) = 0$,
(ii) $H^0(\mathbb{Q}_p, H_p(A)) = 0$, *unless E has potentially ordinary reduction and admits complex multiplication. In this latter case* $\dim_{\mathbb{Q}_p} H^0(\mathbb{Q}_p, H_p(A)) = 1$.

Proof. It will be convenient to introduce the following notation. Let $\chi\colon G_p \longrightarrow \mathbb{Z}_p^\times$ be the character giving the action of G_p on $\mathbb{Q}_p(1)$. By the irreducibility of the cyclotomic equation over $\mathbb{Q}_p$, χ is surjective. Let

$$\rho\colon G_p \longrightarrow \mathrm{GL}(V_p(E)), \qquad \varphi\colon G_p \longrightarrow \mathrm{GL}(\mathrm{Sym}^2(V_p(E)))$$

be the natural representations arising from the action of G_p on $V_p(E)$ and $\mathrm{Sym}^2(V_p(E))$ respectively. By the Weil pairing, we know that $\det(\rho) = \chi$. For each $\sigma \in G_p$, we write $\lambda(\sigma)$, $\mu(\sigma)$ for the eigenvalues in $\overline{\mathbb{Q}}_p$ of $\rho(\sigma)$. Recall that if $\{e_1, e_2\}$ is a $\mathbb{Q}_p$-basis of $V_p(E)$, then a $\mathbb{Q}_p$-basis of $\mathrm{Sym}^2(V_p(E))$ is given by the set $\{e_1 \otimes e_1, e_2 \otimes e_2, e_1 \otimes e_2 + e_2 \otimes e_1\}$.

We first consider case (i). Suppose that, on the contrary, there does exist $z \neq 0$ in $\mathrm{Sym}^2(V_p(E))$ such that $\varphi(\sigma)(z) = z$ for all $\sigma \in G_p$. This means that 1 must be an eigenvalue of $\varphi(\sigma)$ for every σ in G_p. But linear algebra tells us that the eigenvalues with multiplicity of $\varphi(\sigma)$ are $\lambda(\sigma)^2$, $\mu(\sigma)^2$, $\chi(\sigma) = \lambda(\sigma)\mu(\sigma)$. Now choose σ_0 in G_p such that $\chi(\sigma_0)$ is not a root of unity. This is clearly possible because χ is surjective. Since $\chi(\sigma_0) \neq 1$, we must have one of the other eigenvalues of $\varphi(\sigma_0)$ equal to 1, say $\lambda(\sigma_0)^2 = 1$. Taking $\tau = \sigma_0^2$, we conclude that we must have one of the eigenvalues of $\rho(\tau)$ equal to 1, say $\lambda(\tau) = 1$. Since $\chi(\tau) \neq 1$, we can then choose a basis $\{e_1, e_2\}$ of $V_p(E)$ consisting of eigenvectors for the eigenvalues 1, $\chi(\tau)$, respectively. Writing z as a linear combination of the basis elements $e_1 \otimes e_1$, $e_2 \otimes e_2$, $e_1 \otimes e_2 + e_2 \otimes e_1$, and using the fact that $\chi(\tau)^2 \neq 1$, we deduce immediately from the equation $\varphi(\tau)(z) = z$, that z is a scalar multiple of $e_1 \otimes e_1$. Hence $\varphi(\sigma)$ must fix $z = e_1 \otimes e_1$ for all σ in G_p. Now for any such σ we let

$$\rho(\sigma) = \begin{pmatrix} a(\sigma) & b(\sigma) \\ c(\sigma) & d(\sigma) \end{pmatrix} \tag{2.9}$$

be the matrix of $\rho(\sigma)$ relative to the basis $\{e_1, e_2\}$. Then the equation $\varphi(\sigma)(z) = z$ yields immediately $a(\sigma)^2 = 1$ and $c(\sigma) = 0$ for all $\sigma \in G_p$. Then $\sigma \mapsto a(\sigma)$ defines a quadratic character of G_p, and we let L be the fixed field of the kernel of this character. It is then clear that $G(\overline{\mathbb{Q}}/L)$ fixes the 1-dimensional subspace of $V_p(E)$ which is generated by e_1. But this is impossible, since the p-primary subgroup of $E(L)$ is finite. This completes the proof of (i).

In preparation for the proof of (ii), we make some remarks on linear algebra. Let σ_0 be any element of G_p such that the eigenvalues $\lambda(\sigma_0)$, $\mu(\sigma_0)$ of $\rho(\sigma_0)$ are distinct elements of $\mathbb{Q}_p$. Note that such a σ_0 always exists - for example, take σ_0 to be any element of G_p such that $\chi(\sigma_0)$ is not a square in $\mathbb{Q}_p$, which is always possible since $\chi(G_p) = \mathbb{Z}_p^\times$. Let $\{e_1, e_2\}$ be the basis of $V_p(E)$ consisting of eigenvectors for $\lambda(\sigma_0)$, $\mu(\sigma_0)$, respectively. Put $W_i = \mathbb{Q}_p e_i$ $(i = 1, 2)$. Assume now that

$$H^0(G_p, H_p(A)) \neq 0. \tag{2.10}$$

We claim that W_1 and W_2 are then stable under the action of G_p and that $V_p(E) = W_1 \oplus W_2$ as G_p-modules. Indeed (2.10) implies that there exists $z \neq 0$ in $\mathrm{Sym}^2(V_p(E))$ such that

$$\varphi(\sigma)(z) = \chi(\sigma)z \quad \text{for all} \quad \sigma \in G_p. \tag{2.11}$$

Writing z as a $\mathbb{Q}_p$-linear combination of the basis elements $e_1 \otimes e_1$, $e_2 \otimes e_2$, $e_1 \otimes e_2 + e_2 \otimes e_1$, and applying (2.11) with $\sigma = \sigma_0$, we deduce immediately that z is a scalar multiple of $e_1 \otimes e_2 + e_2 \otimes e_1$ since $\lambda(\sigma_0) \neq \mu(\sigma_0)$. Hence we can take $z = e_1 \otimes e_2 + e_2 \otimes e_1$. Applying (2.11) for any σ in G_p, we then deduce that if (2.9) denotes the matrix of $\rho(\sigma)$ relative to $\{e_1, e_2\}$, then necessarily $b(\sigma) = c(\sigma) = 0$. This proves that $V_p(E)$ is a direct sum of the G_p-subspaces W_1 and W_2, as claimed.

Let j_E denote the j-invariant of E. If $\mathrm{ord}_p(j_E) < 0$, there is a quadratic extension L of $\mathbb{Q}_p$ such that E is isomorphic over L to the Tate curve E_q for a suitable q in $\mathbb{Q}_p^\times$. But $\mathrm{Sym}^2(V_p(E))$ does not change if we replace E by a quadratic twist, and so we can assume that $E = E_q$. But then we have the exact sequence of G_p-modules

$$0 \to \mathbb{Q}_p(1) \to V_p(E_q) \to \mathbb{Q}_p \to 0.$$

Hence the eigenvalues of $\rho(\sigma)$ are equal to $\chi(\sigma)$ and 1. Choose σ_0 to be any element of G_p such that $\chi(\sigma_0) \neq 1$. If (2.10) were valid, then it would follow that the above exact sequence would split as an exact sequence of G_p-modules, since we could take e_1 to be any non-zero element of $\mathbb{Q}_p(1)$. But it is well know (see **[15]**, A.1.2) that this exact sequence of G_p-modules does not split, and so (2.10) cannot hold in this case.

Suppose next that E has potential good ordinary reduction at p. This means that there exists a finite extension L of $\mathbb{Q}_p$ such that E has good ordinary reduction over L. Write $\hat{E}$ for the formal group and $\tilde{E}$ for the reduction of E over L, and let $G_L = G(\overline{\mathbb{Q}_p}/L)$. Then, in an obvious notation, we have the exact sequence of G_L-modules

$$0 \to V_p(\hat{E}) \to V_p(E) \to V_p(\tilde{E}) \to 0, \tag{2.12}$$

where both $V_p(\hat{E})$ and $V_p(\tilde{E})$ have $\mathbb{Q}_p$-dimension equal to 1. Moreover, $V_p(\tilde{E})$ is an unramified G_L-module. Thus we can choose σ_0 to be any element of the inertial subgroup of G_L such that $\chi(\sigma_0) \neq 1$, and then take e_1 to be any non-zero element of $V_p(\hat{E})$. Hence, if (2.10) is valid, we conclude that (2.12) must split, even as an exact sequence of G_p-modules. But then by **[15]**, A.2.4, E must have complex multiplication. Conversely, if E admits complex multiplication by an order in an imaginary quadratic field K, and has potential good ordinary reduction at p, then it is well known that p must split in K, and

$$V_p(E) = V_1 \oplus V_2,$$

where V_1 and V_2 are G_p-submodules of $\mathbb{Q}_p$-dimension 1, and where G_p acts on both V_1 and V_2 by characters which are not of finite order. We deduce immediately from this decomposition that $\dim_{\mathbb{Q}_p}(H^0(G_p, H_p(A))) = 1$.

Suppose next that E has potential supersingular reduction at p. This means that there exists a finite Galois extension L of $\mathbb{Q}_p$ such that E has good supersingular reduction over L. Let $\hat{E}$ be the formal group of E over the ring of integers O_L of L. Then $\hat{E}$ has height 2 and $V_p(E) = V_p(\hat{E})$. Now the Weil pairing gives an isomorphism of G_p-modules

$$\mathrm{Hom}(V_p(E), V_p(E)) \xrightarrow{\sim} H_p(A) \oplus \mathbb{Q}_p.$$

Let $\mathrm{End}(\hat{E})$ denote the ring of endomorphisms of $\hat{E}$ which are defined over O_L. By the main result of **[19]**, we have

$$\mathrm{End}(\hat{E}) \otimes \mathbb{Q}_p \xrightarrow{\sim} \mathrm{Hom}(V_p(E), V_p(E))^{G_L} \xrightarrow{\sim} H_p(A)^{G_L} \oplus \mathbb{Q}_p$$

where $G_L = G(\overline{\mathbb{Q}_p}/L)$. But, as E is defined over $\mathbb{Q}_p$, $G(L/\mathbb{Q}_p)$ operates on $\mathrm{End}(\hat{E})$ via its action on the coefficients of the power series defining an endomorphism, and the above maps plainly preserve this action. Hence, taking $G(L/\mathbb{Q}_p)$-invariants, we obtain an isomorphism

$$\mathrm{End}(\hat{E})^{G(L/\mathbb{Q}_p)} \otimes \mathbb{Q}_p \xrightarrow{\sim} H_p(A)^{G_p} \oplus \mathbb{Q}_p.$$

But Lubin [12] has shown that there is an injective homomorphism $c: \mathrm{End}(\hat{E}) \hookrightarrow O_L$ obtained by mapping an endomorphism to the coefficient of t in its power series development, where t denotes any parameter for $\hat{E}$. This map c commutes with the action of $G(L/\mathbb{Q}_p)$. Hence $\mathrm{End}(\hat{E})^{G(L/\mathbb{Q}_p)}$ has rank 1 over $\mathbb{Z}_p$, and we deduce that $H_p(A)^{G_p} = 0$ in this case. This completes the proof of Lemma 2.3. □

Lemma 2.4.

(i) $\dim_{\mathbb{Q}_p}(H^1_f(\mathbb{Q}_p, H_p(A))) = 1$, *unless E has potentially ordinary reduction and admits complex multiplication. In this latter case* $\dim_{\mathbb{Q}_p} H^1_f(\mathbb{Q}_p, H_p(A)) = 2$.

(ii) $\dim_{\mathbb{Q}_p}(H^1_f(\mathbb{Q}_p, H_p(B))) = 2$.

Proof. Let M denote either the motive A or the motive B. Let B_{DR} be Fontaine's field of p-adic periods, and, as usual, we define

$$DR_p(M) = (H_p(M) \otimes_{\mathbb{Q}_p} B_{DR})^{G_p}.$$

We obtain a $\mathbb{Q}_p$-vector space, which is endowed with the filtration $\mathrm{Fil}^k\, DR_p(M)$ ($k \in \mathbb{Z}$) coming from the canonical filtration on B_{DR}. Now it is well known that $H_p(M)$ is a de Rham representation for every prime p, i. e. $DR_p(M)$ has dimension 3 over $\mathbb{Q}_p$. Thus the formula of Bloch-Kato (see [1], p. 355) gives

$$\begin{aligned}(2.13)\quad & \dim_{\mathbb{Q}_p}(H^1_f(\mathbb{Q}_p, H_p(M))) \\ &= \dim_{\mathbb{Q}_p}(DR_p(M)/\mathrm{Fil}^0\, DR_p(M)) + \dim_{\mathbb{Q}_p} H^0(\mathbb{Q}_p, H_p(M)).\end{aligned}$$

Hence, in view of Lemma 2.3, it remains to compute the first dimension on the right of this formula. This is done most simply by using the fact that Fontaine's stronger de Rham conjecture is true for M. We recall that $H_{DR}(M)$ is a 3-dimensional vector space over $\mathbb{Q}$, which is endowed with the Hodge filtration $\mathrm{Fil}^k\, H_{DR}(M)$ ($k \in \mathbb{Z}$). Then there is an isomorphism of $\mathbb{Q}_p$-vector spaces

$$(2.14)\qquad \theta_p: DR_p(M) \xrightarrow{\sim} H_{DR}(M) \otimes_{\mathbb{Q}} \mathbb{Q}_p,$$

which preserves the filtrations. Hence we conclude from (2.13) and (2.14) that

$$\begin{aligned}(2.15)\quad & \dim_{\mathbb{Q}_p}(H^1_f(\mathbb{Q}_p, H_p(M))) \\ &= \dim_{\mathbb{Q}_p}(H_{DR}(M)/\mathrm{Fil}^0\, H_{DR}(M)) + \dim_{\mathbb{Q}_p} H^0(\mathbb{Q}_p, H_p(M)).\end{aligned}$$

Now M is critical at $s = 0$, and it well known (see [3], p. 156) that this implies that the first term on the right hand side of (2.15) is equal to $d^+(M)$. Since $d^+(B) = 2$ and $d^+(A) = 1$, this completes the proof of Lemma 2.4. □

3. Global cohomology over $\mathbb{Q}$. In order to simplify the discussion we will assume for the rest of the paper that the prime p is odd. Let S denote any finite set of non-archimedean primes of $\mathbb{Q}$ containing p and all primes not equal to p where the $G_{\mathbb{Q}}$-representation $H_p(A)$ is ramified (or equivalently the $G_{\mathbb{Q}}$-representation $H_p(B)$ is ramified). We write $\mathbb{Q}_S$ for the maximal extension of $\mathbb{Q}$ unramified outside S and ∞, and put $G_S = \mathrm{Gal}(\mathbb{Q}_S/\mathbb{Q})$. Then $\mathfrak{A}_{p^\infty}$ and $\mathfrak{B}_{p^\infty}$, as defined by (2.6), are discrete G_S-modules, and the aim of this section will be to study their G_S-cohomology. The principal tool will be the Poitou-Tate sequence for finite G_S-modules (see, for example, [13], p. 70), together with an important modification of this sequence, which was first pointed out by Cassels in the case of elliptic curves. This modified sequence is very well known to the experts, but we have included a proof here for lack of a good reference.

Let $\mathcal{M}$ be a finite G_S-module of p-power order and define $\mathcal{M}'$ to be $\mathrm{Hom}(\mathcal{M}, \mu_{p^\infty})$. Of course $\mathcal{M}'$ is endowed with its canonical structure as a G_S-module. Assume that for each place $v \in S$, we are given an arbitrary subgroup $W_v(\mathcal{M})$ of $H^1(\mathbb{Q}_v, \mathcal{M})$. We write $W_v(\mathcal{M})^\perp$ for the orthogonal complement of $W_v(\mathcal{M})$ under the Tate dual pairing

$$H^1(\mathbb{Q}_v, \mathcal{M}) \times H^1(\mathbb{Q}_v, \mathcal{M}') \longrightarrow \mathbb{Q}_p/\mathbb{Z}_p. \tag{3.1}$$

We now define the map $\varphi_{\mathcal{M}}$ to be the restriction map from $H^1(G_S, \mathcal{M})$ to $\bigoplus_{v \in S} H^1(\mathbb{Q}_v, \mathcal{M})$ composed with the canonical surjection

$$\bigoplus_{v \in S} H^1(\mathbb{Q}_v, \mathcal{M}) \longrightarrow \bigoplus_{v \in S} \frac{H^1(\mathbb{Q}_v, \mathcal{M})}{W_v(\mathcal{M})},$$

and we define $\varphi_{\mathcal{M}'}$ in an entirely similar manner, except that we replace $W_v(\mathcal{M})$ by its orthogonal complement $W_v(\mathcal{M})^\perp$. If X is a p-primary abelian group, we write $\hat{X} = \mathrm{Hom}(X, \mathbb{Q}_p/\mathbb{Z}_p)$ for its Pontrjagin dual.

Theorem 3.1. (Cassels-Poitou-Tate). *We have a canonical exact sequence*

(3.2)
$$0 \to \mathrm{Ker}(\varphi_{\mathcal{M}}) \to H^1(G_S, \mathcal{M}) \xrightarrow{\varphi_{\mathcal{M}}} \bigoplus_{v \in S} \frac{H^1(\mathbb{Q}_v, \mathcal{M})}{W_v(\mathcal{M})} \to \mathrm{Ker}(\varphi_{\mathcal{M}'})^\wedge$$
$$\to H^2(G_S, \mathcal{M}) \to \bigoplus_{v \in S} H^2(\mathbb{Q}_v, \mathcal{M}) \to H^0(G_S, \mathcal{M}')^\wedge \to 0.$$

Proof. We start with the classical Poitou-Tate exact sequence (see [**13**], Theorem 4.10)

(3.3)
$$H^1(G_S, \mathcal{M}) \xrightarrow{\beta_{\mathcal{M}}} \bigoplus_{v \in S} H^1(\mathbb{Q}_v, \mathcal{M}) \xrightarrow{\delta_{\mathcal{M}}} H^1(G_S, \mathcal{M}')^\wedge$$
$$\longrightarrow H^2(G_S, \mathcal{M}) \longrightarrow \bigoplus_{v \in S} H^2(\mathbb{Q}_v, \mathcal{M}) \longrightarrow H^0(G_S, \mathcal{M}')^\wedge \longrightarrow 0,$$

here $\beta_{\mathcal{M}}$ is the natural restriction map, and $\delta_{\mathcal{M}}$ can be identified with the dual of $\beta_{\mathcal{M}'}$ using Tate local duality (3.1). Now it is easy to see that the sequence (3.2) exists granted (3.3), and all that has to be checked about exactness is obvious except at the term $\underset{v\in S}{\oplus} H^1(\mathbb{Q}_v, \mathcal{M})/W_v(\mathcal{M})$. Simple diagram chasing from (3.3) yields the exact sequence

$$\begin{aligned}(3.4)\qquad & 0 \to \operatorname{Ker}(\beta_{\mathcal{M}'}) \to \operatorname{Ker}(\varphi_{\mathcal{M}'}) \xrightarrow{\alpha} \underset{v\in S}{\oplus} W_v(\mathcal{M}') \\ & \xrightarrow{\gamma} \operatorname{Coker}(\beta_{\mathcal{M}'}) \to \operatorname{Coker}(\varphi_{\mathcal{M}'}) \to 0.\end{aligned}$$

We are required to prove that $\operatorname{Im}(\alpha) = \operatorname{Ker}(\hat{\varphi}_{\mathcal{M}})$. But the above sequence (3.4) shows that $\operatorname{Im}(\alpha) = \operatorname{Ker}(\gamma)$. But, by (3.3), we have a factorisation of $\hat{\varphi}_M$ as a composition

$$\hat{\varphi}_{\mathcal{M}} : W_v(\mathcal{M}') \xrightarrow{\gamma} \operatorname{Coker}(\beta_{\mathcal{M}'}) \xrightarrow{i} H^1(G_S, \mathcal{M})^\wedge,$$

where i is an injection. Hence $\operatorname{Ker}(\gamma) = \operatorname{Ker}(\hat{\varphi}_{\mathcal{M}})$, and the proof of Theorem 3.1 is complete. □

We now apply Theorem 3.1 where $\mathcal{M}_0$ denotes either $\mathfrak{A}_{p^n}$ or $\mathfrak{B}_{p^n}$ (see (2.6)). It is more elegant in this case to denote $\mathcal{M}$ by $\mathfrak{M}_{p^n}$, and to write $\mathfrak{R}_{p^n} = \operatorname{Hom}(\mathfrak{M}_{p^n}, \mu_{p^n})$. We denote M for the motive corresponding to the choice of A or B made in defining $\mathfrak{M}_{p^n}$, and R for the remaining motive of A or B. We must now suppose that, for each $v \in S$ and each integer $n \geq 1$, we are given a subgroup $W_v(\mathfrak{M}_{p^n})$ of $H^1(\mathbb{Q}_v, \mathfrak{M}_{p^n})$ having the following functorial properties as n varies. Firstly, the map induced by the inclusion of $\mathfrak{M}_{p^n}$ into $\mathfrak{M}_{p^{n+1}}$ sends $W_v(\mathfrak{M}_{p^n})$ into $W_v(\mathfrak{M}_{p^{n+1}})$, and secondly the map induced by multiplication by p from $\mathfrak{M}_{p^{n+1}}$ into $\mathfrak{M}_{p^n}$ sends $W_v(\mathfrak{M}_{p^{n+1}})$ into $W_v(\mathfrak{M}_{p^n})$. For each $v \in S$, let $W_v^\perp(\mathfrak{R}_{p^n})$ be the orthogonal complement of $W_v(\mathfrak{M}_{p^n})$ under the local Tate pairing at v. We can therefore define

$$W_v(\mathfrak{M}_{p^\infty}) = \varinjlim W_v(\mathfrak{M}_{p^n}), \quad W_v^\perp(\mathfrak{R}_{p^\infty})) = \varinjlim W_v^\perp(\mathfrak{R}_{p^n})$$

$$W_v(T_p(M)) = \varprojlim W_v(\mathfrak{M}_{p^n}), \quad W_v^\perp(T_p(R)) = \varprojlim W_v^\perp(\mathfrak{R}_{p^n}).$$

Thus for M we can define the Selmer groups by the exactness of

$$(3.5)\quad 0 \longrightarrow \mathcal{S}(\mathbb{Q}, T_p(M), W) \longrightarrow H^1(G_S, T_p(M)) \longrightarrow \underset{v\in S}{\oplus} \frac{H^1(\mathbb{Q}_v, T_p(M))}{W_v(T_p(M))}$$

$$(3.6)\qquad 0 \longrightarrow \mathcal{S}(\mathbb{Q}, \mathfrak{M}_{p^\infty}, W) \longrightarrow H^1(G_S, \mathfrak{M}_{p^\infty}) \longrightarrow \underset{v\in S}{\oplus} \frac{H^1(\mathbb{Q}_v, \mathfrak{M}_{p^\infty})}{W_v(\mathfrak{M}_{p^\infty})},$$

and for R we can define similar Selmer groups by replacing the local conditions W by the local conditions $W^\perp$. Taking inductive (resp. projective) limits of the exact sequences (3.2) with $\mathcal{M} = \mathfrak{M}_{p^n}$, and noting that both inductive limits and projective limits are exact because the groups are finite, we obtain:-

Theorem 3.2. *We have the following exact sequences: –*

Discrete version:

$$\begin{aligned} & 0 \longrightarrow \mathcal{S}(\mathbb{Q}, \mathfrak{M}_{p^\infty}, W) \longrightarrow H^1(G_S, \mathfrak{M}_{p^\infty}) \longrightarrow \underset{v\in S}{\oplus} \frac{H^1(\mathbb{Q}_v, \mathfrak{M}_{p^\infty})}{W_v(\mathfrak{M}_{p^\infty})} \\ (3.7)\qquad & \longrightarrow \mathcal{S}(\mathbb{Q}, T_p(R), W^\perp)^\wedge \longrightarrow H^2(G_S, \mathfrak{M}_{p^\infty}) \longrightarrow \underset{v\in S}{\oplus} H^2(\mathbb{Q}_v, \mathfrak{M}_{p^\infty}) \\ & \longrightarrow H^o(G_S, T_p(R))^\wedge \longrightarrow 0.\end{aligned}$$

Compact version:

$$0 \longrightarrow \mathcal{S}(\mathbb{Q}, T_p(M), W) \longrightarrow H^1(G_S, T_p(M)) \longrightarrow \bigoplus_{v \in S} \frac{H^1(\mathbb{Q}_v, T_p(M))}{W_v(T_p(M))}$$

$$(3.8) \quad \longrightarrow \mathcal{S}(\mathbb{Q}, \mathfrak{N}_{p^\infty}, W^\perp)^\wedge \longrightarrow H^2(G_S, T_p(M)) \longrightarrow \bigoplus_{v \in S} H^2(\mathbb{Q}_v, T_p(M))$$

$$\longrightarrow H^o(G_S, \mathfrak{N}_{p^\infty})^\wedge \longrightarrow 0.$$

If X is an abelian group, we write $T_p(X) = \varprojlim (X)_{p^n}$ where $(X)_{p^n}$ denotes the kernel of multiplication by p^n on X.

Lemma 3.3. *We have the exact sequence*

$$(3.9) \quad 0 \longrightarrow \frac{\mathfrak{M}_{p^\infty}(\mathbb{Q})}{\mathfrak{M}_{p^\infty}(\mathbb{Q})_{div}} \longrightarrow \mathcal{S}(\mathbb{Q}, T_p(M), W) \longrightarrow T_p(\mathcal{S}(\mathbb{Q}, \mathfrak{M}_{p^\infty}, W)) \longrightarrow 0.$$

Proof. For each $n \geq 1$, we have the exact sequence

$$(3.10) \qquad 0 \longrightarrow \mathfrak{M}_{p^n} \xrightarrow{i(n)} \mathfrak{M}_{p^\infty} \xrightarrow{p^n} \mathfrak{M}_{p^\infty} \longrightarrow 0.$$

Taking G_v-cohomology for any $v \in S$, gives the exact sequence

$$(3.11) \quad 0 \longrightarrow \frac{\mathfrak{M}_{p^\infty}(\mathbb{Q}_v)}{p^n \mathfrak{M}_{p^\infty}(\mathbb{Q}_v)} \longrightarrow H^1(\mathbb{Q}_v, \mathfrak{M}_{p^n}) \longrightarrow (H^1(\mathbb{Q}_v, \mathfrak{M}_{p^\infty}))_{p^n} \longrightarrow 0.$$

Now passing to the projective limit relative to the maps given by multiplication by p gives the exact sequence

$$(3.12) \quad 0 \longrightarrow \frac{\mathfrak{M}_{p^\infty}(\mathbb{Q}_v)}{\mathfrak{M}_{p^\infty}(\mathbb{Q}_v)_{div}} \longrightarrow H^1(\mathbb{Q}_v, T_p(M)) \longrightarrow T_p(H^1(\mathbb{Q}_v, \mathfrak{M}_{p^\infty})) \longrightarrow 0.$$

Now from (3.11), we clearly have the exact sequence

$$0 \longrightarrow i(n)_*^{-1}(0) \longrightarrow i(n)_*^{-1} W_v(\mathfrak{M}_{p^\infty}) \longrightarrow W_v(\mathfrak{M}_{p^\infty})_{p^n} \longrightarrow 0,$$

where $i(n)_*$ denotes the map on the H^1's induced by $i(n)$, and passing to the projective limit gives

$$(3.13) \qquad 0 \longrightarrow \frac{\mathfrak{M}_{p^\infty}(\mathbb{Q}_v)}{\mathfrak{M}_{p^\infty}(\mathbb{Q}_v)_{div}} \longrightarrow W_v(T_p(M)) \longrightarrow T_p(W_v(\mathfrak{M}_{p^\infty})) \longrightarrow 0.$$

Hence

$$(3.14) \qquad \bigoplus_{v \in S} \frac{H^1(\mathbb{Q}_v, T_p(M))}{W_v(T_p(M))} \xrightarrow{\sim} \bigoplus_{v \in S} \frac{T_p(H^1(\mathbb{Q}_v, \mathfrak{M}_{p^\infty}))}{T_p(W_v(\mathfrak{M}_{p^\infty}))}.$$

Next we take G_S-cohomology of (3.10) to obtain the exact sequence

$$0 \longrightarrow \frac{\mathfrak{M}_{p^\infty}(\mathbb{Q})}{p^n \mathfrak{M}_{p^\infty}(\mathbb{Q})} \longrightarrow H^1(G_S, \mathfrak{M}_{p^n}) \longrightarrow (H^1(G_S, \mathfrak{M}_{p^\infty}))_{p^n} \longrightarrow 0,$$

and so passing to the projective limit gives

$$(3.15)\quad 0 \longrightarrow \frac{\mathfrak{M}_{p^\infty}(\mathbb{Q})}{\mathfrak{M}_{p^\infty}(\mathbb{Q})_{div}} \longrightarrow H^1(G_S, T_p(M)) \longrightarrow T_p(H^1(G_S, \mathfrak{M}_{p^\infty})) \longrightarrow 0.$$

The result now follows by simple diagram chasing from (3.7), (3.8), (3.13) and (3.14), the essential point being that (3.14) shows that the natural map

$$\underset{v\in S}{\oplus} \frac{H^1(\mathbb{Q}_v, T_p(M))}{W_v(T_p(M))} \longrightarrow \underset{v\in S}{\oplus} T_p\left(\frac{H^1(\mathbb{Q}_v, \mathfrak{M}_{p^\infty})}{W_v(\mathfrak{M}_{p^\infty})}\right)$$

is injective. □

We recall that $H^1_f(\mathbb{Q}_v, \mathfrak{M}_{p^\infty})$ is defined to be the image of $H^1_f(\mathbb{Q}_v, H_p(M))$ under the map induced by the canonical surjection from $H_p(M)$ to $\mathfrak{M}_{p^\infty}$. From now on, we will always assume that $W_v(\mathfrak{M}_{p^\infty}) = H^1_f(\mathbb{Q}_v, \mathfrak{M}_{p^\infty})$. Now it is well known that $H^1_f(\mathbb{Q}_v, H_p(M))$ is the orthogonal complement of $H^1_f(\mathbb{Q}_v, H_p(R))$ (see [1]). Finally we define

$$(3.16)\qquad H^1_f(\mathbb{Q}, T_p(M)) = \mathrm{Ker}\left(H^1(G_S, T_p(M)) \longrightarrow \underset{v\in S}{\oplus} \frac{H^1(\mathbb{Q}_v, T_p(M))}{H^1_f(\mathbb{Q}_v, T_p(M))}\right)$$

$$(3.17)\qquad H^1_f(\mathbb{Q}, \mathfrak{M}_{p^\infty}) = \mathrm{Ker}\left(H^1(G_S, \mathfrak{M}_{p^\infty}) \longrightarrow \underset{v\in S}{\oplus} \frac{H^1(\mathbb{Q}_v, \mathfrak{M}_{p^\infty})}{H^1_f(\mathbb{Q}_v, \mathfrak{M}_{p^\infty})}\right)$$

We will write $\mathrm{ck}_{\mathbb{Z}_p} X$ for the $\mathbb{Z}_p$-corank of a discrete $\mathbb{Z}_p$-module X.

Proposition 3.4. *For M equal to A or B, we have*

$$\mathrm{ck}_{\mathbb{Z}_p}(H^1(G_S, \mathfrak{M}_{p^\infty})) = d^-(M) + \delta(M),$$

where the integer $\delta(M)$ is given by $H^2(G_S, \mathfrak{M}_{p^\infty}) \cong (\mathbb{Q}_p/\mathbb{Z}_p)^{\delta(M)}$.

Proof. This is an easy consequence of Tate's global Euler characteristic theorem. Indeed, $H^0(G_S, \mathfrak{B}_{p^\infty})$ is finite because $H^0(\mathbb{Q}_p, \mathfrak{B}_{p^\infty})$ is finite. Similarly, if E does not admit complex multiplication, then $H^0(G_S, \mathfrak{A}_{p^\infty})$ is finite because $H^0(\mathbb{Q}_p, \mathfrak{A}_{p^\infty})$ is finite. On the other hand if E admits complex multiplication then a standard linear algebra argument (see, for example, [**18**], Section 7) shows that if $H^0(G_S, \mathfrak{A}_{p^\infty}) \neq 0$ then

$$\tilde{\rho}_p(G_{\mathbb{Q}}) \subset \left\{ \begin{pmatrix} a & 0 \\ 0 & d \end{pmatrix} \middle| \, a, d \in \mathbb{F}_p \right\},$$

where $\tilde{\rho}_p$ is the natural representation

$$\tilde{\rho}_p \colon G_{\mathbb{Q}} \longrightarrow \mathrm{Aut}(E_p).$$

Now the argument of [17], Proposition 21 shows that

$$t_v \equiv 1 + v (\mathrm{mod}\ p) \qquad \text{for all } v \notin S,$$

where $t_v = 1 + v - A_v$ and A_v is the number of points of $\tilde{E}(v)$ defined over $\mathbb{F}_v$. But if E admits complex multiplication then $t_v = 0$ for a set of primes of density 1/2. Hence, in this case, $H^0(G_S, \mathfrak{A}_{p^\infty}) = 0$. Now a standard argument with the global Euler characteristic theorem then shows that

$$\mathrm{ck}_{\mathbb{Z}_p}(H^1(G_S, \mathfrak{M}_{p^\infty})) = d^-(M) + \delta(M), \tag{3.18}$$

where $\delta(M) = \mathrm{ck}_{\mathbb{Z}_p}(H^2(G_S, \mathfrak{M}_{p^\infty}))$. Also $H^2(G_S, \mathfrak{M}_{p^\infty})$ is divisible because G_S has p-cohomological dimension equal to 2 since $p > 2$. This completes the proof of the proposition. □

Proposition 3.5. *We have*

$$\mathrm{ck}_{\mathbb{Z}_p} H^1_f(\mathbb{Q}, \mathfrak{A}_{p^\infty}) = \mathrm{ck}_{\mathbb{Z}_p} H^1_f(\mathbb{Q}, \mathfrak{B}_{p^\infty}).$$

In particular, $H^1_f(\mathbb{Q}, \mathfrak{A}_{p^\infty})$ *is finite if and only if* $H^1_f(\mathbb{Q}, \mathfrak{B}_{p^\infty})$ *is also finite.*

Proof. As usual we let M denote either A or B. Now, for a cofinitely generated $\mathbb{Z}_p$-module X, $T_p(X)$ is a free $\mathbb{Z}_p$-module of rank equal to the $\mathbb{Z}_p$-corank of X. Hence as $\mathfrak{M}_{p^\infty}(\mathbb{Q})/\mathfrak{M}_{p^\infty}(\mathbb{Q})_{div}$ is finite, Lemma 3.3 implies that

$$\mathrm{ck}_{\mathbb{Z}_p} H^1_f(\mathbb{Q}, \mathfrak{M}_{p^\infty}) = \mathrm{rk}_{\mathbb{Z}_p} H^1_f(\mathbb{Q}, T_p(M)).$$

Now as $V = H_p(M)$ is de Rham, $H^1_f(\mathbb{Q}_p, V)^\perp = H^1_f(\mathbb{Q}_p, V^*(1))$. Hence taking the alternate sum of $\mathbb{Z}_p$-coranks of the exact sequence (3.7) with $W_p(\mathfrak{M}_{p^\infty}) = H^1_f(\mathbb{Q}_p, \mathfrak{M}_{p^\infty})$ gives

(3.19)
$$\begin{aligned}
&\mathrm{ck}_{\mathbb{Z}_p} H^1_f(\mathbb{Q}, \mathfrak{M}_{p^\infty}) - \mathrm{ck}_{\mathbb{Z}_p} H^1_f(\mathbb{Q}, \mathfrak{R}_{p^\infty}) \\
&= \mathrm{ck}_{\mathbb{Z}_p} H^1(G_S, \mathfrak{M}_{p^\infty}) - \mathrm{ck}_{\mathbb{Z}_p} H^2(G_S, \mathfrak{M}_{p^\infty}) - \mathrm{ck}_{\mathbb{Z}_p} H^o(G_S, \mathfrak{R}_{p^\infty}) \\
&\quad + \sum_{v \in S} \left(\mathrm{ck}_{\mathbb{Z}_p} H^1_f(\mathbb{Q}_v, \mathfrak{M}_{p^\infty}) - \mathrm{ck}_{\mathbb{Z}_p} H^1(\mathbb{Q}_v, \mathfrak{M}_{p^\infty}) + \mathrm{ck}_{\mathbb{Z}_p} H^2(\mathbb{Q}_v, \mathfrak{M}_{p^\infty}) \right).
\end{aligned}$$

Now Tate local duality gives

$$\sum_{0 \le i \le 2} (-1)^i \mathrm{ck}_{\mathbb{Z}_p} H^i(\mathbb{Q}_v, \mathfrak{M}_{p^\infty}) = \begin{cases} -\dim_{\mathbb{Q}_p}(V), & \text{for } v = p \\ 0, & \text{for } v \ne p. \end{cases} \tag{3.20}$$

Also for $v \ne p$

$$\mathrm{ck}_{\mathbb{Z}_p} H^1_f(\mathbb{Q}_v, \mathfrak{M}_{p^\infty}) = \mathrm{ck}_{\mathbb{Z}_p} H^o(\mathbb{Q}_v, \mathfrak{M}_{p^\infty}). \tag{3.21}$$

Result now follows on combining (3.19)–(3.21) with Proposition 3.4 and the results of Section 2. □

It is conjectured that $H^1_f(\mathbb{Q}, \mathfrak{A}_{p^\infty})$, and so $H^1_f(\mathbb{Q}, \mathfrak{B}_{p^\infty})$ are always finite. Deep progress has been made on this result by Flach [7], Wiles [20], Diamond

[5], and Langer [11]. We note also that when E has complex multiplication and p is relatively prime to the number of roots of unity in $K = \mathrm{End}(E) \otimes \mathbb{Q}$, then the finiteness of $H^1_f(\mathbb{Q}, \mathfrak{A}_{p^\infty})$ can also be shown to be a consequence of Rubin's proof of the Main Conjecture for E. For simplicity, we shall simply state this finiteness result as a hypothesis when we need it in what follows.

Let us put $S' = S\backslash\{p\}$. We then define

$$r(S) = \#\{v \in S' : L_v(A, s) \text{ has a pole at } s = 0\}. \tag{3.22}$$

Proposition 3.6. *Assume that* $H^1_f(\mathbb{Q}, \mathfrak{A}_{p^\infty})$ *is finite. Then*
(i) $\mathrm{ck}_{\mathbb{Z}_p} H^1(G_S, \mathfrak{A}_{p^\infty}) = 2, H^2(G_S, \mathfrak{A}_{p^\infty}) = 0$;
(ii) $\mathrm{ck}_{\mathbb{Z}_p} H^1(G_S, \mathfrak{B}_{p^\infty}) = 1 + r(S), H^2(G_S, \mathfrak{A}_{p^\infty}) \xrightarrow{\sim} (\mathbb{Q}_p/Z_p)^{r(S)}$.

Proof. As usual we let M denote either A or B. Now $H^1_f(\mathbb{Q}, \mathfrak{M}_{p^\infty})$ is finite so by Lemma 3.3 so is $H^1_f(\mathbb{Q}, T_p(M))$. Hence the first exact sequence of Theorem 3.2 shows that

$$\mathrm{ck}_{\mathbb{Z}_p} H^2(G_S, \mathfrak{M}_{p^\infty}) = \sum_{v \in S} \mathrm{ck}_{\mathbb{Z}_p} H^2(\mathbb{Q}_v, \mathfrak{M}_{p^\infty}) - \mathrm{rk}_{\mathbb{Z}_p} H^0(G_S, T_p(R)).$$

Hence the result follows from Lemma 2.2 and Proposition 3.4. □

Recall that $S' = S\backslash\{p\}$ and we let M represent either A or B for the rest of this section. Let

$$\psi'_M : H^1(G_S, \mathfrak{M}_{p^\infty}) \longrightarrow \bigoplus_{v \in S'} \frac{H^1(\mathbb{Q}_v, \mathfrak{M}_{p^\infty})}{H^1_f(\mathbb{Q}_v, \mathfrak{M}_{p^\infty})}$$

be the natural map.

Proposition 3.7. *Assume that* $H^1_f(\mathbb{Q}, \mathfrak{A}_{p^\infty})$ *is finite. Then both* ψ'_A *and* ψ'_B *are surjective.*

Proof. Our finiteness hypothesis clearly implies that, from Lemma 3.2, we have

$$\mathfrak{M}_{p^\infty}(\mathbb{Q}) = H^1_f(\mathbb{Q}, T_p(M)),$$

and by (3.13), if $\mathfrak{M}_{p^\infty}(\mathbb{Q}_p)$ is finite, the map

$$\mathfrak{M}_{p^\infty}(\mathbb{Q}_p) \hookrightarrow H^1_f(\mathbb{Q}_p, T_p(M))$$

is an injection. Thus the natural inclusion of $\mathfrak{M}_{p^\infty}(\mathbb{Q})$ into $\mathfrak{M}_{p^\infty}(\mathbb{Q}_p)$ shows that the restriction map

$$j_p(M) : H^1_f(\mathbb{Q}, T_p(M)) \longrightarrow H^1_f(\mathbb{Q}_p, T_p(M))$$

is injective. On the other hand if $\mathfrak{M}_{p^\infty}(\mathbb{Q}_p)$ is not finite, then by Lemma 2.3, it must be the case that $M = A$, p is potentially ordinary for E and E admits complex multiplication. But then, as was mentioned above, $H^1_f(\mathbb{Q}, T_p(M)) = \mathfrak{M}_{p^\infty}(\mathbb{Q}) = 0$ and so $j_p(M)$ is still injective.

Let

$$j_S(M) : H^1_f(\mathbb{Q}, T_p(M)) \longrightarrow \underset{v\in S}{\oplus} H^1_f(\mathbb{Q}_v, T_p(M))$$

be the map given by restriction at all $v \in S$, which must also be injective. Now $j_S(M)$ is the dual of the map

$$\underset{v\in S}{\oplus} \frac{H^1(\mathbb{Q}_v, \mathfrak{M}^*_{p^\infty}(1))}{H^1_f(\mathbb{Q}_v, \mathfrak{M}^*_{p^\infty}(1))} \longrightarrow H^1_f(\mathbb{Q}, T_p(M))^\wedge \tag{3.23}$$

occurring in the exact sequence (3.7) for $\mathfrak{M}^*_{p^\infty}(1)$. Hence this map is surjective. Simple diagram chasing now gives the exact sequence

$$0 \longrightarrow \mathrm{Coker}(\psi'_R)^\wedge \longrightarrow H^1_f(\mathbb{Q}, T_p(M)) \overset{j_p(M)}{\longrightarrow} H^1_f(\mathbb{Q}_p, T_p(M)),$$

where R denotes A if M equals B and vice versa. Hence the injectivity of $j_p(M)$ shows that $\mathrm{Coker}(\psi'_R) = 0$ as required. □

We now define $S'(\mathbb{Q}, \mathfrak{M}_{p^\infty}) = \mathrm{Ker}(\psi'_M)$.

Proposition 3.8. *Assume that $H^1_f(\mathbb{Q}, \mathfrak{A}_{p^\infty})$ is finite. Then*

$$\mathrm{ck}_{\mathbb{Z}_p} S'(\mathbb{Q}, \mathfrak{M}_{p^\infty}) = d^-(M),$$

unless $M = B$, p is potentially ordinary for E and E admits complex multiplication. In this latter case

$$\mathrm{ck}_{\mathbb{Z}_p} S'(\mathbb{Q}, \mathfrak{B}_{p^\infty}) = 2.$$

Proof. Firstly from the surjectivity of (3.23) for $\mathfrak{M}_{p^\infty}$, we have the exact sequence

$$\begin{aligned} 0 \longrightarrow H^1_f(\mathbb{Q}, \mathfrak{M}_{p^\infty}) \longrightarrow H^1(G_S, \mathfrak{M}_{p^\infty}) \longrightarrow \underset{v\in S}{\oplus} \frac{H^1(\mathbb{Q}_v, \mathfrak{M}_{p^\infty})}{H^1_f(\mathbb{Q}_v, \mathfrak{M}_{p^\infty})} \\ \longrightarrow H^1_f(\mathbb{Q}, T_p(R))^\wedge \longrightarrow 0, \end{aligned}$$

and comparing this with the defining exact sequence for $S'(\mathbb{Q}, \mathfrak{M}_{p^\infty})$ gives

$$\begin{aligned} \mathrm{ck}_{\mathbb{Z}_p} S'(\mathbb{Q}, \mathfrak{M}_{p^\infty}) &= \mathrm{ck}_{\mathbb{Z}_p}(H^1(\mathbb{Q}_p, \mathfrak{M}_{p^\infty})/H^1_f(\mathbb{Q}_p, \mathfrak{M}_{p^\infty})) \\ &= \mathrm{ck}_{\mathbb{Z}_p} H^1_f(\mathbb{Q}_p, \mathfrak{M}^*_{p^\infty}(1)), \end{aligned}$$

with the second equality following from the fact that $H_p(M)$ is de Rham. The result now follows immediately from Lemma 2.4. □

Let v be any prime different from p. By Lemma 2.2, $H^1_f(\mathbb{Q}_v, T_p(B))$ is finite, and hence, by duality, $H^1(\mathbb{Q}_v, \mathfrak{A}_{p^\infty})/H^1_f(\mathbb{Q}_v, \mathfrak{A}_{p^\infty})$ is finite of the same order. We can compute this order as follows, in the spirit of Tamagawa numbers. Following Bloch and Kato, we define

$$H^1_f(\mathbb{Q}_v, T_p(B))^0 = \mathrm{Ker}(H^1(\mathbb{Q}_v, T_p(B)) \to H^1(I_v, T_p(B)) \qquad (v \neq p).$$

If u and w are non-zero elements of $\mathbb{Q}_p$, we will write $u \sim w$ if u/w is a p-adic unit. If follows from the definition of the Euler factor $L_v(B,s)$ that

$$\#(H^1_f(\mathbb{Q}_v, T_p(B))^0) \sim L_v(B,0)^{-1} \qquad (v \neq p). \tag{3.23}$$

Hence, defining

$$c_v^{(p)}(B) = [H^1_f(\mathbb{Q}_v, T_p(B)) : H^1_f(\mathbb{Q}_v, T_p(B))^0] \qquad (v \neq p), \tag{3.24}$$

we conclude that

$$\#(H^1(\mathbb{Q}_v, \mathfrak{A}_{p^\infty})/H^1_f(\mathbb{Q}_v, \mathfrak{A}_{p^\infty})) \sim c_v^{(p)}(B) L_v(B,0)^{-1} \qquad (v \neq p). \tag{3.25}$$

If T is any finite set of finite primes of $\mathbb{Q}$ not containing p, we define

$$H^1_{f,T}(\mathbb{Q}, \mathfrak{A}_{p^\infty}) = \operatorname{Ker}\left(H^1(G_S, \mathfrak{A}_{p^\infty}) \to \bigoplus_{v \in S\setminus T} \frac{H^1(\mathbb{Q}_v, \mathfrak{A}_{p^\infty})}{H^1_f(\mathbb{Q}_v, \mathfrak{A}_{p^\infty})} \right),$$

where S is now chosen to be any finite set of primes of $\mathbb{Q}$ containing p, T, and all primes where E has bad reduction.

Proposition 3.9. *Assume that $H^1_f(\mathbb{Q}, \mathfrak{A}_{p^\infty})$ is finite. Then*

(3.26)
$$[H^1_{f,T}(\mathbb{Q}, \mathfrak{A}_{p^\infty}) : H^1_f(\mathbb{Q}, \mathfrak{A}_{p^\infty})] \sim (\Pi_{v \in T} c_v^{(p)}(B) L_v(B,0)^{-1})/\#(\mathfrak{B}_{p^\infty}(\mathbb{Q}))$$

provided T is non-empty.

Proof. To simplify the notation, let us put

$$U_v = H^1(\mathbb{Q}_v, \mathfrak{A}_{p^\infty})/H^1_f(\mathbb{Q}_v, \mathfrak{A}_{p^\infty}).$$

Our hypothesis that $H^1_f(\mathbb{Q}, \mathfrak{A}_{p^\infty})$ is finite implies, as was shown earlier, that

$$H^2(G_S, \mathfrak{A}_{p^\infty}) = 0 \text{ and } H^1_f(\mathbb{Q}, T_p(B)) = \mathfrak{B}_{p^\infty}(\mathbb{Q}).$$

Hence it follows from (3.7) with $W_v(\mathfrak{A}_{p^\infty}) = H^1_f(\mathbb{Q}_v, \mathfrak{A}_{p^\infty})$ for all $v \in S$ that we have the commutative diagram with exact rows

$$\begin{array}{ccccccccc} 0 \to & H^1_f(\mathbb{Q}, \mathfrak{A}_{p^\infty}) & \to & H^1(G_S, \mathfrak{A}_{p^\infty}) & \to & \bigoplus_{v \in S} U_v & \to & \widehat{\mathfrak{B}_{p^\infty}(\mathbb{Q})} & \to 0 \\ & \downarrow & & \| & & \downarrow \pi & & & \\ 0 \to & H^1_{f,T}(\mathbb{Q}, \mathfrak{A}_{p^\infty}) & \to & H^1(G_S, \mathfrak{A}_{p^\infty}) & \xrightarrow{\theta} & \bigoplus_{v \in S\setminus T} U_v & & & \end{array} \tag{3.27}$$

Since the projection π is surjective, it follows immediately from this diagram that Coker(θ) is finite. In fact, we claim that θ is surjective. Granted this, simple diagram chasing using (3.27), and noting (3.25), establishes the conclusion of Proposition 3.9. To analyse Coker(θ), we apply the exact sequence (3.7) again, but now with $M = A$ and with

$$W_v(\mathfrak{A}_{p^\infty}) = H^1_f(\mathbb{Q}_v, \mathfrak{A}_{p^\infty}) \ (v \in S\setminus T), \quad W_v(\mathfrak{A}_{p^\infty}) = H^1(\mathbb{Q}_v, \mathfrak{A}_{p^\infty}) \ (v \in T).$$

We conclude that we have the exact sequence

$$0 \to \widehat{\mathrm{Coker}(\theta)} \to H^1(G_S, T_p(B)) \xrightarrow{\lambda} \bigoplus_{v \in S \setminus T} \frac{H^1(\mathbb{Q}_v, T_p(B))}{H^1_f(\mathbb{Q}_v, T_p(B))} \bigoplus_{v \in T} H^1_f(\mathbb{Q}_v, T_p(B)).$$

But the torsion subgroup of $H^1(G_S, T_p(B))$ is equal to $\mathfrak{B}_{p^\infty}(\mathbb{Q})$ by Lemma 3.3, and similarly the torsion subgroup of $H^1(\mathbb{Q}_v, T_p(B))$ is equal to $\mathfrak{B}_{p^\infty}(\mathbb{Q}_v)$. Since T is non-empty by hypothesis, it follows that λ must be injective on the torsion subgroup of $H^1(G_S, T_p(B))$. But $\widehat{\mathrm{Coker}(\theta)}$ is contained in this torsion subgroup because it is finite, and so $\mathrm{Coker}(\theta) = 0$. This completes the proof of Proposition 3.9. □

4. Galois cohomology over the cyclotomic $\mathbb{Z}_p$-extension of $\mathbb{Q}$. Let F_∞ denote the cyclotomic $\mathbb{Z}_p$-extension of $\mathbb{Q}$. In other words F_∞ is the unique subfield of $\mathbb{Q}(\mu_{p^\infty})$ whose Galois group over $\mathbb{Q}$ is topologically isomorphic to the additive group of $\mathbb{Z}_p$. Since $p \in S$ we have that $F_\infty \subset \mathbb{Q}_S$, and we set

$$G_{\infty,S} = \mathrm{Gal}(\mathbb{Q}_S/F_\infty), \quad \Gamma = \mathrm{Gal}(F_\infty/\mathbb{Q}).$$

We also write $\Lambda = \mathbb{Z}_p[[\Gamma]]$ for the Iwasawa algebra of Γ. Now Γ acts naturally on the groups $H^i(G_{\infty,S}, \mathfrak{M}_{p^\infty})$, where M denotes either A or B. Hence we can view them as Λ-modules and the aim of this section is to study them as such. We write $X^\wedge$ for the Pontrjagin dual of a Λ-module X, and we will write M for either A or B for the rest of this section. Finally, we will also assume throughout this section that E does not have complex multiplication.

Theorem 4.1. *Assume that $H^1_f(\mathbb{Q}, \mathfrak{A}_{p^\infty})$ is finite. Then*
(i) $H^2(G_{\infty,S}, \mathfrak{M}_{p^\infty}) = 0$,
(ii) $H^1(G_{\infty,S}, \mathfrak{M}_{p^\infty})^\wedge$ *has Λ-rank equal to $d^-(M)$,*
(iii) $H^1(G_{\infty,S}, \mathfrak{M}_{p^\infty})^\wedge$ *has no non-zero finite Λ-submodules.*

We remark immediately that, as we are still assuming that p is odd, (ii) implies (i) and (iii) by Greenberg ([**10**],Propositions 3–5). It is also known ([**10**], Proposition 3) that $\mathrm{ck}_\Lambda\, H^1(G_{\infty,S}, \mathfrak{M}_{p^\infty}) \geq d^-(M)$, so it suffices to prove that

(4.1) $$\mathrm{ck}_\Lambda\, H^1(G_{\infty,S}, \mathfrak{M}_{p^\infty}) \leq d^-(M).$$

Now in order to prove (4.1) we need to study another Iwasawa module attached to $\mathfrak{M}_{p^\infty}$. For each place w of F_∞, let $F_{\infty,w}$ denote the union of the completions at w of all finite extensions of $\mathbb{Q}$ contained in F_∞. We now define $S'(F_\infty, \mathfrak{M}_{p^\infty})$ by the exactness of the sequence

(4.2) $$0 \longrightarrow S'(F_\infty, \mathfrak{M}_{p^\infty}) \longrightarrow H^1(G_{\infty,S}, \mathfrak{M}_{p^\infty}) \xrightarrow{\psi'_{\infty,M}} \bigoplus_{w|S'} H^1(F_{\infty,w}, \mathfrak{M}_{p^\infty}),$$

where $\psi'_{\infty,M}$ is given by restriction at all places w of F_∞ lying above $S' = S \setminus \{p\}$.

Theorem 4.2. *Assume that* $H^1_f(\mathbb{Q}, \mathfrak{A}_{p^\infty})$ *is finite. Then*
(i) $\mathcal{S}'(F_\infty, \mathfrak{M}_{p^\infty})^\wedge$ *has* Λ*-rank equal to* $d^-(M)$,
(ii) $\mathcal{S}'(F_\infty, \mathfrak{M}_{p^\infty})^\wedge$ *has no non-zero finite* Λ*-submodules,*
(iii) $H^1(\Gamma, \mathcal{S}'(F_\infty, \mathfrak{M}_{p^\infty})) = 0$,
(iv) *the map* $\psi'_{\infty,M}$ *is surjective.*

Again we have immediately that (iii) implies (ii), and by Greenberg ([**10**], Proposition 2)

$$\mathrm{ck}_\Lambda\, H^1(F_{\infty,w}, \mathfrak{M}_{p^\infty}) = 0, \qquad \text{for } w \text{ not lying above } p.$$

Hence statement (*i*) of Theorem 4.2 implies statement (*ii*) of Theorem 4.1. Thus it only remains to prove statements (*i*), (*iii*) and (*iv*) of Theorem 4.2. In fact as indicated above, (*i*) would follow if we can show that

$$\mathrm{ck}_\Lambda\, \mathcal{S}'(F_\infty, \mathfrak{M}_{p^\infty}) \le d^-(M). \tag{4.3}$$

We remark in passing that if E has complex multiplication and $p \ge 5$ is ordinary, then using the two variable main conjecture, Flach ([**6**], Proposition 3.14 and Corollary 3.16) proves that
(a) $\mathcal{S}(F_\infty, \mathfrak{M}_{p^\infty})^\wedge$ is Λ-cotorsion,
(b) $\mathcal{S}(F_\infty, \mathfrak{M}_{p^\infty})^\wedge$ has no non-zero finite Λ-submodules,
where

$$0 \longrightarrow \mathcal{S}(F_\infty, \mathfrak{M}_{p^\infty}) \longrightarrow H^1(G_{\infty,S}, \mathfrak{M}_{p^\infty})$$
$$\xrightarrow{\psi_{\infty,M}} \frac{H^1(F_{\infty,p}, \mathfrak{M}_{p^\infty})}{H^1_f(F_{\infty,p}, \mathfrak{M}_{p^\infty})} \oplus \left(\bigoplus_{w|S'} H^1(F_{\infty,w}, \mathfrak{M}_{p^\infty}) \right).$$

Now (*a*) immediately implies Theorem 4.1 in this case, and also shows that $\mathrm{Coker}(\psi_{\infty,M})$ is Λ-cotorsion. We also know that $\mathfrak{M}^*_{p^\infty}(1)(\mathbb{Q}) = 0$ in this case, and so it can easily be seen that the map

$$\psi_M : H^1(G_S, \mathfrak{M}_{p^\infty}) \longrightarrow \bigoplus_{v \in S} \frac{H^1(\mathbb{Q}_v, \mathfrak{M}_{p^\infty})}{H^1_f(\mathbb{Q}_v, \mathfrak{M}_{p^\infty})}$$

is surjective. This implies that $\mathrm{Coker}(\psi_{\infty,M}) = 0$ (exactly as in the case when E does not admit complex multiplication given below), and so diagram chasing both gives the exact sequence

$$0 \longrightarrow \mathcal{S}(F_\infty, \mathfrak{M}_{p^\infty}) \longrightarrow \mathcal{S}'(F_\infty, \mathfrak{M}_{p^\infty}) \longrightarrow \frac{H^1(F_{\infty,p}, \mathfrak{M}_{p^\infty})}{H^1_f(F_{\infty,p}, \mathfrak{M}_{p^\infty})} \longrightarrow 0,$$

and shows that $\psi'_{\infty,M}$ is surjective. The remainder of Theorem 4.2 follows from the above exact sequence and (*b*), on noting that for ordinary p we have that $\mathrm{ck}_\Lambda\, H^1_f(F_{\infty,p}, \mathfrak{M}_{p^\infty}) = d^+(H_p(M))$.

Returning to the case where E does not have complex multiplication, we firstly need a purely local result. If $v \ne p$ we let

$$H^1_f(\mathbb{Q}_v, \mathfrak{M}_{p^\infty})^o = \mathrm{Ker}(H^1(\mathbb{Q}_v, \mathfrak{M}_{p^\infty}) \longrightarrow H^1(I_v, \mathfrak{M}_{p^\infty})),$$

where I_v is the inertia group at v. We now write

$$c_v^{(p)}(R) = \#\left(\frac{H_f^1(\mathbb{Q}_v, \mathfrak{M}_{p^\infty})^o}{H_f^1(\mathbb{Q}_v, \mathfrak{M}_{p^\infty})}\right), \tag{4.4}$$

where R equals A if M equals B and vice versa. Note that $H_f^1(\mathbb{Q}_v, \mathfrak{M}_{p^\infty})^o / H_f^1(\mathbb{Q}_v, \mathfrak{M}_{p^\infty})$ is dual to $H_f^1(\mathbb{Q}_v, T_p(R))/H_f^1(\mathbb{Q}_v, T_p(R))^o$, where

$$H_f^1(\mathbb{Q}_v, T_p(R))^o = \mathrm{Ker}(H^1(\mathbb{Q}_v, T_p(R)) \longrightarrow H^1(I_v, T_p(R))),$$

so $c_v^{(p)}(R)$ is just the pth part of the Tamagawa factor $c_v(R)$. For a more detailed account of the Tamagawa factors for $\mathrm{Sym}^2(E)$ we refer the reader to **[8]**. We also note that from **[18]**, Proposition 4.3 we have that $c_v^{(p)}(A) = c_v^{(p)}(B)$ and if we denote this common value by $c_v^{(p)}$, then we have, for $v \neq p$,

$$c_v^{(p)} = \begin{cases} 1 & \text{if } d_v \leq 0, \\ \#(E_{d_v}(\mathbb{Q}_v)(p)) & \text{if } d_v > 0, \end{cases} \tag{4.5}$$

where $d_v = -\mathrm{ord}_v(j_E)$, and j_E is the modular invariant of E.

Lemma 4.3. *Assume $v \neq p$. Then restriction induces a homomorphism*

$$\mu_{M,v} : \frac{H^1(\mathbb{Q}_v, \mathfrak{M}_{p^\infty})}{H_f^1(\mathbb{Q}_v, \mathfrak{M}_{p^\infty})} \longrightarrow (\underset{w|v}{\oplus} H^1(F_{\infty,w}, \mathfrak{M}_{p^\infty}))^\Gamma,$$

which is surjective, and has finite kernel. Furthermore

$$\#\,\mathrm{Ker}(\mu_{M,v}) = c_v^{(p)}. \tag{4.6}$$

We immediately remark that we do not need the order of this kernel in the proof of Theorems 4.1 and 4.2, but we include it here for completeness.

Proof. Firstly, $H_f^1(\mathbb{Q}_v, \mathfrak{M}_{p^\infty})$ maps to $\underset{w|v}{\oplus} H_f^1(F_{\infty,w}, \mathfrak{M}_{p^\infty})$ under the restriction map. However, as w does not divide p, $F_{\infty,w}$ is the unique unramified $\mathbb{Z}_p$-extension of $\mathbb{Q}_w$. Thus if $\mathbb{Q}_w^{nr}$ denotes the maximal unramified extension of $\mathbb{Q}_w$, the Sylow p-subgroup of $\mathrm{Gal}(\mathbb{Q}_w^{nr}/F_{\infty,w})$ is trivial. Hence

$$H_f^1(F_{\infty,w}, \mathfrak{M}_{p^\infty}) = H^1(\mathrm{Gal}(\mathbb{Q}_w^{nr}/F_{\infty,w}), \mathfrak{M}_{p^\infty}^{I_{\infty,w}}) = 0. \tag{4.7}$$

Thus $H_f^1(\mathbb{Q}_v, \mathfrak{M}_{p^\infty})$ is mapped to zero under the restriction map and so $\mu_{M,v}$ clearly exists.

For each w dividing v, the restriction map from $H^1(\mathbb{Q}_v, \mathfrak{M}_{p^\infty})$ to $H^1(F_{\infty,w}, \mathfrak{M}_{p^\infty})^{\Gamma_w}$, where $\Gamma_w = \mathrm{Gal}(F_{\infty,w}/\mathbb{Q}_v)$, is surjective as Γ_w has cohomological dimension equal to one. The surjectivity of $\mu_{M,v}$ is now immediate.

Finally, we have that

$$\mathrm{Ker}(\mu_{M,v}) = \frac{\mathrm{Ker}(\mu'_{M,v})}{H_f^1(\mathbb{Q}_v, \mathfrak{M}_{p^\infty})},$$

where

$$\begin{aligned}\operatorname{Ker}(\mu'_{M,v}) &:= \operatorname{Ker}(H^1(\mathbb{Q}_v, \mathfrak{M}_{p^\infty}) \longrightarrow \underset{w|v}{\oplus} H^1_f(F_{\infty,w}, \mathfrak{M}_{p^\infty}))\\ &= \operatorname{Ker}(H^1(\mathbb{Q}_v, \mathfrak{M}_{p^\infty}) \longrightarrow \underset{w|v}{\oplus} H^1_f(I_{\infty,w}, \mathfrak{M}_{p^\infty}))\\ &= \operatorname{Ker}(H^1(\mathbb{Q}_v, \mathfrak{M}_{p^\infty}) \longrightarrow H^1_f(I_v, \mathfrak{M}_{p^\infty}))\\ &= H^1_f(\mathbb{Q}_v, \mathfrak{M}_{p^\infty})^o,\end{aligned}$$

and the required result follows. □

Now we clearly have the commutative diagram

$$\begin{array}{ccccccc} 0\to & S'(\mathbb{Q},\mathfrak{M}_{p^\infty}) & \to & H^1(G_S,\mathfrak{M}_{p^\infty}) & \overset{\psi'_M}{\to} & \underset{v\in S'}{\oplus}\dfrac{H^1(\mathbb{Q}_v,\mathfrak{M}_{p^\infty})}{H^1_f(\mathbb{Q}_v,\mathfrak{M}_{p^\infty})} & \to 0\\ & \downarrow{\scriptstyle\alpha_1} & & \downarrow{\scriptstyle\alpha_2} & & \downarrow{\scriptstyle\mu_M} & \\ 0\to & S'(F_\infty,\mathfrak{M}_{p^\infty})^\Gamma & \to & H^1(G_{\infty,S},\mathfrak{M}_{p^\infty})^\Gamma & \to & \underset{w|S'}{\oplus} H^1(F_{\infty,w},\mathfrak{M}_{p^\infty})^\Gamma & \to 0,\end{array}$$

on noting that ψ'_M is surjective by Proposition 3.7, and so the surjectivity of the right-hand map on the bottom row follows from the surjectivity of $\mu_M = \oplus\mu_{M,v}$. We have already seen that $\operatorname{Ker}(\mu_M)$ is finite and α_2 is clearly surjective, so we have from Proposition 3.8 that

$$\operatorname{ck}_{\mathbb{Z}_p} S'(F_\infty, \mathfrak{M}_{p^\infty})^\Gamma \leq \operatorname{ck}_{\mathbb{Z}_p} S'(\mathbb{Q}, \mathfrak{M}_{p^\infty}) = d^-(M).$$

On the other hand, it is clear from the structure theory of Λ-modules that

$$\operatorname{ck}_\Lambda S'(F_\infty, \mathfrak{M}_{p^\infty}) \leq \operatorname{ck}_{\mathbb{Z}_p} S'(F_\infty, \mathfrak{M}_{p^\infty})^\Gamma,$$

and so (4.3) follows immediately.

It only remains to prove (*iii*) and (*iv*) of Theorem 4.2. As $H^2(G_{\infty,S}, \mathfrak{M}_{p^\infty}) = 0$ we have the diagram

$$\begin{array}{ccccc} 0\longrightarrow & H^1(\Gamma, H^1(G_{\infty,S},\mathfrak{M}_{p^\infty})) & \longrightarrow & H^2(G_S,\mathfrak{M}_{p^\infty}) & \longrightarrow 0\\ & \downarrow{\scriptstyle\alpha_1} & & \downarrow{\scriptstyle\alpha_2} & \\ 0\longrightarrow & \underset{w|S}{\oplus} H^1(\Gamma, H^1(F_{\infty,w},\mathfrak{M}_{p^\infty})) & \longrightarrow & \underset{v\in S}{\oplus} H^2(\mathbb{Q}_v,\mathfrak{M}_{p^\infty}) & \longrightarrow 0.\end{array}$$

Now the surjectivity of (3.23) clearly implies that α_2 is injective. Furthermore we have from Theorem 3.1 that $\operatorname{Coker}(\alpha_2) = H^o(G_S, T^*(1))^\wedge = 0$. Hence $\operatorname{Ker}(\alpha_1) = \operatorname{Coker}(\alpha_1) = 0$. Now since we are assuming that E does not have complex multiplication, we know that $H^2(\mathbb{Q}_p, \mathfrak{M}_{p^\infty}) = 0$. This clearly implies that $H^1(\Gamma, H^1(F_{\infty,p}, \mathfrak{M}_{p^\infty})) = 0$, and so

$$H^1(\Gamma, H^1(G_{\infty,S}, \mathfrak{M}_{p^\infty})) \overset{\sim}{\longrightarrow} \underset{w|S}{\oplus} H^1(\Gamma, H^1(F_{\infty,w}, \mathfrak{M}_{p^\infty})). \tag{4.8}$$

However, the surjectivity of

$$H^1(G_{\infty,S}, \mathfrak{M}_{p^\infty})^\Gamma \longrightarrow \underset{w|S'}{\oplus} H^1(F_{\infty,w}, \mathfrak{M}_{p^\infty})^\Gamma \tag{4.9}$$

implies that we have the exact sequence

$$
\begin{aligned}
&0 \to H^1(\Gamma, \mathcal{S}'(F_\infty, \mathfrak{M}_{p^\infty})) \to H^1(\Gamma, H^1(G_{\infty,S}, \mathfrak{M}_{p^\infty})) \\
&\to \underset{w|S'}{\oplus} H^1(\Gamma, H^1(F_{\infty,w}, \mathfrak{M}_{p^\infty})).
\end{aligned} \tag{4.10}
$$

Hence $H^1(\Gamma, \mathcal{S}'(F_\infty, \mathfrak{M}_{p^\infty})) = 0$ and (*iii*) of Theorem 4.2 follows.

We finally establish (*iv*) of Theorem 4.2. The surjectivity of (4.9) shows that

$$\mathrm{Im}(\psi'_\infty)^\Gamma = \underset{w|S'}{\oplus} H^1(F_{\infty,w}, \mathfrak{M}_{p^\infty})^\Gamma,$$

and so we obtain the exact sequence

(4.11)

$$0 \longrightarrow \mathrm{Coker}(\psi'_\infty)^\Gamma \longrightarrow H^1(\Gamma, \mathrm{Im}(\psi'_\infty)) \longrightarrow \underset{w|S}{\oplus} H^1(\Gamma, H^1(F_{\infty,w}, \mathfrak{M}_{p^\infty})).$$

But, by (*iii*) of Theorem 4.2

$$H^1(\Gamma, H^1(G_{\infty,S}, \mathfrak{M}_{p^\infty})) \xrightarrow{\sim} H^1(\Gamma, \mathrm{Im}(\psi'_\infty)), \tag{4.12}$$

and so by (4.8) and (4.11), we have that $\mathrm{Coker}(\psi'_\infty)^\Gamma = 0$. Thus as $\mathrm{Coker}(\psi'_\infty)$ is a discrete Λ-module, it must be zero.

References

[1] S. Bloch and K. Kato, *L-functions and Tamagawa numbers of motives*, in Cartier, P. et al (eds.), The Grothendieck festschrift, Vol. 1, Boston Basel stuttgart, Birkhäuser (1990), 333–400.

[2] J. Coates and C.G. Schmidt, *Iwasawa theory for the symmetric square of an elliptic curve*, J. Reine Angew. Math., **375/376** (1987), 104–156.

[3] J. Coates, *Motivic p-adic L-functions and Arithmetic*, LMS Lecture Notes 153, CUP (1991), 141–172.

[4] J. Coates and G. McConnell, *On the Iwasawa theory of modular elliptic curves of analytic rank* ≤ 1, J. London Math. Soc., **50(2)** (1994), 243–264.

[5] F. Diamond, *On deformation rings and Hecke rings*, Ann. of Math., **144** (1996), 137–166.

[6] M. Flach, *Selmer groups for the symmetric square of an elliptic curve*, Ph. D. dissertation, University of Cambridge (1990).

[7] M. Flach, *A finiteness theorem for the symmetric square of an elliptic curve*, Invent. Math., **109** (1992), 307–327.

[8] M. Flach, M., *On the degree of modular parametrizations*, Séminaire de Theorie des Nombres, Paris (1991–1992), 23–36.

[9] J.-M. Fontaine and B. Perrin-Riou, *Autour des conjectures de Bloch et Kato I. Cohomologie galoisienne*, C.R. Acad. Sci. Paris, **313** (1991), 189–196.

[10] R. Greenberg, *Iwasawa Theory for p-adic Représentations*, in Advanced Studies in Pure math., **17**, Kinokuniya & Academic Press (1989), 97–137.

[11] A. Langer, *Selmer groups and Torsion zero cycles on the selfproduct of a semistable elliptic curve* II, preprint.

[12] J. Lubin, *One parameter formal Lie groups over p-adic integer rings*, Ann. of Math., **80** (1964), 464–484.

[13] J.S. Milne, *Arithmetic Duality Theorems*, (Perspect. Math., Vol. 1), Academic Press (1986).

[14] B. Perrin-Riou, *Fonctions L p-adiques des représentations p-adiques*, to appear.

[15] K. Rubin and A. Silverberg, *Families of Elliptic curves with constant mod p representations*, this colume.

[16] J.-P. Serre, *Abelian l-adic representations and elliptic curves*, New-York, Benjamin (1968).

[17] J.-P. Serre, *Propriétés galoisiennes des points d'ordre fini des courbes elliptiques*, Invent. Math., **15** (1972), 259–331.

[18] A. Sydenham, *Iwasawa theory for the symmetric square of an elliptic curve*, Ph. D. dissertation, University of Cambridge (1996).

[19] J. Tate, *p-divisible groups*, in Proceedings of a Conference on Local fields, Springer (1967), 158–183.

[20] A. Wiles, *Modular elliptic curves and Fermat's Last Theorem*, Ann. of Math., **141** (1995), 443–551.

Conference on Elliptic Curves and Modular Forms
Hong Kong, December 18-21, 1993

The refined conjecture of Serre

FRED DIAMOND
DEPARTAMENT OF MATHEMATICS, MIT, 77 MASSACHUSETTS AVENUE,
CAMRIDGE, MA 02139, USA

1. Introduction. Let ℓ be a prime, and let $\mathbf{F}$ be an algebraic closure of the field $\mathbf{F}_\ell$. Fix a homomorphism $\alpha : \mathcal{O} \to \mathbf{F}$ where $\mathcal{O}$ is the ring of algebraic integers in $\mathbf{C}$. Suppose that

$$\rho : \mathrm{Gal}\,(\overline{\mathbf{Q}}/\mathbf{Q}) \to GL_2(\mathbf{F})$$

is a continuous irreducible representation and that $\det(\rho(\tau)) = -1$ if τ is a complex conjugation. Serre [**16**] has conjectured that any such representation is modular, meaning that there are positive integers k and N such that ρ arises from an eigenform in $S_k(\Gamma_1(N))$ (see [**15**]). Serre also defines integers $k(\rho)$ and $N(\rho)$ (satisfying $k \geq 2$ and N relatively prime to ℓ) and formulates a refinement which predicts that ρ arises from $S_{k(\rho)}(\Gamma_1(N(\rho)))$. In [**15**], Ribet discusses the progress on the problem of establishing the equivalence between Serre's conjecture and this refinement. Combining his work with that of others, Ribet concludes that this problem [**15**], (1.4) will be solved for $\ell > 3$ if the hypothesis $\det(\rho) = \chi_\ell$ is removed from a crucial intermediate result [**15**], (1.5). The main purpose of this article is to explain how to remove this restriction as well as include the case $\ell = 3$. We find that Ribet's methods yield the following:

Theorem 1.1. *Suppose that ℓ is odd and that $\rho : \mathrm{Gal}\,(\overline{\mathbf{Q}}/\mathbf{Q}) \to GL_2(\mathbf{F})$ is an irreducible representation arising from $S_k(\Gamma_1(N))$ for some positive integers k and N. Then ρ arises from $S_{k(\rho)}(\Gamma_1(N(\rho))$.*

The proof so closely follows the arguments of [**15**] and [**14**] that we make no attempt to give an independent treatment. Instead we refer to these articles throughout Sections 3, 4 and 5 and detail only the changes needed to adapt Ribet's proof to the more general situation.

Before explaining those changes, however, we recall that Serre also defines a Dirichlet character $\chi(\rho) : (\mathbf{Z}/N\mathbf{Z})^\times \to \mathbf{C}^\times$ and predicts that ρ arises from $S_{k(\rho)}(\Gamma_1(N(\rho))^{\chi(\rho)}$. Serre subsequently found counterexamples in which $\ell \leq 3$. On the other hand, a lemma of Carayol discussed in Section 2 yields the following corollary to Theorem 1.1.

Corollary 1.2. *Suppose that ρ is irreducible and arises from $S_k(\Gamma_1(N))$ for some positive integers k and N. If $\ell = 3$ then suppose also that ρ is not induced from a character of* $\mathrm{Gal}\,(\overline{\mathbf{Q}}/\mathbf{Q}(\sqrt{-3}))$. *Then ρ arises from $S_{k(\rho)}(\Gamma_1(N(\rho))^{\chi(\rho)}$.*

In Section 6 we explain how to deduce a variant of these results formulated by Wiles [**19**]. We prove that if ρ is as in the corollary, then ρ arises from a certain space of cusp forms of weight two.

Finally we remark that Theorem 1.1 allows us to weaken the hypotheses on the weight and determinant in the main result of [**6**]. Theorem 1 of [**6**] now applies if ρ arises from $S_k(\Gamma_1(N))$ with $2 \le k \le \ell - 1$ and ℓ prime to N, provided that $\ell > 3$, or that $\ell = 3$ and ρ is not induced from a character of $\mathrm{Gal}\,(\overline{\mathbf{Q}}/\mathbf{Q}(\sqrt{-3}))$.

The author is grateful to Ken Ribet and Andrew Wiles for helpful comments during the preparation of this article.

2. Carayol's Lemma. Suppose that $f = \sum_{n=1}^{\infty} a_n(f)e^{2\pi i n z}$ is an eigenform in $S_k(\Gamma_1(N))$, where by "eigenform" we shall always mean a normalized eigenform for the Hecke operators T_n, $n \ge 1$. Let K_f denote the number field generated by the eigenvalues $a_n(f)$, and let $\mathcal{O}_f$ denote the ring of integers of K_f. Recall that we have fixed a homomorphism $\alpha : \mathcal{O} \to \mathbf{F}$ where $\mathcal{O}$ is the ring of algebraic integers in $\mathbf{C}$ and $\mathbf{F}$ is an algebraic closure of $\mathbf{F}_\ell$. Let $\lambda = \ker(\alpha|_{\mathcal{O}_f})$ and $R_f = \mathcal{O}_{f,\lambda}$. Recall that the λ-adic representation $\mathrm{Gal}\,(\overline{\mathbf{Q}}/\mathbf{Q}) \to GL_2(K_{f,\lambda})$ associated to f is unramified outside $N\ell$, and that if p is a prime not dividing $N\ell$ then $a_p(f)$ is the trace of the image of Frob_p. This representation has a model

$$\rho_f : \mathrm{Gal}\,(\overline{\mathbf{Q}}/\mathbf{Q}) \to GL_2(R_f),$$

which is unique up to isomorphism if $\rho_f \otimes_{R_f} \mathbf{F}$ is irreducible. If S is a space of cusp forms stable under the action of the Hecke operators, we say that ρ arises from S if there is an eigenform f in S such that $\rho \cong \rho_f \otimes_{R_f} \mathbf{F}$.

For relatively prime positive integers m and N, we will write $\Gamma_1(m,N)$ for $\Gamma_1(m) \cap \Gamma_0(N)$. Note that $(\mathbf{Z}/N\mathbf{Z})^\times \cong \Gamma_1(m,N)/\Gamma_1(mN)$ acts naturally on $S_k(\Gamma_1(mN))$. For a Dirichlet character χ modulo N, we write $S_k(\Gamma_1(mN))^\chi$ for the associated isotypic component. Thus $S_k(\Gamma_1(mN))^\chi$ consists of the forms f in $S_k(\Gamma_1(mN))$ satisfying $\langle a \rangle f = \chi(a) f$ for all integers $a \equiv 1 \bmod m$ relatively prime to N. Note also that $S_k(\Gamma_1(mN))^\chi$ is stable under the action of the Hecke operators T_n.

To deduce Corollary 1.2 from Theorem 1.1, one can apply the following version of Carayol's Lemma.

Lemma 2.1. *Suppose that χ and ψ are Dirichlet characters modulo N satisfying $\alpha \circ \chi = \alpha \circ \psi$. Suppose that one of the following holds:*

1. $m > 3$,
2. $\ell > 3$, *or*
3. $\ell = 3$ *and ρ is not induced from a character of* $\mathrm{Gal}\,(\overline{\mathbf{Q}}/\mathbf{Q}(\sqrt{-3}))$.

If ρ is irreducible and arises from $S_k(\Gamma_1(mN))^\chi$ with $k \ge 2$ and $(m,N) = 1$, then ρ arises from $S_k(\Gamma_1(mN))^\psi$.

Note that if $m > 3$, then $\Gamma_1(m,N)$ has no elliptic elements, and $(\mathbf{Z}/N\mathbf{Z})^\times$ acts without fixed points on $\Gamma_1(mN)\backslash\mathcal{Z}$ where $\mathcal{Z}$ is the upper-half plane. So under the first condition, the conclusion is immediate from [**3**], Lemme 1 (see

[**3**], Proposition 3). Under the second condition, this is [**3**], Proposition 3). Under the third condition, one can apply a variant of [**3**], Lemme 1 explained in [**3**], (4.4).

If f is an eigenform in $S_k(\Gamma_1(mN))^\chi$ such that $\rho \cong \rho_f \otimes_{R_f} \mathbf{F}$, then under the hypotheses of the lemma, we see that there is an eigenform g in $S_k(\Gamma_1(mN))^\psi$ such that $\alpha(a_p(f)) = \alpha(a_p(g))$ for all primes p not dividing ℓmN. We will need a stronger version of the above lemma in the case $k = 2$; we will show that g can be chosen so that $\alpha(a_n(f)) = \alpha(a_n(g))$ for all $n \geq 1$. The argument is essentially the one found in [**3**] or in [**6**], Section 3, except that we consider the action of the Hecke operators on group cohomology rather than the Galois action on etale cohomology. Though a similar argument works for $k > 2$, we state, prove and use the result only for $k = 2$.

Before doing so, we need a few definitions. Let $\tilde{\mathbf{T}}$ denote the polynomial ring over $\mathbf{Z}$ generated by the variables T_n for $n \geq 1$. Let us call a maximal ideal of $\tilde{\mathbf{T}}$ *Eisenstein* if it contains $T_r - (r+1)$ for all primes $r \equiv 1 \bmod mN$. We say that a $\tilde{\mathbf{T}}$-module is *Eisenstein* if all maximal ideals in its support are Eisenstein. Let $\mathbf{m}$ denote the kernel of $\alpha \circ \theta_f$ where $\theta_f : \tilde{\mathbf{T}} \to \mathcal{O}$ is the homomorphism defined by $T_n \mapsto a_n(f)$. Note that $\mathbf{m}$ is Eisenstein if and only if $\rho_f \otimes_{R_f} \mathbf{F}$ is reducible; hence we assume that $\mathbf{m}$ is not Eisenstein.

Lemma 2.2. *Suppose that $k = 2$ and let ℓ, m, N, χ, ψ and ρ be as in Lemma 2.1. Let f be an eigenform in $S_2(\Gamma_1(mN))^\chi$ such that $\rho \cong \rho_f \otimes_{R_f} \mathbf{F}$. Then there is an eigenform g in $S_2(\Gamma_1(mN))^\psi$ such that $\alpha(a_n(f)) = \alpha(a_n(g))$ for all $n \geq 1$.*

Proof. Suppose first that $m > 3$. Let $K \subset \mathbf{C}$ be a finite extension of $\mathbf{Q}$ containing the values of χ and ψ. Let $\mathcal{O}'$ be the localization of $\mathcal{O}_K$ at the kernel of $\alpha|_{\mathcal{O}_K}$, λ' a uniformizer in $\mathcal{O}'$ and $\mathbf{F}' = \alpha(\mathcal{O}_K) \cong \mathcal{O}'/\lambda'\mathcal{O}'$. For an $\mathcal{O}'$-module M, write $\mathcal{L}^i(M, \chi)$ for the cohomology group $H^i(\Gamma_1(m,N), M)$ where $\Gamma_1(m,N)$ acts on M by χ^{-1}. Using for example that $\Gamma_1(m,N)$ is a free group on a finite set of generators, we see that $\mathcal{L}^i(\mathcal{O}', \chi)$ is finitely generated over $\mathcal{O}'$, that $\mathcal{L}^i(M, \chi) = 0$ if $i > 1$, and that $\mathcal{L}^i(M, \chi) \cong \mathcal{L}^i(\mathcal{O}', \chi) \otimes_{\mathcal{O}'} M$ if M is flat over $\mathcal{O}'$. As in [**17**], Chapter 8, $\mathcal{L}^i(M, \chi)$ is endowed with a natural action of the Hecke operators T_n and there is a $\tilde{\mathbf{T}}$-equivariant inclusion

$$\mathrm{Sh}_\chi : S_2(\Gamma_1(mN))^\chi \oplus S_2(\Gamma_1(mN))^\chi \hookrightarrow \mathcal{L}^1(\mathbf{C}, \chi).$$

We can replace χ by ψ in the above definitions and assertions.

Since $\mathbf{m}$ contains the annihilator of $S_2(\Gamma_1(mN))^\chi$ and $\mathcal{L}^1(\mathcal{O}', \chi) \otimes_{\mathcal{O}'} \mathbf{C} \cong \mathcal{L}^1(\mathbf{C}, \chi)$, we see that $\mathbf{m}$ contains the annihilator of $\mathcal{O}'$-module $\mathcal{L}^1(\mathcal{O}', \chi)$. Since $\mathcal{L}^1(\mathcal{O}', \chi)_{\mathbf{m}}$ is finitely generated over $\mathcal{O}'$, $\mathbf{m}$ is in the support of

$$\mathcal{L}^1(\mathcal{O}', \chi) \otimes_{\mathcal{O}'} \mathbf{F}' \cong \mathcal{L}^1(\mathbf{F}', \chi) \cong \mathcal{L}^1(\mathbf{F}', \psi),$$

where the first isomorphism is gotten from the vanishing of $\mathcal{L}^2(\mathcal{O}', \chi)$ and the second from the hypothesis that $\alpha \circ \chi = \alpha \circ \psi$. To reverse the implications and conclude that $\mathbf{m}$ contains the annihilator of $S_2(\Gamma_1(mN))^\psi$, it suffices to check that the following $\tilde{\mathbf{T}}$-modules are Eisenstein:

- the torsion subgroup of $\mathcal{L}^1(\mathcal{O}', \psi)$;

- the cokernel of Sh_{ψ}.

That the former module is Eisenstein follows from the fact that $T_r = r+1$ on $\mathcal{L}^0(\mathbf{F}', \psi)$ for all primes r not dividing mN. The latter module is Eisenstein as it is isomorphic to a space of modular forms spanned by Eisenstein series of weight two and level mN. (See **[10]**, Section 5 for an explicit description.)

We have now proved that $\mathfrak{m}$ contains the annihilator of $S_2(\Gamma_1(mN))^{\psi}$. It follows that $\alpha \circ \theta_f = \alpha \circ \theta$ for some homomorphism $\theta : \tilde{\mathbf{T}} \to \mathcal{O}$ which factors through the image of $\tilde{\mathbf{T}}$ in End $(S_2(\Gamma_1(mN))^{\psi})$. Such a homomorphism is necessarily of the form θ_g for some eigenform g in $S_2(\Gamma_1(mN))^{\sigma \circ \psi}$ with $\sigma \in \mathrm{Gal}(\overline{\mathbf{Q}}/\mathbf{Q})$. Since this implies that $\alpha \circ \sigma \circ \psi = \alpha \circ \psi$, we may assume that σ is in the inertia subgroup for $\ker \alpha$ and replace g by $g^{\sigma^{-1}}$ which is in $S_2(\Gamma_1(mN))^{\psi}$. Then $\alpha(a_n(f)) = \alpha(a_n(g))$ as desired for all $n \geq 1$.

Now suppose that either of the last two conditions in the lemma are satisfied. Choose a prime q as in **[6]**, Lemma 2 so that q does not divide $6mN$. By what we have already proved, there is an eigenform g' in $S_2(\Gamma_1(mqN))^{\psi}$ so that $\alpha(a_n(f)) = \alpha(a_n(g'))$ for all n relatively prime to q. But since q does not divide the conductor of the newform associated to g', there is an eigenform g in $S_2(\Gamma_1(mN))^{\psi}$ so that $a_n(g) = a_n(g')$ for all n relatively prime to q. Considering the trace and determinant of the image of Frob$_q$ in $\rho_f \otimes_{R_f} \mathbf{F} \cong \rho_g \otimes_{R_g} \mathbf{F}$, we see also that $\alpha(a_{q^i}(f)) = \alpha(a_{q^i}(g))$ for all $i > 0$. So in fact $\alpha(a_n(f)) = \alpha(a_n(g))$ for all $n \geq 1$. □

3. Character groups. A key case of "lowering the level" was treated by Ribet in **[15]** and **[14]** under the assumption that ρ arises from an eigenform on $\Gamma_0(N)$. An important ingredient in Ribet's proof is a certain relationship between character groups arising from Jacobians of modular curves and Shimura curves. We now undertake the analysis of character groups in Sections 3 and 4 of **[14]**, but in the context of modular forms on $\Gamma_1(N)$ rather than $\Gamma_0(N)$.

For relatively prime positive integers N and c, the modular curve $X_1(N,c)$ over $\mathbf{Q}$ arises from the moduli problem parametrizing triples (E, P, C) where E is an elliptic curve, P is a point of E of order N and C is a cyclic cubgroup of E of order c. Suppose that N and d are relatively prime and let q be a prime not dividing Nd. Applying the results of Raynaud discussed in **[14]**, Section 2 to the model for $X_1(N, dq)$ over $\mathbf{Z}_q$ described by Deligne and Rapoport **[5]**, we consider the connected component J^0 of the special fiber of the Neron model over $\mathbf{Z}_q$ of the Jacobian $J_1(N, dq)$ of $X_1(N, dq)$. We thus obtain an exact sequence

$$1 \to T \to J^0 \to J_1(N,d) \times J_1(N,d) \to 0,$$

where T is a torus whose character group X we describe as follows. The analogue of **[14]**, (3.1) is that X is isomorphic to the group of degree-0 divisors on the set $\Sigma_1(N,d)$ of supersingular points of $X_1(N,d)(\overline{\mathbf{F}}_q)$. This set may be identified with the set of isomorphism classes of "enhanced elliptic curves" (E, P, C) over $\overline{\mathbf{F}}_q$, where E is a supersingular elliptic curve, P is a point of E of order N and C is a cyclic subgroup of E of order d. Fix such a point $\mathbf{E}_0 = (E_0, P_0, C_0)$, and let $R = \mathrm{End}\,(E_0)$ and $H = R \otimes \mathbf{Q}$. Then R is a maximal order in the quaternion algebra H over of discriminant q. (Note that R is used to denote an Eichler order in [14].) We again have a natural notion of an "enhanced Tate module" of $\mathbf{E} = (E, P, C)$, which we denote $T(\mathbf{E})$. Let U denote the open compact subroup of $R_{\mathbf{f}}^{\times}$ consisting of those elements which induce automorphisms of $T(\mathbf{E}_0)$ (where $R_{\mathbf{f}}$ denotes $R \otimes \hat{\mathbf{Z}}$). If $\mathbf{E}$ defines an element of $\Sigma_1(N, d)$, then we may choose an isogeny $\lambda : E \to E_0$ and an element g of $H_{\mathbf{f}}^{\times}$ so that $g\lambda$ induces an isomorphism of enhanced Tate modules. This construction, similar to [14], (3.3), defines a bijection between $\Sigma_1(N, d)$ and the double coset space

$$U \backslash H_{\mathbf{f}}^{\times} / H^{\times}.$$

The curve $X_1(N, dq)$ comes with the usual Hecke correspondences T_n for integers $n \geq 1$ and $\langle m \rangle$ for integers m relatively prime to N. For example, if r is a prime not dividing Ndq, then T_r is described by

$$(E, P, C) \mapsto \sum_{B} (E/B, P \bmod B, (C + B)/B)$$

where the sum is over subgroups B of E of order r. These correspondences are defined over $\mathbf{Q}$, as is the operator w_q defined by

$$(E, P, C) \mapsto (E/C_q, P \bmod C_q, (C + E[q])/C_q).$$

We use the same symbols to denote the endomorphisms of $J_1(N, dq)$, T and X these induce by Pic functoriality, and we find that [14], (3.7),(3.8) hold with "involution" replaced by "automorphism", $X_0(N)$ replaced by $X_1(N, dq)$ and $\Sigma(M)$ replaced by $\Sigma_1(N, d)$. (See [15], Section 8.)

We let $\mathbf{T} = \mathbf{T}_1(N, dq)$ be the subring of End$(J_1(N, dq))$ generated by the operators T_n for $n \geq 1$. We recall that $\langle m \rangle$ is in $\mathbf{T}$ for m relatively prime to N and that $\mathbf{T}$ also acts faithfully on $S = S_2(\Gamma_1(N, dq))$. Decompose $S = S_0 \oplus S_1$ where S_0 is the "q-old" subspace of S and S_1 is the "q-new" subspace. Thus S_0 is isomorphic to the direct sum of two copies of $S_2(\Gamma_1(N, d))$ and S_1 is its orthogonal complement. The action of $\mathbf{T}$ on S respects the decomposition, and we may consider the q-old quotient $\mathbf{T}_0$ of $\mathbf{T}$ and the q-new quotient $\mathbf{T}_1$ of $\mathbf{T}$. No changes are then needed for [14], (3.10), (3.11). We postpone the discussion of monodromy pairings and component groups until the next section.

For distinct primes p and q not dividing N, we consider the two natural degeneracy maps

$$\alpha_p, \beta_p : X_1(N, pq) \to X_1(N, q).$$

The maps induced on the character groups are described explicitly by the corresponding maps $\Sigma_1(N, p) \to \Sigma_1(N, 1)$ defined by $\alpha_p(E, P, C) = (E, P, 0)$ and $\beta_p(E, P, C) = (E/C, P \bmod C, 0)$. These combine to produce a map

$$\delta_p : L_q \to X_q$$

where L_q is the character group X defined with $d = p$ and X_q is the product of two copies of the corresponding character group defined with $d = 1$. Thus L_q replaces the group L in [14] and X_q replaces $X \oplus X$. To prove that δ_p is surjective ([14], (3.15)), we appeal directly to the strong approximation theorem rather than the arithmetic of Eichler orders. Fix a supersingular enhanced elliptic curve $\mathbf{E}_0 = (E_0, P_0, 0)$, and let $R = \mathrm{End}(E_0)$ and $H = R \otimes \mathbf{Q}$. Define $U \subset R_{\mathbf{f}}^{\times}$ as above and note that $\nu(U) = \hat{\mathbf{Z}}$ where ν is the reduced norm. If $\mathbf{E} = (E, P, 0)$ defines an element of $\Sigma_1(N, 1)$, choose an isogeny $\lambda : E_0 \to E$ and an element x in $H_{\mathbf{f}}^{\times}$ so that λx induces an isomorphism of enhanced Tate modules. Using the strong approximation theorem we find an element β in $xUH_p^{\times} \cap H^{\times}$ so that $\nu(\beta)^{-1}$ is the degree of λ. If n is sufficiently large and $p^n \equiv 1 \bmod N$, then $p^n\lambda\beta$ is an isogeny $\mathbf{E}_0 \to \mathbf{E}$ whose degree is an even power of p. We then proceed as in [**14**], page 452 to conclude that δ_p is surjective. (Note that $\mathbf{E}_{2j}$ gets replaced by $\langle p\rangle^{-j}\mathbf{E}_{2j}$.)

Next we consider the action of $\mathbf{T} = \mathbf{T}_1(N, pq)$ on L_q. Write Y for the kernel of δ_p. We distinguish T_p in $\mathbf{T}$ from the pth Hecke operator in $\mathbf{T}_1(N, q)$, denoting the latter by τ_p. As in [**14**], (3.19), we find that T_p preserves Y, that $T_p = -w_p$ on Y and that T_p on X_q is represented by the matrix

$$\begin{pmatrix} \tau_p & -1 \\ \langle p\rangle p & 0 \end{pmatrix}$$

of endomorphisms of the character group obtained from $J_1(N, q)$. Define in the obvious way various "old" and "new" subspaces of $S = S_2(\Gamma_1(N, pq))$ and quotients of $\mathbf{T}$. With the evident changes in notation, we have [**14**], (3.20), (3.21).

We now turn our attention to the Shimura curve analogous to the one considered in [**14**], Section 4. We fix a quaternion algebra B over $\mathbf{Q}$ of discriminant pq and a maximal order L in B. Also fix a cyclic left L-submodule $Q \subset L/NL$ of exponent N and order N^2. Let Γ denote the set of $g \in L$ of reduced norm 1 satisfying $xg = x$ for all x in Q. Choosing an isomorphism $B \otimes \mathbf{R} \cong M_2(R)$ we obtain an action of Γ on the upper half-plane $\mathcal{Z}$. The Shimura curve $\Gamma\backslash\mathcal{Z}$ has a canonical model over $\mathbf{Q}$ parametrizing triples (A, i, α) where A is an abelian surface, i is an action $L \hookrightarrow \mathrm{End}(A)$ and and α is an L-linear embedding $Q \hookrightarrow A$. We have also the usual Hecke correspondences T_n for integers $n \geq 1$ and $\langle m\rangle$ for integers m relatively prime to N.

The theory of Cerednik and Drinfeld [**7**], Section 4 provides a model $\mathcal{C}$ over $\mathbf{Z}_p$ for this Shimura curve. The formal scheme associated to this model is as in [**14**], pages 457-8, except that in the description of the p-adic space X, an "enhancement" should now be an L-linear embedding of Q. Let A be an abelian surface over $\overline{\mathbf{F}}_p$ with an action of L defined by i, an enhancement $\alpha : Q \hookrightarrow A$, and an $L \otimes \mathbf{Z}_p$-isomorphism ι between the formal group of A and the formal module Φ defined in [**7**]. We let $\mathcal{M} = \mathrm{End}_L(A)$ and $\mathcal{H} = \mathcal{M} \otimes \mathbf{Q}$. Thus $\mathcal{H}$ is a definite quaternion algebra over $\mathbf{Q}$ with discriminant q, $\mathcal{M}$ is a maximal order in $\mathcal{H}$ and ι induces an isomorphism $\mathcal{M}_p \cong M_2(\mathbf{Z}_p)$. Let K denote the subgroup of prime-to-p ideles in $\mathcal{M}_{\mathbf{f}}^{\times}$ which induce automorphisms of the enhanced Tate module of (A, α). The construction described in [**14**] then gives a bijection between X and the double coset space

$$K\backslash\mathcal{H}_{\mathbf{f}}^{\times}/\mathcal{H}^{\times}$$

compatible with the action of $GL_2(\mathbf{Q}_p) \cong \mathcal{H}_p^{\times}$.

Let r be the order of p in $(\mathbf{Z}/N\mathbf{Z})^{\times}/\{\pm 1\}$, and let G denote the subgroup of $GL_2(\mathbf{Q}_p)$ consisting of those g for which $v_p(\det g)$ is divisible by $2r$. By the method of [**11**], Section 4 we find that the dual graph $\mathcal{G}$ associated to the special fiber of $\mathcal{C}$ is isomorphic to $G\backslash(\Delta\times X)$, where Δ is the tree associated with $SL_2(\mathbf{Q}_p)$. Now let $V = K\mathcal{M}_p^\times$ and let $W = KW_p$ where W_p is the subgroup of $\mathcal{M}_p^\times$ corresponding to

$$\left\{ \begin{pmatrix} a & b \\ c & d \end{pmatrix} \in GL_2(\mathbf{Z}_p) \,\middle|\, c \equiv 0 \bmod p \right\}$$

under the isomorphism induced by ι. Setting $\mathcal{V} = V\backslash\mathcal{H}_\mathbf{f}^\times/\mathcal{H}^\times$, $\mathcal{E} = W\backslash\mathcal{H}_\mathbf{f}^\times/\mathcal{H}^\times$ and defining two degeneracy maps α and β from $\mathcal{E}$ to $\mathcal{V}$ as in [**14**], we obtain the description of $\mathcal{G}$ in [**14**], (4.4). (Use that $G\mathbf{Q}_p^\times = GL_2(\mathbf{Q}_p)^+$.) This gives the description in [**14**], (4.5) of the character group of the connected component of the special fiber of the Neron model over $\mathbf{Z}_p$ of the Jacobian J of the Shimura curve. We will denote this group, analogous to Z in [**14**], by Y_p. Instead of [**14**], (4.7) we have

Proposition 3.1 *There is a supersingular elliptic curve E over $\overline{\mathbf{F}}_q$, with a point P of order N, a subgroup C of order p and an isomorphism $\mathcal{M} \cong \mathrm{End}\,(E)$ such that $W \subset \mathcal{M}_\mathbf{f}^\times$ corresponds to the stabilizer of (P, C) in $(\mathrm{End}\,(E) \otimes \hat{\mathbf{Z}})^\times$.*

The existence of the curve E and the isomorphism $\mathcal{M} \cong \mathrm{End}\,(E)$ is proved in [**14**], (3.6). To choose P note that if r is a prime dividing N, then there is an isomorphism $\mathcal{M}_r \cong M_2(\mathbf{Z}_r)$ such that W_r corresponds to

$$\left\{ \begin{pmatrix} a & b \\ c & d \end{pmatrix} \in GL_2(\mathbf{Z}_r) \,\middle|\, a - 1 \equiv c \equiv 0 \bmod N\mathbf{Z}_r \right\}.$$

We construct C similarly.

Using (E, P, C) as $\mathbf{E}_0$ in the double coset description of $\Sigma_1(N,p)$, the isomorphism $\mathcal{M} \cong R$ induces a bijection $\iota : \Sigma_1(N,p) \to \mathcal{E}$. Similarly using $(E, P, 0)$ we construct $\lambda : \Sigma_1(N,1) \to \mathcal{V}$. We have the same compatibilities as in [**14**], (4.8). This establishes an isomorphism $Y \cong Y_p$. (Recall that Y is the kernel of the map $\delta_p : L_q \to X_q$, whereas Y_p is defined as a character group.) That this isomorphism commutes with the Hecke operator T_p follows from the argument of [**14**], pages 463-4; one need only take care to distinguish between the various automorphisms and their inverses. (In particular T_p coincides with Frob_p on $\mathcal{C}(\overline{\mathbf{F}}_p)$. This induces the inverse of the automorphism of $\mathcal{G}$ denoted $F(\mathcal{C}/\mathbf{Z}_p)$ in [**11**]. This in turn coincides with the automorphism of $G\backslash(\Delta \times X)$ induced by an element m of $GL_2(\mathbf{Q}_p)$ with $\nu_p(\det m) = 1$.) It is straighforward to verify the compatibility with T_r for primes $r \neq p$, and with $\langle m \rangle$ for integers m prime to N. Letting $\tilde{\mathbf{T}}$ denote the polynomial ring over $\mathbf{Z}$ generated by the variables T_n for $n \geq 1$, we conclude that $\tilde{\mathbf{T}}$ acts on Y_p via its quotient $\mathbf{T}$ and that

Theorem 3.1. *There is a $\mathbf{T}$-isomorphism $Y_p \to Y$.*

4. Lowering the level. We first follow Ribet's argument in Sections 6-8 of [**15**] to use the above results on character groups to prove

Theorem 4.1. *Suppose that ℓ is odd and that $\rho : \mathrm{Gal}(\overline{\mathbf{Q}}/\mathbf{Q}) \to GL_2(\mathbf{F})$ is an irreducible representation arising from $S_2(\Gamma_1(N,p))$ for some prime p not dividing $N\ell$. If ρ is unramified at p, then ρ arises from $S_2(\Gamma_1(N))$.*

With $\Gamma_1(N,p)$ replaced by $\Gamma_0(Np)$ and $\Gamma_1(N)$ replaced by $\Gamma_0(N)$, this is [**15**], (1.5). Ribet's proof builds on Mazur's Principle and the work in [**14**] where the theorem is proved under the additional hypothesis that $p \not\equiv 1 \bmod \ell$ or that ℓ does not divide N. Note that [**15**], (1.5) already gives Theorem 4.1 in the case $N \leq 3$, so we assume that $N > 3$ until we have completed the proof of Theorem 4.1.

Before explaining the proof of Theorem 4.1, we continue the discussion of the Jacobians and character groups from the preceding section. Recall that we denote the pth Hecke operator on $J_1(N,pq)$ by T_p and that on $J_1(N,q)$ by τ_p. We use δ_p to denote the homomorphism of Jacobians

$$J_1(N,q)^2 \to J_1(N,pq),$$

as well as that on the character groups $L_q \to X_q$, induced by the two degeneracy maps $X_1(N,pq) \to X_1(N,q)$. Thus the maps denoted δ_p become $\mathbf{T}$-equivariant if we define the action of T_p on $J_1(N,q)^2$ by $\begin{pmatrix} \tau_p & \langle p\rangle p \\ -1 & 0 \end{pmatrix}$ and on X_q by $\begin{pmatrix} \tau_p & -1 \\ \langle p\rangle p & 0 \end{pmatrix}$. We let $\eta_p = T_p^2 - \langle p\rangle$. The proof of [**15**], (6.1) gives now a $\mathbf{T}$-equivariant homomorphism

$$\sigma_p : J_1(N,pq) \to J_1(N,q)^2$$

such that $\sigma_p\delta_p = \eta_p$, and we let κ_p denote the dual of σ_p. Theorem 3.1 gives $\mathbf{T}$-equivariant exact sequences

$$0 \to Y_q \xrightarrow{\iota_q} L_p \xrightarrow{\delta_q} X_p \to 0, \quad 0 \to Y_p \xrightarrow{\iota_p} L_q \xrightarrow{\delta_p} X_q \to 0$$

analogous to [**15**], (6.2), (6.3).

Next we recall that there are natural monodromy pairings on the above character groups. In the case of L_p and L_q these are the restrictions of the standard Euclidean pairings on $\mathbf{Z}^{\Sigma_1(N,q)}$ and $\mathbf{Z}^{\Sigma_1(N,p)}$ (the former is a group of divisors on supersingular $\overline{\mathbf{F}}_p$ points, the latter on supersingular $\overline{\mathbf{F}}_q$ points). In the case of Y_p and Y_q the pairings are induced by the standard ones on $\mathbf{Z}^{\mathcal{E}}$ where $\mathcal{E}$ is the corresponding set of edges. (This follows from the fact that $H^\times \cap xUx^{-1}$ is trivial for $x \in H_{\mathbf{f}}^\times$ under the assumption $N > 3$.) Hence ι_p and ι_q are compatible with the monodromy pairings.

For an abelian variety A over $\mathbf{Q}$ with an action of $\mathbf{T}$, we denote the dual abelian variety with its functorial $\mathbf{T}$-action by A'. If Z_p is the character group associated to the mod p reduction of A, then we write Z'_p the corresponding character group obtained from A', and we use the same convention for component groups. (Thus for example there is a canonical isomorphism of the underlying groups $L_p \to L'_p$ which is not necessarily $\mathbf{T}$-equivariant.) If A has semistable reduction mod p, then the associated component group is canonically isomorphic as a T-module to the cokernel of the map $Z'_p \to \mathrm{Hom}(Z_p, \mathbf{Z})$

induced by the monodromy pairing. We write Φ_p and Θ_p for the component groups obtained from $J_1(N,p) \times J_1(N,p)$ and $J_1(N,pq)$ mod p, and Ψ_q for the one obtained from J mod q where J is the Jacobian of the Shimura curve considered above. In the statements of **[15]**, (6.4),(6.5) we should subscript ι and κ with q. As in [15], (6.6) we have a **T**-equivariant exact sequence

$$0 \to K' \to X_p/\eta_p X_p \to \Psi'_q \to C' \to 0$$

where K' and C' the kernel and cokernel of a **T**-equivariant map $\Phi'_p \to \Theta'_p$. The argument of [15], pages 672-3 shows that Φ_p, Θ_p, Φ'_p and Θ'_p are Eisenstein in the sense that $T_r = r+1$ and $\langle r \rangle = 1$ for primes r not dividing Npq. No changes are needed in the statements of **[15]**, (6.7),(6.8).

Recall that to a maximal ideal $\mathbf{m}$ of $\mathbf{T}$ such that $\mathbf{T}/\mathbf{m}$ has characteristic ℓ, we associate a semisimple representation

$$\rho_{\mathbf{m}} : \mathrm{Gal}\,(\overline{\mathbf{Q}}/\mathbf{Q}) \to GL_2(\mathbf{T}/\mathbf{m}).$$

If r is a prime not dividing $Npq\ell$, then $\rho_{\mathbf{m}}$ is unramified at r and $\rho_{\mathbf{m}}(\mathrm{Frob}_r)$ has characteristic polynomial $X^2 - T_r X + r\langle r \rangle$ mod $\mathbf{m}$. This characterizes $\rho_{\mathbf{m}}$ up to isomorphism. The representations $\rho : \mathrm{Gal}\,(\overline{\mathbf{Q}}/\mathbf{Q}) \to GL_2(\mathbf{F})$ arising from $S_2(\Gamma_1(N,pq))$ are those isomorphic to $\rho_{\mathbf{m}} \otimes \mathbf{F}$ for some maximal ideal $\mathbf{m}$ and some embedding $\mathbf{T}/\mathbf{m} \to \mathbf{F}$. If $\mathbf{m}$ is Eisenstein in the sense that it contains $T_r - (r+1)$ and $\langle r \rangle - 1$ for all but finitely primes r, then $\rho_{\mathbf{m}}$ is reducible. (As noted in Section 2, a weaker notion of Eisenstein implies reducibility of $\rho_{\mathbf{m}}$.) Thus we have [15], (6.9) as stated except that we are writing η_q rather than η.

Now let us assume that ρ satisfies the hypotheses of Theorem 4.1, namely that $\rho : \mathrm{Gal}\,(\overline{\mathbf{Q}}/\mathbf{Q}) \to GL_2(\mathbf{F})$ is irreducible, arises from $S_2(\Gamma_1(N,p))$ for some prime p not dividing $N\ell$, and is unramified at p. We assume also that ℓ is odd. As in **[15]**, page 654 consider the three-dimensional representation $\sigma = \rho \times \chi$ where χ is the mod ℓ cyclotomic character, and choose a prime q not dividing $Np\ell$ so that the conjugacy class of $\sigma(\mathrm{Frob}_q)$ coincides with that of $\sigma(c)$ where c is complex conjugation. We then find a maximal ideal $\mathbf{m}$ of $\mathbf{T} = \mathbf{T}_1(N,pq)$ and an embedding $\mathbf{T}/\mathbf{m} \to \mathbf{F}$ so that ρ is obtained from $\rho_{\mathbf{m}}$ by extension of scalars. Moreover $q \equiv -1$ mod ℓ and

$$T_q^2 \equiv \langle q \rangle \equiv 1 \bmod \mathbf{m},$$

so that in particular $\eta_q \in \mathbf{m}$.

Next consider the kernel of $\mathbf{m}$ in $J_1(N,pq)$, denoted $J_1(N,pq)[\mathbf{m}]$. For our contravariant Hecke operators and arithmetic Frobenius Frob_r, the Eichler-Shimura relation gives

$$\mathrm{Frob}_r^2 - \langle r \rangle^{-1}\mathrm{Frob}_r + \langle r \rangle^{-1} r = 0$$

on $J_1(N,pq)[\ell]$ for primes r not dividing $Npq\ell$. It therefore follows from **[2]** that $J_1(N,pq)[\mathbf{m}]$ is isomorphic as a $\mathbf{T}/\mathbf{m}[\mathrm{Gal}\,(\overline{\mathbf{Q}}/\mathbf{Q})]$-module to a direct sum of copies of $W = \mathrm{Hom}(V, \mu_l)$ where V is a model for $\rho_{\mathbf{m}}$. We define λ as the number of copies, so that $\lambda > 0$ and $\dim_{\mathbf{T}/\mathbf{m}}(J_1(N,pq)[\mathbf{m}]) = 2\lambda$. We have similarly that $J[\mathbf{m}]$ is a direct sum of copies of μ copies of W where $\dim_{\mathbf{T}/\mathbf{m}}(J[\mathbf{m}]) = 2\mu$.

Under the assumption that ρ does not arise from $S_2(\Gamma_1(N,q))$, we have equalities and inequalities identical to [**15**], (7.4)-(7.8). (For the analogues of [**15**], (7.7),(7.8), use that Frob_q acts as T_q on Y_q and L_q. Since $T_q^2 = \langle q\rangle$ on these groups it follows that Frob_q acts as $q\langle q\rangle^{-1}T_q$ on $\mathrm{Hom}(Y_q,\mu_\ell)$ and $\mathrm{Hom}(L_q,\mu_\ell)$, and as $\langle q\rangle^{-1}T_q$ on Y_q'.) The resulting contradiction shows that ρ arises from $S_2(\Gamma_1(N,q))$.

To complete the proof of Theorem 4.1, we appeal to Mazur's Principle [**15**], (8.1) to conclude that ρ arises from $S_2(\Gamma_1(N))$. We remark on a potentially confusing point in the proof of Mazur's Principle in [**15**]. Ribet considers a maximal ideal $\mathbf{m}$ in $\mathbf{T}_1(N,q)$ such that $\rho_{\mathbf{m}}$ provides a model V for ρ and asserts that $J_1(N,q)[\mathbf{m}]$ is isomorphic to a direct sum of copies of V. Implicit in this assertion is a choice of conventions different from the one used above. For example, one could use the above definitions for the curve $X_1(N,q)$ and the Hecke correspondences T_n, but define the action of T_n on $J_1(N,q)$ using Albanese functoriality. In that case, Frob_q would act on the character group X as $\langle q\rangle^{-1}T_q$ rather than T_q as stated on [**15**], page 674. Here X denotes the character group of the torus T in the reduction mod q of $J_1(N,q)$, so Frob_q would then act as qT_q on $T(\overline{\mathbf{F}}_q)$. With our conventions, however, $J_1(N,q)[\mathbf{m}]$ is isomorphic to a direct sum of copies of $W = \mathrm{Hom}(V,\mu_\ell)$, Frob_q coincides with T_q on X, and with $q\langle q\rangle^{-1}T_q$ on $T(\overline{\mathbf{F}}_q)$. The square of this last scalar is $q^2\langle q\rangle^{-1}$ while the determinant of Frob_q on W is $q\langle q\rangle^{-1}$, yielding Mazur's Principle just the same. Yet another set of conventions is used in [**9**] where the modular curve parametrizes elliptic curves together with embeddings of μ_N and the Hecke operators on the Jacobian are defined using Albanese functoriality. (Note in particular [**9**], (3.12).)

5. Proof of Theorem 1.1. As explained in [**15**], Theorem 1.1 follows from Theorem 4.1 if $\ell > 3$. Very few changes are needed to Sections 2, 3 and 4 of [**15**] to include the case $\ell = 3$. We record a slightly stronger version of [**15**], (2.2):

Theorem 5.1. *Let f be an eigenform in $S_k(\Gamma_1(N))$. Suppose that ℓ does not divide N, that $2 \le k \le \ell+1$ and that $\rho = \rho_f \otimes_{R_f} \mathbf{F}$ is irreducible. Then there is an eigenform g in $S_2(\Gamma_1(N\ell))$ such that $\alpha(a_n(f)) = \alpha(a_n(g))$ for all $n \ge 1$. In particular $\rho \cong \rho_g \otimes_{R_g} \mathbf{F}$ arises from $S_2(\Gamma_1(N\ell))$.*

For $\ell > 3$, this follows from [**1**], (3.5), and for $N > 4$ from [**9**], (9.3). Note that the eigenform g must be cuspidal since ρ is irreducible. Note also that $S_k(\Gamma_1(N)) = 0$ if $2 \le k \le 4$ and $N \le 4$.

Appealing to the work of Gross [**9**], Edixhoven [**8**] and Coleman and Voloch [**4**] on the weight in Serre's conjecture, we see that [**15**], (3.3) is valid for odd ℓ so that Theorem 1.1 will be a consequence of the following.

Theorem 5.2. *Let ℓ be an odd prime and N a positive integer. If $\ell = 3$, then suppose also that 9 divides N. Let ρ be an irreducible representation arising from $S_2(\Gamma_1(N))$. Suppose that $p \ne \ell$ is a prime dividing $N/N(\rho)$. Then ρ arises from $S_2(\Gamma_1(N/p))$.*

We proceed as in [**15**], Section 4, combining methods and results of Carayol [**3**] with Theorem 4.1. We first regard ρ as arising from an automorphic representation π of $GL_2(\mathbf{A})$ associated to a weight two newform of conductor dividing N. We may assume that the power of p dividing N coincides with the conductor of π_p. We distinguish three cases according to whether π_p is principal series, special or cuspidal.

Suppose that π_p is principal series. Let d be the highest power of ℓ dividing N. Now apply Lemma 2.1 to conclude that ρ arises from $S_k(\Gamma_1(N))^\chi$ where χ is a mod N/d Dirichlet character of order prime to ℓ. The conclusion of Theorem 5.2 follows from Carayol's argument [**15**], (4.4).

Suppose that π_p is special. The discussion in [**15**] explains why the desired conclusion follows from Theorem 4.1. Indeed Theorem 4.1 provides a solution to [**15**], (4.6). The discussion applies also to the case $\ell = 3$.

Suppose that π_p is cuspidal, in which case p^2 divides N. We first remark that if $k = 2$, then the hypothesis that ℓ does not divide N is not needed in [**15**], (5.1). Indeed [**13**], (4.2) provides the exact sequence [**15**], (5.3) of free $\mathbf{Z}_\ell$-modules. So we may choose a prime q not dividing $N\ell$ such that ρ arises from a newform of weight two and conductor dividing Nq and for which the associated automorphic representation π' is special at q. In view of the cases considered above, we need only prove that if π'_p is cuspidal, then ρ arises from $S_2(\Gamma_1(Nq/p))$. For $\ell > 3$ this case is treated by Carayol [**3**], Section 5. If $\ell = 3$, then the hypothesis that 9 divides N ensures that the stabilizers in [**3**], Lemme 2 are trivial so that Carayol's argument applies here as well. This completes the proof of Theorem 1.1.

As a final remark we mention an alternative to the above argument treating the cuspidal case. Recall that Carayol uses the Jacquet-Langlands correspondence to switch to an automorphic representation on $(B \otimes \mathbf{A})^\times$, where B is a quaternion algebra over $\mathbf{Q}$ of discriminant pq, and he then applies an analogue of Lemma 2.1. One can instead switch to a definite quaternion algebra of discriminant p without introducing an auxiliary prime q. As an analogue of Carayol's lemma in the setting of a definite quaternion algebra, we have [**6**], Theorem 7. (The hypothesis that ℓ does not divide N appears in the statement, but in fact it is never used in the proof.) To obtain the desired conclusion, one can apply [**6**], Theorem 7 with $T = S(D) = \{p\}$ and ψ'_p as the trivial character.

6. Wiles' Variant. We now explain how to deduce a variant of Theorem 1.1 proposed by Wiles [**19**]. We wish to prove that if

$$\rho : \mathrm{Gal}\,(\overline{\mathbf{Q}}/\mathbf{Q}) \to GL_2(\mathbf{F})$$

is irreducible and modular, then ρ arises from an eigenform of weight two for which the associated ℓ-adic representation is of the same "deformation type" as ρ. This means that there is a weight two eigenform f such that $\rho \cong \rho_f \otimes_{R_f} \mathbf{F}$, subject to some additional constraints on the local representations $\rho_f|_{D_p}$. Here D_p denotes a decomposition group at p and the constraints are determined by $\rho|_{D_p}$. For primes p not dividing $\ell N(\rho)$ we will require simply that ρ_f be unramified at p. To describe the restrictions imposed at the ramified primes, we will need more notation and definitions.

Let $\sigma : G \to GL_2(R)$ be a continuous homomorphism, where G is a profinite group and R is a complete local Noetherian ring with finite residue field of characteristic ℓ. Let $\mathfrak{m}$ denote the maximal ideal of R and write σ_n for $\sigma \otimes_R R/\mathfrak{m}^{n+1}$. Then σ is called a lifting of σ_0. Two such liftings are equivalent if they are conjugate by an element of the kernel of $GL_2(R) \to GL_2(R/\mathfrak{m})$, and an equivalence class of liftings is called a deformation of σ_0 (see [**12**]).

Consider now the case of $G = D_p$. Let I_p denote the inertia subgroup of D_p and H_p the wild inertia subgroup. We will write χ_ℓ for the composition of the ℓth cyclotomic character with the natural homomorphism $\mathbf{Z}_\ell^\times \to R^\times$. We recall the types of local representations considered by Wiles. For $p \neq \ell$ we make the following definitions:

- σ is *type A* if σ_0 is ramified and $\sigma \cong \begin{pmatrix} \psi\chi_\ell & * \\ 0 & \psi \end{pmatrix}$ for some unramified character $\psi : D_p \to R^\times$.
- σ is *type B* if $\sigma|_{I_p} \cong \begin{pmatrix} \psi & 0 \\ 0 & 1 \end{pmatrix}$ for some nontrivial character $\psi : I_p \to R^\times$ of finite order prime to ℓ.

For $p = \ell$ we say

- σ is *finite* if $\det \sigma|_{I_p} = \chi_\ell|_{I_p}$ and for each $n \geq 0$, σ_n arises from a finite flat group scheme over $\mathbf{Z}_p$. (This means that there is a finite flat group scheme G_n over $\mathbf{Z}_p$ with an action of R such that $G_n(\overline{\mathbf{Q}}_p)$ is free of rank two over $R/\mathfrak{m}^{n+1}$ and σ_n is isomorphic to $D_p \to \mathrm{Aut}_R(G_n(\overline{\mathbf{Q}}_p))$.)
- σ is *Selmer* if $\sigma \cong \begin{pmatrix} \phi & * \\ 0 & \psi \end{pmatrix}$ for characters $\phi, \psi : D_p \to R^\times$ with ψ unramified and $\phi|_{H_p} = \chi_\ell|_{H_p}$.

In this last case we say also that σ is *ψ_0-Selmer* (where for a homomorphism $\psi : D_p \to R^\times$, we write ψ_0 for $\psi \bmod \mathfrak{m}$). Note that if σ is Selmer, then ψ_0 is determined by σ_0 unless σ_0 is the direct sum of two distinct unramified characters.

We remark on some slight differences between the above terminology and that of Wiles. Firstly, we use the adjectives "type A", "type B", etc., to describe local Galois representations rather than deformations of a fixed global residual representation. For $p = \ell$ there are further distinctions. Note that a representation may be both finite and Selmer, and that Wiles calls a deformation "flat" if the local representation is (according to the definitions above) finite, but not Selmer. Note also that our notion of Selmer is more general than that of [**19**] in that we do not require $\phi_0 \neq \psi_0$. Wiles considers also "ordinary" deformations for which there is no restriction on $\phi|_{H_p}$. We make no mention of these as the results analogous to those below are immediate from the ones in the Selmer case. Finally, Wiles calls a deformation "strict" if the local representation σ is Selmer with $\phi = \psi\chi_\ell$ and σ_0 is not finite. We make no further mention of these either since the following proposition shows that if σ_0 is strict, then "strict" and "Selmer" are equivalent.

Proposition 6.1. *Suppose that $p = \ell$ and $\sigma \cong \begin{pmatrix} \phi\chi_\ell & * \\ 0 & \psi \end{pmatrix}$ for unramified characters ϕ and ψ. If $\phi \neq \psi$, then σ_0 is finite.*

Proof. We may assume that R is Artinian and $\psi = 1$. Now $\sigma \cong \begin{pmatrix} \phi\chi_\ell & t \\ 0 & 1 \end{pmatrix}$ for a cocycle $t : D_p \to M(1)$ where M is a free R-module of rank one on which D_p acts via ϕ. Let u denote the cohomology class in $H^1(D_p, M(1))$ defined by t, and let u_0 denote its image in $H^1(D_p, M_0(1))$ where $M_0 = M/\mathfrak{m}M$. Let $G = \ker\phi$ and let F be the fixed field of G (so F is a finite unramified extension of $\mathbf{Q}_p$). Choose n so that $\ell^n R = 0$. Since $H^2(G, \mu_{\ell^r}) \to H^2(G, \mu_{\ell^s})$ is injective for $r \leq s$, we see that the natural map of $R[D_p/G]$-modules $H^1(G, \mu_{\ell^n}) \otimes_{\mathbf{Z}_\ell} M \to H^1(G, M(1))$ is an isomorphism. So by Kummer theory we have $H^1(G, M(1)) \cong F^\times/(F^\times)^{\ell^n} \otimes_{\mathbf{Z}_\ell} M$. Now consider the commutative diagram

$$\begin{array}{ccccc} H^1(G, M(1))^{D_p} & \tilde{\to} & ((F^\times/(F^\times)^{\ell^n}) \otimes_{\mathbf{Z}_\ell} M)^{D_p} & \to & M^{D_p} \\ \downarrow & & \downarrow & & \downarrow \\ H^1(G, M_0(1)) & \tilde{\to} & (F^\times/(F^\times)^{\ell}) \otimes_{\mathbf{F}_\ell} M_0 & \to & M_0, \end{array}$$

where the rightmost horizontal maps are induced by $v_p : F^\times \to \mathbf{Z}$. If $\phi \neq 1$, then $M^{D_p} \subset \mathfrak{m}M$, so that the element $\operatorname{res} u_0$ of $H^1(G, M_0(1))$ is in the image of $(\mathcal{O}_F^\times/(\mathcal{O}_F^\times)^\ell) \otimes_{\mathbf{F}_\ell} M_0$. But this means that σ_0 is "peu ramifié" in the sense of **[16]** and therefore σ_0 is finite (see **[8]**, (8.2)). □

We remark that the same argument for $p \neq \ell$ proves the following:

Proposition 6.2. *Suppose $p \neq \ell$, $\sigma|_{I_p} \cong \begin{pmatrix} 1 & * \\ 0 & 1 \end{pmatrix}$ and σ_0 is ramified. Then σ is type A.*

We now return to the setting of ℓ-adic representations arising from modular forms. We let f be a newform of weight k and conductor N. Suppose that $\rho = \rho_f \otimes_{R_f} \mathbf{F}$ is irreducible. Write $N(\rho) = \prod_p p^{\alpha_p}$ and $N = \prod_p p^{\beta_p}$. We consider first $\sigma = \rho_f|_{D_p}$ for primes $p \neq \ell$. Recall that p^{β_p} is the conductor of $\sigma \otimes_{R_f} \overline{K}_{f,\lambda}$, that p^{α_p} is the conductor of $\sigma_0 \otimes_{R_f} \mathbf{F}$ and that $\alpha_p \leq \beta_p$. Also note that if σ is type A then $\alpha_p = 1$ and that if σ is type B then p^{α_p} is the conductor of $\det\sigma_0$.

Proposition 6.3. *Suppose that $\alpha_p = \beta_p$, $\det\sigma|_{I_p}$ has order prime to ℓ and σ_0 is type A (resp. type B). Then σ is type A (resp. type B).*

Proof. For ℓ odd, it is immediate from the classification in **[6]**, Section 1 that $\tau_p = \sigma \otimes_{R_f} \overline{K}_{f,\lambda}$ has the desired form. It follows easily that so does σ.

For $\ell = 2$, the classification of **[6]**, pp. 255-6 is valid except for the conclusion in case 3 that F is unramified if τ_p is induced from a character whose reduction is equal to its conjugate. Note in this case however that the semisimplification of $\sigma \otimes_{R_f} \mathbf{F}$ is scalar so that σ_0 cannot be type B. The argument is then the same as for ℓ odd. □

We remark that the above proposition is implicit in **[19]** where results of Langlands are used to obtain information about the structure at p of the automorphic representation corresponding to f, and consequently information about σ.

Now we consider $\sigma = \rho_f|_{D_p}$ for $p = \ell$.

It is well-known that σ is finite if $k = 2$ and $\beta_\ell = 0$. Indeed $\det\sigma$ is the product of χ_ℓ with an unramified character, and the σ_n occur in the ℓ^n-torsion of an abelian variety with good reduction at $p = \ell$.

Recall ([**18**], Theorem 2) that if $k = 2$, $\beta_\ell \leq 1$ and $a_\ell(f)$ is a unit in R_f, then ρ_f is ψ_0-Selmer where ψ_0 is the unramified character $D_\ell \to (\mathcal{O}_f/\lambda)^\times$ defined by $\mathrm{Frob}_\ell \mapsto a_\ell(f) \mod \lambda$. Recall also that if $2 \leq k \leq \ell+1$, $\beta_\ell = 0$ and σ_0 is Selmer, then the eigenvalue $a_\ell(f)$ is a unit in R_f and σ_0 is ψ_0-Selmer where ψ_0 is again defined by $\mathrm{Frob}_\ell \mapsto a_\ell(f) \mod \lambda$. This is a consequence of results of Deligne and Fontaine stated in Edixhoven's article [**8**], (2.5), (2.6) using Katz's definition of modular forms over $\mathbf{F}$. If ℓ does not divide N then the reduction of an eigenform f gives (according to the definitions in [**8**]) an eigenform f' over $\mathbf{F}$ of type (N,k), and f' is characterized by $a_n(f') = \alpha(a_n(f))$. Conversely, if $k \geq 2$, $N\ell > 4$ and ℓ does not divide N, then an eigenform over $\mathbf{F}$ of type (N,k) is the reduction of a characteristic zero eigenform of weight k and level N. Recall also that if $\rho_{f,0}$ is irreducible then f must be a cusp form.

We now prove

Theorem 6.4. *Let $\mathbf{F}_0$ be a finite subfield of $\mathbf{F}$ and let $\rho_0 : \mathrm{Gal}(\overline{\mathbf{Q}}/\mathbf{Q}) \to GL_2(\mathbf{F}_0)$ and $\psi_0 : D_\ell \to \mathbf{F}_0^\times$ be continuous homomorphisms. Suppose that $\rho = \rho_0 \otimes_{\mathbf{F}_0} \mathbf{F}$ is irreducible and modular. Assume also that $\ell > 3$ or that $\ell = 3$ and ρ is not induced from a character of $\mathrm{Gal}(\overline{\mathbf{Q}}/\mathbf{Q}(\sqrt{-3}))$. Then there is a newform f of weight two and conductor N dividing $\ell^2 N(\rho)$ so that $\rho \cong \rho_f \otimes_{R_f} \mathbf{F}$ and $\det\rho_f = \chi\chi_\ell$ for a character χ of order prime to ℓ. Moreover*

- *if $\rho_0|_{D_\ell}$ is finite, then f can be chosen so that ℓ does not divide N;*
- *if $\rho_0|_{D_\ell}$ is ψ_0-Selmer, then f can be chosen so that $\alpha(a_\ell(f)) = \psi_0(\mathrm{Frob}_\ell)$ and ℓ^2 does not divide N.*

Proof. The first assertion is immediate on combining Theorem 1.1, Theorem 5.1 and Lemma 2.1.

For the second assertion, let g be an eigenform in $S_{k(\rho)}(\Gamma_1(N(\rho)))$ such that $\rho \cong \rho_g \otimes_{R_g} \mathbf{F}$. Recall that if $\rho_0|_{D_\ell}$ is finite, then $k(\rho) = 2$, and that if $\rho_0|_{D_\ell}$ is Selmer, then $2 \leq k(\rho) \leq \ell+1$. We claim that if $\rho_0|_{D_\ell}$ is ψ_0-Selmer, then g can be chosen so that $\alpha(a_\ell(g)) = \psi_0(\mathrm{Frob}_\ell)$. This is clear unless $\rho_0|_{D_\ell}$ is the direct sum of distinct unramified characters ϕ_0 and ψ_0 and $\alpha(a_\ell(g)) = \phi_0(\mathrm{Frob}_\ell)$. But in that case $k(\rho) = \ell$ and [**8**], (2.9) gives a "companion form" g' over $\mathbf{F}$ of type $(N(\rho),1)$. We also see from the proof of [**8**], (2.8) that $X^2 - a_\ell(g')X + \det(\rho_0(\mathrm{Frob}_\ell))$ has distinct roots $\psi_0(\mathrm{Frob}_\ell)$ and $\phi_0(\mathrm{Frob}_\ell)$ and that there is an eigenform g'' of type $(N(\rho),\ell)$ satisfying

- $a_p(g'') = \alpha(a_p(g))$ for $p \neq \ell$;
- $a_\ell(g'') = \psi_0(\mathrm{Frob}_\ell)$.

Since $S_\ell(\Gamma_1(N(\rho))) \neq 0$, we see that $N(\rho)\ell > 4$, so g'' is the reduction of an eigenform with which we replace g.

If $\rho_0|_{D_\ell}$ is finite, then let $f = g$. If $\rho_0|_{D_\ell}$ is Selmer but not finite, then by Theorem 5.1, there is an eigenform f in $S_2(\Gamma_1(\ell N(\rho))$ satisfying $\alpha(a_n(f)) = \alpha(a_n(g))$ for all $n \geq 1$. Therefore $\rho \cong \rho_f \otimes_{R_f} \mathbf{F}$ and $\alpha(a_\ell(f)) = \psi_0(\mathrm{Frob}_\ell)$. Finally by the variant (Lemma 2.2) of Carayol's result, we see that f can be chosen with character of order prime to ℓ. □

Corollary 6.5. *Under the hypotheses of the theorem, there is a weight two* *newform f such that*

- ρ_f *is unramified outside* $\ell N(\rho)$;
- $\rho \cong \rho_f \otimes_{R_f} \mathbf{F}$;
- *for each prime p such that $\rho_0|_{D_p}$ is type A (resp. type B, resp. finite, resp. ψ_0-Selmer), $\rho_f|_{D_p}$ is also type A (resp. type B, resp. finite, resp. ψ_0-Selmer).*

References

[1] A. Ash and G. Stevens, *Modular forms in characteristic ℓ and special values of their L-functions*, Duke Math. J. **53** (1986), 849-868.

[2] N. Boston, H.W. Lenstra and K. Ribet, *Quotients of group rings arising from two-dimensional representations*, C. R. Acad. Sci. Paris, Série I **312** (1991), 323-328.

[3] H. Carayol, *Sur les représentations galoisiennes modulo l attachées aux formes modulaires*, Duke Math. J. **59** (1989), 785-801.

[4] R.F. Coleman and J.F. Voloch, *Companion forms and Kodaira-Spencer theory*, Invent. Math. **110** (1992), 263-281.

[5] P. Deligne and M. Rapoport, *Les schémas modulaires de courbes elliptiques*, Lecture Notes in Math. **349** (1973), 143-316.

[6] F. Diamond and R. Taylor, *Lifting modular mod l representations*, Duke Math. J. **74** (1994), 253-269.

[7] V.G. Drinfeld, *Coverings of p-adic symmetric regions* (in Russian), Funkts. Anal. Prilozn. **10** (1976), 29-40. Translation in Funct. Anal. Appl. **10** (1976), 107-115.

[8] B. Edixhoven, *The weight in Serre's conjectures on modular forms*, Invent. Math. **109** (1992), 563-594.

[9] B.H. Gross, *A tameness criterion for Galois representations associated to modular forms mod p*, Duke math. J. **61** (1990), 445-517.

[10] H. Hida, *Galois representations into $GL_2(\mathbf{Z}_p[[X]])$ attached to ordinary cusp forms*, Invent. Math. **85** (1986), 545-613.

[11] B. Jordan and R. Livné, *Local diophantine properties of Shimura curves*, Math. Ann. **270** (1985), 235-248.

[12] B. Mazur, *Deforming Galois representations*, in Galois Groups over **Q**, MSRI Publ. **16** (1989), 385-437.

[13] K. Ribet, *Congruence relations between modular forms*, Proc. I.C.M. (1983), 503-514.

[14] K. Ribet, *On modular representations of* Gal $(\overline{\mathbf{Q}}/\mathbf{Q})$ *arising from modular forms*, Invent. Math. **100** (1990), 431-476.

[15] K. Ribet, *Report on mod l representations of* Gal $(\overline{\mathbf{Q}}/\mathbf{Q})$, in Motives, Proc. Symp. Pure Math. **55**:2 (1994), 639-676.

[16] J.-P. Serre, *Sur les représentations modulaires de degré* 2 *de* Gal $(\overline{\mathbf{Q}}/\mathbf{Q})$, Duke Math. J. **54** (1987), 179-230.

[17] G. Shimura, *Introduction to the arithmetic theory of automorphic functions*, Iwanami Shoten and Princeton University Press, 1971.

[18] A. Wiles, *On ordinary λ-adic representations associated to modular forms*, Invent. Math. **94** (1988), 529-573.

[19] A. Wiles, Course at Princeton University, February–April, 1994.

Conference on Elliptic Curves and Modular Forms
Hong Kong, December 18-21, 1993

Wiles minus epsilon implies Fermat

Noam D. Elkies
Dept. of Mathematics, Harvard University, 1 Oxford Street,
Cambridge, MA 02138, USA

For an elliptic curve E/Q and for each prime l define the group $G_l(E) \subseteq GL_2(Z/l)$ as the image of the representation of $\mathrm{Gal}(\bar{Q}/Q)$ on $E[l]$, that is, $G_l(E)$ is the Galois group of the extension of Q obtained by adjoining the coordinates of all the l-torsion points of E. By "Wiles minus epsilon" I mean the following recently re-announced result [**4, 3**]:

Theorem. *Let E/Q be a semistable elliptic curve. Assume that $G_3(E) = GL_2(Z/3)$. Then E is modular.*

The "epsilon" is then the concluding part of Wiles' monumental paper, wherein he removes the hypothesis on $G_3(E)$ by cleverly switching between G_3 and G_5 on E and an auxiliary semistable curve. Since Ribet had already shown [**1**], building on observations of Serre [**2**], that the semistable Frey curve associated to a counterexample to Fermat's Last Theorem (FLT) could not be modular, that "epsilon" also completed the proof of FLT.

In this note we show that the above theorem together with the ideas of Frey-Serre-Ribet already implies FLT. The argument is unabashed ad-hockery; in our defense we note that the same is true of Wiles' G_3/G_5 trick (which among other things depends crucially on the fact that the modular curve $X(5)$ has genus zero!), and that though our aims are more modest (we cannot conclude that every semistable curve admitting a rational 3-isogeny is modular), our approach is almost entirely elementary.

Recall that the Frey curve $E = E(A, B, C)$ associated to a solution of

$$A + B + C = 0 \tag{1}$$

in relatively prime nonzero integers A, B, C is the elliptic curve

$$Y^2 = X(X + A)(X - B) \tag{2}$$

or its $\sqrt{-1}$-twist $Y^2 = X(X - A)(X + B)$.

Now fix a prime $p > 3$. Then it is a corollary of the work of Serre and Ribet [**1, 2**] that if (A, B, C) is a solution to (1) in integers, each of the form

$$2^r 3^s m^p \quad \textit{with} \quad \textit{either} \quad r = 0 \textit{ or } r \geq 4, \tag{3}$$

then the Frey curve $E = E(A, B, C)$ is semistable but <u>not</u> modular. This is because if E were modular then the p-torsion points of E would yield a weight-2 cusp form mod p on $X_0(N)$ for $N = 1, 2, 3$, or 6, and no such form exists. [The only possibly new wrinkle here is that I must allow $r = 4$. Fortunately in that case E actually has good reduction at 2, with odd j-invariant. The examples $1 + 1 = 2$ (or $1 + 2 = 3$), $1 + 3 = 4$, and $1 + 8 = 9$ show that we cannot allow $r = 1, 2$, or 3. There are infinitely many examples such as $3 + 125 = 128$ that preclude taking $p = 3$. In [**2**] the condition $p > 3$ assures that $G_p[E]$ acts irreducibly on $E[p]$, i.e. that E does not admit a rational p-isogeny, and indeed the Frey curve associated to $3 + 125 = 128$ admits a rational 3-isogeny; this is curve #30F in the Antwerp tables. Our argument also requires $p > 3$ as will be seen later.]

Thus by the above Theorem $G_3(E)$ must be strictly smaller than $\mathrm{GL}_2(Z/3)$. We shall show that this is impossible for such a Frey curve, or rather (which is good enough), that given a putative counterexample we shall construct a possibly different counterexample whose associated Frey curve has maximal G_3. This will prove that (1) has no solutions with each of A, B, C of the form (3). In particular this will prove FLT while circumventing the G_3/G_5 trick.

Note that $G_3(E)$ cannot be contained in $\mathrm{SL}_2(Z/3)$ because Q does not contain the cube roots of unity. Thus if $G_3(E)$ is properly contained in $\mathrm{GL}_2(Z/3)$ then either it is a 2-group, and so contained in a 2-Sylow subgroup of $\mathrm{GL}_2(Z/3)$, or it acts reducibly on $E[3]$, whence E admits a rational 3-isogeny.

We dispense with the first possibility by noting that, for any elliptic curve E, if $G_3(E)$ is a 2-group then the j-invariant of E is a cube (because $j^{1/3}$ parametrizes the modular curve corresponding to the 2-Sylow subgroup of $\mathrm{GL}_2(Z/3)$). But the j-invariant of the Frey curve $E(A, B, C)$ is $2^8(A^2 + AB + B^2)^3/(ABC)^2$, which is a cube if and only if $ABC = 2D^3$ for some integer D. This, however, is incompatible with $A + B + C = 0$ unless (A, B, C) is proportional to a permutation of $(1, 1, -2)$: the elliptic curve $ABC = 2D^3, A + B + C = 0$ has Weierstrass model $y^2 = x^3 + 1$, conductor 36 and rank zero; its six rational points yield the permutations of $(1, 0, -1)$ and $(1, 1, -2)$. But $(1, 1, -2)$ does not satisfy the condition on the 2-valuations of A, B, C, and anyway yields a Frey curve with complex multiplication ($j = 12^3$) which is thus modular.

We may thus assume that E admits a 3-isogeny. In that case some twist of E has a subgroup of rational points isomorphic to $(Z/2)^2 \times (Z/3)$. Writing that twist of E as $Y^2 = (X - x_1)(X - x_2)(X - x_3)$, and translating X to put the 3-torsion points at $X = 0$, we calculate that $(x_1 : x_2 : x_3)$ must lie on the rational plane quartic nicely parametrized as

$$(4) \qquad (x_1 : x_2 : x_3) :: \left((ab)^2 : (bc)^2 : (ca)^2\right) \quad \textit{with} \quad a + b + c = 0.$$

[In effect[1] $(a : b : c)$ is a uniformizer for $X_0(12)$; the modular automorphisms of that curve coming from $\Gamma(1)/\Gamma(2)$ are visible in the permutations of a, b, c, and the involution w_3 comes from $(a : b : c) \leftrightarrow (b - c : c - a : a - b)$.] Thus A, B, C are proportional to $a^2(c^2 - b^2)$, $b^2(a^2 - c^2)$, and $c^2(b^2 - a^2)$, or equivalently to $a^3(b-c), b^3(c-a), c^3(a-b)$, for some rational (a, b, c) with $a+b+c = 0$. We may

[1]The arithmetic subgroup $\Gamma(2) \cap \Gamma_0(3)$ of $\mathrm{PSL}_2(Z)$ is conjugate in $\mathrm{PGL}_2^+(Q)$ with $\Gamma_0(12)$ (because the groups $\Gamma(2)$ and $\Gamma_0(4)$ are conjugate); thus the corresponding modular curves are isomorphic.

take a, b, c to be coprime integers, and assume without loss of generality that a is even and b, c are odd. Then $a, b, c, b-c, c-a, a-b$ are pairwise coprime except for $\gcd(a, b-c) = 2$, and possibly a common factor of 3 among $b-c, c-a, a-b$. Thus each of $a, b, c, b-c, c-a, a-b$ is a p-th power times some powers of 2 and 3. (This is where we use $p > 3$; our earlier example of $3 + 125 = 128$ with $p = 3$ yields $(a, b, c) = (4, 1, -5)$ with c not of the form (3).)

So A is even, say with 2-valuation $v_2(A) = r \geq 4$, and B, C are odd. Furthermore, either a is singly even and $v_2(b-c) = r-3$, or $b-c$ is singly even and $v_2(a) = (r-1)/3$ — but not both, since a and $b-c$ cannot both be singly even under the hypotheses of the previous paragraph. This immediately excludes the case $r = 4$. (That case could also be ruled out by noting that the (good) reduction mod 2 of E would have both a 2- and a 3-torsion point, but an elliptic curve mod 2 cannot have 6 rational points!). When r is larger, we'll use the fact that $a, b, c, c-a, a-b, b-c$ are each a p-th power times some powers of 2 and 3 to produce a new A, B, C as above, except with smaller r, and in a finite number of steps reach $r = 4$ and obtain our contradiction.

This can probably be done in several ways. Here's one: the identity

$$a(b-c) + b(c-a) + c(a-b) = 0 \tag{5}$$

gives new A, B, C with r replaced by either $r-2$ or $(r+2)/3$, both $< r$. The only problem is that if r was 5 or 7 then the new r will be 3, which is forbidden. In the first case, I use instead the identity

$$(b-c)^2 - (c-a)^2 = (b-a)(b+a-2c) = 3c(a-b), \tag{6}$$

and in the second case the identity

$$a^2 - b^2 = (a+b)(a-b) = c(b-a) \tag{7}$$

(the w_3 image of (6)), to get new A, B, C with $r = 4$. This completes the proof of the impossibility of nonzero coprime integers A, B, C satisfying (1) and (3), and thus in particular of FLT, given of course that "Wiles minus epsilon" is confirmed.

Acknowledgements. Thanks to Ken Ribet for explaining and discussing parts of [4], and to Barry Mazur for several corrections and improvements of exposition.

I am grateful to the National Science Foundation and the Packard Foundation for partial support during the preparation of this note.

References

[1] Ribet, K.A., *On modular representations of Gal($\bar{Q}/Q$) arising from modular forms.* Invent. Math. **100** (1990), 431–476.

[2] Serre, J.-P., *Sur les représentations modulaires de degré 2 de Gal($\bar{Q}/Q$)*, Duke Math. J. **54** (1987), 179–230.

[3] Taylor, R., Wiles, A., *Ring Theoretic Properties of Certain Hecke Algebras*, Preprint, 1994.

[4] Wiles, A., *Modular elliptic curves and Fermat's Last Theorem*, Preprint, 1994.

Conference on Elliptic Curves and Modular Forms
Hong Kong, December 18-21, 1993

Geometric Galois Representations

JEAN-MARC FONTAINE
UNIVERSITÉ DE PARIS-SUD
MATHÉMATIQUE-BÂTIMENT 425, 91405-ORSAY CEDEX FRANCE

BARRY MAZUR
HARVARD UNIVERSITY
DEPARTMENT OF MATHEMATICS
1 OXFORD STREET, CAMBRIDGE, MA 02138

The main point of this paper [1] is to present in detail some of our conjectures concerning p-adic representations of the Galois groups of number fields (cf. [**10**] for a brief synopsis). The motivating idea behind our conjectures is simply the hope that, given an irreducible p-adic representation ρ of G_K, the Galois group of a number field K (assumed unramified except at a finite number of places), one can find necessary and sufficient local conditions (on the restriction of the representation to the decomposition groups G_{K_V} at the finite set of places v of K of residual characteristic p) for the representation ρ to "come from algebraic geometry". The local condition we have in mind is that ρ restricted to G_{K_V} be *potentially semi-stable.* This will be made precise in Conjecture 1 of § 1 below.

There is a well known conjecture which, vaguely put, asserts that a pure motive of rank 2 and of Hodge type $(0,r),(r,0)$ "comes from" a modular newform of weight $k = r+1$ (for related conjectures, see [**28**] and the large literature concerning the connection between elliptic curves and modular curves; for spectacular recent work in the way of proving such conjectures, see [**31**] and [**29**]). Combining our Conjecture 1 with this "well known conjecture" leads to a conjectural necessary and sufficient condition, stated only in terms of a local condition at p, for an irreducible representation $\rho : G_{\mathbb{Q}} \to GL_2(\overline{\mathbb{Q}}_p)$ to be the representation associated to a cuspidal newform.

Conjecture: *Let*

$$\rho : G_{\mathbb{Q}} \to GL_2(\overline{\mathbb{Q}}_p)$$

be an irreducible representation which is unramified except at a finite number of primes and which is not the Tate twist of an even representation which factors through a finite quotient group of $G_{\mathbb{Q}}$.

[1] Research supported in part by the NSF, Contract No. DMS-89-05205;by the Institut Universitaire de France and the CNRS (URA D 0752).

Then ρ is associated [2] *to a cuspidal newform f if and only if ρ is potentially semi-stable at p.*

For a refinement of the above conjecture, see Conjecture 3c below.

We also include certain "finiteness conjectures" for irreducible representations which "come from algebraic geometry", which have a fixed level of ramification and a fixed Hodge-Tate type. These "finiteness conjectures" are akin to the assertion (in view of the Conjecture displayed above) that there are only a finite number of newforms of given level and weight. This discussion is collected in §3. In §7 and §8 we examine Conjecture 1 in the case where the representation is potentially abelian, and potentially everywhere unramified.

Part II of our paper is devoted to a deformation - theoretic study of the condition of *potential semi-stability* for degree two, p-adic representations of $G_{\mathbb{Q}_p}$ which are residually absolutely irreducible. The main results here are due to Ramakrishna [**23**], but we revisit Ramakrishna's theory to give a slightly more detailed picture, particularly in the case when the representations are assumed to be potentially Barsotti-Tate.

Here is our motivation for doing such a study. Fixing an absolutely irreducible representation

$$\overline{\rho}_p : G_{\mathbb{Q}_p} \to GL_2(\mathbf{F}_p),$$

Ramakrishna has shown that the universal deformation ring $R(\overline{\rho}_p)$ is isomorphic to a power series ring in five variables $\mathbb{Z}_p[[T_1, T_2, T_3, T_4, T_5]]$ (and therefore the p-adic variety $X(\overline{\rho}_p) := \mathrm{Hom}(R(\overline{\rho}_p), \mathbb{Z}_p)$ which classifies lifts of $\overline{\rho}_p$ to $GL_2(\mathbb{Z}_p)$ is smooth on five parameters). Fix a finite set of primes S including p, let $G_{\mathbb{Q},S}$ denote the Galois group of the maximal algebraic extension of $\mathbb{Q}$, unramified outside S, and fix an imbedding of $\overline{\mathbb{Q}}$ in $\overline{\mathbb{Q}}_p$ and hence a homomorphism $i : G_{\mathbb{Q}_p} \to G_{\mathbb{Q},S}$. Let $\overline{\rho}_{\mathbb{Q},S} : G_{\mathbb{Q},S} \to GL_2(\mathbf{F}_p)$ be any representation (if such exists) which extends $\overline{\rho}_p$. That is, $\overline{\rho}_p = \overline{\rho}_{\mathbb{Q},S} \circ i$. One knows that the universal deformation ring $R(\overline{\rho}_{\mathbb{Q},S})$ is of Krull dimension ≥ 4, and if smooth, is a power series ring over $\mathbb{Z}_p$ on three parameters. It is conjectured always to have Krull dimension 4.

Now, since $5 - 3 = 2$, for our "finiteness conjectures" to be feasible it would be nice to prove that all *potential semi-stable* liftings

$$\rho_p : G_{\mathbb{Q}_p} \to GL_2(\mathbb{Z}_p),$$

of $\overline{\rho}$ with fixed Hodge-Tate weights (r, s) lie in a finite union of (at most!) *two*-parameter subspaces of the universal deformation space $X(\overline{\rho}_p)$. Indeed, the best would be if there were a finite quotient ring, call it $R(\overline{\rho}_p)_{r,s}$, of $R(\overline{\rho}_p)$ through which all homomorphisms $R(\overline{\rho}_p) \to \mathbb{Z}_p$ classifying all potentially semi-stable liftings of $\overline{\rho}_p$ (with Hodge-Tate weights r, s) factors, and such that $R(\overline{\rho}_p)_{r,s}$ is of Krull dimension ≤ 3, and whose generic fiber over $\mathbb{Q}_p$ is formally smooth of dimension two.

[2]recall that "associated" is a technical term which means that one can find an isomorphism between $\overline{\mathbb{Q}}_p$ and $\mathbb{C}$ such that for all but a finite number of prime numbers l, the above isomorphism brings the trace of $\rho(\mathrm{Frob}_l)$ to the eigenvalue of the Hecke operator T_l on the newform f. Here, Frob_l denotes the Frobenius element at l, and of course, one need only consider prime numbers l which do not divide the level of f, and at which the representation ρ is unramified.

This would indeed be nice, for then we could entertain the hope that our "finiteness conjecture" is a phenomenon related to a possible transversality question; namely, do the generic fibers of Spec$R(\overline{\rho}_p)_{r,s}$ and of $R(\overline{\rho}_{\mathbb{Q},S})$ have images in the generic fiber of Spec$R(\overline{\rho}_p)$ which intersect transversally?

We do not know whether rings $R(\overline{\rho}_p)_{r,s}$ as described above exist for all Hodge-Tate weights r, s. But Ramakrishna established that the locus of crystalline liftings possessing Hodge- Tate weights contained in the interval $[0, p-1]$ indeed forms a subscheme of $\mathrm{Spec}(R(\overline{\rho}_p))$, smooth on two formal parameters over $\mathrm{Spec}(\mathbb{Z}_p)$. A Hodge-Tate lifting with Hodge-Tate weights r, s contained in $[0, p-1]$ will be said to be of **moderate Hodge-Tate type**. We complement Ramakrishna's theory by classifying (at least for $p \geq 5$) all weakly admissible two-dimensional irreducible potentially semi-stable filtered $(\varphi, N, G_{\mathbb{Q}_p})$ -modules (Thm. A of § 11) and then by constructing explicitly the universal such module which classifies crystalline liftings of $\overline{\rho}_p$ to $GL_2(\mathbb{Z}_p)$ of moderate Hodge-Tate type (Thm. B2 of § 12).

We also discuss the interesting case of **potential Barsotti-Tate liftings** of $\overline{\rho}_p$ to $GL_2(\mathbb{Z}_p)$. The aim here is, firstly, to preview a theory (classifying Barsotti-Tate groups "strictly of slope 1/2") which will be expounded more fully in further publications by one of us, and, secondly, to formulate a consequence of this theory which gives a classification of potentially Barsotti-Tate liftings of $\overline{\rho}_p$ to $GL_2(\mathbb{Z}_p)$. This classification is put forward in Thm. C3 of § 13 below; we give a series of hints towards its proof in Appendix C. Specifically, the locus of all liftings of $\overline{\rho}_p$ to $GL_2(\mathbb{Z}_p)$ which are potentially Barsotti-Tate (PBT, for short) is either empty, or else it is a union of two disjoint smooth two-parameter spaces in $X(\overline{\rho}_p)$. More exactly, there is a quotient-ring R_{PBT} of $R(\overline{\rho}_p)$ though which any homomorphism $R(\overline{\rho}_p) \to \mathbb{Z}_p$ which classifies a PBT lifting factors, and

$$R_{\mathrm{PBT}} \cong \mathbb{Z}_p[[Y_1, Y_2]] \times_{\mathbf{F}_p} \mathbb{Z}_p[[Y'_1, Y'_2]],$$

i.e., the fiber product of two power series rings over $\mathbb{Z}_p$ each in two variables, the fiber product being taken over their common residue field $\mathbf{F}_p$.

The PBT liftings whose classifying homomorphisms factor through $\mathbb{Z}_p[[Y_1, Y_2]]$ are quite different from those which factor through $\mathbb{Z}_p[[Y_1', Y_2']]$. For certain residual representations $\overline{\rho}_p$ (specifically, for those $\overline{\rho}_p$ whose invariants [3] j_1, j_2 have the property that $j_2 - j_1 = 1$) this difference can readily be seen by considering their associated admissible pst modules:

Those PBT liftings whose classifying homomorphisms factor through $\mathbb{Z}_p[[Y_1, Y_2]]$ have associated pst modules occuring in our "type I" series (cf. Thm. A of §11) and the two p-adic parameters Y_1 and Y_2 correspond to a two-parameter variation of (the two) coefficients of the characteristic polynomial $X^2 - aX + d$ of the "Frobenius" operator φ acting on the associated pst modules of these liftings. The datum of the filtration on these pst modules contributes, however, no further variation.

Those PBT liftings whose classifying homomorphisms factor through $\mathbb{Z}_p[[Y'_1, Y'_2]]$ have associated pst modules occurring in our "type IV" series (cf. Thm. A of §11) and here the two p-adic parameters Y_1' and Y_2' correspond to somewhat different features of the associated pst modules. The characteristic

[3] cf. Thm. C3 of §13

polynomial of the "Frobenius" operator φ acting on the associated pst modules of these liftings have the form $X^2 + d$, and the p-adic parameter Y_1' corresponds to the possibility of varying the coefficient d. The datum of the filtration on these pst modules is important, and contributes another one-parameter of p-adic variation corresponding to the second variable Y_2'.

Full proofs of Thm. C1-3 will be given in later publications. Beyond this PBT case, however there still remains a good deal of work to do to get a completely satisfactory, and completely general, local picture of pst liftings, even in the special context of two-dimensional representations of $G_{\mathbb{Q}_p}$.

Notation:
K := number field; $\overline{K}$:= an algebraic closure of K, $G_K = \mathrm{Gal}(\overline{K}/K)$. S : a finite set of finite places of K; $G_{K,S}$:= the Galois group of the maximal algebraic extension of K in $\overline{K}$ unramified outside S; p := a fixed prime number; E := a field containing $\mathbb{Q}_p$; N := a fixed positive integer.

For each finite place v of K, we denote by K_v the completion of K at v. We choose an algebraic closure $\overline{K}_v$ of K_v and an embedding of $\overline{K}$ into $\overline{K}_v$. This gives us an identification of $G_v = \mathrm{Gal}(\overline{K}_v/K_v)$ with a decomposition subgroup of G_K. We denote by I_v the inertia subgroup of G_v.

If G is a profinite group, a **p-adic representation of** G is a finite dimensional $\mathbb{Q}_p$-vector space equipped with a continuous and linear action of G. Similarly, an **E-representation of** G is a $\mathbb{Q}_p$-vector space V of finite dimension (called the **degree of the representation**) equipped with a linear action of G which has the property that, if we choose a basis of V, the image of the corresponding map

$$\rho : G \to GL_N(E)$$

is contained in $GL_N(E_0)$ where E_0 is a suitable finite $\mathbb{Q}_p$-algebra contained in E and the map $G \to GL_N(E_0)$ is continuous (this doesn't depend on the choice of the basis).

Part I. The conjectures

§1. Geometric representations and representations coming from algebraic geometry.

A p-adic representation of G_K is called **geometric** if

(a) it is unramified outside a finite set of places of K,

(b) its restriction to every decomposition group G_v (for v ranging through all non-archimedean places of K) is *potentially semi-stable* in the sense of **[13]** (see also **[12]** for v's dividing p).

Equivalently, we may ask that its restriction to the decomposition groups for $v|p$ are potentially semi-stable, since, by a theorem of Grothendieck, the restriction to the other decomposition groups are automatically potentially semi-stable.

Remark: We do not know whether the condition (a) is satisfied for all semi-simple representations.

A continuous *irreducible* $\mathbb{Q}_p$-representation of G_K is said to **come from algebraic geometry** if it is isomorphic to a subquotient of an etale cohomology group with coefficients in $\mathbb{Q}_p(r)$ for some Tate twist $r \in \mathbb{Z}$, of an algebraic variety over K (equivalently [4] : of a projective smooth algebraic variety over K).

Conjecture 1: *An irreducible p-adic representation is geometric if and only if it comes from algebraic geometry.*

Evidence: The part of conjecture 1 saying that "irreducible representations coming from algebraic geometry are "geometric" is in no way original to this article, and goes in a direction where much more precise results are expected to be true about varieties defined over local fields. See, for example, Conjecture $\underline{C}_{\text{pst}}$ of **[12]** (6.2.1) and the further conjectures in (6.2.4) there. This part of Conjecture 1 has been known for a long time to hold for abelian varieties, and, more recently, **[6]** for varieties which have good reduction at all places dividing p (in which case the representations of G_v are crystalline) and also in slightly more general instances (see **[15]** and **[19]** for a survey, see also the forthcoming work of Tsuji).

The part of Conjecture 1 saying that irreducible "geometric" representations "come from algebraic geometry" is presently known for irreducible potentially abelian representations (see §6 below). The recent work of Wiles **[31]** and Taylor-Wiles **[29]** establishes this also for a very significant class of irreducible p-adic representations V of $G_{\mathbb{Q}}$, of dimension two over $\text{End}_{\mathbb{Q}_p[G_{\mathbb{Q}}]}(V)$.

§2. Hodge-Tate representations.

If the extension $E/\mathbb{Q}_p$ is finite, an E-representation of degree N of G_K is called **geometric** if it is geometric as a p-adic representation of degree $N \cdot \dim_{\mathbb{Q}_p} E$. Generally, an E-representation V of degree N of G_K is called **geometric** if there exist a finite $\mathbb{Q}_p$-algebra E_0 contained in E, a geometric E_0-representation V_0 of G_K and an isomorphism of E-representations $E \otimes_{E_0} V_0 \simeq V$.

Let V be an E-representation of G_K of degree N which is geometric. If v is a place of K dividing p and if the representation is potentially semi-stable at v then V, as a representation of G_v, is Hodge-Tate. Let us recall what this means. Put $\mathbb{C}_v :=$ the completion of $\overline{\mathbb{Q}}_v$ (on which G_v acts by continuity), and $B_{HT,v} :=$ the ring $\oplus_{r \in \mathbb{Z}} \mathbb{C}_v(r)$, with $\mathbb{C}_v(r)$ the usual Tate twist. The ring $B_{\text{HT},v}$ is the ring of Laurent polynomials in the indeterminate t, a generator of $\mathbb{Z}_p(1)$, with coefficients in $\mathbb{C}_v$ and $(B_{\text{HT},v})^{G_v} = K_v$. The G_v-representation V is **Hodge-Tate** in the sense that, if $\underline{D}^*_{\text{HT},v}(V) = \text{Hom}_{\mathbb{Q}_p[G_v]}(V, B_{\text{HT},v})$, the natural map

$$B_{\text{HT},v} \otimes_{K_v} \underline{D}^*_{\text{HT},v}(V) \to \text{Hom}_{\mathbb{Q}_p}(V, B_{\text{HT},v})$$

is an isomorphism. This implies that $\underline{D}^*_{\text{HT},v}(V)$ is a free $E \otimes_{\mathbb{Q}_p} K_v$-module of rank N, equipped with a $\mathbb{Z}$-gradation by sub-$E \otimes_{\mathbb{Q}_p} K_v$-modules

$$\text{gr}^r \underline{D}^*_{\text{HT},v}(V) = \text{Hom}_{\mathbb{Q}_p[G_v]}(V, \mathbb{C}_v(r)).$$

By the (E, v)-**Hodge-Tate type** $h(v)$ of V, we mean the isomorphism class of the graded $E \otimes_{\mathbb{Q}_p} K_v$-module $\underline{D}^*_{\text{HT},v}(V)$, and we call N the **degree** of $h(v)$. For instance, if $K_v = \mathbb{Q}_p$, $\underline{D}^*_{\text{HT},v}(V)$ is just a graded E-vector space and to know $h(v)$ amounts to the same as to know the **Hodge-Tate numbers**, that is, the

[4] using resolutions of singularities.

non-negative integers $h_r(v) = \dim_E(\mathbb{C}_v(r) \otimes_{\mathbb{Q}_p} V)^{G_v} = \dim_E \mathrm{gr}^{-r} \underline{D}^*_{\mathrm{HT},v}(V)$ (these are almost all 0 and $\sum h_r(v) = N$).

Remark: If v divides p, the **trivial (E, v)-Hodge-Tate type of degree** N is the isomorphism class of $D = (E \otimes_{\mathbb{Q}_p} K_v)^N$ with $\mathrm{gr}^0 D = D$. An E-representation V of G_{K_v} is Hodge-Tate of trivial type if and only if the image of I_v in $\mathrm{Aut}_E(V)$ is finite ([**24**], th. 1 and [**25**], th. 11, cor.). In particular, if V is a geometric representation of $G_{K,S}$ and if v is not in S but divides p, the (E, v)-Hodge-Tate type of V is trivial.

By an **E-Hodge-Tate type of K**, we mean a function h assigning to each place v dividing p an (E, v)-Hodge-Tate type $h(v)$, all those $h(v)$ having the same degree N. Let **Geom**$(K, S, h; E)$ denote the set of isomorphism classes of geometric irreducible E-representations of $G_{K,S}$ with the indicated Hodge-Tate type $h(v)$ for each v dividing p (such a representation is necessarily of dimension N).

§3. The finiteness conjectures.

If V is a geometric E-representation of G_K, for any finite place v of K, there is an unique invariant open subgroup $\mathfrak{L}_v(V)$ of I_v which is such that, if L is a finite extension of K_v contained in $\overline{K}_v$, then V, when viewed as a representation of $\mathrm{Gal}(\overline{K}_v/L)$, is semi-stable if and only if $\mathrm{Gal}(\overline{K}_v/L) \cap I_v \subset \mathfrak{L}_v(V)$. If $v \notin S$, we have $\mathfrak{L}_v(V) = I_v$.

By an **inertial level for** S, we mean a rule $\mathfrak{L}$ which assigns to each $v \in S$ an open invariant subgroup $\mathfrak{L}_v$ of I_v.

Let **Geom**$(K, S, \mathfrak{L}, h; E)$ denote the set of isomorphism classes of geometric irreducible E-representations V of $G_{K,S}$ in the set **Geom**$(K, S, h; E)$ such that, for each $v \in S, \mathfrak{L}_v \subset \mathfrak{L}_v(V)$.

Conjecture 2a: *For any finite set of places S of K, inertial level $\mathfrak{L}$ for S, $\overline{\mathbb{Q}}_p$-Hodge-Tate type h of K, the set* ***Geom***$(K, S, \mathfrak{L}, h; \overline{\mathbb{Q}}_p)$ *is finite.*

Example [**14**], prop. 1 (see [**1**] for other examples of this kind): If $K = \mathbb{Q}, S = \{7\}, \mathfrak{L}_7 = I_7$ and h, given by the Hodge numbers $(h_r)_{r \in \mathbb{Z}}$, is such that $h_r h_s \neq 0 \Rightarrow s - r \leq 3$, then **Geom**$(K, S, Z, h; \overline{\mathbb{Q}}_7)$ is empty, unless there is $i \in \mathbb{Z}$ such that $h_{-i} = 1$ and $h_r = 0$ if $r \neq i$, in which case **Geom**$(K, S, Z, h; \overline{\mathbb{Q}}_7)$ has one element which is the class of $\overline{\mathbb{Q}}_7(i)$.

Conjecture 2b: *For any finite set of places S, inertial level $\mathfrak{L}$ for S, finite extension field E of $\mathbb{Q}_p$, E-Hodge-Tate type h of K, the set* ***Geom***$(K, S, \mathfrak{L}, h; \overline{\mathbb{Q}}_p)$ *is finite.*

Conjecture 2c: *For any finite set of places S, finite extension field E of $\mathbb{Q}_p$, E-Hodge-Tate type h of K, the set* ***Geom***$(K, S, h; E)$ *is finite.*

§4. Remarks:

(a) Concerning the finiteness conjectures.

Conjecture 2c is in the spirit of the finiteness conjectures of Shafarevich (proved by Faltings [**5**]).

Obviously, *conjecture* 2a *implies conjecture* 2b. *Conjecture* 2b *and* 2c are equivalent: the implication $2c \Rightarrow 2b$ is obvious. Conversely, attached to a potentially semi-stable representation of degree N (at v) one has (for this theory, see

[**12**]) a representation D_v of the Weil-Deligne group of K_v of degree N with coefficients in a field E' which is E if v does not divide p and a finite unramified extension of E if v divides p. In particular, if we choose a basis of D_v, we have an homomorphism $\rho_v : I_v \to GL_N(E')$ whose kernel is the invariant open subgroup $\mathfrak{L}_v(V)$.

If v does not divide $p, E' = E$, and, because the order of the finite subgroups of $GL_N(E)$ is bounded, there is an integer $M_v(E, N)$ such that $[I_v : \mathfrak{L}_v(V)]$ divides $M_v(E, N)$. If v divides p, one can check (cf. [**12**]) that, for any $g \in I_v$, the characteristic polynomial of g acting on D_v has coefficients in E. This implies [**7**] that there exists a linear representation of degree N of $I_v/\mathfrak{L}_v(V)$ with coefficients in the field $E_1 = E(^{p-1}\sqrt{1})$ if $p \neq 2$ (resp. $E(^4\sqrt{1})$ if $p = 2$) which has the same character as ρ_v. Therefore, because the order of the finite subgroups of $GL_N(E_1)$ is bounded, in this case as well, there is an integer $M_v(E, N)$ such that $[I_v : \mathfrak{L}_v(V)]$ divides M_v.

For any $v \in S$, it is easy to see that, given the integer $M_v(E, N)$, there is an open invariant subgroup $\mathfrak{L}_v(E, N)$ of I_v which is contained in all open invariant subgroups whose order divides $M_v(E, N)$. Therefore $\mathbf{Geom}(K, S, h; E) \subset \mathbf{Geom}(K, S, \mathfrak{L}, h; E)$ where $\mathfrak{L}$ is given by $v \mapsto \mathfrak{L}_v(E, N)$.

(b) Changing K. Let L be a finite extension of K contained in $\overline{K}$. It is easy to check that an E-representation V of G_K is geometric if and only if its restriction $\mathrm{Res}_K^L V$ to G_L is geometric; similarly an E-representation W of G_L is geometric if and only if the induced representation $\mathrm{Ind}_L^K W$ of G_K is. It is also easy to compute the Hodge-Tate types and the inertial levels of $\mathrm{Res}_K^L V$ (resp. $\mathrm{Ind}_L^K W$) from those of V (resp. W). From that, using adjunction, we see easily that Conjecture 1 (resp. 2a, 2b) is true for K if and only if it is true for L.

(c) Semi-stable representations. A geometric E-representation V of G_K is said to be *semi-stable* if, for each finite place v of K, V is semi-stable as a representation of G_v, i.e. if $\mathfrak{L}_v(V) = I_v$. Let $\mathbf{Geom}_{\mathrm{st}}(K, S, h; \overline{\mathbb{Q}}_p)$ denote the set of isomorphism classes of geometric irreducible $\overline{\mathbb{Q}}_p$-representations V of $G_{K,S}$ in the set $\mathbf{Geom}(K, S, h; \overline{\mathbb{Q}}_p)$ which are semi-stable. The following is a special case of conjecture 2a (corresponding to choosing $\mathfrak{L}$ such that $\mathfrak{L}_v = I_v$ for all $v \in S$):

Conjecture 2a′: *For any finite set of places S of K, $\overline{\mathbb{Q}}_p$-Hodge-Tate type h of K, the set $\mathbf{Geom}_{st}(K, S, h; \overline{\mathbb{Q}}_p)$ is finite.*

Conversely, it is easy to check that conjecture $2a'$ for all number fields implies conjecture 2a.

(d) Concerning geometric p-adic representations, and "compatible families" of representations.

The curious implication of Conjecture 1, taken together with standard conjectures concerning the étale cohomology of algebraic varieties of number fields, is that any irreducible p-adic geometric representation V of $G_{K,S}$ has these ("motivic") properties:

(1) For all nonarchimedean places $v \notin S$, let $\Phi_v \in G_v \subset G_K$ be a lifting of the geometric Frobenius $f_v \in G_v/I_v$ and let $P_v(V; T) \in \overline{\mathbb{Q}}_p[T]$ denote the

characteristic polynomial of Φ_v acting on V. Then, there exists $r \in \mathbb{Z}$ and a finite extension E of $\mathbb{Q}$ contained in $\mathbb{Q}_p$ such that, for all $v \notin S, P_v(V,T) \in E[T]$, the complex roots of $P_v(V,T)$ (for a chosen embedding of E into $\mathbb{C}$) have their complex absolute values equal to $q_v^{r/2}$ where q_v is the cardinality of the residue field of v.
(2) There exists a finite extension E' of the above E and, for any finite place λ of E', a geometric E'_λ-representation V_λ of G_{K,S_λ} where $S_\lambda = S \cup \{v|v$ has the same residual characteristic as $\lambda\}$ such that, if $v \notin S_\lambda, P_v(V_\lambda,T)(:=$ the characteristic polynomial of Φ_v acting on the E'_λ-vector space $V_\lambda) = P_v(V,T)$.

(e) *L*-functions and weights.

Assume $K = \mathbb{Q}$ and let V be an irreducible geometric $\overline{\mathbb{Q}}_p$-representation of $G_\mathbb{Q}$ of degree N. For each prime l, one can associate to V an N-dimensional $\overline{\mathbb{Q}}_p$-linear representation of the Weil-Deligne group W'_l of $\mathbb{Q}_l$ (cf. [**13**] or [**17**]). Hence if we choose an imbedding of $\overline{\mathbb{Q}}_p$ into $\mathbb{C}$, one can define in the usual way a local factor $L_l(V,s)$ for each prime l and one can define the global L-function $L(V,s)$ as being the formal product of all those local factors. One conjectures that this Dirichlet series converges for $\Re(s) >> 0$ and admits a meromorphic continuation in the whole complex plane; moreover, one can give a conjectural interpretation of the order of the zero or pole at $s = 0$ in terms of Galois cohomology (see [**17**], n°3.4).

One can define the **weight** $w(V)$ of V : if $\dim_{\overline{\mathbb{Q}}_p} V = 1$, there is an unique integer i such that the action of $G_\mathbb{Q}$ on $V(i)$ is finite and $w(V) = 2i$; if $\dim_{\overline{\mathbb{Q}}_p} V = N, w(V) = w(\wedge^N V)/N$; hence this is a rational number. For $r,s \in \mathbb{Z}$, define the **Hodge numbers** $h_{r,s}(V)$ of V as being 0 unless $r+s = w(V)$ in which case $h_{r,s}(V)$ is the Hodge-Tate number $h_r(V)$.

If we choose an imbedding of $\overline{\mathbb{Q}}$ into $\mathbb{C}$, this defines the complex conjugation $c \in G_\mathbb{Q}$, hence we can define

$$h^+(V) = \dim_{\overline{\mathbb{Q}}_p}\{v \in V|cv = v\} \text{ and } h^-(V) = \dim_{\overline{\mathbb{Q}}_p}\{v \in V|cv = -v\}.$$

Define also $h^+_{r,s}(V) = h^-_{r,s}(V) = h_{r,s}(V)/2$ if $r \neq s$ and, if $j = w(V)/2 \in \mathbb{Z}$,

$$h^+_{j,j}(V) = h^+(V) - \sum_{r\neq s} h^+_{r,s}(V), \quad h^-_{j,j}(V) = h_{j,j}(V) - h^+_{j,j}(V).$$

Conjecture 3a: *Let V be an irreducible geometric $\overline{\mathbb{Q}}_p$-representation of $G_\mathbb{Q}$. Then the weight $w(V)$ of V is an integer. Moreover, for $r,s \in \mathbb{Z}$ such that $r+s = w(V)$, the numbers $h^+_{r,s}(V)$ and $h^-_{r,s}(V)$ are non negative integers and $h_{r,s}(V) = h_{s,r}(V)$.*

Assuming this conjecture, these numbers define in the usual way an isomorphism class of a linear representation of the Weil group $W_\mathbb{R}$, hence a Γ-factor $L_\infty(V,s)$ and we can consider the complete L-function $\wedge(V,s) = L_\infty(V,s) \cdot L(V,s)$.

Because we have a representation of W'_l for each prime number l and a representation of $W_\mathbb{R}$, we can define in the usual way the **conductor** N_v and the **ϵ-factor** $\epsilon(V,s)$ (cf. e.g. [**3**]). We thus have the conjectural functional equation:

Conjecture 3b: *Let V be as in* 3a. *Then $\wedge(V,s)$ converges for $\Re(s) >> 0$ and admits a meromorphic continuation in the whole complex plane, satisfying*

$$\wedge(V,s) = \epsilon(V,s) \cdot L(V^*, 1-s).$$

(f) Modular forms.

Combining conjecture 1 with classical conjectures (e.g. [**26**]), we obtain

Conjecture 3c: *Let V be an irreducible geometric $\overline{\mathbb{Q}}_p$-representation of $G_{\mathbb{Q}}$ of degree two which is not a Tate twist of a finite representation. Then there is an integer $i \in \mathbb{Z}$ such that $V(-i)$ is isomorphic to the representation associated to a "new" modular form.*

Using the previous discussion, one can be more precise: the integer i must be the smallest integer such that $h_j(V) \neq 0$. Twisting by $\mathbb{Z}_p(-i)$ if necessary, we can assume $i = 0$. Then, if w is the biggest integer such that $h_w(V) \neq 0$, the weight of V is w and we see that the weight of the corresponding modular form f must be $w+1$. The conductor of f must be the conductor N_V of V. The nebentypus can be also computed in the usual way using the representations of the Weil-Deligne groups. To prove this conjecture one "need only" to prove that the L-function of V is the Mellin transform of a modular form.

Observe also

i) that, because of the finiteness of the dimension of the space of modular forms of fixed weight and level, conjecture 2a and conjecture 2b in the two dimensional case follows from conjecture 3c;

ii) that, because of the fact that, if X is an elliptic curve over $\mathbb{Q}$, if $T_p(X)$ is its Tate module and if $V = \mathbb{Q}_p \otimes_{\mathbb{Z}_p} T_p(X), V$ is geometric [**12**], conjecture 3c implies the Shimura-Taniyama-Weil conjecture.

(g) Tannakian categories. (for a discussion of tannakian categories, see, for instance, [**12**]). Let K be a number field, and let $\underline{\mathrm{Rep}}_{\mathbb{Q}_p}(G_K)$ denote the tannakian category of p-adic representations of G_K. A tannakian subcategory is a full subcategory containing an object of positive dimension, and which is stable under passage to sub-object, quotient, direct sum, tensor product and dual.

Let $\underline{\mathrm{Rep}}_{\mathbb{Q}_p,g}(G_K)$ (resp. $\underline{\mathrm{Rep}}^{ss}_{\mathbb{Q}_p,g}(G_K)$) denote the tannakian subcategory of $\underline{\mathrm{Rep}}_{\mathbb{Q}_p}(G_K)$ whose objects are geometric (resp. semi-simple geometric) representations of G_K.

Conjecture 4a. *The category $\underline{Rep}^{ss}_{\mathbb{Q}_p,g}(G_K)$ is the smallest tannakian subcategory of $\underline{Rep}_{\mathbb{Q}_p}(G_K)$ containing all representations of the form $H^m_{\acute{e}t}(X_{/\overline{K}}, \mathbb{Q}_p)$ where X ranges through all proper smooth varieties over K.*

Note that Conjecture 4a is equivalent to the union of Conjecture 1 and the classical conjecture, due to Tate, that if X is smooth and proper over K, the G_K-representation $H^m_{\acute{e}t}(X_{/\overline{K}}, \mathbb{Q}_p)$ is semi-simple.

Although we don't have a precise definition of "mixed motives" in mind, it is tempting to us to ask the somewhat vague question:

Question 4b: *Is the category $\underline{Rep}_{\mathbb{Q}_p,g}(G_K)$ the smallest tannakian subcategory of $\underline{Rep}_{\mathbb{Q}_p}(G_K)$ containing the p-adic realization of all mixed motives over K.*

One concrete interpretation of the above question is the following:

Question 4c: *Is the category $\underline{Rep}_{\mathbb{Q}_p,g}(G_K)$ the smallest tannakian subcategory of $\underline{Rep}_{\mathbb{Q}_p}(G_K)$ containing all representations of the form $H^m_{\acute{e}t}(X._{/\overline{K}}, \mathbb{Q}_p)$ where X ranges through all simplicial schemes of finite type over K ?*

§5. "Universal" geometric deformation rings.

A preliminary remark on deformations: Let G be a profinite group and Λ a local complete noetherian ring, of residue field k. Let GSW be the category of local complete noetherian Λ-algebras, with residue field k. For short, by Λ-*algebra* we mean an object of $\mathfrak{C}_\Lambda$.

If A is any Λ-algebra, an A-**representation** of G is an A-module of finite type equipped with a linear and continuous action of G. We say that such a representation is *flat* if the underlying A-module is flat.

We give ourselves a finite dimensional k-representation $\overline{V}$ of G, which is such that the natural map $k \to \mathrm{End}_{k[G]}(\overline{V})$ is an isomorphism[5].

For each Λ-algebra A, an A-**deformation of** $\overline{V}$ is a flat A-representation V of G such that $k \otimes_A V$ is isomorphic to $\overline{V}$. Denote by $\Psi_{\overline{V}}(A)$ the set of isomorphism classes of A-deformations of $\overline{V}$. We get in this way a functor

$$\Psi_{\overline{V}} : \mathfrak{C}_\Lambda \to \underline{\text{Sets}}.$$

Then [**22**][6], if $\dim_k H^1(G, \mathrm{End}_k(\overline{V})) < +\infty$, this functor is representable. Therefore, one can define the universal Λ-deformation ring $R_\Lambda(\overline{V})_G$ together with a $R_\Lambda(\overline{V})_G$-deformation of $\overline{V}$, well defined up to isomorphism, with the obvious universal property.

Remark: If we choose a basis of $\overline{V}$ over k, the action of G on $\overline{V}$ gives us an homomorphism $\overline{\rho} : G \to GL_d(k)$. Usually, one considers continuous homomorphisms from $G \to GL_d(A)$ lifting $\overline{\rho}$ rather than deformations of $\overline{V}$. Recall that if $\rho, \rho' : G \to GL_d(A)$ are two such homomorphisms, one says that ρ and ρ' are isomorphic (resp. strictly isomorphic) if there exists $\alpha \in GL_d(A)$ (resp. $\alpha \in GL_d(A)$ lifting the identity in $GL_d(k)$) such that $\rho'(g) = \alpha\rho(g)\alpha^{-1}$, for all $g \in G$. From the surjectivness of A^* onto k^* and the fact that $\mathrm{End}_{k[G]}(\overline{V}) = k$, we see that the two equivalence relations are actually the same. If we denote by $\psi_{\overline{\rho}}(A)$ the set of equivalence classes, we get also in this way a functor $\psi_{\overline{\rho}} : \mathfrak{C}_\Lambda \to \underline{\text{Sets}}$. It is an easy exercise to build a natural isomorphism between the functors ψ_ρ and $\psi_{\overline{V}}$ and we will use this natural isomorphism to identify $\Psi_{\overline{\rho}}$ and $\psi_{\overline{V}}$.

Now let $\mathfrak{D}$ be a strictly full subcategory of the category $\underline{\mathrm{Rep}}^f_\Lambda(G)$ of Λ-modules of finite length equipped with a linear and continuous action of G which is stable under subobjects, quotients and direct sums. Assume that $\overline{V}$ is

[5]actually, we will be mostly interested in the absolutely irreducible case.

[6]at least in the absolutely irreducible case, but the same proof works in general.

an object of $\mathfrak{D}$. Then we can define the subfunctor $\psi_{\overline{V},\mathfrak{D}}$ of $\psi_{\overline{V}}$ by the condition that, for any Λ-algebra A, $\psi_{\overline{V},\mathfrak{D}}(A)$ consists of elements which are represented by representations V such that, for any artinian quotient A' of $A, A' \otimes_A V$ is an object of $\mathfrak{D}$. One sees [**23**] that, if $\dim_k H^1(G, \mathrm{End}_k(\overline{V})) < +\infty$, then $\psi_{\overline{V},\mathfrak{D}}$ is representable by a quotient $R_\Lambda(\overline{V})_\mathfrak{D}$ of $R_\Lambda(\overline{V})_G$. We call $R_\Lambda(\overline{V})_\mathfrak{D}$ the **universal ring of deformations of $\overline{V}$ lying in $\mathfrak{D}$.**

We can apply that to $G = G_{K,S}$ and $\Lambda = \mathcal{O}_E$ the ring of the integers of a finite extension of $\mathbb{Q}_p$. Hence, given $\overline{V}$, we can consider the noetherian and complete $\mathcal{O}_E$-algebra $R_{\mathcal{O}_E}(\overline{V})_S := R_{\mathcal{O}_E}(\overline{V})_{G_{K,S}}$. Now, if moreover, we fix an inertia level $\mathfrak{L}$ for S and two integers $a, b \in \mathbb{Z}$ with $a \leq b$, we can consider the full subcategory $\mathfrak{D} = \underline{\mathrm{Rep}}^f_{\mathcal{O}_E}(G_{K,S})_{\mathfrak{L},[a,b],\mathrm{st}}$ (resp. $\underline{\mathrm{Rep}}^f_{\mathcal{O}_E}(G_{K,S})_{\mathfrak{L},[a,b],\mathrm{cris}}$) of $\underline{\mathrm{Rep}}^f_{\mathcal{O}_E}(G_{K,S})$ whose objects are the T's such that, for each $v \in S$,
i) if v doesn't divides p, $\mathfrak{L}_v$ acts trivially on T,
ii) if v divides p, one can find a semi-stable (resp. crystalline) p-adic representation V of $\mathfrak{L}_v$ satisfying $(\mathbb{C}_v(r) \otimes_{\mathbb{Q}_p} V)^{\mathfrak{L}_v} = 0$ such that T is isomorphic to a sub-quotient of V as a $\mathbb{Z}_p[\mathfrak{L}_v]$-module.

We denote by $R_{\mathcal{O}_E}(\overline{V})_{S,\mathfrak{L},[ab],\mathrm{st}}$ (resp. $R_{\mathcal{O}_E}(\overline{V})_{S,\mathfrak{L},[ab],\mathrm{cris}}$) the corresponding quotient of $R_{\mathcal{O}_E}(\overline{V})_S$. Observe that the second is a quotient of the first.

Conjecture 5: *For any finite set of places S of K, finite extension E of $\mathbb{Q}_p$ of residue field k, inertial level $\mathfrak{L}$, integers $a \leq b$, finite dimensional k-representation $\overline{V}$ of $G_{K,S}$ belonging to $R_{\mathcal{O}_E}(\overline{V})_{S,\mathfrak{L},[ab],st}$, which is such that the natural map $k \to \mathrm{End}_{k[G]}(\overline{V})$ is an isomorphism, the ring $R_{\mathcal{O}_E}(\overline{V})_{S,\mathfrak{L},[ab],st}$ is a finite $\mathcal{O}_E$-algebra.*

This conjecture implies conjecture 2b: if $S, \mathfrak{L}$ and h are given as in conjecture 2a, we can choose $a \leq b$ such that, for all v's dividing p, if D is a graded $(\overline{\mathbb{Q}}_p \otimes_{\overline{\mathbb{Q}_p}} K_v)$-module representing the class $h(v)$, we have $gr^{-r}D = 0$ if $r \notin [a, b]$. Then, if V is an E-representation of $G_{K,S}$, say of degree N, representing an element of $\mathbf{Geom}(K, S, \mathfrak{L}, h; E)$, the fact that V is irreducible implies that one can find an $\mathcal{O}_E$-lattice T of V stable under $G_{K,S}$ and such that, if $\overline{V} = k \otimes_{\mathcal{O}_E} T$, the map $k \to \mathrm{End}_{k[G]}(\overline{V})$ is an isomorphism. Obviously, each finite quotient of T lies in $\underline{\mathrm{Rep}}^f_{\mathcal{O}_E}(G_{K,S})_{\mathfrak{L},[a,b],\mathrm{st}}$, hence the natural homomorphism from $R_{\mathcal{O}_E}(\overline{V})_S$ to $\mathcal{O}_E$ factors through $R_{\mathcal{O}_E}(\overline{V})_{S,\mathfrak{L},[a,b],\mathrm{st}}$. Therefore the conjecture implies that, given $\overline{V}$, there are only finitely many elements of $\mathbf{Geom}(K, S, \mathfrak{L}, h; E)$ within a representative of which we can find a $G_{K,S}$-stable $\mathcal{O}_E$-lattice T such that $k \otimes_{\mathcal{O}_E} T \cong \overline{V}$. The result follows from the fact, easy to check, that there are only finitely many isomorphism classes of k-representations $\overline{V}$ of degree N of $G_{K,S}$ such that the map $k \to \mathrm{End}_{k[G]}(\overline{V})$ is an isomorphism.

Remarks: a) Let V be an E-representation of $G_{K,S}$ which admits a lattice T stable under $G_{K,S}$ such that each finite quotient lies in $\underline{\mathrm{Rep}}^f_{\mathcal{O}_E}(G_{K,S})_{\mathfrak{L},[a,b],\mathrm{st}}$ (resp. $\underline{\mathrm{Rep}}^f_{\mathcal{O}_E}(G_{K,S})_{\mathfrak{L},[a,b],\mathrm{cris}}$). It should not be very hard to prove that V is geometric. Up to now, we have checked this property only in special cases (for instance, this is OK for $\underline{\mathrm{Rep}}^f_{\mathcal{O}_E}(G_{K,S})_{\mathfrak{L},[a,b],\mathrm{cris}}$ if $b - a \leq p - 1$ and $K/\mathbb{Q}$ is unramified at p or if $b - a \leq 1$ and, for all v dividing p, the ramification index of $K_v/\mathbb{Q}_p$ is $\leq p - 1$).

(b) Universal geometric deformation rings and the Langlands program: Let R be either $R_{\mathcal{O}_E}(\overline{V})_{S,\mathfrak{L},[a,b],\mathrm{st}}$ or $R_{\mathcal{O}_E}(\overline{V})_{S,\mathfrak{L},[a,b],\mathrm{cris}}$ as above and T_R the corresponding free R-module with action of $G_{K,S}$ (well defined up to isomorphism). For any place v of K not contained in S, there is an element which might be called a **"Hecke element"** τ_v in R given by taking the trace of Frob_v acting on T_R, where Frob_v is any choice of Frobenius element at v in $G_{K,S}$.

Can one "reconstuct" the rings $R_{\mathcal{O}_E}(\overline{V})_{S,\mathfrak{L},[a,b],\mathrm{st}}$ and $R_{\mathcal{O}_E}(\overline{V})_{S,\mathfrak{L},[a,b],\mathrm{cris}}$ and their systems of Hecke elements $v \mapsto \tau_v$ as completions (at maximal ideals of residual characteristic p) of rings which are generated by Hecke operators and which act (faithfully) on automorphic representations spaces for specific reductive groups (notably $GL_{N/K}$) ?

§6. The special case where ρ is potentially abelian.

A p-adic representation V of G_K is **potentially abelian** if there is an open subgroup of G_K which operates on V through an abelian quotient group.

For any abelian variety A over K, let $A[p^n](\overline{K})$ denote the group of $\overline{K}$-rational points of A annihilated by multiplication by p^n, let $T_p(A)$=lim.proj. $A[p^n](\overline{K})$
and put $V_p(A) = \mathbb{Q}_p \otimes_{\mathbb{Z}_p} T_p(A)$. Recalling the terminology of Remark (g) of § 4, let $\underline{\mathrm{Rep}}_{\mathbb{Q}_p,CM}(G_K)$ denote the smallest Tannakian subcategory of $\underline{\mathrm{Rep}}_{\mathbb{Q}_p}(G_K)$ containing all representations of G_K which factor through finite groups, and also containing the $V_p(A)$ for A ranging through all abelian varieties over K which are potentially of CM type.

Proposition: *Let V be a potentially abelian p-adic representation of G_K and, for each place v of K, let V_v be the underlying representation of G_v. The following are equivalent:*
1) For every place v of K dividing p, V_v is of Hodge-Tate type and the action of I_v is semi-simple;
2) for every place v of K dividing p, V_v is of Hodge-Tate type;
3) for every place v of K dividing p, V_v is of de Rham type;
4) for every place v of K dividing p, V_v is potentially semi-stable;
5) for every place v of K dividing p, V_v is potentially crystalline;
6) the representation V is geometric;
7) the representation V is an object in $\underline{Rep}_{\mathbb{Q}_p,CM}(G_K)$.

Proof. Since potentially abelian representations are unramified outside a finite set of places ([**27**], Cor. of p. III.11) we have (4) $\Leftrightarrow$ (6). One knows that every abelian variety A which is potentially of CM type has potentially good reduction. It follows that if $V = V_p(A)$, then V satisfies (5). Moreover, any finite representation satisfies (5), and (5) is stable under tannakian operations. Therefore (7) $\Rightarrow$ (5).

The implications (5) $\Rightarrow$ (4) $\Rightarrow$ (3) $\Rightarrow$ (2) and (1) $\Rightarrow$ (2) are all trivial.

From ([**27**], Thm. 3 p. III-52) one has that (1) implies that V is "locally algebraic" which, in turn, implies (7) by ([**4**], Prop. D1).

It remains to establish the implication (2) $\Rightarrow$ (1) which is an exercise, using the Theorem of Sen [**25**] that a representation, which is Hodge-Tate and whose only Hodge-Tate weight is 0 (with any multiplicity) must factor through a finite quotient group. □

§ 7. **The special case where p is potentially unramified.**

After a finite base change we can make p unramified, i.e., we can view our representation as being a representation of $G_{K,S}$ where S is a finite set of primes, none of which have residual characteristic p.

Conjecture 5a: *If p is distinct from all of the residual characteristics of S, then any p-adic representation of $G_{K,S}$ factors through a finite quotient group of $G_{K,S}$.*

Remark: Conjecture 5a for all p-adic representations follows from the same conjecture, but stated only for semi-simple geometric p-adic representations. This latter conjecture follows from Conjecture 1 in conjunction with the Tate Conjecture about the subspace of étale cohomology generated by algebraic cycles.

An equivalent way of stating Conjecture 5a is that if p is distinct from all of the residual characteristics of S, then any quotient group of $G_{K,S}$ which is a p-adic analytic group, is finite. Conjecture 5a for S empty bears on the structure of the Galois group of a Golod-Shafarevich p-tower: Let $\Gamma(K,p)$ denote the Galois group of the maximal everywhere unramified pro-p-extension of the number field K. Conjecture 5a implies.

Conjecture 5b: *Any quotient group of $\Gamma(K,p)$ which is a p-adic analytic group, is finite.*

For some partial corroboration of this conjecture, see the recent work of N.Boston ([**2**], Theorem 1), and F.Hajir [**18**].

Part II. Representations of dimension 2 of $\mathrm{Gal}(\overline{\mathbb{Q}}_p/\mathbb{Q}_p)$.

In this part, $G_p = G_{\mathbb{Q}_p} = \mathrm{Gal}(\overline{\mathbb{Q}}_p/\mathbb{Q}_p)$ and I_p is the inertia subgroup. For any $\mathbb{Z}_p$-module V and any $r \in \mathbb{Z}$, we denote by $V(r) = V \otimes_{\mathbb{Z}_p} \mathbb{Z}_p(r)$ the usual Tate twist.

For $\epsilon \in \overline{\mathbb{F}}_p$, the residue field of $\overline{\mathbb{Q}}_p$, we denote $[\epsilon]$ its Teichmüller representative in $\overline{\mathbb{Q}}_p$.

For simplicity, we will assume $p \neq 2$ (and sometimes $p \geq 5$).

§8. **The representations of G_p of dimension 1.**

Let A be a complete noetherian local ring with finite residue field. For any profinite group G, a *character with values in A* is a continuous homomorphism $\eta : G \to A^*$. If N is any closed invariant subgroup of G contained in the kernel of η, we still denote $\eta : G/N \to A^*$ the character we get by factoring, and conversely.

If V is an A-module equipped with an action of $G, V(\eta)$ denotes the same A-module with the action of G twisted by η.

If $a \in A^*$, we denote by $\eta_a : G_p/I_p \to A^*$ the unique character such that $\eta_a(\text{arith.frob.}) = a$.

Let $\xi_1 : G_p \to \mathbb{F}_p^*$ be the character giving the action of G_p on the p-th roots of 1. For $i \in \mathbb{Z}/(p-1)\mathbb{Z}$ and $\epsilon \in \mathbb{F}_p^*$, denote by $\overline{U}_{i,\epsilon}$ the one dimensional $\mathbb{F}_p$-representation of G_p which is $\mathbb{F}_p$ on which G_p acts through $\eta_\epsilon \xi_1^i$. Then any one dimensional $\mathbb{F}_p$-representation of G_p is isomorphic to one and only one of the $\overline{U}_{i,\epsilon}$.

Define $\hat{\xi}_1(g) = [\xi_1(g)]$ and $\chi : G_p \to \mathbb{Z}_p^*$ the cyclotomic character which we can write as $\chi = \hat{\xi}_1 \cdot \chi_0$ with χ_0 taking values in $1 + p\mathbb{Z}_p$.

If we denote $R(\overline{U}_{i,\epsilon}) = R_{\mathbb{Z}_p}(\overline{U}_{i,\epsilon})_{G_p}$ the universal $\mathbb{Z}_p$-deformation ring, one sees that one can write $R(\overline{U}_{i,\epsilon}) = \mathbb{Z}_p[\![T_1, T_2]\!]$, in such a way that the corresponding character $\eta : G_p \to \mathbb{Z}_p[\![T_1, T_2]\!]^*$ giving the universal deformation has the property that if, for $r \in \mathbb{Z}_p$

$$e_r : \mathbb{Z}_p[[T_1, T_2]] \longrightarrow \mathbb{Z}_p[[T_1]]$$

is the ring-homomorphism to $\ker_\rho T_1$ to T_1 and T_2 to $(1+p)^r - 1$, then

$$e_r \cdot \eta = \eta_{[\epsilon](1+T_1)} \cdot \hat{\xi}_1^i \cdot \chi_0^{T_2}$$

for all $r \in \mathbb{Z}_p$.

Now, let U be a one dimensional $\mathbb{Q}_p$ representation of G_p. To give U up to isomorphism is the same as to give the unique $(i, \varepsilon, t_1, t_2) \in ((\mathbb{Z}/(p-1)\mathbb{Z}) \times \mathbb{F}_p^* \times p\mathbb{Z}_p \times p\mathbb{Z}_p$ such that G_p acts on U via $\eta_{[\epsilon](1+t_1)} \cdot \hat{\xi}_1^i \cdot \chi_0^r$ where $(1-p)^r - 1 = t_{2r}$. If T is a $\mathbb{Z}_p$-lattice and if $\overline{U} = T/pT$, then $\overline{U} \simeq \overline{U}_{i,\varepsilon}$; moreover U is potentially semi-stable if it is Hodge-Tate, which amounts to requiring that r be in $\mathbb{Z}$. In this case r is the (unique) Hodge-Tate weight of U.

Now fix $\overline{U}_{i,\varepsilon}$ and $r \in \mathbb{Z}$. We observe that , if U is potentially semi-stable of Hodge-Tate weight r and such that $\overline{U} \simeq \overline{U}_{i,\varepsilon}$, then there is a unique $t_1 \in p\mathbb{Z}_p$ such that G_p acts on U via $\eta_{[\epsilon](1+t_1)} \cdot \hat{\xi}_1^i \cdot \chi_0^r = \hat{\xi}_1^{i-\bar{r}} \cdot \eta_{[\epsilon](1+t_1)} \cdot \chi^r$ (where $\bar{r}$ denotes the image of r in $\mathbb{Z}/(p-1)\mathbb{Z}$), hence $U(\hat{\xi}_1^{\bar{r}-i})$ is a crystalline representation with r as unique Hodge-Tate weight. Now, if we consider the full subcategory $\mathfrak{D}$ of the category of finite $\mathbb{Z}_p$-representations T of G_p which are such that $T(\hat{\xi}_1^{\bar{r}-i})$ is isomorphic to a subquotient of a crystalline representation of unique Hodge-Tate weight r, $\mathfrak{D}$ is stable under subobjects, quotients and direct sums, so we can speak of the universal deformation of $\overline{U}_{i,\varepsilon}$ "lying in $\mathfrak{D}$". The corresponding ring $R_{\mathbb{Z}_p}(\overline{U}_{i,\varepsilon})_{\mathfrak{D}}$ is isomorphic to $\mathbb{Z}_p[\![T_1]\!]$ with the corresponding character $\eta_{[\epsilon](1+T_1)} \cdot \hat{\xi}_1^i \cdot \chi_0^r$.

§9. The absolutely irreducible $\mathbb{F}_p$-representations of G_p of dimension 2 and their deformations.

By convention, for any positive integer $n, \mathbb{Q}_{p^n} \subset \overline{\mathbb{Q}}_p$ will denote the unramified extention of $\mathbb{Q}_p$ in $\overline{\mathbb{Q}}_p$ of degree n and $\mathbb{F}_{p^n}$ the corresponding residue field. We choose $\pi_1, \pi_2 \in \overline{\mathbb{Q}}_p$ such that $\pi_1^{p-1} = -p$ and $\pi_2^{p+1} = \pi_1$ and, for $j = 1, 2$, let $F_j =$ the Galois closure over $\mathbb{Q}_p$ of $\mathbb{Q}_p(\pi_j)$; let $L = \mathbb{Q}_{p^{2(p-1)}}(\pi_2)$ and $L_0 = \mathbb{Q}_{p^{2(p-1)}}$ so that we have the diagram

$$\begin{array}{ccccccc}
 & & \mathbb{Q}_p(\pi_2) & - & F_2 = \mathbb{Q}_{p^2}(\pi_2) & - & L = \mathbb{Q}_{p^{2(p-1)}}(\pi_2) \\
 & & | & & | & & | \\
F_1 & = & \mathbb{Q}_p(\sqrt[p-1]{1}) = \mathbb{Q}_p(\pi_1) & - & \mathbb{Q}_{p^2}(\pi_1) & - & \mathbb{Q}_{p^{2(p-1)}}(\pi_1) \\
 & & | & & | & & | \\
 & & \mathbb{Q}_p & - & \mathbb{Q}_{p^2} & - & L_0 = \mathbb{Q}_{p^{2(p-1)}}
\end{array}$$

The $\mathrm{Gal}(L/\mathbb{Q}_p(\pi_2))(\simeq \mathrm{Gal}(L_0/\mathbb{Q}_p))$ is cyclic of order $2(p-1)$ generated by the "Frobenius element" τ satisfying $\tau x = x^p$ if $x \in \mathbb{F}_{p^{2(p-1)}}$, the residue field of L. The group $\mathrm{Gal}(L/L_0)$ is canonically isomorphic to the multiplicative group of $\mathbb{F}_{p^2}$, via the **"fundamental character"** $\xi_2 : \mathrm{Gal}(L/L_0) \to \mathbb{F}_{p^2}^*$ associated to $g \in \mathrm{Gal}(L/L_0)$, the image of $g\pi_2/\pi_2$ in $\mathbb{F}_{p^2}$. The group $\mathrm{Gal}(L/\mathbb{Q}_p)$ is a semi-direct product of $\mathrm{Gal}(L/\mathbb{Q}_p(\pi_2))$ and the normal subgroup $\mathrm{Gal}(L/L_0)$, with $\tau g \tau^{-1} = g^p$ for any $g \in \mathrm{Gal}(L/L_0)$.

Consider (ι, ε) where $\iota \in \mathbb{Z}/(p^2-1)\mathbb{Z}$ and $\varepsilon \in \mathbb{F}_p^*$. Given such data, *we choose* $\zeta \in \mathbb{F}_{p^2}$ such that $\zeta^{p+1} = -\varepsilon$. Attached to the above data, *and this choice*, we have a unique homomorphism

$$\overline{\rho}_{\iota,\varepsilon} : \mathrm{Gal}(L/\mathbb{Q}_p) \to \mathrm{Aut}_{\mathbb{F}_p}(\mathbb{F}_{p^2})$$

such that the restriction to $\mathrm{Gal}(L/L_0)$ consists in $\mathbb{F}_{p^2}$-linear automorphisms given by the one-dimensional $\mathbb{F}_{p^2}$-character ξ_2^ι, and τ acts on $\mathbb{F}_{p^2}$ via the $\mathbb{F}_{p^2}$-semilinear automorphism $x \mapsto \zeta \cdot x^p$. We denote also $\overline{\rho}_{\iota,\varepsilon} : G_p \to Aut_{\mathbb{F}_p}(\mathbb{F}_{p^2})$ the natural lifting and $\overline{V}_{\iota,\varepsilon}$ the 2-dimensional $\mathbb{F}_p$-representation of G_p which is $\mathbb{F}_{p^2}$ on which G_p acts through $\overline{\rho}_{\iota,\varepsilon}$.

Proposition . *i) The isomorphism class of the $\mathbb{F}_p[G_p]$-module $\overline{V}_{i,\varepsilon}$ is independent of the choice of ζ such that $\zeta^{p+1} = -\varepsilon$. The "determinant" $\wedge^2 \overline{V}_{\iota,\varepsilon}$ is isomorphic to $\overline{U}_{2\bar{\iota},\varepsilon}$ (where $\bar{\iota}$ is the image of ι in $\mathbb{Z}/(p-1)\mathbb{Z}$). One has $\overline{V}_{\iota',\varepsilon'} \simeq \overline{V}_{\iota,\varepsilon}$ if and only if $\varepsilon' = \varepsilon$ and $\iota' \in \{l, p\iota\}$. The representation $\overline{V}_{\iota,\varepsilon}$ is absolutely irreducible if and only if $p\iota \neq \iota$ (in $\mathbb{Z}/(p^2-1)\mathbb{Z}$).*

ii) Let $\overline{V}$ be a 2-dimensional absolutely irreducible $\mathbb{F}_p$-representation of G_p. Then $\overline{V} \simeq \overline{V}_{\iota,\varepsilon}$ for a suitable $(\iota, \varepsilon) \in (\mathbb{Z}/(p^2-1)\mathbb{Z}) \times \mathbb{F}_p^*$.

Proof. Exercise. □

Let $\iota \in \mathbb{Z}/(p^2-1)\mathbb{Z}$ such that $p\iota \neq \iota$. The residual representation $\overline{V}_{\iota,\varepsilon}$ being absolutely irreducible possesses a universal deformation ring $R_{\iota,\varepsilon} = R(\overline{V}_{\iota,\varepsilon})$ as in [**22**], and we have the theorem due to R. Ramakrishna ([**23**] Th.4.1):

Theorem . *For any absolutely irreducible $\overline{V} = \overline{V}_{\iota,\varepsilon}$, the "deformation problem is smooth" and the ring $R_{\iota,\varepsilon}$ is isomorphic to a power series ring in five variables over $\mathbb{Z}_p$, i.e. to $\mathbb{Z}_p[\![T_1, T_2, \cdots, T_5]\!]$.*

§10. Potentially semi-stable p-adic representations and pst-modules.

It is known[**12**] that, using the ring B_{st} [**11**], one gets an equivalence of categories between potentially semi-stable p-adic representations of $G_p = \mathrm{Gal}(\overline{\mathbb{Q}}_p/\mathbb{Q}_p)$ and the category of "admissible filtered (φ, N, G_p)-modules". We want to recall a bit of this theory.

(a) **Filtered modules** : Let F be a finite Galois extension of $\mathbb{Q}_p$ contained in $\overline{\mathbb{Q}}_p$ and F_0 the maximal unramified extension of $\mathbb{Q}_p$ contained in F. A **filtered** $(\varphi, N,\ \mathrm{Gal}(F/\mathbb{Q}_p))$**-module** is a finite dimensional F_0-vector space D endowed with

- a "Frobenius" endomorphism $\varphi : D \to D$ bijectiv and semi-linear via the action of the absolute Frobenius on F_0,

- a "monodromy" operator $N : D \to D$ linear and satisfying $N\varphi = p\varphi N$,
- an action of $\mathrm{Gal}(F/\mathbb{Q}_p)$ semi-linear with respect to the natural action of this group (via its quotient $\mathrm{Gal}(F_0/\mathbb{Q}_p)$) on F_0 and commuting with φ and N,
- a decreasing filtration $(\mathrm{Fil}^r D_F)_{r\in\mathbb{Z}}$ of the F-vector space $D_F = F \otimes_{F_0} D$ stable under the natural action of $\mathrm{Gal}(F/\mathbb{Q}_p)$ (acting via $(g, \lambda \otimes d) \mapsto g\lambda \otimes gd$) verifying $\mathrm{Fil}^r D_F = 0$ for $f \gg 0$ and $= D_F$ for $r \ll 0$. Observe that, if we put $D_{dR} = (D_F)^{\mathrm{Gal}(F/\mathbb{Q}_p)}$, the obvious map $F \oplus_{\mathbb{Q}_p} D_{dR} \to D_F$ is an isomorphism and the condition that the $\mathrm{Fil}^r D'_F$s are stable under $\mathrm{Gal}(F/\mathbb{Q}_p)$ means exactly that the filtration comes from a filtration of the $\mathbb{Q}_p$-vector space D_{dR}.

Whenever $F = \mathbb{Q}_p, D$ is just a $\mathbb{Q}_p$-vector space together with two linear endomorphisms φ and N (with φ bijective and $N\varphi = p\varphi N$) and a filtration and we call D a **filtered** (φ, N)**-module**.

The filtered $(\varphi, N, \mathrm{Gal}(F/\mathbb{Q}_p))$-modules are the objects of an additive $\mathbb{Q}_p$-linear category (which is not abelian) where "morphisms" are defined in the evident way. One can define in a natural way the dual D^* of an object of this category and the tensor product $D_1 \otimes D_2$ of two objects.

Recall ([**9**], [**12**]) that for such a D, one can define its Hodge polygon (associated to the Hodge-Tate type, i.e. to the number $h^r = \dim gr^r D_F$) and its Newton polygon (associated to the slope of Frobenius). One says that D is **weakly admissible** (or **w.a.** for short) if
i) for any sub-$(\varphi, N, \mathrm{Gal}(F/\mathbb{Q}_p))$-module D' of$D$, with $D'_F \subset D_F$ equipped with the induced filtration, the Hodge polygon of D' lies above the Newton polygon of D';
ii) the Hodge polygon and the Newton polygon of D ends up at the same point.

The weakly admissible modules form an abelian category [**12**]. If D is an object of this category, sub-objects of D are sub-$(\varphi, N, \mathrm{Gal}(F/\mathbb{Q}_p))$-modules such that the Hodge polygon and Newton polygon of D end up at the same point.

Changing F : Call $\mathfrak{F}$ the set of finite Galois extensions of $\mathbb{Q}_p$ contained in $\overline{\mathbb{Q}}_p$. Given $F, F' \in \mathfrak{F}$ with $F \subset F'$, we say that a $(\varphi, N, \mathrm{Gal}(F'/\mathbb{Q}_p))$-module D' is F**-semi-stable** if the natural map $F'_0 \otimes_{F_0} (D')^{\mathrm{Gal}(F'/F)} \to D'$ is an isomorphism. To any filtered $(\varphi, N, \mathrm{Gal}(F/\mathbb{Q}_p))$-module D, one can associate a filtered $(\varphi, N, \mathrm{Gal}(F'/\mathbb{Q}_p))$-module $D_{/F'}$ in an obvious way (the underlying F'_0-vector space is $F'_0 \otimes_{F_0} D$). We get in this way an equivalence of categories between filtered $(\varphi, N, \mathrm{Gal}(F/\mathbb{Q}_p))$-modules and the full subcategory of filtered $(\varphi, N, \mathrm{Gal}(F'/\mathbb{Q}_p))$-modules consisting of F-semi-stable ones. Thus we can form the inductive limit over $\mathfrak{F}$ of the category of filtered $(\varphi, N, \mathrm{Gal}(F/\mathbb{Q}_p))$-modules; we call an object of this $\mathbb{Q}_p$-linear category a **pst-module**.

Each time we have a pst-module Δ, we may choose an $F \in \mathfrak{F}$ such that this object is F-semi-stable and we can speak of its F**-realisation** which is a filtered $(\varphi, N, \mathrm{Gal}(F/\mathbb{Q}_p))$-module D. On this D we have an F_0-linear action of the inertia subgroup of $\mathrm{Gal}(F/\mathbb{Q}_p)$ that we can view as a continuous action of I_p whose kernel contains $I_p \cap \mathrm{Gal}(\overline{\mathbb{Q}}_p/F)$. If $F' \in \mathfrak{F}$, Δ is F'-semi-stable if and only if $I_p \cap \mathrm{Gal}(\overline{\mathbb{Q}}_p/F')$ acts trivially on D. To know Δ is the same as to know D; the choice of F doesn't matter and we will sometimes speak of the pst-module D.

If D is a filtered $(\varphi, N, \mathrm{Gal}(F/\mathbb{Q}_p))$-module and if $F \subset F', D_{F'}$ is weakly admissible if and only if D is. Then it makes sense to speak of a **weakly admissible pst-module**. These modules form an *abelian* $\mathbb{Q}_p$*-linear category.*

(b) **Representations** : If $F \in \mathfrak{F}$, we have a natural functor $\underline{D}_{\mathrm{st},F}$ from the category of p-adic representations of G_p to the category of filtered $(\varphi, N,\ \mathrm{Gal}(F/\mathbb{Q}_p))$-modules:

$$V \mapsto D = \underline{D}_{\mathrm{st},F}(V) = (B_{\mathrm{st}} \otimes_{\mathbb{Q}_p} V)^{\mathrm{Gal}(\overline{\mathbb{Q}_p}/F)}.$$

We have $\dim_{F_0} \leq \dim_{\mathbb{Q}_p} V$ and V is said to be F**-semi-stable** if equality holds. The property to be F-semi-stable is stable under taking subobject, quotient, direct sum, tensor product, dual. Say a filtered $(\varphi, N,\ \mathrm{Gal}(F/\mathbb{Q}_p))$-module is **admissible** if it is isomorphic to $\underline{D}_{\mathrm{st},F}(V)$ for an F-semi-stable V. *It is known that admissible implies weakly admissible and conjectured that the converse is true.* The restriction of $\underline{D}_{\mathrm{st},F}$ to F-semi-stable representations is fully faithful, hence induces an *equivalence of categories between p-adic F-semi-stable representations of* G_p *and* **admissible filtered** $(\varphi, N,\ \mathrm{Gal}(F/\mathbb{Q}_p))$**-modules**. There is a natural quasi-inverse $\underline{V}_{\mathrm{st}}$ of $\underline{D}_{\mathrm{st},F}$ ($\underline{V}_{\mathrm{st}}(D)$ is defined as a suitable sub-$\mathbb{Q}_p$-vector space of $B_{\mathrm{st}} \otimes_{F_0} D$). This is an equivalence of tannakian categories, i.e. $\underline{D}_{\mathrm{st},F}$ is compatible in a natural way with duality and tensor product.

If D is admissible and if D' is a sub-$(\varphi, N,\ \mathrm{Gal}(F/\mathbb{Q}_p))$-module of D, then D' is admissible if and only if it is weakly admissible.

It is sometimes convenient to use a contravariant version of this equivalence. If V is F-semi-stable, one can define $\underline{D}^*_{\mathrm{st},F}(V)$ in three different ways (there are canonical isomorphisms between them [**12**])

$$\underline{D}^*_{\mathrm{st},F}(V) = Hom_{\mathbb{Q}_p[\ \mathrm{Gal}(\overline{\mathbb{Q}}_p/F)]}(V, B_{\mathrm{st}}) = \underline{D}_{\mathrm{st},F}(V^*) = (\underline{D}_{\mathrm{st},F}(V))^*$$

Let V be F-semi-stable and $D = \underline{D}_{\mathrm{st},F}(V)$. Then V is *de Rham* and the filtered $\mathbb{Q}_p$-vector space $\underline{D}_{dR}(V) = (B_{dR} \otimes_{\mathbb{Q}_p} V)^{G_p}$ can be identified with $D_{dR} = (D_F)^{\mathrm{Gal}(F/\mathbb{Q}_p)}$. Hence V is also Hodge-Tate and the Hodge-Tate multiplicities $h^r(V) = \dim_{\mathbb{Q}_p} \mathrm{Hom}_{\mathbb{Q}_p[G_p]}(V, \mathbb{C}_p(r))$ satisfy

$$h^r(V) = \dim_F \mathrm{Fil}^{-r} D_F / \mathrm{Fil}^{-r+i} D_F = \dim_F \mathrm{Fil}^r D^*_F / \mathrm{Fil}^{r+1} D^*_F.$$

Tate twists : For any filtered $(\varphi, N,\ \mathrm{Gal}(F/\mathbb{Q}_p))$-module D and for $i \in \mathbb{Z}$, we denote by $D\{i\}$ the filtered $(\varphi, N,\ \mathrm{Gal}(F/\mathbb{Q}_p))$-module which is D as a F_0-vector space with the same action of N and of $\mathrm{Gal}(F/\mathbb{Q}_p)$, and

$$\varphi_{\text{ new}} = p^{-i} \varphi_{\text{ old}} \text{ and } \mathrm{Fil}^r D_{F,\text{ new}} = \mathrm{Fil}^{r+1} D_{F,\text{ old}}.$$

If V is any p-adic representation of G_p and if $i \in \mathbb{Z}$, the usual Tate twist $V(i)$ is F-semi-stable if and only if V is. In this case, $\underline{D}_{\mathrm{st},F}(V(i))$ can be identified with $\underline{D}_{\mathrm{st},F}(V)\{i\}$ and $\underline{D}^*_{\mathrm{st},F}(V(i))$ with $\underline{D}^*_{\mathrm{st},F}(V)\{-i\}$.

Changing F : A p-adic representation V of G_p is said to be **potentially semi-stable** if it is F-semi-stable for some $F \in \mathcal{F}$. If V is F-semi-stable, with D $= \underline{D}_{\mathrm{st},F}(V)$, and if $F' \in \mathcal{F}$ contains F, V is also F'-semi-stable and

$\underline{D}_{st,F'}(V)$ can be identified with $D_{/F'}$. Hence we have an evident notion of **admissible** pst-**modules** and we get an equivalence between potentially semi-stable representations and the full subcategory of weakly admissible pst-modules whose objects are admissible ones (and conjecturally they are all).

Remark: A p-adic representation V of G_p is **semi-stable** if it is $\mathbb{Q}_p$-semi-stable, F-**crystalline** if it is F-semi-stable and $N = 0$ on $\underline{D}_{st,F}(V)$, **potentially crystalline** if it is F-**crystalline** for some F, **crystalline** if it is $\mathbb{Q}_p$-crystalline.
The one dimensional case : Consider a one dimensional pst-module Δ, choose $F \in \mathcal{F}$ such that Δ is F-semi-stable and let D be its F-realisation. This is a one dimensional F_0-vector space on which I_p acts linarly via a character of finite order with values in $\mathcal{O}_{F_0}$. The fact that this action commutes with φ implies that it takes values in $\mathbb{Z}_p$, hence its order divides $p-1$. This means that Δ is F_1-semi-stable (recall $F_1 = \mathbb{Q}_p(\sqrt[p-1]{1})$), hence we can choose $F = F_1$ and D is a one dimensional $\mathbb{Q}_p$-vector space; it is an easy exercise to check that i) there is a unique $(r, a, i) \in \mathbb{Z} \times \mathbb{Q}_p^* \times (\mathbb{Z}/p-1)\mathbb{Z}$ such that $D \simeq D(r; a; i)$, where $D(r; a; i)$ is defined as follows. Its underlying $\mathbb{Q}_p$-vector space we take to be $\mathbb{Q}_p$, and hence $D_{F_1} = F_1$. Put

$$\varphi 1 = a, N1 = 0, g1 = \hat{\xi}_1(g)^i \text{ if } g \in \text{ Gal}(F_1/\mathbb{Q}_p),\ \text{Fil}^r F_1 = F_1,\ \text{Fil}^{r+1} F_1 = 0.$$

ii) the module $D(r; a; i)$ is weakly admissible if and only if $v_p(a) = r$; in this case it is also admissible and G acts on $\underline{V}_{st}(D(r; a; i))$ via the character $\chi^{-r} \cdot \eta_{(a/p^r)^{-1}} \cdot \hat{\xi}_1^i$.
All the one dimensional potentially semi-stable p-adic representations of G_p are F_1-crystalline.

§11. The irreducible 2-dimensional weakly admissible pst-modules.

If $F \in \mathcal{F}$ and D is a 2-dimensional filtered $(\varphi, N,\ \text{Gal}(F/\mathbb{Q}_p))$-module, D_F is a 2-dimensional filtered F-vector space and there are well defined integers r, s satisfying $r \leq s$ such that, if $i \in \mathbb{Z}$, $\text{Fil}^i D_F = D_F \Leftrightarrow i \leq r$ and $\text{Fil}^i D_F = 0 \Leftrightarrow i > s$. We call (r, s) the Hodge-Tate type of D.

We fix (r, s) and we want to describe some 2-dimensional filtered $(\varphi, N,\ \text{Gal}(F/\mathbb{Q}_p))$-modules, D of Hodge-Tate type (r, s). If $r = s$, there is no choice for the filtration. If $r < s$, the filtration of D_F is determined by the knowledge of the F-line $\text{Fil}^{r+1} D_F = \text{Fil}^s D_F$ and we will denote it by Δ. The underlying F_0-vector space of each D will be taken to be $(F_0)^2$ and we denote $\{e_1, e_2\}$ the canonical basis.

(a) Let's first choose $F = F_1$, hence $F_0 = \mathbb{Q}_p, \Gamma_1 = \text{Gal}(F/\mathbb{Q}_p)$ and the characters of Γ_1 with values in $\mathbb{Z}_p$ are the $\hat{\xi}_1^i$ for $0 \leq i < p-1$.
Type I : Choose $(a, d, i) \in \mathbb{Q}_p \times \mathbb{Q}_p^* \times \mathbb{Z}$ with $0 \leq i < p-1$; define $D = D_I(r, s; a, d, i)$ by $\varphi e_1 = e_2, \varphi e_2 = -de_1 + ae_2, Ne_1 = e_2 = 0, gx = \hat{\xi}_1^i(g).x \quad (x \in D, g \in \Gamma_1)$, and if $r < s$, $\quad \Delta = F_1.1 \otimes e_1$.
Type II : Choose $(b, c, i) \in \mathbb{Q}_p^* \times \mathbb{Q}_p \times \mathbb{Z}$ with $0 \leq i < p-1$; define $D = D_{II}(r, s; b, c; i)$ by $\varphi e_1 = pbe_1, \varphi e_2 = be_2, Ne_1 = e_1, Ne_2 = 0, gx = \hat{\xi}_1^i(g).x \quad (x \in D, g \in \Gamma_1)$, and if $r < s, \Delta = F_1.(1 \otimes e_1 + c \otimes e_2)$.
Type III ; Choose $(a_1, a_2, i_1, i_2) \in (\mathbb{Q}_p^*)^2 \times \mathbb{Z}^2$, with $0 \leq i_1 < i_2 \leq p-1$; define $D = D_{III}(r, s; a_1, a_2; i_1, i_2)$ by $\varphi e_1 = ae_1, \varphi e_2 = a_2 e_2, Ne_1 = Ne_2 = 0, ge_1 = \hat{\xi}_1^{i_1}.e_1, \quad ge_2 = \hat{\xi}_1^{i_2}.e_2 \quad (g \in \Gamma_1)$, and, if $r < s, \Delta = F_1(\pi_1^{i_2} \otimes e_1 + \pi_1^{i_1} \otimes e_1)$.

(b) We now choose $F = F_2$ with, as in §9, $F_2 = \mathbb{Q}_{p^2}(\pi_2)$ where we have chosen $\pi_2 \in \overline{\mathbb{Q}}_p$ such that $\pi_2^{p^2-1} = -p$. Then $F_0 = \mathbb{Q}_{p^2}$. Let $\Gamma_2 =$ Gal$(F_2/\mathbb{Q}_p)$ and $I\Gamma_2$ the inertia subgroup. We denote by $\hat{\xi}_2 : I\Gamma_2 \to \mu_{p^2-1}(\mathbb{Q}_{p^2})$ the isomorphism defined by $\hat{\xi}_2(g) = g\pi_2/\pi_2$. The group Γ_2 is the semi-direct product of the invariant subgroup $I\Gamma_2$ by the subgroup of order 2 generated by the unique nontrivial element $\overline{\tau}$ of Γ_2 such that $\overline{\tau}\pi_2 = \pi_2$; moreover, if $g \in I\Gamma_2, \overline{\tau}g\overline{\tau} = g^p$.
Type IV ; Choose $d \in \mathbb{Q}_p^*$, integers i_1, i_2 satisfying $0 \leq i_1 < i_2 \leq p-1$. If $s > r$, choose also $\alpha \in \mathbb{P}^1(\mathbb{Q}_p)$. If $s = r$ (resp. $s > r$) define $D = D_{IV}(r,r;d;i_1,i_2)$(resp. $D_{IV}(r,s;d;i_1,i_2;\alpha)$) by $\varphi e_1 = e_2, \varphi e_2 = -de_1, Ne_1 = Ne_2 = 0, \overline{\tau}e_1 = e_1, \overline{\tau}e_2 = e_2, ge_1 = \hat{\xi}_2^{i_1+pi_2}(g)e_1$ and $ge_2 = \hat{\xi}_2^{i_2+pi_1}(g)e_2$ for $g \in I\Gamma_2$ and , if $r < s, \Delta = F_2.(\pi_2^{(p-1)i_1} \otimes e_1 + \alpha.\pi_2^{(p-1)i_2} \otimes e_2)$ if $\alpha \neq \infty$ and $\Delta = F_2.1 \otimes e_2$ if $\alpha = \infty$.

Theorem A : i) *The following pst-modules are weakly admissible and irreducible:*
1) $D_I(r,r;a,d,i)$ *if* $v_p(a) \geq r, v_p(d) = 2r$ *and* $X^2 - aX + d$ *irreducible over* $\mathbb{Q}_p$,
1') $D_I(r,s;a,d,i)$ *for* $r < s$, *if* $v_p(a) > r$ *and* $v_p(d) = r+s$,
2) $D_{II}(r,s;b,c;i)$ *for* $s - r \geq 3$ *and odd, if* $v_p(b) = (s-r-1)/2$,
3) $D_{III}(r,s;a_1,a_2;i_1,i_2)$ *for* $s - r \geq 2$, *if* $v_p(a_1) > r, v_p(a_2) > r, v_p(a_1a_2) = r+s$,
4) $D_{IV}(r,r;d;i_1,i_2)$*(resp.* $D_{IV}(r,s;d;i_1,i_2;\alpha)$ *if* $v_p(d) = 2r$*(resp.* $r+s$*).*
ii) The above objects are all absolutely irreducible exept $D_I(r,r;a,d,i)$ *for which the ring of endomorphisms is isomorphic to* $\mathbb{Q}_p[X]/(X^2 - aX + d)$.
iii) Any irreducible 2-dimensional w.a. pst-module, which is F_2*-semi-stable, is isomorphic to one and only one object of the lists 1)-4).*
iv) If $p \geq 5$, *any irreducible 2-dimensional w.a. pst-module is* F_2*-semi-stable.*

Proof. See App., §A. □

This theorem reduces the problem of finding the complete list of isomorphism classes of irreducible two dimensional p-adic potentially semi-stable representations of G_p to finding out which are the pst-modules in the above list which are admissible (and conjecturally they all are). Each time that we discover that one of those D's is admissible we will denote the corresponding representation $\underline{V}_{st}(D)$ (resp. the dual representation $\underline{V}_{st}^*(D) = \underline{V}_{st}(D)^*$) using the same notation but replacing D with V (resp. V^*) (e.g. $V_{IV}(r,s;d;i_1,i_2;\alpha) = \underline{V}_{st}(D_{IV}(r,s;d;i_1,i_2;\alpha))$ and $V_{IV}^*(r,s;d;i_1,i_2;\alpha)$ is the dual of this representation).

Given a two dimensional p-adic representation V of G_p and two integers $r \leq s$, we say that V **is of Hodge-Tate type** (r,s) if V is Hodge-Tate and $\mathrm{Hom}_{\mathbb{Q}_p[G_p]}(V, \mathbb{C}_p(i)) \neq 0$ if and only if $i \in \{r,s\}$. We see that if $V = \underline{V}_{st}^*(D)$ for some admissible pst-module D, than the Hodge-Tate type of V is the same as the Hodge-Tate type of D. Using this fact and checking the list of D's occurring in the previous theorem, we get the following result:

Proposition 1. *Let V be a two dimensional irreducible p-adic representation of G_p which is potentially semi-stable of Hodge-Tate type (r,s). Then, we* *are in one and only one of the following cases:*

1) *There is a character of finite order ν such that $V(\nu)$ is crystalline, in which case there is a unique integer i satisfying $0 \leq i < p-1$ such that one can choose $\nu = \hat{\xi}_1^i$ and a unique $(a,d) \in \mathbb{Q}_p \times \mathbb{Q}_p^*$ such that $D_I(r,s;a,d;i)$ is admissible and $V \simeq V_I^*(r,s;a,d;i)$;*
2) *The representation is not potentially crystalline, in which case there is a unique integer i satisfying $0 \leq i < p-1$ such that $V(\hat{\xi}_1^i)$ is semi-stable and a unique $(b,c) \in \mathbb{Q}_p^* \times \mathbb{Q}_p$ such that $D_{II}(r,s;b,c;i)$ is admissible and $V \simeq V_{II}^*(r,s;b,c;i)$;*
3) *The representation is F_1-crystalline, but there is no character ν such that $V(\nu)$ is crystalline, in which case there is a unique couple (i_1,i_2) of integers satisfying $0 \leq i_1 < i_2 \leq p-1$ and a unique $(a_1,a_2) \in (\mathbb{Q}_p^*)^2$ such that $D_{III}(r,s;a_1,a_2;i_1,i_2)$ is admissible and $V \simeq V_{III}^*(r,s;a_1,a_2;i_1,i_2)$;*
4) *The representation is not F_1-semi-stable, in which case it is F_2-crystalline, there is a unique couple (i_1,i_2) of integers satisfying $0 \leq i_1 < i_2 \leq p-1$, a unique $d \in \mathbb{Q}_p^*$ and , if $r < s$ a unique $\alpha \in \mathbb{P}^1(\mathbb{Q}_p)$, such that, if $r = s$ (resp. $r < s$), $D_{IV}(r,r;d;i_1,i_2)$ (resp. $D_{IV}(r,s;d;i_1,i_2;\alpha)$) is admissible and $V \simeq V_{IV}^*(r,r;d;i_1,i_2)$ (resp. $V \simeq V_{IV}^*(r,s;d;i_1,i_2;\alpha)$).*

From the fact that we have an equivalence of tannakian categories, one sees that, if D is admissible, then $\wedge^2 D$ is also and that $\underline{V}_{\mathrm{st}}(\wedge^2 D) \simeq \wedge^2 \underline{V}_{\mathrm{st}}(D)$. An easy computation gives us the following result:

Proposition 2. *Let D be one of the w.a. pst-modules listed in the previous theorem. Assume it is admissible and let $V = \underline{V}_{st}^*(D)$. Then G_p acts on $\det V$, via $\eta_u \chi^{r+s} \hat{\xi}_1^{-j}$,*

1) *with $u = d/p^{r+s}$ and $j = 2i$ if $V = V_I^*(r,s;a,d,i)$,*
2) *with $u = b^2/p^{r+s-1}$ and $j = 2i$ if $V = V_{II}^*(r,s;b,c;i)$,*
3) *with $u = a_1 a_2/p^{r+s}$ and $j = i_1 + i_2$, if $V = V_{III}^*(r,s;a_1,a_2;i_1,i_2)$,*
4) *with $u = d/p^{r+s}$ and $j = i_1 + i_2$, if $V = V_{IV}^*(r,s;d;i_1,i_2;\alpha)$ or $V_{IV}^*(r,s;d;i_1,i_2)$ with $s = r$.*

§12. Ramakrishna's theorem : crystalline representations of dimension 2 and their deformations.

We say that a p-adic representation V of G_p **is absolutely irreducible mod** p if there exists a $\mathbb{Z}_p$-lattice T of V stable under G_p such that the G_p-module T/pT is absolutely irreducible; we put $\overline{V} = T/pT$ and call it the **reduction mod p** of V. This is well defined, because the other lattices stable under G_p are the $p^n T$, for $n \in \mathbb{Z}$. Of course, any p-adic representation which is absolutely irreducible mod p is absolutely irreducible, but the converse is not true.

Theorem B1. *The D's of type IV which $s - r \leq p-1$ listed in theorem A are admissible. Moreover*

i) *If d is a p-adic unit, if $X^2 - aX + d$ is irreducible over $\mathbb{Q}_p$ and if λ is a root of this polynomial, $V_I^*(r,r;a,d;i)$ is isomorphic to a one dimensional $\mathbb{Q}_p(\lambda)$-vector space on which G_p acts through the character $\eta_\lambda . \chi^r . \hat{\xi}_1^{-i}$;*

ii) *If* $1 \leq s-r \leq p-1$, $V_I^*(r,s;a,d;i)$ *is absolutely irreducible mod p and its reduction mod p is isomorphic to* $\overline{V}_{\iota,\varepsilon}$ *where* ι *is the image of* $rp+s-(p+1)i$ *in* $\mathbb{Z}/(p^2-1)\mathbb{Z}$ *and* ε *the image of* d/p^{r+s} *in* $\mathbb{F}_p^*$.

Proof. See App., §B. □

Let $\underline{\mathrm{Rep}}_{\mathbb{Q}_p}(G_p)$ be the category of (finite dimensional) p-adic representations of G_p and $\underline{\mathrm{Rep}}^f_{\mathbb{Z}_p}(G_p)$ the category of $\mathbb{Z}_p$-modules of finite length equipped with a linear and continuous action of G_p. If $a \leq b$ are integers, we denote by $\underline{\mathrm{Rep}}_{\mathbb{Q}_p}(G_p)_{cr,[a,b]}$ the full subcategory of $\underline{\mathrm{Rep}}_{\mathbb{Q}_p}(G_p)$ consisting of those V which are crystalline and such that $h^r(V)(= \dim_{\mathbb{Q}_p} gr^r \underline{D}^*_{HT}(V)) = 0$ if $r \notin [a,b]$. Denote also by $\underline{\mathrm{Rep}}^f_{\mathbb{Z}_p}(G_p)_{cr,[a,b]}$ the full subcategory of $\underline{\mathrm{Rep}}^f_{\mathbb{Z}_p}(G_p)$ whose objects are T's for which one can find an object V of $\underline{\mathrm{Rep}}_{\mathbb{Q}_p}(G_p)_{cr,[a,b]}$ such that T is isomorphic to a subquotient of V. This is stable under subobjects, quotients, direct sums. Hence, for any (ι,ε) as in §9 such that $\overline{V}_{\iota,\varepsilon}$ is absolutely irreducible one can speak of the ring $R_{\iota,\varepsilon}(cr,[a,b])$ of the universal $\mathbb{Z}_p$-deformation of $\overline{V}_{\iota,\varepsilon}$ lying in $\underline{\mathrm{Rep}}^f_{\mathbb{Z}_p}(G_p)_{cr,[a,b]}$ as soon as $\overline{V}_{\iota,\varepsilon}$ itself is an object of this category.

Theorem B2. *Let* $\overline{V} = \overline{V}_{\iota,\varepsilon}$ *be absolutely irreducible.*

i) *The representation* $\overline{V}$ *lies in* $\underline{Rep}^f_{\mathbb{Z}_p}(G_p)_{cr,[0,p-1]}$;

ii) *if* r,s *are the unique integers such that* $0 \leq r < s \leq p-1$ *and the image of* $r+ps$ *mod* p^2-1 *is equal to* ι *or* $p\iota$, *there is a unique isomorphism* $R_{\iota,\varepsilon}(cr,[0,p-1]) \simeq \mathbb{Z}_p[\![Y_1,Y_2]\!]$ *such that, for any morphism* $Y_1 \mapsto y_1, Y_2 \mapsto y_2$ *of* $R_{\iota,\varepsilon}(cr,[0,p-1])$ *to* $\mathbb{Z}_p$, *if* T *is representative of the corresponding isomorphism class of* $\mathbb{Z}_p$*-deformation, then*

$$\mathbb{Q}_p \otimes_{\mathbb{Z}_p} T \simeq V_I^*(r,s;p^r y_2, p^{r+s}[\varepsilon](1+y_1),0).$$

Proof. See App., §B. This theorem is a slight refinement of a result of Ramakrishna [**23**] who proved the fact that $R_{\iota,\varepsilon}(cr,[0,p-1]) \simeq \mathbb{Z}_p[\![Y_1,Y_2]\!]$. □

Remark If $0 < s-r \leq p-1, u \in \mathbb{Z}$ and if $V_I^*(r,s;a,d;i)$ is as in theorem B1, then $V_I^*(r+u,s+u;a,d;i) \simeq V_I^*(r,s;p^u a,p^{2u}d;0)(\chi^u \hat{\xi}_1^{-i})$. If $\overline{u}$ denotes the remainder of euclidean division of u by $p-1$, the character $\chi^u \hat{\xi}_1^{-\overline{u}} = \chi_0^u$ takes values in $1+p\mathbb{Z}_p$. For each $u \in \mathbb{Z}$, twisting the representation on $\mathbb{Z}_p[\![Y_1,Y_2]\!]$ which is the "universal deformation of $\overline{V}_{\iota,\varepsilon}$ for which all finite quotients lie in $\underline{\mathrm{Rep}}^f_{\mathbb{Z}_p}(G_p)_{cr,[0,p-1]}$" by χ_0^u, we get the "universal deformation of $\overline{V}_{\iota,\varepsilon}$ for which all finite quotients T are such that $T(\hat{\xi}_1^{\overline{u}})$ lie in $\underline{\mathrm{Rep}}^f_{\mathbb{Z}_p}(G_p)_{cr,[u,u+p-1]}$". Any $V = \underline{V}^*_{\mathrm{st}}(D)$ for D absolutely irreducible of type I and Hodge-Tate type (r',s') with $s'-r' \leq p-1$ such that $\overline{V} \simeq \overline{V}_{\iota,\varepsilon}$ can be obtained by a suitable specialization of one of those universal deformations.

§13. Potentially Barsotti-Tate representations of dimension 2 and their deformations.

We will only give a sketch of the proof of the main results of this section (see Appendix, § C). Details will be published elsewhere.

Consider (ι, ε) where $\iota \in \mathbb{Z}/(p^2-1)\mathbb{Z}$ and $\varepsilon \in \mathbb{F}_p^*$. We denote by $V_{\iota,\varepsilon}$ the p-adic representation of G_p which is the "canonical lifting" of $\overline{V}_{\iota,\varepsilon}$: this is $\mathbb{Q}_{p^2}$ on which G_p acts through the homomorphism

$$\rho_{\iota,\varepsilon} : \mathrm{Gal}(L/\mathbb{Q}_p) \to \mathrm{Aut}_{\mathbb{Q}_p}(\mathbb{Q}_{p^2})$$

such that the restriction to $\mathrm{Gal}(L/L_0)$ consists in $\mathbb{Q}_{p^2}$-linear automorphisms given by the one-dimensional $\mathbb{Q}_{p^2}$-character $\hat{\xi}_2^{\iota}$, and τ acts on $\mathbb{Q}_{p^2}$ via the $\mathbb{Q}_{p^2}$-semilinear automorphism $x \mapsto [\zeta]\cdot\sigma x$ (where, as in §9, $\zeta^{p+1} = -\varepsilon$).

Let $\mathbb{Q}_{p^2}^{ab}$ be the maximal abelian extension of $\mathbb{Q}_{p^2}$ contained in $\overline{\mathbb{Q}}_p$ and $\theta : \mathbb{Q}_{p^2}^* \to \mathrm{Gal}(\mathbb{Q}_{p^2}^{ab}/\mathbb{Q}_{p^2})$ the inverse of the local reciprocity map[7]. The restriction θ_0 of θ to the group of units is an isomorphism onto the inertia subgroup $\mathrm{In}(\mathbb{Q}_{p^2}^{ab}/\mathbb{Q}_{p^2})$ of $\mathrm{Gal}(\mathbb{Q}_{p^2}^{ab}/\mathbb{Q}_{p^2})$; as the natural map $I_p \to \mathrm{In}(\mathbb{Q}_{p^2}^{ab}/\mathbb{Q}_{p^2})$ is onto, the inverse of θ_0 can be viewed as a character

$$\nu : I_p \to \mathbb{Q}_{p^2}^*.$$

Given $d \in \mathbb{Q}_p$ with $v_p(d) = 1$ and $\iota \in \mathbb{Z}/(p^2-1)\mathbb{Z}$, one sees easily that, up to conjugacy, there is one and only one continuous homomorphism

$$\rho_{LT,d,\iota} : G_p \to \mathrm{Aut}_{\mathbb{Q}_p}(\mathbb{Q}_{p^2})$$

("*LT*" for "Lubin-Tate") such that the restriction of $\rho_{LT,d,\iota}$ to I_p is the character $\nu\cdot\hat{\xi}_2^{\iota}$ and the deteminant of $\rho_{LT,d,\iota}$ is the character $\eta_d/p \cdot \chi \cdot \hat{\xi}_1^{\iota}$. We denote by $V_{LT,d,\iota}$ a chosen representative of the corresponding isomorphism class of p-adic representations. One sees that, if $(d', \iota',) \neq (d, \iota)$, then $V_{LT,d',\iota'} \not\simeq V_{LT,d,\iota}$ and that $V_{LT,d,\iota}$ is absolutely irreducible mod p if and only if $p+1$ doesn't divide $\iota+1$ in which case $\overline{V}_{LT,d,\iota} \simeq \overline{V}_{\iota+1,\varepsilon}$ with ε the image of d/p in $\mathbb{F}_p^*$. also $V_{LT,d,0} \simeq V_I^*(0,1;0,d;0) \simeq \mathbb{Q}_p \otimes_{\mathbb{Z}_p} T_p(J_d)$ where J_d is a p-divisible group over $\mathbb{Z}_p$ which viewed as a p-divisible over the ring $\mathbb{Z}_{p^2}$ of the integers of $\mathbb{Q}_{p^2}$ is a Lubin-Tate formal group for $\mathbb{Q}_{p^2}$.

Observe that $D_{IV}(r,r;d;i_1,i_2) \simeq D_{IV}(0,0;p^{-2r}d;i_1,i_2)\{-r\}$, hence one of those modules is admissible if and only if the other is, in which case $V_{IV}^*(r,r;d;i_1,i_2) \simeq V_{IV}^*(0,0;p^{-2r}d;i_1,i_2)(r)$. When $r < s$, we have a similar statement for $D_{IV}(r,s;d;i_1,i_2;\alpha) \simeq D_{IV}(0,s-r;p^{-2r}d;i_1,i_2;p^r\alpha)\{-r\}$ with, in case of admissiblity, $V_{IV}^*(r,s;d;i_1,i_2;\alpha) \simeq V_{IV}^*(0,s-r;p^{-2r}d;i_1,i_2;p^r\alpha)(r)$. In particular, the first representation is absolutely irreducible mod p if and only if the second is.

Theorem C1. *The D's of type 4 with $s - r \leq 1$ listed in theorem A are admissible. Moreover, if $0 \leq i_1 < i_2 \leq p-1$ and if $\bar{i}$ denotes the image of $i_1 + pi_2$ in $\mathbb{Z}/(p^2-1)\mathbb{Z}$, then*

i) if $d = [\varepsilon]u^2$ with $\varepsilon \in \mathbb{F}_p^*$ and $u \in 1 + p\mathbb{Z}_p$, $V_{IV}^*(0,0;d;i_1,i_2) \simeq V_{-\bar{i},\varepsilon}(\eta_u)$ (hence this representation is absolutely irreducible mod p and its reduction mod p is isomorphic to $\overline{V}_{-\bar{i},\varepsilon}$),

[7]hence the image of $\theta(p)$ in $\mathrm{Gal}(\overline{\mathbb{F}}_{p^2}/\mathbb{F}_{p^2})$ is the arithmetic Frobenius

ii) *if* $d = p[\varepsilon]u^2$ *with* $\varepsilon \in \mathbb{F}_p^*, u \in 1 + p\mathbb{Z}_p$ *and if* $\alpha \in \mathbb{P}^1(\mathbb{Q}_p)$, *then* $V = V_{IV}^*(0,1;d;i_1,i_2;\alpha)$ *is absolutely irreducible mod p if and only if - either* $v_p(\alpha) \geq 0$, *in which case* $\overline{V} \simeq \overline{V}_{1-\bar{i},\varepsilon}$ *- or* $v_p(\alpha) \leq -2$ *and* $i_2 - i_1 > 1$, *in which case* $\overline{V} \simeq \overline{V}_{1-p\bar{i},\varepsilon}$;

iii) *if* $v_p(d) = 1, V_{IV}^*(0,1;d;i_1,i_2;0) \simeq V_{LT,d,1-\bar{i}}$ *and* $V_{IV}^*(0,1;d;i_1,i_2;\infty) \simeq V_{LT,d,1-p\bar{i}}$.

We observed that all w.a. pst modules listed in theorem A with $s - r \leq 1$ are admissible because they are either of type I or of type IV. Say that a p-adic representation V of G_p is **potentially Barsotti-Tate** if one can find a finite extension E of $\mathbb{Q}_p$ contained in $\overline{\mathbb{Q}}_p$ and a Barsotti-Tate group J defined over the ring of integers of E such that $V \simeq V_p(J) = \mathbb{Q}_p \otimes_{\mathbb{Z}_p} T_p(J)$ as a $\mathbb{Q}_p[\,\mathrm{Gal}(\overline{\mathbb{Q}}_p/E)]$-module, where $T_p(J)$ is the Tate module of J. If this is the case, one knows that V is potentially crystalline with Hodge-Tate weights $\in \{0,1\}$. One conjectures the converse is true and this is clearly the case whenever 0 is the only weight (this means that the image of inertia is finite, hence that J is étale) or whenever 1 is the only weight (this means that the image of inertia on $V(-1)$ is finite, hence that J is of multiplicative type, i.e. is the Cartier dual of an étale Barsotti-Tate group).

Assume V is irreducible of dimension 2. Then the Hodge-Tate type can be $(0,0), (1,1)$ or $(0,1)$. If it is $(0,0)$ this means that $V \simeq \underline{V}_{\mathrm{st}}^*(D)$ with D equal to one of the w.a. modules $D_I(0,0;a,d,i)$ or $D_{IV}(0,0;d;i_1,i_2)$ listed in theorem A. If it is $(1,1)$ this means that $V(-1)$ is of type $(0,0)$. The problem for type $(0,1)$ is solved by the next theorem, for which we need two more definitions:

Definitions: Let J_0 be a Barsotti-Tate group over $\mathbb{F}_p$. Then J_0 is equipped with two endomorphisms, the Frobenius $\varphi = F$ and the Verschiebung V satisfying $\varphi V = V\varphi = p$. We say that J_0 **is strictly of slope** $1/2$ if there is an automorphism u of J_0 (necessarily unique) such that $\varphi = Vu$.

Let $\mathcal{O}_E$ be the ring of the integers of a finite, totally ramified extension E of $\mathbb{Q}_p$. We say that a Barsotti-Tate group J over $\mathcal{O}_E$ **is strictly of slope** $1/2$ if its special fiber is strictly of slope $1/2$ and if moreover, given any invariant differential form ω on J, one can find differential forms ω_1, ω_2 with ω_1 exact and $\lambda \in \mathcal{O}_E$ with $v_p(\lambda) \geq 1/2$ such that $\omega = \omega_1 + \lambda\omega_2$.

Theorem C2. *Assume* $p \geq 5$. *Let* V *be an irreducible two dimensional p-adic representation of* G_p *which is Hodge-Tate of Hodge-Tate type* $(0,1)$. *Then the following are equivalent:*

i) V *is potentially Barsotti-Tate,*

ii) V *is potentially crystalline,*

iii) V *is potentially semi-stable,*

iv) *There is a w.a. pst module* D *which is either one of the modules* $D_I(0,1;a,d,i)$ *or one of the modules* $D_{IV}(0,1;d,i_1,i_2;\alpha)$ *listed in Theorem A such that* $V \simeq \underline{V}_{st}^*(D)$.

Moreover,

a) *if* $V \simeq V_I^*(0,1;a,d,i)$, *there is a Barsotti-Tate group* J *defined over* $\mathbb{Z}_p$, *strictly of slope 1/2, such that* $V \simeq V_p(J)(\hat{\xi}_1^{-i})$ *when viewed as a* G_p*-module;*

b) Let $\pi = (\pi_2)^{p-1}$ *(hence* $\pi^{p+1} = -p$*) and* $\hat{\xi}_\pi : Gal(\mathbb{Q}_p(\pi_2)/\mathbb{Q}_p(\pi)) \to \mathbb{Z}_p^*$ *the character defined by* $\hat{\xi}_\pi(g) = g\pi_2/\pi_2$*. If* $V \simeq V_{IV}^*(0,1;d;i_1,i_2;\alpha)$ *there is a Barsotti-Tate group* J *defined over the integers of* $\mathbb{Q}_p(\pi)$*, strictly of slope 1/2, such that* $V \simeq V_p(J)(\hat{\xi}_\pi^{-i_1-i_2})$ *when viewed as a* $Gal(\overline{\mathbb{Q}}_p/\mathbb{Q}_p(\pi))$*-module.*

From now on, $\mathcal{O}_E = \mathbb{Z}_p[\pi]$ and $E = \mathbb{Q}_p(\pi)$. We denote by $\underline{\mathrm{Rep}}^f_{\mathbb{Z}_p}(G_p)_{BT,E,1/2}$ the full subcategory of $\underline{\mathrm{Rep}}^f_{\mathbb{Z}_p}(G_p)$ consisting of those T's for which we can find a Barsotti-Tate group J over $\mathcal{O}_E$ strictly of slope 1/2 such that T, when viewed as a $\mathrm{Gal}(\overline{\mathbb{Q}}_p/E)$-module, is isomorphic to a quotient of $T_p(J)$). This category is stable under passage to subobjects, quotients, direct sums. Hence for any (ι,ε) as in §9 such that $\overline{V}_{\iota,\varepsilon}$ is absolutely irreducible, one can speak of the ring $R_{\iota,\varepsilon}(BT,E,1/2)$ of the universal $\mathbb{Z}_p$-deformation of $\overline{V}_{\iota,\varepsilon}$ lying in $\underline{\mathrm{Rep}}^f_{\mathbb{Z}_p}(G_p)_{BT,E,1/2}$ as soon as $\overline{V}_{\iota,\varepsilon}$ itself is an object of this category.

Theorem C3. *Let* j_1, j_2 *be integers satisfying* $0 \le j_1 < j_2 \le p-1$, ι *the image of* $j_1 + pj_2$ *in* $\mathbb{Z}/(p^2-1)\mathbb{Z}$, $\varepsilon \in \mathbb{F}_p^*$ *and* $\overline{V} = \overline{V}_{\iota,\varepsilon}$*. Then* $\overline{V}$ *belongs to* $\underline{Rep}^f_{\mathbb{Z}_p}(G_p)_{BT,E,1/2}$*. One can build a* $\mathbb{Z}_p[\![Y_1,Y_2]\!] \times_{\mathbb{F}_p} \mathbb{Z}_p[\![Y_1',Y_2']\!]$*-deformation* $T_{\iota,\varepsilon}(BT,E,1/2)$ *of* $\overline{V}$ *in such a way that all finite quotients lie in* $\underline{Rep}^f_{\mathbb{Z}_p}(G_p)_{BT,E,1/2}$ *and that*

i) if $y_1, y_2 \in p\mathbb{Z}_p$ *and if* T *is the* $\mathbb{Z}_p$*-deformation of* $\overline{V}$ *obtained from* $T_{\iota,\varepsilon}(BT,E,1/2)$ *via the map* $Y_1 \mapsto y_1, Y_2 \mapsto y_2, Y_1' \to 0, Y_2' \to 0$*, then*

$$\mathbb{Q}_p \otimes_{\mathbb{Z}_p} T \simeq V_I^*(0,1;p[\varepsilon](1+y_1);j_2) \text{ if } j_2 - j_1 = 1 \text{ [8]}$$

$$(\text{ resp. } V_{IV}^*(0,1;p[\varepsilon](1+y_1);j_1+1,j_2,p^{-1}y_2) \text{ if } j_2 - j_1 > 1);$$

ii) if $y_1', y_2' \in p\mathbb{Z}_p$ *and if* T *is the* $\mathbb{Z}_p$*-deformation of* $\overline{V}$ *obtained from* $T_{\iota,\varepsilon}(BT,E,1/2)$ *via the map* $Y_1 \mapsto 0, Y_2 \mapsto 0, Y_1' \mapsto y_1', Y_2' \mapsto y_2'$*, then*

$$\mathbb{Q}_p \otimes_{\mathbb{Z}_p} T \simeq V_{IV}^*(0,1;p[\varepsilon](1+y_1');0,j_1+1,p^{-1}y_2') \text{ if } j_2 = p-1$$

$$(\text{ resp. } V_{IV}^*(0,1;p[\varepsilon](1+y_1');j_1,j_2+1,1/y_2') \text{ if } j_2 < p-1).$$

Remarks :

1) We observe that each isomorphism class of p-adic representation V as in Theorem C2 which is absolutely irreducible mod p is isomorphic to $\mathbb{Q}_p \otimes_{\mathbb{Z}_p} T$ for one and only one T obtained via the construction referred to in Theorem C3.

2) Assume ι and ε are as in Theorem C3. From this theorem, we get a natural homomorphism

$$R_{\iota,\varepsilon}(BT,E,1/2) \to \mathbb{Z}_p[\![Y_1,Y_2]\!] \times_{\mathbb{F}_p} \mathbb{Z}_p[\![Y_1',Y_2']\!]$$

It is likely that this map is onto with kernel killed by p.

[8]with the convention that $V_I^*(r,s;a,d;p-1) = V_I^*(r,s;a,d;0)$.

3) With the above conventions, we see that the determinant of the considered representation is $\eta_{[\varepsilon](1+y_1)}\cdot\chi\cdot\hat{\xi}_1^{-i}$ in case (i) and $\eta_{[\varepsilon](1+y_1')}\cdot\chi\cdot\hat{\xi}_1^{-i}$ in case (ii). Thus, if we consider $\mathbb{Z}_p$-deformations of $\overline{V}_{\iota,\varepsilon}$ with a given determinant, we have one less parameter.

4) The families of type I and of type IV are of different nature : if y_1, y_2 or y_1', y_2' are as above, we can look at the characteristic polynomial of Frobenius acting on the Dieudonné module of the Barsotti-Tate group J defined over $\mathcal{O}_E$: if $j_2 - j_1 = 1$, for $V_I^*(0,1;y_2,p[\varepsilon](1+y_1);j_2)$, this is $X^2 - y_2X + p[\varepsilon](1+y_1)$, hence if we fix this polynomial, there is only one representation; if we require this polynomial to have coefficients in $\mathbb{Q}$ with roots Weil numbers, there are only finitely many representations. For all the other cases, the characteristic polynomial is $X^2 + p[\varepsilon](1+y_1)$ (or $X^2 + p[\varepsilon](1+y_1')$) and y_2 (or y_2') may vary.

Appendix : Proof of results on potentially semi-stable representations

§A : Proof of theorem A.

By twisting, we see that it is enough to consider the case where $r = 0$, i.e. the case of type $(0, s)$ with $s \in \mathbb{N}$.

(a) *The case of twisted $\mathbb{Q}_p$-semi-stable modules (type I and II)*: Assume D is a two dimensional filtered (φ, N, Γ_1)-module of type $(0, s)$ which has the property that the action of $\Gamma_1 = \mathrm{Gal}(F_1/\mathbb{Q}_p)$ on it is diagonal. There is a unique integer i satisfying $0 \leq i < p-1$ such that the action is given by $\hat{\xi}_1^i$. Twisting by $\hat{\xi}_1^{-i}$, we can assume the action of Γ_1 is trivial i.e. we can suppose that D is just a (φ, N)-module, i.e. there is no action of Γ_1 and the filtration is defined on D. This D is a two dimensional $\mathbb{Q}_p$-vector space and the Frobenius φ acts linearly on it. The fact that the Newton polygon of its characteristic polynomial $X^2 - aX + d$ must be above the Hodge polygon and ends up at the same point means $v_p(d) = s$ and $a \in \mathbb{Z}_p$.

Suppose $s = 0$. We must have $N = 0$. Such a D is determined by (a, d), is weakly admissible and is irreducible if and only if there is no line in D stable under φ, i.e. if $X^2 - aX + d$ is irreducible, putting us in case 1 of the statement of Theorem A, part (i).

Suppose now $s \geq 1$ and $N = 0$. Observe that $\Delta = \mathrm{Fil}^s D$ can't be stable under φ. Otherwise Δ would be stable under φ and N and one could find $c \in \mathbb{Z}_p$ with $0 \leq v_p(c) \leq s$ such that $\varphi x = cx$ if $x \in \Delta$. But,

-if $v_p(c) < s$ the Newton polygon (of φ acting on Δ) is not above its Hodge polygon and D woould not be weakly admissible;

-if $v_p(c) = s, D$ would be weakly admissible but Δ would be a proper weakly admissible sub-object and D would not be irreducible.

Hence, if we choose a basis e of Δ, $\{e, \varphi e\}$ forms a basis of D. Now D is weakly admissible. As Δ is not stable under φ, the possible proper weakly admissible sub-objects are lines Δ' stable under φ such that $\varphi x = ux$ for $x \in \Delta'$ and u a unit. This doesn't occur if and only if p divides a and we are in case $1'$ of the statement of Theorem A, part (i).

The last case to consider is $s \geq 1$ and $N \neq 0$. The condition $N\varphi = p\varphi N$ implies, that φ must have two distinct eigenvalues $b, pb \in \mathbb{Z}_p$ with $v_p(b) + v_p(pb) = s$, hence s must be odd and $v_p(b) = (s-1)/2$. If e is a non-zero eigenvector corresponding to the eigenvalue pb, then e and Ne form a basis of D, and we have $\varphi e = pbe, \varphi Ne = b \cdot Ne, N^2 e = 0$. The unique proper subobject of D is the $\mathbb{Q}_p$-subspace Δ' spanned by Ne and weak admissibility amounts to requiring that $\Delta \neq \Delta'$. There is a unique $c \in \mathbb{Q}_p$ such that Δ is the line spanned by $e + c \cdot Ne$. Now D is weakly admissible; it is irreducible if and only if Δ' is not weakly admissible, i.e. if $v_p(b) > 0$ which amounts to saying that $s \geq 3$ and we are in case 2 of the statement of Theorem A, part(i).

Hence we have proved that *the D's of type I or II listed in theorem A are irreducible weakly admissible and that they exhaust the possibilities of D's which are F_1-semi-stable with a diagonal action of* Γ_1.

(b) *The case of F_1-semi-stable modules which are not twists of $\mathbb{Q}_p$-semi-stable modules (type III)* : Let D be a two dimensional (φ, N, Γ_1)-module of type $(0, s)$. The group Γ_1 acts linearly on D and because it is cyclic of degree $p-1$, this action is diagonalisable. To ask that it is not diagonal amounts to requiring that there are interes $0 \leq i_1 < i_2 < p-1$ and two lines D_1 and D_2 in D such that $gx = (\hat{\xi}_1(g))^{i_j} \cdot x$ for all $g \in \Gamma_1$ and all $x \in D_j, j \in \{1, 2\}$.

The Frobenius φ acts linearly on D. The fact that φ commutes with the action of Γ_1 means that D_1 and D_2 are stable under φ, hence there are elements $a_1, a_2 \in \mathbb{Q}_p$ such that $\varphi x = a_j \cdot x$ for all $x \in D_j$. The fact that the Newton polygon lies above the Hodge polygon and ends at the same point implies $a_1, a_2 \in \mathbb{Z}_p$ and $v_p(a_1) + v_p(a_2) = s$. As N is nilpotent, the fact that it commutes with the action of Γ_1 implies $N = 0$.

Now, if $s = 0$, one sees easily that such a D is weakly admissible, but D_1 and D_2 are proper sub-objects and D is not irreducible. Assume $s \geq 1$. For $j = 1, 2$, the line $F_1 \otimes D_j$ is stable under the action of Γ_1, but if Δ were this line, either $v_p(a_j) < s$ and D would not be weakly admissible or $v_p(a_j) = s$ and D would be weakly admissible but D_j would be a proper sub-object and D would not be irreducible. Therefore Δ must be a line stable under Γ_1 and $\neq D_1, D_2$. Then it is easy to see that one can choose a basis e_1 of D_1 and a basis e_2 of D_2 in such a way that Δ is generated by $\pi_1^{i_2} \otimes e_1 + \pi_1^{i_1} \otimes e_1$. It is now easy to check that such a D is weakly admissible. It will be irreducible if there is no proper sub-object. A proper sub-object must be stable under φ, hence must be D_1 or D_2. But, one sees that D_j is a proper subobject if and only if $v_p(a_j) = 0$.

Hence we have proved that *the D's of type III listed in theorem A are irreducible weakly admissible and that they exhaust the possibilities of D's which are F_1-semi-stable with a non-diagonal action of* Γ_1.

(c) *The case of F_2−, but not F_1-semi-stable modules (type IV)* : Let D be a two dimensional (φ, N, Γ_2)-module of type $(0, s)$. Let D_0 the sub-$\mathbb{Q}_p$-vector space of D consisting of those x such that $\overline{\tau} x = x$. One deduces easily from the fact that $\overline{\tau}$ acts semi-linearly on the $\mathbb{Q}_{p^2}$-vector space D that D_0 is of dimension 2 over $\mathbb{Q}_p$ and spans D as a $\mathbb{Q}_{p^2}$-vector space. The fact that φ commutes with the action of $\overline{\tau}$ is then equivalent to the fact that the restriction of φ to D_0 is a $\mathbb{Q}_p$-linear automorphism of D_0.

The group $I\Gamma_2$ is cyclic of order $p^2 - 1$ and acts linearly on D and, if we want D not to be semi-stable, the inertia group of F_2/F_1, which is the subgroup of index $p+1$, cannot act trivially. Therefore $I\Gamma_2$ acts on D through characters and at least one of them is of order not dividing $p-1$. This means that one can find a line D' of D and an integer i satisfying $0 \leq i < p^2 - 1$ and i not divisible by $p+1$ such that $gx = \hat{\xi}_2^i(g) \cdot x$ if $g \in I\Gamma_2$ and $x \in D'$. Then, $g(\varphi x) = \varphi(gx) = \varphi(\hat{\xi}_2^i(g) \cdot x) = \hat{\xi}_2^{pi}(g) \cdot \varphi x$; but i not divisible by $p+1$ means $pi \not\equiv i \pmod{p^2-1}$ which implies that $\varphi x \in D'$; we see that $D = D' \oplus D''$ with $gy = \hat{\xi}_2^{pi}(g) \cdot y$ if $g \in |\Gamma_2$ and $y \in D''$. Permuting i and pi mod p^2-1 if necessary, we can assume that $i = i_1 + pi_2$ and $pi \equiv i_2 + pi_1 \pmod{p^2-1}$ with integers i_1, i_2 satisfying $0 \leq i_1 < i_2 \leq p-1$.

Now if $x \in D'$, we have $g\overline{\tau}x = \overline{\tau}g^p x = \overline{\tau}(\hat{\xi}_2^{pi}(g) \cdot x) = \hat{\xi}_2^i(g) \cdot \overline{\tau}x$, hence $\overline{\tau}$ leaves D' fixed; similarly it leaves also D'' fixed. If $x \in D_0$ and if $x = x' + x''$, with $x' \in D'$ and $x'' \in D''$, we then see that $\overline{\tau}x' = x'$ and $\overline{\tau}x'' = x''$, hence we can write $D_0 = D_0' \oplus D_0''$ with $D_0' = D' \cap D_0$ and $D_0'' = D'' \cap D_0$. If $x' \in D_0'$, we must have $g\varphi x' = \varphi g x' = \varphi(\hat{\xi}_2^i(g) \cdot x') = \hat{\xi}_2^{pi}(g) \cdot \varphi x'$, hence $\varphi x' \in D_0''$. Similarly $\varphi^2 x' \in D_0'$. Summarizing, we see that we can find $d \in \mathbb{Q}_p$ and a basis e_1, e_2 of D such that

$$\overline{\tau}e_1 = e_1, \overline{\tau}e_2 = e_2, ge_1 = \hat{\xi}_2^{i_1+pi_2}(g)e_1 \text{ and } ge_2 = \hat{\xi}_2^{i_2+pi_1}(g)e_2$$

$$\text{for } g \in I\Gamma_2, \quad \varphi e_1 = e_2, \quad \varphi e_2 = -de_1.$$

As N is nilpotent, the fact that it commutes with the action of $I\Gamma_2$ implies $N = 0$. Now we have actions of φ, N and $\mathrm{Gal}(F_2/\mathbb{Q}_p)$ satisfying the required properties. If we want D to be weakly admissible, the fact that the Hodge polygon lies below the Newton polygon and that they end up at the same point amounts to requiring that $v_p(d) = s$.

If $s = 0$, we then have a weakly admissible module. There is no line stable under φ and Γ_2, hence D is irreducible and we get $D \simeq D_{IV}(0,0;d;i_1,i_2)$.

If $s \geq 1, \Delta$ can be any line of D_{F_2} stable under the action of Γ_2. One sees immediately that there is a unique $\alpha \in \mathbb{P}^1(\mathbb{Q}_p)$ such that this is the line generated by $\pi_2^{(p-1)i_1} \otimes e_1 + \alpha \cdot \pi_1^{(p-1)i_2} \otimes e_2$. Again because there is no line of D stable under φ and Γ_2, this D is weakly admissible and does not contain any proper weakly admissible subobject. Hence we have proved that *the D's of type IV listed in theorem A are irreducible weakly admissible and that they exhaust the possibilities of D's which are F_2-semi-stable but not F_1-semi-stable.*

(d) The only thing which is left to prove is the fact that, if $p \geq 5$, *any irreducible 2-dimensional weakly admissible pst-module is F_2-semi-stable.* : One can assume that this module is F-semi-stable, for F a finite Galois extension of $\mathbb{Q}_p$ containing F_2 and contained in $\overline{\mathbb{Q}_p}$. The inertia group $\mathrm{In}(F/\mathbb{Q}_p)$ of the extension $F/\mathbb{Q}_p$ acts linearly on the corresponding two dimensional F_0-vector space D. If $g \in \mathrm{In}(F/\mathbb{Q}_p)$, the fact that the action of g commutes with the action of φ implies that the characteristic polynomial of g has coefficients in $\mathbb{Q}_p$. Because $p \geq 5, [\mathbb{Q}_p(\sqrt[p]{1}) : \mathbb{Q}_p] > 2$ and any element of $\mathrm{In}(F/\mathbb{Q}_p)$ of order a power of p acts trivially; so the p-Sylow group P of $\mathrm{In}(F/\mathbb{Q}_p)$ acts trivially. Replacing F by F^P, one can assume $F/\mathbb{Q}_p$ tame, i.e. that $\mathrm{In}(F/\mathbb{Q}_p)$ is cyclic of order prime to p; if g is a generator of this group, the roots of the polynomial characteristic of g acting on D are in $\mathbb{Q}_{p^2}$, hence of order dividing p^2-1. Therefore g^{p^2-1} acts trivially and the result follows from the fact that this element generates the inertia group of F/F_1. □

§B : Proof of theorems B1 and B2.

(a)*Proof of theorem* B1: By twisting, it is enough to prove the theorem for $r = 0$ and $i = 0$. In this case we observe that $D_I(0, s; a, d; 0)$ can be viewed as a weakly admissible filtered φ-module over $\mathbb{Q}_p$ ($N = 0$ and the Galois group action is trivial) which satisfies $\mathrm{Fil}^0 D = D$ and $\mathrm{Fil}^p D = 0$ hence is admissible ([**15**],[**30**]).

To prove (i) we are reduced to checking that $V_I^*(0, 0; a, d; 0)$ is isomorphic to a one dimensional $\mathbb{Q}_p(\lambda)$-vector space on which G_p acts through the unramified character η_λ. But $\mathrm{Fil}^0 B_{\mathrm{cris}}$ and $\mathrm{Fil}^0 B_{\mathrm{st}}$ contain the completion $\hat{\mathbb{Q}}_p^{nr}$ of the maximal unramified extension of $\mathbb{Q}_p$ contained in $\overline{\mathbb{Q}}_p$ and hence we have an injective map $\mathrm{Hom}_{\mathbb{Q}_p[\varphi]}(D_I(0, 0; a, d; 0), \hat{\mathbb{Q}}_p^{nr}) \to V_I^*(0, 0; a, d; 0)$, which is an isomorphism for reasons of dimensions. The assertion is then an easy exercise.

(ii) is a consequence of theorem B2 which will be prove below. □

(b)*The category* $\underline{MF}^f_{]-p+1,0]}$ *and the functor* $\overline{V}_{cr}$: Let $\underline{\mathrm{Rep}}_{\mathbb{Q}_p}(G_p)_{\mathrm{cr},[0,p-1[}$ be the full subcategory of $\underline{\mathrm{Rep}}_{\mathbb{Q}_p}(G_p)_{\mathrm{cr},[0,p-1]}$ consisting of those V which have no non-trivial subobject V' such that $V'(-p+1)$ is unramified. Similarly denote by $\underline{\mathrm{Rep}}^f_{\mathbb{Z}_p}(G_p)_{\mathrm{cr},[0,p-1[}$ the full subcategory of $\underline{\mathrm{Rep}}^f_{\mathbb{Z}_p}(G_p)_{\mathrm{cr},[0,p-1]}$ consisting of those T which are isomorphic to a subquotient of an object of $\underline{\mathrm{Rep}}_{\mathbb{Q}_p}(G_p)_{\mathrm{cr},[0,p-1[}$. It is easy to see that, for any V in $\underline{\mathrm{Rep}}_{\mathbb{Q}_p}(G_p)_{\mathrm{cr},[0,p-1]}$, one can find a short exact sequence

$$0 \to V' \to V \to V'' \to 0$$

such that $V'(-p+1)$ is unramified and V'' is in $\underline{\mathrm{Rep}}_{\mathbb{Q}_p}(G_p)_{\mathrm{cr},[0,p-1[}$.

Let $\overline{V}$ be a 2-dimensional $\mathbb{F}_p$-vector space on which the action of the inertia group is irreducible. We see that, for any local artinian $\mathbb{Z}_p$-algebra A of residue field $\mathbb{F}_p$, any A-deformation of $\overline{V}$ which lies in $\underline{\mathrm{Rep}}^f_{\mathbb{Z}_p}(G_p)_{\mathrm{cr},[0,p-1]}$ actually lies in $\underline{\mathrm{Rep}}_{\mathbb{Q}_p}(G_p)_{\mathrm{cr},[0,p-1[}$.

For $a, b \in \mathbb{Z}$ satisfying $a \leq b$, let $\underline{MF}^f_{[a,b]}$ be the following category:
-the objects are $\mathbb{Z}_p$-modules M of finite length, equipped with
i) a filtration of M by sub-$\mathbb{Z}_p$-modules

$$M = \mathrm{Fil}^a M \supset \mathrm{Fil}^{a+1} M \supset \cdots \supset \mathrm{Fil}^j M \supset \cdots \supset \mathrm{Fil}^b M \supset \mathrm{Fil}^{b+1} M = 0;$$

ii) for each j, a $\mathbb{Z}_p$-linear map $\varphi^j : \mathrm{Fil}^j M \to M$ such that $\varphi^{j+1}(x) = p\varphi^j(x)$ if $x \in \mathrm{Fil}^{j+1} M$, and such that $M = \Sigma_{a \leq j \leq b} \varphi^j(\mathrm{Fil}^j M)$.
-the morphisms are $\mathbb{Z}_p$-linear maps compatible with all the structures.

Recall [**16**] that this is an abelian category and that for any object M the $\mathrm{Fil}^j M$'s are direct summands (as $\mathbb{Z}_p$-modules). If $a < b$, we denote by $\underline{MF}_{]a,b]}$ the full subcategory of $\underline{MF}_{[a,b]}$ consisting of those M which have no non-trivial subobject N such that $\mathrm{Fil}^{a+1} N = 0$.

Let $\overline{k}$ be the residue field of $\overline{\mathbb{Q}}_p$ and σ the absolute Frobenius acting on $\overline{k}$ (via $x \mapsto x^p$) and on the ring $W(\overline{k})$ of Witt vectors with coefficients in $\overline{k}$ by functoriality. Let A_{cris} be the ring constructed in [**11**]. Recall that this is a $W(\overline{k})$-algebra which is a domain equipped with i) an action of G_p compatible with the ring structure and with the obvious action on $W(\overline{k})$, ii) a Frobenius $\varphi : A_{\mathrm{cris}} \to A_{\mathrm{cris}}$ compatible with the ring structure, commuting with the action of G_p and σ-semi-linear, iii) a decreasing filtration $(\operatorname{Fil}^j A_{\text{cris}})_{j\in\mathbb{N}}$ by ideals, which are direct summands as $W(\overline{k})$-modules, stable under G_p.

Moreover, for $0 \leq j \leq p-1$, we have $\varphi(\operatorname{Fil}^j M) \subset p^j \operatorname{Fil}^j M$, therefore, one can define $\varphi^j : \operatorname{Fil}^j A_{\text{cris}} \to A_{\text{cris}}$ by $\varphi^j x = p^{-j}\varphi x$.

For any object M of $\underline{MF}^f_{[0,p-1]}$ and $0 \leq j \leq p-1$ we can use the natural map $\operatorname{Fil}^j A_{\text{cris}} \otimes_{\mathbb{Z}_p} \operatorname{Fil}^{-j} M \to M$, which is injective, to identify $\operatorname{Fil}^j A_{\text{cris}} \otimes_{\mathbb{Z}_p} \operatorname{Fil}^{-j} M$ with a sub-A_{cris}-module of $A_{\text{cris}} \otimes_{\mathbb{Z}_p} M$ and define $\operatorname{Fil}^0(A_{\text{cris}} \otimes_{\mathbb{Z}_p} M) = \Sigma_{0\leq j\leq p-1} \operatorname{Fil}^j A_{\text{cris}} \otimes_{\mathbb{Z}_p} \operatorname{Fil}^{-j} M$. It is easy to see that there is a unique σ-semi-linear map

$$\varphi : \operatorname{Fil}^0(A_{\text{cris}} \otimes_{\mathbb{Z}_p} M) \to A_{\text{cris}} \otimes_{\mathbb{Z}_p} M$$

such that, for $0 \leq j \leq p-1$, $\varphi(\lambda \otimes x) = \varphi^j\lambda \otimes \varphi^{-j}x$ if $\lambda \in \operatorname{Fil}^j A_{\text{cris}}$ and $x \in \operatorname{Fil}^j M$.

Recall ([**16**], see also [**30**]) that if for M in $\underline{MF}_{[-p+1,0]}$, we define

$$\underline{V}_{\text{cr}}(M) = \{v \in \operatorname{Fil}^0(A_{\text{cris}} \otimes_{\mathbb{Z}_p} M) | \varphi v = v\},$$

then $\underline{V}_{\text{cr}}(M)$ is a finite $\mathbb{Z}_p$-representation of G_p. If M is in $\underline{MF}_{]-p+1,0]}$, this is an object of $\underline{\operatorname{Rep}}^f_{\mathbb{Z}_p}(G_p)_{\text{cr},[0,p-1[}$, which has the same length as M as a $\mathbb{Z}_p$-module. The functor

$$\underline{V}_{\text{cr}} : \underline{MF}^f_{]-p+1,0]} \to \underline{\operatorname{Rep}}^f_{\mathbb{Z}_p}(G_p)_{\text{cr},[0,p-1[}$$

defined in this way induces an equivalence between those two categories[9]. We denote by $\underline{D}_{\text{cr}}$ a quasi-inverse.

(c) *The mod p representation*: Recall that $r, s \in \mathbb{Z}$ with $0 \leq r < s \leq p-1$ and $\varepsilon \in \mathbb{F}_p^*$. We give to the two dimensional $\mathbb{F}_p$-vector space $(\mathbb{F}_p)^2$ (with $\overline{u}_1, \overline{u}_2$ the canonical basis) the structure of an object $\overline{M} = \overline{M}(r, s; \varepsilon)$ of $\underline{\operatorname{Rep}}^f_{\mathbb{Z}_p}(G_p)_{\text{cr},[0,p-1[}$ by defining

$$\operatorname{Fil}^{-s}\overline{M} = \overline{M},\ \operatorname{Fil}^{-s+1}\overline{M} = \operatorname{Fil}^{-r}\overline{M} = \mathbb{F}_p\overline{u}_1,\ \operatorname{Fil}^{-r+1}\overline{M} = 0,$$
$$\varphi^{-r}\overline{u}_1 = \overline{u}_2, \varphi^{-s}\overline{u}_2 = -\varepsilon^{-1} \cdot \overline{u}_1.$$

It is not hard to check (compare with [**16**]): i) that one can identify $\underline{V}_{\text{cr}}(\overline{M})$ with the residue field $\mathbb{F}_{p^2}$ of $\mathbb{Q}_{p^2}$ in such a way that the inertia subgroup I_p acts via the character ξ_2^{r+ps}; ii) that the determinant of the action of G_p is the character $\eta_\varepsilon\chi^{r+ps}$; all together this proves that $\underline{V}_{\text{cr}}(\overline{M}) \simeq \underline{V}_{i,\varepsilon}$ and this proves the first part of the assertion (ii) of Theorem B1.

(d) Let's prove the following lemma:

Lemma . *Let A be a local artinian ring, $\mathfrak{S}$ an abelian A-linear category, F, G two covariant functors from $\mathfrak{S}$ to the category of A-modules of finite length which are A-linear, exact, faithful and such that, for any object N of $\mathfrak{S}$, the A-modules $F(N)$ and $G(N)$ have the same length. Then, if $d \in \mathbb{N}$ and M is an object of $\mathfrak{S}$, $F(M)$ is free of rank d over A if and only $G(M)$ is.*

Indeed, for $a \in A$ and N an object of $\mathfrak{S}$, call $[a]_N$ the endomorphism of N which is the action of a. Let m_A be the maximal ideal of A and $k = A/m_A$. The

[9] actually, what is defined in [**16**] is a contravariant version of this construction but the passage from one construction to the other is straightforward.

faithfulness of F implies that, for any object N of $\mathfrak{G}$, one has $[a]_N = 0$ for all $a \in m_A$ if and only if $F(N)$ is a k-vector space. Now let $\overline{M}$ be the biggest quotient of M such that $[a]_{\overline{M}} = 0$ for all $a \in m_A$. It is clear that $F(\overline{M}) = F(M)/m_A F(M)$ and $G(\overline{M}) = G(M)/m_A G(M)$. Assume that $F(M)$ is free of rank d. Then $\dim_k F(\overline{M}) = d$. Hence we have also $\dim_k G(\overline{M}) = d$ and if $e_1, e_2, \cdots, e_d$ are lifting in $G(M)$ of a basis of $G(\overline{M})$ over k, the e_j generate $G(M)$. But they are linearly independent over A, for otherwise, $\text{length}_A G(M) < d{\cdot}\text{length}_A A = \text{length}_A F(M)$.

(e) *Deformations*: Let A be a local artinian $\mathbb{Z}_p$-algebra. Any A-representation T of V can be viewed as a $\mathbb{Z}_p$-representation together with an imbedding of A into the ring of the endomorphisms of T. Therefore, if T is an object of $\underline{\text{Rep}}^f_{\mathbb{Z}_p}(G_p)_{\text{cr},[0,p-1[}$, $\underline{D}_{\text{cr}}(T)$ is an **A-object of** $\underline{MF}^f_{]-p+1,0]}$; that is, there is a natural structure of A-module on it, the filtration is given by sub-A-modules and the maps φ^r are A-linear. Conversely, given any A-object M Of $\underline{MF}^f_{]-p+1,0]}$, $\underline{V}_{\text{cr}}(M)$ is an A-representation of G_p. Therefore, for any A, the functor $\underline{V}_{\text{cr}}$ induces an equivalence between the category of A-objects of $\underline{MF}^f_{]-p+1,0]}$ and the category of A-objects in $\underline{\text{Rep}}^f_{\mathbb{Z}_p}(G_p)_{\text{cr},[0,p-1[}$. Applying the lemma to the first of those two categories and to the functors which associate to M respectively the underlying A-module and the A-module underlying $\underline{V}_{\text{cr}}(M)$, we see, that for any A-object M of $\underline{MF}^f_{]-p+1,0]}$, $\underline{V}_{\text{cr}}(M)$ is flat as an A-module if and only if M is.

Now let V be an A-deformation of $\overline{V}$ and $M = \underline{D}_{\text{cr}}(V)$; if m_A is the maximal ideal of A, one can identify $M/m_A M$ with $\overline{M}$; as M must be a free-A-module, any couple $\{u_1, u_2\}$ of elements of M lifting $\{\overline{u}_1, \overline{u}_2\}$ is a basis of M over A; one knows **[16]** that any morphism in $\underline{MF}^f_{]-p+1,0]}$ is strictly compatible with the filtrations. This implies that $\text{Fil}^{-s}M = M$ and that one can choose $u_1 \in \text{Fil}^{-r}M$; as $\varphi^{-r}\overline{u}_1 = \overline{u}_2$, one can choose $u_2 = \varphi^{-r}u_1$; then we must have $\varphi^{-s}u_2 = -[\varepsilon^{-1}](1+x_1)u_1 + x_2u_2$ with $x_1, x_2 \in m_A$.

Now $\text{Fil}^{-r}M$ contains the free-A-module of rank one spanned by u_1. *It can't be bigger*, for otherwise one could find a nonzero $\lambda \in A$ such that $\lambda u_2 \in \text{Fil}^{-r}M$; multiplying λ by a suitable power of p, one could assume $p\lambda = 0$; but then $p^{s-r} \cdot \varphi^{-r}(\lambda u_2) = \varphi^{-s}(\lambda u_2) = \lambda(-[\varepsilon^{-1}](1+x_1)u_1 + x_2u_2) \neq 0$ because λ is a unit in A; on the other hand, $p^{s-r}\varphi^{-r}(\lambda u_2) = p^{s-r-1}\varphi^{-r}(p\lambda \cdot u_2) = 0$ giving us a contradiction. So we see that

$$\text{Fil}^{-s}M = M,\ \text{Fil}^{-s+1}M = \text{Fil}^{-r}M = Au_1,\ \text{Fil}^{-r+1}M = 0,$$

$$\varphi^{-r}u_1 = u_2, \quad \varphi^{-s}u_2 = -[\varepsilon^{-1}](1+x_1)u_1 + x_2u_2.$$

Now if we change the lifting of u_1, it must be to the element of the form λu_1 with λ a unit in A; then $\varphi^{-r}(\lambda u_1) = \lambda u_2$ and $\varphi^{-s}u_2 = -[\varepsilon^{-1}](1+x_1) \cdot \lambda u_1 + x_2 \cdot \lambda u_2$ and we see that x_1 and x_2 depends only on the isomorphism class of M.

Conversely it is clear that if we now identify $\{u_1, u_2\}$ with the canonical basis of A^2 we have equipped A^2, with the structure of an A-object $M = M_A(r,s,\varepsilon;x_1,x_2)$ of $\underline{MF}^f_{]-p+1,0]}$ and that

$$V_A(r,s,\varepsilon;x_{,1},x_2) = \underline{V}_{\text{cr}}(M(r,s,\varepsilon;x_{,1},x_2))$$

is an A-deformation of $\overline{V}$ and that for any A-deformation V of $\overline{V}$, there is a unique $(x_1, x_2) \in m_A$ such that $V \simeq V(r, s, \varepsilon; x_{,1}, x_2)$.

Now let $\mathfrak{a}$ be the set of artinian quotients of the ring $\mathbb{Z}_p[\![X_1, X_2]\!]$. For any $A \in \mathfrak{a}$, consider $M_A = M_A(r, s, \varepsilon; x_{,1}, x_2)$ and $V_A = V_A(r, s, \varepsilon; x_1, x_2)$ with x_i = the image of x_i in A. Then the M_A's form a projective system of objects of $\underline{MF}^f_{]p+1,0]}$ and therefore the V_A's form a projective system of finite representations of G_p which are in $\underline{\mathrm{Rep}}^f_{\mathbb{Z}_p}(G_p)_{\mathrm{cr},[0,p-1[}$. Then it is clear that $\text{lim-proj}\cdot_{A\in\mathfrak{a}} V_A$ is a free $\mathbb{Z}_p[\![X_1, X_2]\!]$-module of rank two equipped with an action of G_p which gives an identification of $\mathbb{Z}_p[\![X_1, X_2]\!]$ with $R_{i,\varepsilon}(\mathrm{cr}, [0, p-1])$. Now if we put $Y_1 = (1 + x)^{-1} - 1$ and $Y_2 = [\varepsilon](1 + x_1)^{-1}x_2$, we have $\mathbb{Z}_p[\![X_1, X_2]\!] = \mathbb{Z}_p[\![Y_1, Y_2]\!]$.

(f) *End of the proof of* (ii): It is easy to see that if we send $\mathbb{Z}_p[\![X_1, X_2]\!]$ to $\mathbb{Z}_p$ via $X_1 \mapsto x_1, X_2 \mapsto x_2$ if $y_1 = (1 + x_1)^{-1} - 1$ and $y_2 = -[\varepsilon](1 + x_1)^{-1}x_2$ and if T and V are as in the statement of the theorem, then V is crystalline and that the dual D of $D^* = \underline{D}^*_{\mathrm{cris}}(V) = \underline{D}_{\mathrm{st},\mathbb{Q}_p}(V)$ is the two dimensional $\mathbb{Q}_p$-vector space $D = \mathbb{Q}_p u_1 + \mathbb{Q}_p u_2$ with

$$\mathrm{Fil}^{-s}D = D,\ \mathrm{Fil}^{-s+1}D = \ \mathrm{Fil}^{-r}D = \mathbb{Q}_p u_1,\ \mathrm{Fil}^{-r+1}D = 0,$$

$$\text{and } \varphi u_1 = p^{-r}u_2, \quad \varphi u_2 = p^{-s}(-[\varepsilon^{-1}](1 + x_1)u_1 + x_2u_2).$$

Hence D is of Hodge-Tate type $(-s, -r)$ and the characteristic polynomial of φ acting on D is $X^2 - p^{-s}x_2 \cdot X + p^{-r-s}[\varepsilon]^{-1}(1 + x_1)$. Therefore, D^* is of Hodge-Tate type (r, s) and the characteristic polynomial of φ acting on D^* is $X^2 - p^r x_2[\varepsilon](1 + x_1)^{-1} \cdot X + p^{r+s}[\varepsilon](1 + x_1)^{-1}$; hence $D^* \simeq D_I(r, s; p^r x_2[\varepsilon](1 + x_1)^{-1}, p^{r+s}[\varepsilon](1 + x_1)^{-1} = D_I(r, s; p^r y_2, p^{r+s}[\varepsilon](1 + y_1)\text{'}0)$.

§C : About the proof of theorems C1, C2 and C3:

a) *Proof of theorem C1 for $s = r$*: Twisting if necessary, we are reduced to case $r = s = 0$ and this is an easy exercise on representations of G_p on which the action of I_p is finite. □

b) *About theorem C2* : Let V be as in theorem C2. The inplications i) $\Rightarrow$ ii) $\Rightarrow$ iii) are well known. If V is potentially semi-stable, there is a D listed in theorem A which is of Hodge-Tate type $(0, 1)$ and admissible such that $V \simeq \underline{V}^*_{\mathrm{st}}(D)$. Looking at the list, we see that D is either one of the modules $D_I(0, 1; a, d; i)$ or one of the modules $D_{IV}(0, 1; d, i_1, i_2, \alpha)$, hence iii) $\Rightarrow$ iv).

Remark : Before giving a sketch of proof of the other statements, let us explain how one could easily deduce theorem C2 and the admissibility statement of theorem C1 from results of Laffaille : We already know (th.B1) that any $D = D_I(0, 1; a, d; i)$ is admissible; moreover $V_I^*(0, 1; a, d; i)(\hat{\xi}_1^i) \simeq V_I^*(0, 1; a, d; 0)$ is a crystalline representation of G_p of Hodge-Tate type $(0, 1)$ and **[21]** gives that this representation comes from a Barsotti-Tate group defined over $\mathbb{Z}_p$; using the fact that the p-adic valuation of each root of the characteristic polynomial of φ acting on D, which is $X^2 - aX + d$ is $1/2$, it is easy to see that Γ is strictly of slope $1/2$, hence we get *the assertion (a) of theorem C2.*

Similarly, let D be any of the $D_{IV}(0,1;d,i_1,i_2,\alpha)$'s of theorem A. Denote by D_0 the sub-$\mathbb{Q}_p$-vector space fixed under $\overline{\tau}$. Observe that the filtration on D_{F_2} is actually defined on E and that D_0 equipped with the action of φ and the filtration on $E \otimes D_0$ can be viewed as a "weakly admissible filtered φ-module over E" of dimension 2 and of Hodge-Tate type $(0,1)$. Laffaille's **[20]** gives that this module is admissible and that there is a Barsotti-Tate group J defined over $\mathcal{O}_E$ such that the associated p-adic representation of $\mathrm{Gal}(\overline{\mathbb{Q}}_p/\mathbb{Q}_p)$, $\mathrm{Hom}_{\mathrm{filt}\cdot\varphi-\mathrm{mod}}(D_0, B_{\mathrm{cris}})$, is isomorphic to $V_p(J)$. From that, we easily get the admissibility of D and *the assertion (b) of theorem C2*, except for the fact that one can choose J strictly of slope 1/2; this latter fact can be shown by a direct computation. By twisting, we get the *admissibility of all* $D_{IV}(r,r+1;d,i_1,i_2,\alpha)$'s. From the above discussion we have that all the w.a. modules $D_I(0,1;a,d;I)$ and $D_{IV}(0,1;d,i_1,i_2,\alpha)$ are actually admissible and that the corresponding representations are potentially Barsotti-Tate, hence iv) $\Rightarrow$ i).

Sketch of the proof of the other statements:

1. - If S is any scheme, let us call a **p-group scheme over S** any inductive system $(J_n)_{n\in\mathbb{N}}$ of finite and flat commutative group schemes over S such that map $J_n \to J_{n+1}$ identifies J_n with the kernel of the multiplication by p^n in J_{n+1}. Thus, if the map $J_n \to J_{n+1}$ is also an isomorphism for n big enough, we have a **finite p-group scheme**; that is, a finite and flat commutative group scheme killed by a power of p. If there is an integer h such that J_n is free of rank p^{nh} for all n, we get a **Barsotti-Tate group**.

2. - Any formal group, and therefore also any p-group scheme, over $\mathbb{F}_p$ is equipped with two endomorphisms, the Frobenius $F=\varphi$ and the Verschiebung V satisfying $\varphi V = V\varphi = p$.

A **slope 1/2 structure** on a p-group scheme J over $\mathbb{F}_p$ is an automorphism u of J such that $\varphi = Vu$. If J is a Barsotti-Tate group over $\mathbb{F}_p$, there is at most one slope 1/2 structure u on J and there is such a u if and only if J is strictly of slope 1/2.

The p-group schemes over $\mathbb{F}_p$ with slope 1/2 structure form, in an obvious way an additive category $\underline{pGS}_{\mathbb{F}_p,1/2}$ which turns out to be abelian.

Forgetting u we get an additive functor from $\underline{pGS}_{\mathbb{F}_p,1/2}$ to the category $\underline{pGS}_{\mathbb{F}_p}$ of p-group schemes over $\mathbb{F}_p$ which is exact and faithful; if (J_1,u) and (J_2,u) are two objects of $\underline{pGS}_{\mathbb{F}_p,1/2}$, the cokernel of the injective map

$$\mathrm{Hom}_{\underline{pGS}_{\mathbb{F}_p,1/2}}((J_1,u)(J_2,u)) \to \mathrm{Hom}_{\underline{pGS}_{\mathbb{F}_p}}(J_1,J_2)$$

is killed by p.

3. - Let CW be the formal group of covectors over $\mathbb{F}_p$ (**[8]**, chap.II, §4). Let us recall that, for any finite $\mathbb{F}_p$-algebra A, $CW(A)$ consists of covectors $a = (\cdots, a_{-n}, \cdots, a_{-1}, a_0)$ with $a_{-n} \in A$ for all n and $a_{-n} \in r_A$ for almost all n, where r_A denotes the radical of A; with the obvious notation, we have $a+b=c$, with

$$c_{-n} = S_m(a_{-n-m}, \cdots, a_{-n-1}, a_{-n}; b_{-n-m}, \cdots, b_{-n-1}, b_{-n}) \text{ for } m >> 0,$$

where the S_m's are the polynomials which define the addition on Witt vectors. We have also

$$\varphi a = (\cdots, (a_{-n})^p, \cdots, (a_{-1})^p, (a_0)^p) \text{ and } V_a = (\cdots, a_{-n-1}, \cdots, a_{-2}, a_{-1}).$$

To any p-group scheme J over $\mathbb{F}_p$, one can associate its "contravariant Dieudonné-module" $\underline{M}(J) = \mathrm{Hom}(J, CW)$. One can view $\underline{M}$ as a contravariant additive functor from the category of p-group schemes over $\mathbb{F}_p$ to the category of $\mathbb{Z}_p[\varphi, V]$-modules which are $\mathbb{Z}_p$-modules of finite type. This is an anti-equivalence of categories. A quasi-inverse to $\underline{M}$ is given by associating to such a $\mathbb{Z}_p[\varphi, V]$-module M the p-group scheme $\underline{J}(M)$ defined by $\underline{J}(M)(A) = \mathrm{Hom}_{\mathbb{Z}_p,[\varphi,V]}(M, CW(A))$ for all finite $\mathbb{F}_p$-algebras A ([**8**], chap.III).

Denote by $\mathcal{O}_{1/2}$ the ring $\mathbb{Z}_p[u, u^{-1}, \varphi]/(\varphi^2 - pu)$. Setting $V = \varphi u^{-1}$, we can view $\mathbb{Z}_p[F, V]$ as a subring of $\mathcal{O}_{1/2}$. If (J, u) is an object of $\underline{pGS}_{\mathbb{F}_p,1/2}$, then $\underline{M}(J)$ has a natural structure of $\mathcal{O}_{1/2}$-module; we get in this way an anti-equivalence between $\underline{pGS}_{\mathbb{F}_p,1/2}$ and the category of $\mathcal{O}_{1/2}$-modules which are $\mathbb{Z}_p$-modules of finite type.

4. - For any finite $\mathbb{F}_p$-algebra A, denote $BW_{1/2}(A)$ the set of the $a = (a_n)_{n\in\mathbb{Z}}$ with $a_n \in r_A$ for all n. With obvious the notation, one sees that, for all fixed $n \in \mathbb{Z}$, if $a, b \in BW_{1/2}(A)$, the sequence

$$S_m((a_{n-m})^{p^m}, \cdots, (a_{n-1})^p, a_n; (b_{n-m})^{p^m}, \cdots, (b_{n-1})^p, b_n) \text{ for } m \in \mathbb{N}$$

is stationary; if we denote by c_n its limit and if we set

$$a + b = c = (c_n)_{n\in\mathbb{Z}},$$

$BW_{1/2}(A)$ becomes an abelian group and $BW_{1/2}$ may be viewed as a commutative formal group over $\mathbb{F}_p$ which is equipped with an automorphism u defined by

$$u((a_n)_{n\in\mathbb{Z}}) = (a_{n+1})_{n\in\mathbb{Z}}.$$

We have $\varphi(a_{n\in\mathbb{Z}}) = ((a_n)^p)_{n\in\mathbb{Z}}$ and $V = \varphi u^{-1}$.

We have also a natural morphism from $BW_{1/2}$ to CW sending $(a_n)_{n\in\mathbb{Z}}$ to $(\cdots, (a_{-n})^{p^n}, \cdots, (a_{-1})^p, a_0)$. If (J, u) is an object of $\underline{pGS}_{\mathbb{F}_p,1/2}$, this map induces an isomorphism from $\mathrm{Hom}((J, u), BW_{1/2})$ to the $\mathcal{O}_{1/2}$-module $\underline{M}(J)$ and we use it to identify these two modules.

5. - We denote by $\underline{pGS}_{\mathcal{O}_E}$ the additive category of p-group schemes over $\mathcal{O}_E$. The couples (J, u) consisting of a p-group scheme J over $\mathcal{O}_E$ and a slope $1/2$ structure u on the special fiber $J_{\mathbb{F}_p}$ are in an obvious way the objects of an additive category $\underline{pGS}_{\mathcal{O}_E,1/2}$. We denote by $\underline{pGS}_{\mathcal{O}_E,1/2,\mathrm{strict}}$ the full subcategory of $\underline{pGS}_{\mathcal{O}_E,1/2}$ consisting of those (J, u), for which one can find two Barsotti-Tate groups J' and J'' over $\mathcal{O}_E$, strictly of slope $1/2$ and morphisms $J \to J'$, $J' \to J''$ of p-group schemes such that the sequence

$$0 \to J \to J' \to J'' \to 0$$

is exact and the diagram

$$\begin{array}{ccccccccc} 0 & \to & J_{\mathbb{F}_p} & \to & J'_{\mathbb{F}_p} & \to & J''_{\mathbb{F}_p} & \to & 0 \\ & & \downarrow u & & \downarrow u' & & \downarrow u'' & & \\ 0 & \to & J_{\mathbb{F}_p} & \to & J'_{\mathbb{F}_p} & \to & J''_{\mathbb{F}_p} & \to & 0 \end{array}$$

(where u' (resp. u'') is the unique slope $1/2$ structure which exists on $J'_{\mathbb{F}_p}$ (resp. $J''_{\mathbb{F}_p}$)) is commutative.

6. - Set $\mathcal{O}_{E,1/2} = \mathcal{O}_E[v, v^{-1}]$. The ring $\mathcal{O}_E \otimes_{\mathbb{Z}_p} \mathcal{O}_{1/2}$ is a domain and, setting $1 \otimes \varphi = \pi^{(p+1)/2} . v$ and $1 \otimes u = -v^2$, we identify $\mathcal{O}_{E,1/2}$ with the normalization of $\mathcal{O}_E \otimes_{\mathbb{Z}_p} \mathcal{O}_{1/2}$ in its field of fractions. Also $\mathcal{O}_{E,1/2}$ is a faithfully flat $\mathcal{O}_{1/2}$-algebra.

Define the **category** $\underline{MF}_{E,1/2}$ as follows:

- an object is a couple $(M, \wedge)$ where M is an $\mathcal{O}_{1/2}$-module, which is of finite type as a $\mathbb{Z}_p$-module, and $\wedge$ is a sub $\mathcal{O}_E$-module of $\mathcal{O}_{E,1/2} \otimes_{\mathcal{O}_{1/2}} M$ such that the map from $\wedge \oplus \wedge$ to $\mathcal{O}_{E,1/2} \otimes_{\mathcal{O}_{1/2}} M$ sending (x, y) to $x + vy$, is an isomorphism;

- a morphism $(M, \wedge) \to (M', \wedge')$ is an $\mathcal{O}_{1/2}$-linear map from M to M' such that the $\mathcal{O}_{E,1/2}$-linear map induced by scalar extension sends $\wedge$ into $\wedge'$.

7. - Now, for any finite and flat $\mathcal{O}_E$-algebra $\mathfrak{a}$, there is a unique $\mathcal{O}_E$-linear map

$$\lambda_{\mathfrak{a}} : \mathcal{O}_{E,1/2} \otimes_{\mathcal{O}_{1/2}} BW_{1/2}(\mathfrak{a}/\pi\mathfrak{a}) \to \mathfrak{a}[1/p]/\pi\mathfrak{a}$$

such that, if $a = (a_n)_{n \in \mathbb{Z}} \in BW_{1/2}(\mathfrak{a}/\pi\mathfrak{a})$ and if $\hat{a}_n$ is a lifting of a_n in $\mathfrak{a}$, then $\lambda_{\mathfrak{a}}(1 \otimes a) = \Sigma_{m \in \mathbb{N}} p^{-m} (\hat{a}_{-m})^{p^{2m}} (\bmod \pi\mathfrak{a})$ and $\lambda_{\mathfrak{a}}(v \otimes a) = \pi^{-(p+1)/2} \cdot \Sigma_{m \in \mathbb{N}} p^{-m} (\hat{a}_{-m})^{p^{2m+1}} (\bmod \pi\mathfrak{a})$.

Let $(M, \wedge)$ be an object of $\underline{MF}_{E,1/2}$. Then M defines an object $(J_{\mathbb{F}_p}, u)$ of $\underline{pGS}_{\mathbb{F}_p,1/2}$: for any finite $\mathbb{F}_p$-algebra A,

$$J_{\mathbb{F}_p}(A) = \mathrm{Hom}_{\mathcal{O}_{1/2}}(M, BW_{1/2}(A)).$$

If for any finite and flat $\mathcal{O}_E$-algebra $\mathfrak{a}$, we define

$$J(\mathfrak{a}) = \{\alpha \in \mathrm{Hom}_{\mathcal{O}_{1/2}}(M, BW_{1/2}(\mathfrak{a}/\pi\mathfrak{a})) | \wedge \subset \mathrm{Ker}\ \lambda_{\mathfrak{a}} \circ (id \otimes \alpha)\},$$

we get a functor J from the category of finite and flat $\mathcal{O}_E$-algebras to abelian groups; one can check that J is actually a p-group scheme over $\mathcal{O}_E$. Moreover (J, u) is an object of $\underline{pGS}_{\mathcal{O}_E,1/2,\mathrm{strict}}$.

The correspondence $(M, \wedge) \mapsto (J = \underline{J}(M, \wedge), u = \underline{u}(M))$ defines a contravariant additive functor

$$\underline{Ju} : \underline{MF}_{E,1/2} \to \underline{pGS}_{\mathcal{O}_E,1/2,\mathrm{strict}},$$

which turns out to be an anti-equivalence of categories. If the $\mathbb{Z}_p$-module underlying M is free, J is a Barsotti-Tate group.

8. - Let $\mathcal{O}'_{1/2}$ be the noncommutative ring generated over the commutative ring $\mathbb{Z}_{p^2}[u, u^{-1}]$ by an element φ with relations $\varphi^2 = pu, \varphi u = u\varphi$ and $\varphi a = \sigma a \cdot \varphi$, if $a \in \mathbb{Z}_{p^2}$ (where σ is the Frobenius). This ring contains $\mathbb{Z}_{p^2}$ and $\mathcal{O}_{1/2}$ as commutative subrings and can be identified as a $\mathbb{Z}_p$-module with

$\mathbb{Z}_{p^2} \otimes_{\mathbb{Z}_p} \mathcal{O}_{1/2}$. For any finite $\mathbb{F}_{p^2}$-algebra A, there is unique structure of $\mathbb{Z}_{p^2}$-module on $BW_{1/2}(A)$ such that $[\varepsilon]\cdot(a_n)_{n\in\mathbb{Z}} = (\varepsilon a_n)_{n\in\mathbb{Z}}$ if $\varepsilon \in \mathbb{F}_{p^2}$ and $(a_n)_{n\in\mathbb{Z}} \in BW_{1/2}(A)$. Together with the structure of $\mathcal{O}_{1/2}$-module, $BW_{1/2}(A)$ becomes an $\mathcal{O}'_{1/2}$-module. Moreover, for any $\mathcal{O}_{1/2}$-module M, $\mathbb{Z}_{p^2} \otimes_{\mathbb{Z}_p} M$ has a natural structure of $\mathcal{O}'_{1/2}$-module and the obvious map

$$\mathrm{Hom}_{\mathcal{O}_{1/2}}(M, BW_{1/2}(A)) \to \mathrm{Hom}_{\mathcal{O}'_{1/2}}(\mathbb{Z}_{p^2} \otimes_{\mathbb{Z}_p} M, BW_{1/2}(A))$$

is an isomorphism: we use it to identify these two groups.

9. - Let $F = \mathbb{Q}_{p^2}(\pi)$ be the Galois closure of E in $\overline{\mathbb{Q}}_p$; this is a subfield of $F_2 = \mathbb{Q}_{p^2}(\pi_2)$ and $\Gamma_2 = \mathrm{Gal}(F/\mathbb{Q}_p)$ acts on it. Recall (§11) that we have defined an isomorphism $\hat{\xi}$ from the inertia subgroup $I\Gamma$, of Γ_2 onto the group $\mu_{p^2-1}(\mathbb{Q}_{p^2})$ and called $\overline{\tau}$ the only nontrivial element of $\mathrm{Gal}(F_2/\mathbb{Q}_p(\pi_2))$.

Consider now a couple (M, gr) where M is an $\mathcal{O}_{1/2}$-module which is a $\mathbb{Z}_p$-module of finite type, and gr is a gradation on M indexed by $\mathbb{Z}/(p+1)\mathbb{Z}$,

$$M = \bigoplus_{s\in\mathbb{Z}/(p^2-1)\mathbb{Z}} gr^s M,$$

by sub-$\mathbb{Z}_p[u, u^{-1}]$-modules, such that $\varphi(gr^s M) \subset gr^{ps} M$ for all s. The group Γ_2 acts naturally on $\mathcal{O}'_{1/2} = \mathbb{Z}_{p^2} \otimes_{\mathbb{Z}_p} \mathcal{O}_{1/2}$ (via $g(a\otimes b) = a \otimes b$ if $g \in I\Gamma_2$ and $\overline{\tau}(a\otimes b) = \sigma a \otimes b$). We can define a semi-linear action of Γ_2 on the $\mathcal{O}'_{1/2}$-module $\mathbb{Z}_{p^2} \otimes_{\mathbb{Z}_p} M$ by setting, if $a \in \mathbb{Z}_{p^2}$ and $x \in gr^s M$,

$$\overline{\tau}(a \otimes x) = \sigma a \otimes x \text{ and } g(a \otimes x) = (\hat{\xi}_2(g))^s \cdot a \otimes x \text{ for all } g \in I\Gamma_2.$$

Let $\overline{\mathbb{Z}}_p$ be the integral closure of $\mathbb{Z}_p$ in the chosen algebraic closure $\overline{\mathbb{Q}}_p$ of $\mathbb{Q}_p$. Define $BW_{1/2}(\overline{\mathbb{Z}}_p/\pi)$ to be the inductive limit of the $BW_{1/2}(\mathcal{O}_{F'}/\pi\mathcal{O}_{F'})$ for F' running through finite Galois extensions of $\mathbb{Q}_p$ contained in $\overline{\mathbb{Q}}_p$ and containing F. Set

$$J_{\mathbb{F}_p}(\overline{\mathbb{Z}}_p/\pi) = \mathrm{Hom}_{\mathcal{O}_{1/2}}(M, BW_{1/2}(\overline{\mathbb{Z}}_p/\pi)) = \mathrm{Hom}_{\mathcal{O}'_{1/2}}(\mathbb{Z}_{p^2} \otimes_{\mathbb{Z}_p} M, BW_{1/2}(\overline{\mathbb{Z}}_p/\pi)).$$

This abelian group is equipped with an action of G_p: if

$$u : \mathbb{Z}_{p^2} \otimes_{\mathbb{Z}_p} M \to BW_{1/2}(\overline{\mathbb{Z}}_p/\pi)$$

is an $\mathcal{O}'_{1/2}$-linear map and if $\gamma \in G_p$, then $\gamma(u) = \gamma \circ u \circ \gamma^{-1}$.

10. - The group Γ_2 acts naturally on $\mathcal{O}_F$ and on $\mathbb{Z}_{p^2} \otimes_{\mathbb{Z}_p} \mathcal{O}_{E,1/2} = \mathbb{Z}_{p^2} \otimes_{\mathbb{Z}_p} \mathcal{O}_E[v, v^{-1}] = \mathcal{O}_F[v, v^{-1}]$ (with $gv = v$ if $g \in I\Gamma_2$ and $\overline{\tau}v = -v$). If (M, gr) is as above, the action of Γ_2 on $\mathbb{Z}_{p^2} \otimes_{\mathbb{Z}_p} M$ extends uniquely to a semi-linear action on $\mathbb{Z}_{p^2} \otimes_{\mathbb{Z}_p} (\mathcal{O}_{E,1/2} \otimes_{\mathcal{O}_{1/2}} M)$. Define the category $\underline{MF}_{E/\mathbb{Q}_p,1/2}$ as follows:

-an object is a triple (M, Λ, gr) with (M, Λ) an object of $\underline{MF}_{E/\mathbb{Q}_p,1/2}$ and (M, gr) as above such that $\mathbb{Z}_{p^2} \otimes_{\mathbb{Z}_p} \Lambda \subset \mathbb{Z}_{p^2} \otimes_{\mathbb{Z}_p} (\mathcal{O}_{E,1/2} \otimes_{\mathcal{O}_{1/2}} M)$ is stable under Γ_2;

- a morphism is a morphism of the underlying objects of $\underline{MF}_{E,1/2}$ which is compatible with the gradations.

This category is abelian.

If $(M, \wedge, gr)$ is an object of $\underline{MF}_{E/\mathbb{Q}_p,1/2}$, and if J is the p-group scheme over $\mathcal{O}_E$ associated to $(M, \wedge)$, one can view $J(\overline{\mathbb{Z}}_p)$ (= the inductive limit of the $J(\mathcal{O}_{F'})$ for F' describing the finite Galois extensions of $\mathbb{Q}_p$ contained in $\overline{\mathbb{Q}}_p$ and containing F) as a subgroup of the group $J_{\mathbb{F}_p}(\overline{\mathbb{Z}}_p/\pi)$ defined above and the condition that $\wedge$ is stable under Γ implies that $J(\overline{\mathbb{Z}}_p)$ is stable under G_p.

If $\underline{\mathrm{Rep}}_{p-\mathrm{tor}}(G_p)$ denotes the category of p-torsion abelian groups V such that the kernel of multiplication by p is finite, equipped with a linear and continuous action of G_p, we can see $J(\overline{\mathbb{Z}}_p)$ as an object of this category.

The correspondence $(M, \wedge, gr) \mapsto J(\overline{\mathbb{Z}}_p)$ can be viewed as a contravariant additive functor

$$\underline{J}_{1/2} : \underline{MF}_{E/\mathbb{Q}_p,1/2} \to \underline{\mathrm{Rep}}_{p-\mathrm{tor}}(G_p).$$

11. - The function $\underline{J}_{1/2}$ is exact and faithful. Moreover, if $(M, \wedge, gr)$ and $(M', \wedge', gr')$ are two objects of $\underline{MF}_{E/\mathbb{Q}_p,1/2}$, the cokernel of the map

$$\mathrm{Hom}((M, \wedge, gr), (M', \wedge', gr')) \to \mathrm{Hom}(\underline{J}_{1/2}(M, \wedge, gr), \underline{J}_{1/2}(M', \wedge', gr'))$$

is killed by p.

If the $\mathbb{Z}_p$-module underlying M is free and if $J = \underline{J}(M, \wedge)$ is the corresponding Barsotti-Tate group over $\mathcal{O}_E$, $\underline{J}_{1/2}(M, \wedge, gr) = J(\overline{\mathbb{Z}}_p)$, when viewed as a $\mathrm{Gal}(\overline{\mathbb{Q}}_p/F_2)$-module. Therefore,

$$V = V_p((\underline{J}_{1/2})(M, \wedge, gr)) = \mathbb{Q}_p \otimes_{\mathbb{Z}_p} \text{lim.proj} \cdot \underline{J}_{1/2}(M, \wedge, gr)_{p^n}$$

is a potentially Barsotti-Tate p-adic representation of G_p, hence a fortiori is potentially crystalline. It is F_2-semi-stable; the corresponding admissible filtered (φ, N, Γ_2)-module $D = \underline{D}^*_{\mathrm{st},F_2}(V)$ can be identified with $\underline{D}(M, \wedge, gr)$ defined as follows: the underlying $\mathbb{Q}_{p^2}$-vector space is $\mathbb{Q}_{p^2} \otimes_{\mathbb{Z}_p} M = \mathbb{Q}_p \otimes_{\mathbb{Z}_p} (\mathbb{Z}_{p^2} \otimes_{\mathbb{Z}_p} M)$ with the given action of φ and of Γ_2 and with $N = 0$; the filtration on

$$D_{F_2} = F_2 \otimes_{\mathbb{Q}_{p^2}} D = F_2 \otimes_E (E \otimes_{\mathbb{Q}_{p^2}} D) = F_2 \otimes_{\mathcal{O}_E} (\mathcal{O}_{E,1/2} \otimes_{\mathcal{O}_{1/2}} M)$$

is given by $\mathrm{Fil}^0 D_{F_2} = D_{F_2}$, $\mathrm{Fil}^1 D_{F_2} = F_2 \otimes_{\mathcal{O}_E} \wedge$, $\mathrm{Fil}^2 D_{F_2} = 0$.

12. - With these results in mind, the proof of the theorems becomes an exercise in the category $\underline{MF}_{E/\mathbb{Q}_p,1/2}$:

a) For each w.a. pst-module $D = D_I(0, 1; a, d, i)$ or $D = D_{IV}(0, 1; d, i_1, i_2; \alpha)$ one exhibits an object $(M, \wedge, gr)$ of $\underline{MF}_{E/\mathbb{Q}_p,1/2}$ such that $D \simeq \underline{D}(M, \wedge, gr)$; this gives us statement a) and b) of the theorem C2, hence also, as we already explained in the above remark, the admissibility statement of theorem C1 and the implication iv) $\Rightarrow$ i) of theorem C2, whose proof is completed. □

b) To this $(M, \wedge, gr)$, we can associate

$$T_p((\underline{J}_{1/2})(M, \wedge, gr)) = \text{lim.proj} \cdot \underline{J}_{1/2}(M, \wedge, gr)_{p^n}$$

which can be identified with a lattice T of $V = \underline{V}^*_{\mathrm{st}}(D)$ stable under G_p. Reducing $(M, \wedge, gr)$ mod p we can compute explicitly the two dimensional $\mathbb{F}_p$-representation $\overline{V} = T/pT = \underline{J}_{1/2}((M, \wedge, gr) \bmod p)$ of G_p and check the assertions of theorem C1 for $(r, s) = (0, 1)$. By twisting we deduce the assertions for $s - r = 1$ and the proof of theorem C1 is completed. □

c) To prove the theorem C3, we consider the category $\mathfrak{M}$ opposite to the full subcategory of $\underline{MF}_{E/\mathbb{Q}_p,1/2}$ whose objects are torsion objects. We can view an object of $\mathfrak{M}$ as a $\mathbb{Z}_p$-module N of finite length and an object $(M, \wedge, gr)$ of $\underline{MF}_{E/\mathbb{Q}_p,1/2}$ such that the $\mathbb{Z}_p$-module underlying M is $N^\wedge = \mathrm{Hom}_{\mathbb{Z}_p}(N, \mathbb{Q}_p/\mathbb{Z}_p)$ the Pontrjagin dual of N. Now, $\underline{J}_{1/2}$ can be viewed as a *covariant* additive exact functor from $\mathfrak{M}$ to $\underline{\mathrm{Rep}}^f_{\mathbb{Z}_p}(G_p)$. The proof then consists of playing the same game on $\mathfrak{M}$ that we played with the category $\underline{MF}^f_{]-p+1,0]}$ to prove the theorem B2 (see Appendix, §B). □

References

[1] V.A. Abrashkin. *Modular representations of the Galois group of a local field and generalization of Shafarevich conjecture*, Math. USSR Izv., 35 (1990), pp. 469-518.

[2] N. Boston. *Some Cases of the Fontaine-Mazur Conjecture*, J.Number Theory 42 (1992), pp. 285-291.

[3] P. Deligne.*Les constantes de l'équation fonctionnelle des fonctions L*, pp. 501-595 in *Modular functions of one variable II*, Lecture Notes in Math. 349, Springer-Verlag, Berlin (1973).

[4] P. Deligne. *Motifs et groupes de Taniyama*. pp. 261-279 in *Hodge cycles, motives and Shimura varieties*, Lecture Notes in Math. 900, Spring-Verlag, Berlin (1982).

[5] G. Faltings. *Endlichkeitssätze für abelsche Varietäten über Zahlkörpern*, Inv. Math. 73 (1983), pp. 349-366.

[6] G. Faltings. *Crystalline cohomology and p-adic Galois representations*, pp. 25-80 in *Algebraic Analysis, Geometry and Number Theory*, the Johns Hopkins University Press (1989).

[7] J.-M. Fontaine. *Sur la décomposition des algèbres de groupe*, Annales Scient. E.N.S., 4 (1971), pp. 121-180.

[8] J.-M. Fontaine. *Groupes p-divisibles sur les corps locaux*, Astérisque 47-48, Soc. Math. de France, Paris (1977).

[9] J.-M. Fontaine. *Modules galoisiens, modules filtrés et anneaux de Barsotti-Tate*, pp. 3-80 in *Journées de géométrie algébrique de Rennes (III)*, Astérisque 65, Soc. Math. de France, Paris (1979).

[10] J.-M. Fontaine. *Arbeitstagung report 1988*, Geometric l-adic Galois representations.

[11] J.-M. Fontaine. *Le corps des périodes p-adiques*, pp. 59-101 in *Périodes p-adiques*, Astérisque 223, Soc. Math. de France, Paris (1994).

[12] J.-M. Fontaine. *Représentations p-adiques semi-stables*, pp. 113-184, in *Périodes p-adiques*, Astérisque 223, Soc. Math. de France, Paris (1994).

[13] J.-M. Fontaine. *Représentations l-adiques potentiellement semi-stables*, pp.321-347, in *Périodes p-adiques*, Astérisque 223, Soc. Math. de France, Paris (1994).

[14] J.-M. Fontaine. *Schémas propres et lisses sur* $\mathbb{Z}$, pp. 43-56, in Proceedings of the Indo-French Conference on Geometry, Hindustan Book Agency, New Dehli (1993).

[15] J.-M. Fontaine, L. Illusie. *p-adic periods: a survey*, pp. 57-93, in Proceedings of the Indo-French Conference on Geometry, Hindustan Book Agency, New Dehli (1993).

[16] J.-M. Fontaine, G. Laffaille. *Construction de représentations p-adiques*, Ann. Scient. E.N.S., 15 (1982), pp. 547-608.

[17] J.-M. Fontaine, B. Perrin-Riou. *Autour des conjectures de Bloch et Kato: cohomologie galoisienne et valeurs de fonctions L*, pp. 599-706 in *Motives*, Proceedings of Symposia in Pure Maths. 55, part I (1994).

[18] F. Hajir. *On the growth of p-class groups in p-class field towers*, preprint, Calif. Inst. Tech. (1995).

[19] L. Illusie. *Cohomologie de de Rham et cohomologie étale p-adique*, Séminaire Bourbaki, exp. 726, juin 1990.

[20] G. Laffaille. *Construction de groupes p-divisibles: le cas de dimension 1*, pp.103-123 in Journées de géométrie algébrique de Rennes (III), Astérisque 65, Soc. Math. de France, Paris (1979).

[21] G. Laffaille. *Groupes p-divisibles et modules filtrés: le cas peu ramifié*, Bull. Soc. Math. 108 (1980), pp. 187-206.

[22] B. Mazur. *Deforming Galois representations* pp. 385-437 in *Galois group over* $\mathbb{Q}$, M.S.R.I. Publications, Spring-Verlag, Berlin 1989.

[23] R. Ramakrishna. *On a variation of Mazur's deformation functor*, Compositio Math., 87 (1993), pp. 269-286.

[24] S. Sen. *Lie algebras of Galois groups arising from Hodge-Tate modules*, Ann. of Math. (1973), pp. 160-170.

[25] S. Sen *Continuous cohomology and p-adic galois representtions*, Inv. Math. 82 (1980), pp. 89-116.

[26] J.-P. Serre. *Représentations l-adiques*, pp. 177-193 in Kyoto Int. Symposium on Algebraic Number Theory, Japan Soc. for the Promotion of Science (1977) (oeuvres, t.III, pp. 384-400).

[27] J.-P. Serre. *Abelian l-adic representations and elliptic curves* (2° éd.) Addison-Wesley, Redwood City (1989).

[28] J.-P. Serre. *Sur les représentations modulaires de degré 2 de* $Gal(\overline{\mathbb{Q}}/\mathbb{Q})$). Duke Math. J. 54 (1987) pp. 179-230.

[29] R.Taylor, A. Wiles. *Ring Theoretic Properties of Certain Hecke Algebras*, Preprint 1994.

[30] N.Wach. *Représentations p-adiques cristallines du groupe de Galois d'un corps local*, thèse, univ. Paris-Sud, Orsay (1994).

[31] A. Wiles. *Modular Elliptic Curves and Fermat's Last Theorem*, Preprint 1994.

Conference on Elliptic Curves and Modular Forms
Hong Kong, December 18-21, 1993

On elliptic curves with isomorphic torsion structures and corresponding curves of genus 2

GERHARD FREY
INSTITUTE FOR EXPERIMENTAL MATHEMATICS
UNIVERSITY OF ESSEN
ELLERNSTRASSE 29 D-45326 ESSEN, GERMANY

After stating conjectures about elliptic curves E_1 and E_2 which have Galois isomorphic submodules in their torsion groups and discussing relations with conjectures about solutions of equations of Fermat type we describe properties of curves of genus 2 which are related to such isomorphisms. These discussions include the case that E_1 is isogenous to E_2, and they are used to find a possibility to write down explicit equations for twisted modular curves of level p where p is an odd prime.

Introduction. Solutions of ternary diophantine equations are closely related to questions about arithmetical properties of elliptic curves and corresponding Galois representations induced by torsion points of such curves. Especially it is of interest whether different elliptic curves can give rise to equivalent representations.

In the case that the ground field is $\mathbb{Q}$ it is well known that this leads to very interesting questions about congruence primes of modular forms, and the celebrated theorem of Ribet [**9**] shows how closely for instance Fermat's conjecture is related to Taniyama's conjecture or Serre's conjecture (cf. [**1**]), and for general ground fields K the height conjecture for elliptic curves would answer many questions about solutions of ternary diophantine equations. In the first section we formulate some conjectures arising in this area and discuss their relations.

Among these conjectures the height conjecture for elliptic curves plays a prominent role. It is well known that it would follow from a conjecture of Parshin-Bogomolov-Miyaoka-Yau type about the self intersection of the relative canonical sheaf of arithmetic surfaces with generic fiber equal to a curve of genus ≥ 2 (cf. [**8**]). In [**2**] it is discussed how curves of genus two with elliptic differentials can possibly be used to come closer to the height conjecture.

The aim of this paper is to discuss such curves C under the additional assumption that they are already defined over K. This is equivalent to the condition that there are two elliptic curves E_1 and E_2/K and a number n such that $E_{1,n}$ (which is the G_K-module of the points of order dividing n in $E(\bar{K})$) is isomorphic to $E_{2,n}$ under an "anti-isometric" isomorphism (definition in §1) satisfying certain conditions. In section 2 we examine ramification properties of C/E and study its Galois closure. This enables us to translate conjectures of "asymptotic Fermat type" into conjectures about geometric fundamental groups of curves of genus 2 over global fields (Corollary 2.6).

In section 3 we treat the special case that the Jacobian of C is isogenous to $E \times E$; a special result is that there are curves of genus 2 over finite fields whose geometric fundamental group is infinite.

It is easily seen that covers of elliptic curves E by curves of genus 2 induce related covers of projective lines with very restricted ramification patterns. This is used in section 4 to determine systems of equations for twisted modular curves, they follow from the result of theorem 4.1 which characterizes such covers with the help of solutions of twists by polynomials of degree 3 of an elliptic curve defined over $K(U)$ where U is an indeterminate.

It should be remarked that in order to simplify the exposition we did not formulate the conjectures and results in the greatest possible generality. For instance nearly everywhere the condition "p is an odd prime" could be replaced by "p is an odd number".

The following paper is strongly influenced by common work with E. Kani, and many suggestions from and many discussions with him were of great value for me.

Parts of the results were presented in the lecture I gave during the conference on "Elliptic Curves and Modular Forms"; it is a great pleasure for me to thank the organizers and the Chinese University of Hong Kong for their generous support, for the stimulating scientific atmosphere they succeeded to create and for their warm hospitality which made my visit to Hong Kong an unforgettable experience.

1. Conjectures. Let K be a field with absolute Galois group G_K. For simplicity assume that char $(K) \neq 2, 3$. Throughout the whole paper we shall assume that p is a prime not dividing $2 \cdot \text{char}(K)$.

Let E/K be an elliptic curve. With E_p we denote the group of points of $E(\bar{K})$ whose order divides p. Let $\rho_{E,p}$ be the representation of G_K induced by its action on E_p.

Now fix an elliptic curve E_0 and denote by $\mathcal{R}_K(E_0)$ the following set:

$$\mathcal{R}_K(E_0) = \{E/K; \exists p \in \mathbb{P}\backslash\{2,3,5, \text{char } K\} \text{such that } E_p \cong E_{0,p} \text{ as } G_K - \text{module and that } E \text{ is semistable at all } \mathfrak{p}|p\}.^1$$

K is called a global field if either K is a finite number field or K is a function field of one variable over a perfect field K_0 as field of constants. Let $h(E)$ be the Faltings height of elliptic curves over K.

Now we can state

Conjecture $\underline{(R_{E_0,K})}$: *There is a number M only depending on K and on the conductor of E_0 such that for $E \in \mathcal{R}_K(E_0)$ we have:*

$$h(E) < M.$$

In particular, $\mathcal{R}_K(E_0)$ is finite if K is a number field, or K_0 is finite.

---

[1] The condition that E is semistable at divisors of p should not be necessary for the conjecture stated below.

Let S_0 be a finite set of primes (including the archimedean primes) such that E_0 has good reduction outside of S_0. Assume that E/K is an elliptic curve such that for a prime $p > 3$ we have an G_K-isomorphism

$$\alpha_p : E_{0,p} \to E_p.$$

It follows that for places $\mathfrak{q} \notin S_0 \cup \{\mathfrak{p}|p\}$ the curve E is semistable at $\mathfrak{q}$ and $\min\{0, v_{\mathfrak{q}}(j_E)\} \equiv 0 \bmod p$.

Now assume in addition that $E \in \mathcal{R}_K(E_0)$. Let $j_E \in K$ be the absolute invariant of E and $h(j_E)$ its height. Since E is semistable outside of S_0, we get

$$h(j_E) = h_{S_0}(j_E) + \tilde{h}(j_E) \geq h_{S_0}(j_E) + p\deg(\tilde{N}_E)$$

where h_{S_0} is the contribution of places in S_0, $\tilde{h}$ is the contribution of places not in S_0 and $\tilde{N}_E$ is the part of the conductor N_E of E which is prime to S_0.

So

$$h(j_E) \geq \sum_{\substack{\mathfrak{q}\in S_0 \\ \mathfrak{q}|N_E}} \deg \mathfrak{q} + p \sum_{\substack{\mathfrak{q}\notin S_0 \\ \mathfrak{q}|N_E}} \deg \mathfrak{q},$$

which implies a related formula for the Faltings height of E (cf. [**1**]).

Let us recall that the **height conjecture** $\underline{(H_E)}$ for elliptic curves (cf. [**1**]) states:

Let S_0 be a finite set of places of K. Then there exist constants c, d depending on S_0 and on K such that for all elliptic curves E/K which are semistable outside of S_0 one has

$$h(E) \leq d\deg N_E + c.$$

Assuming the truth of this conjecture we get that

$$\sum_{\substack{\mathfrak{q}\in S_0 \\ \mathfrak{q}|N_E}} \deg \mathfrak{q} + p \sum_{\substack{\mathfrak{q}\notin S_0 \\ \mathfrak{q}|N_E}} \deg \mathfrak{q} \leq d\deg N_E + c \leq \tilde{d}(\sum_{\mathfrak{q}|N_E} \deg \mathfrak{q}) + c$$

with $\tilde{d}$ depending on K and d only.

Hence for almost all p the support of N_E must lie inside of S_0, and so E has to have bounded height. Since for $p > 5$ the modular curve $X(p)$ has genus ≥ 2 there are only finitely many elliptic curves corresponding to $X(p)(K)$, and so we get

Proposition 1.1. *The height conjecture* $\underline{(H_E)}$ *implies* $\underline{(R_{E_0,K})}$ *for all* E_0/K.

Corollary 1.2. *If* K *is a function field then* $\underline{(R_{E_0,K})}$ *is true.*

Proof. The height conjecture for elliptic curves over function fields is true (cf. [**1**]). □

Next we weaken Conjecture $\underline{(R_{E_0,K})}$ to

Conjecture $\underline{(R_{E_0,K})(2)}$: *There is a number* M *depending on* K *and the conductor of* E_0 *such that for all* $E \in \mathcal{R}_K(E_0)$ *with the additional property that* $E_{0,2} \cong E_2$ *we have*

$$h(E) \leq M.$$

Let us discuss this conjecture.
Assume that $E \in \mathcal{R}_K(E_0)$ and that $E_2 \cong E_{0,2}$. Without loss of generality we can assume that $E_{0,2} \subset E_0(K)$, and so E can be given by an equation

$$E : Y^2 = X(X-A)(X-B)$$

with $A, B \in K$. Moreover there is a finite set S_0 of places of K depending on E_0 and on K only, such that the ring of S_0-integers O_{S_0} is a principal ideal domain and such that A, B are relatively prime elements in O_{S_0} (as seen above E has to be semistable outside a finite set of places depending on E_0).

So for $\mathfrak{p} \notin S_0$ we get: $v_{\mathfrak{p}}(A^2B^2(A-B)^2) = v_{\mathfrak{p}}(j_E)$ if $v_{\mathfrak{p}}(j_E) < 0$, and hence we have for $\mathfrak{p} \notin S_0$:

$$v_{\mathfrak{p}}(A^2B^2(A-B)^2) \equiv 0 \mod p$$

if $E \in \mathcal{R}_K(E_0)$ and $\alpha_p : E_{0,p} \to E_p$ is a G_K-isomorphism for $p \geq 5$.

So we can find relatively prime elements $z_1, z_2, z_3 \in O_{S_0} \backslash \{0\}$, with $A = az_1^p$, $B = bz_2^p, A - B = cz_3^p$ and with $a, b, c \in O_{S_0}^*$. Hence the equation:

$$aZ_1^p - bZ_2^p = cZ_3^p$$

has the solution (z_1, z_2, z_3) in O_{S_0}.

Next we state the

Asymptotic Fermat conjecture $\underline{F_{as}}$: *Let S_0 be a finite set of primes of K. Define*

$$L_{S_0} = \{(z_1, z_2, z_3) \in O_{S_0}, \exists a, b, c \in O_{S_0}^*$$

$$and p \geq 5 \; such \; that az_1^p + bz_2^p = cz_3^p\}.$$

Then there exists a number M depending on K and S_0 such that for all $(z_1, z_2, z_3) \in L_{S_0}$ the projective height

$$h((z_1, z_2, z_3)) \leq M.$$

Proposition 1.3. *The conjecture $\underline{(F_{as})}$ implies $\underline{(R_{E_0,K}(2))}$.*

Remark. It is obvious that $\underline{(F_{as})}$ follows from the height conjecture for elliptic curves as well (cf [**1**]).

Proof. We assume that E is given by

$$Y^2 = X(X - az_1^p)(X - bz_2^p) \text{ for } p \geq 5.$$

Conjecture $(F_{\underline{as}})$ implies that there are only finitely many places dividing $\Delta_E = (abc)^2(z_1z_2z_3)^{2p}$, and so E has good reduction outside a finite set of places S_1 depending on K and S_0 only, and the proposition follows. □

The main concern of the paper is to investigate another weakened version of $\underline{(R_{E_0,K})}$.

But before doing this we want to make some remarks concerning the special case that $K = \mathbb{Q}$.

We assume that E_0 is a modular elliptic curve, i.e. there is a $\mathbb{Q}$-rational non constant morphism

$$\varphi : X_0(N_E) \to E_0.$$

Let $\mathcal{R}_{\text{mod}}(E_0)$ be the set of elliptic curves in $\mathcal{R}_{\mathbb{Q}}(E_0)$ which are modular curves and $\mathcal{R}_{\text{mod}}(E_0)(2)$ the set of elliptic curves E in $\mathcal{R}_{\text{mod}}(E_0)$ with $E_2 \cong E_{0,2}$.

For a finite set S_0 of primes define

$$L_{S_0,\text{mod}} = \{(z_1, z_2, z_3) \in \mathbb{Z}_{S_0}; \exists a, b, c \in \mathbb{Z}^*_{S_0} \text{ and } p \geq 5 \text{ with } az_1^p - bz_2^p = cz_3^p \text{ and } E_{\underline{z}} : Y^2 = X(X - az_1^p)(X - bz_2^p) \text{ is modular}\}.$$

Then we get

Proposition 1.4. *$\mathcal{R}_{mod}(E_0)(2)$ is finite for all elliptic curves E_0 whose conductor is in $\mathbb{Z}^*_{S_0}$ if and only if $L_{S_0,mod}$ is finite.*

Proof. If $\mathcal{R}_{\text{mod}}(E_0)(2)$ is infinite for E_0 modular and $N_{E_0} \in \mathbb{Z}^*_{S_0}$ then $L_{S_0,\text{mod}}$ is shown to be infinite as in the proof of Proposition 1.3. Conversely, assume that there are infinitely many $(z_1, z_2, z_3) \in \mathbb{Z}_{S_0}$ and primes $p_i \geq 5$ with $a_1 z_1^{p_i} - bz_2^{p_i} = cz_3^{p_i}$ and $E_{\underline{z}}$ modular. Using Ribet's theorem about minimal levels of cusp forms related to $\rho_{E,p}$, an easy argument due to Mazur (cf. **[1]**) proves that there is a modular elliptic curve E_0 with $E_{0,2} \subset E_0(\mathbb{Q})$ and with conductor $N_{E_0} \in O^*_{S_0}$ such that $\rho_{E_0,p_i} = \rho_{E_{\underline{z}},p_i}$ and hence $\mathcal{R}_{\text{mod}}(E_0)(2)$ is infinite. □

$\mathcal{R}_{\text{mod}}(E_0)$ is closely related to the question how often there are elliptic quotients of the modular curves $X_0(N)$ with equivalent Galois representation induced by submodules of the group of torsion points. On the other hand it is known (cf. **[1]**) that the height conjecture for modular elliptic curves is equivalent to the following

Conjecture: *Let S_0 be a finite set of primes. There exist $c, d \in \mathbb{R}$ such that the following holds:*

Let E be a modular elliptic curve which is semistable outside of S_0 with conductor N_E. Let E^* be the unique elliptic curve in $J_0(N_E)$ which is isogenous to E, and A its complementary abelian variety in $J_0(N_E)$. Then $|(E^* \cap A)| \leq c \cdot N_E^{2d}$.

So it becomes obvious from the representation theoretical point of view as well, how closely the question of congruence primes and solutions of equations of Fermat type are related if we are working over $\mathbb{Q}$ and if we assume Taniyama's conjecture.

The conjecture $(R_{E_0,K})$ is rather coarse in the sense that it doesn't care about the symplectic structure of points of order p of elliptic curves induced by the Weil pairing e_p.

DEFINITION. *Let E_0 and E be two elliptic curves, $\alpha_p : E_{0,p} \to E_p$ a G_K-isomorphism. α_p is anti-isometric if for $P, Q \in E_{0,p}$ we have $e_p(P,Q) = e_p(\alpha_p(P), \alpha_p(Q))^{-1}$. In other words: α_p is anti-isometric if and only if $\Delta_{\alpha_p} := \{(P, \alpha_p P)\} \subset E_{0,p} \times E_p$ is (with respect to the Weil pairings) an isotropic G_K-submodule.*

Let $\overline{\mathcal{R}_K(E_0)}$ be the set of elliptic curves $E \in \mathcal{R}_K(E_0)$ for which there exists an anti-isometric G_K-isomorphism from $E_{0,p}$ to E_p for some prime $p > 5$. The main concern of the rest of the paper is to study properties of elements in $\overline{\mathcal{R}_K(E_0)}$.
We state an obvious weakening of conjecture $\underline{(R_{E_0,K})}$:

CONJECTURE. $\overline{R_{(E_0),K}}$: *There is a number M depending on K and on the conductor of E_0 such that for all $E \in \overline{\mathcal{R}_K(E_0)}$ we have: $h(E) \leq M$.*

First we want to discuss the relation between $\underline{\mathcal{R}_K(E_0)}$ and $\overline{\mathcal{R}_K(E_0)}$.
So assume that $\alpha_p : E_{0,p} \to E_p$ is a G_K-isomorphism.
Let α'_p be another isomorphism.
Then $\alpha'^{-1}_p \circ \alpha_p = \beta_p \in \mathrm{Aut}_{G_K}(E_{0,p})$. So β_p is in the centralizer of $\mathrm{Im}\,(\rho_{E_0,p})$ in $\mathrm{Aut}\,(E_{0,p})$. Let (P_1, P_2) be a base of $E_{0,p}$ and use it to map $\mathrm{Aut}\,(E_{0,p})$ into $\mathrm{Gl}\,(2, \mathbb{Z}/p\mathbb{Z})$. Since $e_p(\beta_p P_1, \beta_p P_2) = e_p(P_1, P_2)^{\det(\beta_p)}$ we get:

Proposition 1.5. *Let $\alpha : E_{0,p} \to E_p$ be a G_K-isomorphism. Assume that $e_p(\alpha P_1, \alpha P_2) = e_p(P_1, P_2)^{k_\alpha}$. Then there are k anti-isometric isomorphisms*

$$\alpha' : E_{0,p} \to E_p$$

with $k = |\{\beta \in Z_{Gl(2,\mathbb{Z}/p\mathbb{Z})}(Im(\rho_{E,p})); \det(\beta) = -k_\alpha^{-1}\}|$.

Corollary 1.6. *Assume that $Z_{Gl(2,\mathbb{Z}/p\mathbb{Z})}(Im\rho_{E,p})) = (\mathbb{Z}/p\mathbb{Z})^* \cdot id$ and that $\alpha \in Iso_{G_K}(E_{0,p}, E_p)$ with $e_p(\alpha P_1, \alpha P_2) = e_p(P_1, P_2)^{k_\alpha}$. Then there is an anti-isometric element α' in $Iso_{G_K}(E_{0,p}, E_p)$ if and only if $-k_\alpha \in (\mathbb{Z}/p\mathbb{Z})^{*2}$, and in this case α' is, up to sign, uniquely determined.*

Corollary 1.7. *Assume that $\alpha \in Iso_{G_K}(E_{0,p}, E_p)$, and that $\eta : E_0 \to E'_0$ is a K-rational isogeny of prime degree l with $\left(\frac{l}{p}\right) = -1$. Then there is an anti-isometric element either in $Iso_{G_K}(E_{0,p}, E_p)$ or in $Iso_{G_K}(E'_{0,p}, E_p)$.*

Proof. If $e_p(\alpha P_1, \alpha P_2) = e_p(P_1, P_2)^{k_\alpha}$ then $e_p(\eta P_1, \eta P_2) = e_p(P_1, P_2)^l = e_p(\alpha P_1, \alpha P_2)^{l \cdot k_\alpha}$ and either $-k_\alpha$ or $-k_{\alpha'} l$ is a square modulo p. □

EXAMPLE. Assume that E_0 contains a K-rational point P of order 2.
Put $E'_0 = E_0/\langle P\rangle$. Let $\mathcal{R}'_K(E_0)$ be the set of curves $E \in \underline{\mathcal{R}_K(E_0)}$ with $\alpha_p \in \mathrm{Iso}_{G_K}(E_{0,p}, E_p)$ and $p \not\equiv \pm 1 \bmod 8$. Then $E \in \mathcal{R}'_K(E_0)$ is either in $\overline{\mathcal{R}_K(E_0)}$ or in $\overline{\mathcal{R}_K(E'_0)}$.

Corollary 1.8. *Assume that $Im(\rho_{E_0,p})$ is contained in a Cartan subgroup A of $Gl(2, \mathbb{Z}/p\mathbb{Z})$ e.g. assume that E_0 has complex multiplication. Let $Iso_{G_K}(E_{0,p}, E_p) \neq \emptyset$. Then there are at least $p-1$ anti-isometric elements $\alpha \in Iso_{G_K}(E_{0,p}, E_p)$ if A is split, and $p+1$ such elements if A is non split.*

Proof. The centralizer of $\mathrm{Im}(\rho_{E_0,p})$ contains A, and $|A \cap \mathrm{Sl}\,(2, \mathbb{Z}/p\mathbb{Z})| = p - 1$ if A is split, and $p + 1$ if A is not split. Hence $A/C \cap \mathrm{Sl}\,(2, \mathbb{Z}/p\mathbb{Z})$ has $p - 1$ elements, and so $\det : C \to \mathbb{F}_p^*$ is surjective, and the assertion follows. □

2. Curves of genus 2 covering elliptic curves. In this section we shall relate curves of genus 2 to elements in $\mathrm{Iso}_K(E_{0,p}, E_p)$ which are anti-isometric, and we shall show that Conjecture $\overline{(R_{E_0,K})}$ is closely connected to a conjecture about unramified geometric Galois coverings of such curves.
We begin by recalling some facts which can be found in [**2**] and [**4**].

Assume that E_0 and E_1 are elliptic curves over K and that for an odd prime p different from $\mathrm{char}(K)$ there is a G_K-isomorphism $\alpha : E_{0,p} \to E_p$ with $e_p(P,Q) = e_p(\alpha P, \alpha Q)^{-1}$ for all $P, Q \in E_{0,p}$ i.e., α is anti-isometric and $\Delta_\alpha = \{(P, \alpha P)\} \subset E_{0,p} \times E_{1,p}$ is isotropic in $E_{0,p} \times E_{1,p}$.

Then

$$J_\alpha := E_0 \times E_1 / \Delta_\alpha$$

is a principally polarized abelian surface over K (cf. [**2**]).

J_α is the Jacobian of a curve C_α/K of genus 2 if and only if there is <u>no</u> isogeny

$$\eta : E_0 \to E_1$$

with $\deg \eta = k(p-k)$ for some $1 \leq k \leq p-1$ such that

$$k \circ \alpha = \eta | E_{0,p}$$

(cf. [**6**]).
We assume that this is the case.

Then there is a map

$$\pi_\alpha : C_\alpha \to E_0$$

with $\deg(\pi_\alpha) = p$, and $wC_\alpha | E_0 = -\mathrm{id}$ where w is the hyperelliptic involution of C_α.
Let $\overline{K}$ be the algebraic closure of K. Hurwitz genus formula yields:
Either there are two points P_1, P_2 on $E_0(\bar{K})$ with $P_1 = -P_2 \neq P_2$ and $Q_i \in C_\alpha(\overline{K})$ with $\pi_\alpha Q_i = P_i (i = 1,2)$, the ramification order of Q_i/P_i is equal to two and all points Q of C different from Q_1, Q_2 are unramified in C_α/E_0. (**Type I**) (which is the generic case)
or, as degeneration there is a Weierstrass point Q of C_α which ramifies of order 3 over E_0 and $C_\alpha \backslash \{Q\}$ is unramified, or there is a Weierstrass point of which E_0 has two ramified extensions which ramify of order 2. (**Type II**).

NOTATION. *For curves X defined over K let $F(X)$ be its function field, and $\bar{F}(X)$ the composite of $F(X)$ with $\bar{K}$.*

Let $\tilde{F_\alpha}$ be the Galois closure of $F(C_\alpha)/F(E_0)$.

Proposition 2.1. *Assume that C_α is of type I. Then $\tilde{F}_\alpha/F(E_0)$ is a geometric extension with Galois group S_p. Hence $\tilde{F}_\alpha$ is the function field of a curve $\tilde{C}_\alpha/K$ which is a Galois cover of E_0 with Galois group S_p.*

Proof. To prove the proposition it is enough to show that the Galois group of $\bar{\tilde{F}}_\alpha | \bar{F}(E_0)$ is equal to S_p. Let $\mathfrak{p}_1$ be one of the places of $\bar{F}(E_0)$ which ramify in $\bar{\tilde{F}}_\alpha$.

Then conorm $_{\bar{F}(E_0)/\bar{F}(C_\alpha)}(\mathfrak{p}_1) = \mathfrak{q}_1^2 \cdot \mathfrak{q}_1 \dots \mathfrak{q}_{p-1}$ where $\mathfrak{q}_1, \dots \mathfrak{q}_{p-1}$ are different places. It follows that the inertia group of an extension of $\mathfrak{q}_1$ to $\bar{\tilde{F}}_\alpha$ is generated by a transposition. But this implies already that $G(\bar{\tilde{F}}_\alpha/\bar{F}(E_0))$ is equal to S_p (cf. [**13**]). □

Now assume that C_α is of type II and that there is a place $\mathfrak{p} \in F(E_0)$ such that
conorm $_{\bar{F}(E_0)/\bar{F}(C_\alpha)}$ $(\mathfrak{p}) = \mathfrak{q}^3 \cdot \mathfrak{q}_1 \dots \mathfrak{q}_{p-3}$ where $\mathfrak{q}_1, \dots, \mathfrak{q}_{p-3}$ are different places. So we have a 3-cycle inside of $G(\bar{\tilde{F}}_\alpha/\bar{F}(E_0))$ and hence this group contains A_p. Since the hyperelliptic involution w of C_α maps E_0 into itself the function field of C_α/w (which is a rational field $K(U)$) is isomorphic to $K(X)[Y]/(f(Y))$ where $K(X)$ is the function field of E_0/w.

We can describe the ramification of $K(U)/K(X)$ (see §4) and get:

The discriminant of $K(U)/K(X)$ is either equal to

i) $\frac{p-1}{2}(\mathfrak{p}_1' + \mathfrak{p}_2') + (\frac{p-3}{2})(\mathfrak{p}_3' + \mathfrak{p}_4') + 2\mathfrak{p}_4$

or equal to

ii) $\frac{p-1}{2}(\mathfrak{p}_1' + \mathfrak{p}_2' + \mathfrak{p}_3' + \mathfrak{p}_4')$

where $\mathfrak{p}_1', \dots, \mathfrak{p}_4'$ correspond to the places of $K(X)$ which ramify in $F(E_0)/K(X)$.

If i) occurs then Δ_f is not a square in $\bar{K}(X)$ and $\bar{K}(X)(\sqrt{\Delta_f})$ is a function field of genus 0. Hence Δ_f is not a square in the function field of E_0 and so $G(\bar{\tilde{F}}_\alpha/\bar{F}(E_0))$ is equal to S_p.

If case ii) occurs then $\bar{\tilde{F}}_\alpha/\bar{F}(E_0)$ has Galois group A_p, and the algebraic closure of K in $\tilde{F}_\alpha$ is an extension of K of degree ≤ 2.

The last possibility is that a point of order 2 of E_0 has two extensions to C_α which ramify of order 2. (case II iii)). Hence $G(\bar{\tilde{F}}_\alpha/\bar{F}(E_0))$ contains a product of two transpositions and so it contains A_p if $p > M$, [2] and the algebraic closure of K in $\tilde{F}_\alpha$ is an extension of degree ≤ 2.

Proposition 2.2. *If $\pi_\alpha : C_\alpha \to E_0$ is of type II of degree $\geq M$ then the Galois group of $\tilde{F}_\alpha/F(E_0)$ contains A_p. $\tilde{F}_\alpha$ is geometric over K with Galois group equal to S_p if case i) occurs.*

If case ii) occurs $\tilde{F}_\alpha$ is geometric over K if and only if $G(\tilde{F}_\alpha/F(E_0))$ is equal to A_p.

Define $F_\alpha' = \tilde{F}_\alpha^{A_p}$, and let C_α' be the curve defined over $\bar{K}$ with function field $\bar{F}_\alpha'$.

Corollary 2.3. *$\tilde{F}_\alpha$ is a geometric extension of F_α'. If C_α is of type I then C_α' is a curve of genus 2 defined over K and $\tilde{C}_\alpha/C_\alpha'$ is unramified. If C_α is of type II then C_α' is a curve of genus 1 which is $\bar{K}$-isogenous to E_0.*

In the type II-case C_α/E_0 is only ramified in the point P; it follows that P is K-rational and so we can assume that P is the ramification point of $E_0/(E_0/w)$ which corresponds to the place $\mathfrak{p}_\infty$ of the function field $K(X)$ of E_0/w.

To simplify the exposition we shall assume from now on that $E_0(K)_2 = \mathbb{Z}/2 \times \mathbb{Z}/2$ if type II occurs. It will be shown in section 4 that there is up to

[2]The author has to thank W. Lempken for pointing out this result.

isomorphism exactly one curve C_3 of genus 2 covering E_0 of degree 3 which is ramified of order 3 at P (cf. proposition 4.2). If case II iii) occurs we can find a curve C_4 of genus 2 covering E_0 of degree 4 which ramifies at P of order 2, and so the composite of its function field with $\tilde{F}_\alpha$ is unramified over $F(C_4)$. It follows that for $p > M$ we have:

Corollary 2.4. *$\tilde{F}_\alpha \cdot F(C_3)/F(C_3)$ resp. $\tilde{F}_\alpha F(C_4)/F(C_4)$ is an unramified Galois extension whose Galois group contains A_p and the algebraic closure of K in $\tilde{F}_\alpha \cdot F(C_3)$ is of degree ≤ 2 over K. In other words: For every elliptic curve E_0 with $E_{0,2} \subset E_0(K)$ there exists an extension K_1 of K with $[K_1 : K] \leq 2$ and a curve C' of genus 2 covering E_0 of degree 3 or 4 such that every covering of degree $p > M$ of type II of E_0 by a curve of genus 2 gives rise to an unramified geometric Galois cover of C' defined over K_1 whose Galois group contains A_p.*

Now assume conversely that $\tilde{C}/E_0$ is a geometric Galois cover with group S_p defined over K. Assume moreover that there are exactly 2 points P_1 and P_2 of $E_0(\bar{K})$ which ramify in $\tilde{C}$ with ramification order 2, or equivalently, assume that $g(\tilde{C}) = \frac{p!}{2} + 1$, and that the inertia group of P_i is generated by a transposition.

Let C' be equal to $\tilde{C}/A_p$.

It follows that C'/E_0 is ramified in P_1 and P_2 and that $\tilde{C}/C'$ is unramified.

Let Q_1 be an extension of P_1 to $\tilde{C}$ with inertia field K_{Q_1} which has index 2 in the function field of $\tilde{C}$ and whose Galois group is generated by a transposition τ. Hence there are $p-2$ copies of S_{p-1} inside of S_p which contain τ, and so Q_1 is ramified in exactly two extensions of degree p over the function field of E_0. It follows that in each of the fields of degree p over E_0 exactly $(p-1)!$ extensions of P_1 to $\tilde{C}$ ramify, or, in other words, there is exactly one extension of P_1 to $C = \tilde{C}/S_{p-1}$ (take some embedding of S_{p-1} into S_p) which is ramified. Since for P_2 the same result holds we get that C is a curve of genus 2 covering E_0 over K of degree p which is of type I.

By using analogous arguments one can characterize coverings of type II: Assume that $\tilde{C}/E_0$ is a Galois cover whose Galois group is contained in S_p and contains A_p, and that exactly one point P of E_0 is ramified in $\tilde{C}$, assume furthermore that the inertia group of an extension of P to $\tilde{C}$ is generated by a 3-cycle resp. by a product of two transpositions. Let C be a subcover of E_0 of degree p. Then C is a curve of genus 2 defined over K covering E_0 of degree p which is of type II.

Having $\pi : C \to E_0$ it follows that there is an elliptic curve E_1/K such that the Jacobian J_C of C is equal to $(E_0 \times E)/\Delta_\alpha$ where α is an anti-isometric element α in $\mathrm{Iso}_{G_K}(E_{0,p}, E_p)$.

Summarizing our results we get

Theorem 2.5. *Assume that is a prime $p > u$ not dividing $2 \cdot char(K)$. Let $\alpha : E_{0,p} \to E_p$ be an anti-isometric G_K-isomorphism between the points of order p of the elliptic curves E_0 and E. Then either there is a geometric Galois cover $\tilde{C}/K$ of E_0 with Galois group S_p with ramifies of order 2 in two points of E_0 and which is unramified over the curve $\tilde{C}/A_p$ (which has genus 2), or there is a Galois cover $\tilde{C}/E_0$ in which exactly one point P of E_0 is ramified of order 2 or 3, the Galois group of $\tilde{C}/E_0$ contains A_p and is contained in S_p, $\tilde{C}/E_0$*

is geometric if and only if $(\tilde{C}/A_p)/E_0$ *is geometric, and there is a curve* C' *of genus* 2 *covering* E_0 *over* $K(E_{0,2})$ *such that* $\tilde{C} \times_{E_0} C'/C'$ *is unramified.*

Conversely to every Galois cover $\pi : \tilde{C} \to E_0$ *whose Galois group is contained in* S_p *and contains* A_p*, which is ramified either in* 2 *points of* E_0 *of order* 2 *or in one point of* E_0 *of order* 2 *or* 3 *such that the ramification groups are generated by* 2*- or resp.* 3*-cycles resp. a product of two two-cycles there corresponds an anti-canonical* G_K*-isomorphism* $\alpha : E_{0,p} \to E_p$*, where* E/K *is an elliptic curve such that a subextension* C *of* E_0 *in* $\tilde{C}$ *of degree* p *is a curve* C *of genus* 2 *with* $J_C = E_0 \times E_1/\Delta_\alpha$.

Type II forces us to formulate Theorem 2.5 in a rather complicated way. By using arguments from **[2]** one can show that if type II occurs and if K is a number field then the height of the corresponding curve C is bounded. Using this we can reformulate conjecture $(\overline{R_{E_0,K}})$ in terms which involve arithmetical properties of the "K-rational fundamental group" of curves of genus 2 with elliptic differentials.

Corollary 2.6. *Let* K *be a number field. The conjecture* $(\overline{R_{E_0,K}})$ *is true if the following holds:*
There is a constant M *depending on* K *and the conductor of* N_{E_0} *such that for all curves* C' *of genus* 2 *with the properties:*
i) C' *covers* E_0 *of degree* 2,
ii) there is a prime $p > 5$ *such that* C' *has a geometric* K*-rational unramified Galois cover* $\tilde{C}$ *with Galois group* A_p *which is normal over* E_0 *with Galois group* S_p *and for all embeddings of* S_{p-1} *to* $G(\tilde{C}/E_0)$ *we have:*
The modular height of $\tilde{C}/S_{p-1}$ *is bounded by* M.

3. Covers induced by isogenies between elliptic curves. The height conjectures stated in the previous sections don't take care of the case that E_0 is isogenous to E_1. In this section we want to discuss this situation more closely.

So let

$$\eta : E_0 \to E$$

be a K-isogeny of degree l_0.

Let p be a prime not dividing $2 \cdot l_0 \cdot \mathrm{char}(K)$. The restriction of η to $E_{0,p}$ induces a K-rational isomorphism

$$\alpha_{0,p} : E_{0,p} \to E_p,$$

and so $E_1 \in \mathcal{R}_K(E_0)$.

Can we use $\alpha_{0,p}$ to construct an anti-isometric isomorphism α such that $(E_0 \times E_1/\Delta_\alpha)$ is the Jacobian of a curve C_α covering E_0 of degree p ?

Proposition 3.1. *Let* $\rho_{E_{0,p}}$ *be the representation of* G_K *induced by the action of* G_K *on* $E_{0,p}$. *We assume that* E_0 *has no complex multiplication and that the center of* $Im(\rho_{E_{0,p}}) \in Gl(E_{0,p})$ *is equal to* $(\mathbb{Z}/p\mathbb{Z})^* \cdot id$.

Then there is a curve C/K *of genus* 2 *covering* E_0 *of degree* p *with* J_C *isogenous to* $E_0 \times E_1$ *if and only if* p *is split in* $\mathbb{Q}(\sqrt{-l_0})$ *but for all divisors* l'

of l_0 *we get:* $l' \cdot p$ *is not norm of an element in* $\mathbb{Z} + \mathbb{Z} \cdot \sqrt{-l_0'}$ *where* l_0' *is the square free part of* l_0.

Proof. In order to find an anti-isometric isomorphism α from $E_{0,p}$ to E_p we have to find an $l \in \mathbb{Z}$ with

$$\deg(l \cdot \eta) \equiv -1 \bmod p$$

or:

$$l^2 \cdot l_0 \equiv -1 \bmod p.$$

So $-l_0$ has to be a square mod p, and hence p is split in $\mathbb{Q}(\sqrt{-l_0})$.

Now assume that p is split in $\mathbb{Q}(\sqrt{-l_0})$, and that $l^2 l_0 \equiv -1 \bmod p$, and define $\alpha := l \cdot \eta_{|E_{0,p}}$.

We know that $E_0 \times E_1/\Delta_\alpha$ is principally polarized and that this polarization is an irreducible curve C_α of genus 2 if and only if there is no isogeny $\eta' : E_0 \to E_1$ with $\deg \eta' = k(p-k)$ for $1 \le k \le p-1$ and $k \cdot \alpha = \eta'_{|E_{0,p}}$.

Let η' be any isogeny from E_0 to E_1. Let $\hat{\eta}$ be the dual isogeny of η from E_1 to E_0.

Then $\eta' \circ \hat{\eta}$ is an endomorphism of E_1, and since E_1 has no complex multiplication there is $n \in \mathbb{Z}$ with $\eta' \circ \hat{\eta} = n \cdot \mathrm{id}_{E_1}$.

It follows that

$$\begin{aligned} l_0 \cdot \eta' &= n \cdot \eta, \qquad \text{and so} \\ \deg \eta' &= \tfrac{n^2}{l_0}. \end{aligned}$$

So if

$$\deg \eta' = k(p-k) \quad \text{with} \quad 1 \le k \le p-1$$

we have:

$$\frac{n^2}{l_0} = k(p-k).$$

So

$$k = \frac{n_1^2}{l_1} \text{ and } (p-k) = \frac{n_2^2}{l_2}$$

with $l_1 \cdot l_2 = l_0$.

Hence

$$p = \frac{n_1^2}{l_1} + \frac{n_2^2}{l_2} = \frac{l_2 n_1^2 + l_1 n_2^2}{l_0}.$$

Since $\frac{n_1^2}{l_1}$ and $\frac{n_2^2}{l_2}$ are integers we have:

$$\frac{n_i^2}{l_i} = l_i' \cdot (n_i')^2$$

where l_i' is the square free part of l_i, and so

$$p = l_1' n_1'^2 + l_2' n_2'^2$$

with $l_1' \cdot l_2' = l_0'$ the square free part of l_0.

By multiplying with l_1' we see that $l_1' \cdot p$ is norm of an element of $\mathbb{Z} + \mathbb{Z}\sqrt{-l_0'}$, and so one direction of the proposition is proved.

Conversely assume that for a square free divisor l' of l_0 we have:

$$l'p = n_1^2 + l_0' n_2^2.$$

Since l' divides n_1 we get: $k := \frac{n_1^2}{l'} \in \mathbb{N}$, and $p = k + \frac{(n_2')^2}{l_2}$ with $l_2 = \frac{l_0}{l'}$.
Then

$$\eta' := \frac{(n_1 n_2')}{l_0} \cdot \eta$$

is an isogeny of E_0 to E_1 of degree $k(p-k)$ with $k = \frac{n_1^2}{l'}$.
It induces an isomorphism $\alpha' \in \mathrm{Iso}_{G_K}(E_{0,p}, E_p)$ with $e_p(\alpha' P, \alpha' Q) = e_p(P,Q)^{-k^2}$. Since $k \cdot \alpha$ has the same property and the center of $\mathrm{Im}\,(\rho_{E_{0,p}}) \in \mathrm{Gl}\,(E_{0,p})$ is equal to $(\mathbb{Z}/p\mathbb{Z})^* \cdot \mathrm{id}$ we must have: $\alpha' = \pm\alpha$, and by replacing η' by $-\eta'$ if necessary we get equality and hence $E_0 \times E_1/\Delta_\alpha$ is not a Jacobian of a curve of genus 2. □

EXAMPLE. Let E_0 be an elliptic curve over a number field K without complex multiplication but with a K-rational isogeny of degree 5.

We can apply the result of prop. 3.1 to construct covers of degree p of E_0 by curves of genus 2 whose Jacobian is equal to $(E_0 \times \eta E_0)/\Delta_\alpha$ for all primes p which are split in $\mathbb{Q}(\sqrt{-5})$ into factors which are not principal ideals.

Next let us assume that E_0 has complex multiplication over K and that O_{-d} the ring of integers in $\mathbb{Q}(\sqrt{-d})$, is equal to its ring of endomorphisms.

Let p be a prime not dividing d.

$\mathrm{Im}\,(\rho_{E_{0,p}})$ is contained in a Cartan subgroup A of $\mathrm{Gl}\,(2, \mathbb{Z}/p\mathbb{Z})$, and so its center contains A.

Using Corollary 1.8 we get:

To each isogeny $\eta : E_0 \to E$ defined over K of degree l_0 prime to p there exist $p+1$ (if A is non split) resp. $p-1$ (if A is split) anti-isometric isomorphism α_p of the form

$$\eta_{|E_{0,p}} \circ \beta$$

with $\beta \in A$.

We recall: α_p does not belong to a cover $\pi_p : C_{\alpha p} \to E_0$ with $\deg \pi_p = p$ if and only if there is an isogeny $\eta' : E_0 \to E_1$ of degree $k(p-k)$ $(1 \le k \le p-1)$ with $\eta'_{|E_{0,p}} = k \circ \alpha_p$.

By applying the dual map $\hat{\eta}$ of η we get an endomorphism $\tilde{\eta} = \hat{\eta} \circ \eta'$ of E_0 with degree $(\hat{\eta}) = k(p-k)l_0'$ with $l_0' | l_0$, and in order to be sure that there is an anti-isometric isomorphism $\alpha : E_{0,p} \to E_{1,p}$ inducing a cover $\pi_p : C_\alpha \to E_0$ of degree p and with $g(C_\alpha) = 2$ we have to bound the number of endomorphisms of E_0 with degree $l_0' k(p-k)$ by a number smaller than $p-1$.

Let us assume that $E_0 = E$ and hence $l_0' = 1$. I. Kiming has proved the following result:

Proposition 3.2. *Let D_0 be a square free natural number and let D be the discriminant of $\mathbb{Q}(\sqrt{-D_0})$. Let $h(D_0)$ be the class number of $\mathbb{Q}(\sqrt{-D_0})$ and w be the number of roots of unity in $\mathbb{Q}(\sqrt{-D_0})$. Define* $\alpha(D_0) = \prod_{\substack{l \in \mathbb{P}\setminus\{2\} \\ l | D_0}} (l + (\frac{-1}{l}))$.

For a prime p let $\beta(p)$ be the number of elements α in the ring of integers of $\mathbb{Q}(\sqrt{-D_0})$ with $N(\alpha) = k(p-k)$ for some $1 \le k \le p-1$.

Define

$$\epsilon(D_0) = \begin{cases} 24 & \text{if } D_0 \equiv 1 \bmod 4 \\ 8 & \text{if } D_0 \equiv 3 \bmod 4 \\ 4 & \text{if } D_0 \equiv 2 \bmod 4. \end{cases}$$

Then

$$\beta(p) = \frac{\epsilon(D_0)h(D_0)}{w \cdot \alpha(D_0)}(p+1) + O(p^{1/2}).$$

For the proof of this proposition we refer to [7]. In the same paper one finds: $h(D_0)/\alpha(D_0) = O(\log D_0 \log\log D_0/\sqrt{D_0})$. Using this we get the

Corollary 3.3. *Assume that D_0 is a square free number such that $\frac{\epsilon(D_0)h(D_0)}{w\cdot\alpha(D_0)} < 1$, for instance assume that D_0 is sufficiently large. Assume moreover that E/K is an elliptic curve with $End_{\bar{K}}(E) = O_{-D_0}$, the ring of integers of $\mathbb{Q}(\sqrt{-D_0})$. Then for p large enough there are anti-isometric G_K-isomorphisms*

$$\alpha_p : E_p \to E_p$$

such that $E \times E/\Delta_{\alpha_p}$ is the Jacobian of a curve C which covers E of degree p.

Proof. By proposition 3.2 it follows that for large p the union of the set of all endomorphisms of E of degree $k(p-k)$ for some $1 \le k \le p-1$ has less than $p-1$ elements. □

This result can be used to prove the existence of curves of genus 2 defined over finite fields k whose geometric fundamental group over k is infinite:

Corollary 3.4. *Let k_0 be a finite field and let E_0/k_0 be an elliptic curve with $E_{0,2} \subset E_0(k_0)$ and $End_{\bar{k}_0}(E_0) = O_{-D_0}$ where D_0 is a square free natural number with $\dfrac{\epsilon(D_0)h(D_0)}{w\alpha(D_0)} < 1$. Let k_1 be the quadratic extension of k_0.*
Then there is (at least) one curve C of genus 2 defined over k_0 such that for infinitely many primes p_i there exists an unramified Galois cover $\tilde{C}_i$ of C defined over k_0 whose Galois group contains A_{p_i}. Hence the composite F of the function fields $F_i = F(C_i)$ is an unramified Galois extension of $F(C)$ with infinite Galois group and the algebraic closure of k_0 in F is contained in k_1.

The proof of this corollary follows immediately from Corollary 3.3 and theorem 2.5 and the fact that there are only finitely many curves of genus 2 defined over k_0.

4. Parametrization of anti-isometric level structures. In Section 2 we discussed the relation between curves of genus 2 covering elliptic curves and anti-isometric isomorphisms between groups of torsion points of elliptic curves. In this section we'll use the ramification pattern of these coverings in order to "parametrize" them.

To simplify the exposition we shall assume that the field K is quadratically closed, i.e. $K^{*2} = K^*$. (Recall that we always assume that char $(K) \neq 2, 3$.)

Next we fix an elliptic curve E_0/K and again we simplify by assuming that $E_{0,2} \subset E(K)$.

Let

$$\pi : C \to E_0$$

be a cover of degree p with $g(C) = 2$ and let w be the hyperelliptic involution which acts as -id on E_0.

Then $K(U) := F(C)^w$ is a rational function field which contains $K(X) = F(E_0)^w$, and $\deg[K(U) : K(X)] = p$. We'll describe the ramification of $K(U)/K(X)$.

The Weierstrass points of C are the six places of $\bar{K}(U)$ which ramify in $\bar{F}(C)/\bar{K}(U)$, these places are extensions of the four places of $\bar{K}(X)$ which ramify in $\bar{F}(E_0)/\bar{K}(X)$ and which correspond to points in $E_{0,2}$. Since the projection under π of the Weierstrass points of C maps three of these points to one point in $E_{0,2}$ and the remaining three Weierstrass points to different points in $E_{0,2}$ (cf. [**2**]) we can choose X such that the places $\mathfrak{p}_0(X = 0), \mathfrak{p}_1(X = 1), \mathfrak{p}_\infty(\frac{1}{X} = 0)$ and $\mathfrak{p}_a(X = a)$ for some $a \in K^*\backslash\{1\}$ ramify in $F(E_0)/K(X)$ and such that $\mathfrak{p}_0, \mathfrak{p}_1, \mathfrak{p}_\infty$ have exactly one extension whose ramification order is prime to 2, and that $\mathfrak{p}_a$ has three such extensions.

If the cover $C \to E_0$ is of type I then only ramification of order two occurs, and so we get: $\mathfrak{p}_0, \mathfrak{p}_1, \mathfrak{p}_\infty$ have exactly one unramified extension and $\frac{p-1}{2}$ ramified extensions with ramification order 2, and $\mathfrak{p}_a$ has three unramified extensions and $\frac{p-3}{2}$ ramified extensions with ramification order 2. Moreover there is exactly one place $\mathfrak{p}$ of $\bar{K}(X)$ different from $\mathfrak{p}_0, \mathfrak{p}_1, \mathfrak{p}_\infty, \mathfrak{p}_a$ which has one extension ramified of order 2, and $\bar{K}(U)/\bar{K}(X)$ is unramified outside of $\{\mathfrak{p}_0, \mathfrak{p}_1, \mathfrak{p}_\infty, \mathfrak{p}_a, \mathfrak{p}\}$.

If $C \to E_0$ is of type II then one of the extensions of $\mathfrak{p}_0, \mathfrak{p}_1, \mathfrak{p}_\infty, \mathfrak{p}_a$ is ramified of order 3 or 4, and $\bar{K}(U)/\bar{K}(X)$ is unramified outside of $\{\mathfrak{p}_0, \mathfrak{p}_1, \mathfrak{p}_\infty, \mathfrak{p}_a\}$.

We can assume that $\mathfrak{p}_0$ and $\mathfrak{p}_1$ have only extensions with ramification order dividing 2, and so there are $\frac{p-1}{2}$ ramified extensions of $\mathfrak{p}_0$ and $\mathfrak{p}_1$ in $\bar{K}(U)/\bar{K}(X)$.

If $\mathfrak{p}_\infty$ has a ramified extension with ramification order 3 then $\mathfrak{p}_\infty$ and $\mathfrak{p}_a$ have $\frac{p-3}{2}$ extensions with ramification order 2, we get case i) of the discussion in Section 2.

In case II ii) $\mathfrak{p}_\infty$ has $\frac{p-1}{2}$ extensions and $\mathfrak{p}_a$ has $\frac{p-5}{2}$ extensions which ramify of order 2, $\mathfrak{p}_\infty$ has one extension which is unramified, and $\mathfrak{p}_a$ has two unramified extensions.

In case II iii) one of the places $\mathfrak{p}_\infty$ and $\mathfrak{p}_a$ has an extension which is ramified of order 4.

Now choose U such that $\mathfrak{q}_0(U = 0), \mathfrak{q}_1(U = 1)$ and $\mathfrak{q}_\infty(\frac{1}{U} = 0)$ are the extensions of $\mathfrak{p}_0, \mathfrak{p}_1$ and $\mathfrak{p}_\infty$ whose ramification orders are odd. We note that U is completely determined by this choice, and that $\mathfrak{q}_0$, $\mathfrak{q}_1$ and $\mathfrak{q}_\infty$ are of degree 1 over K.

The curve C is given by the equation

$$Z^2 = U(U-1)f_3(U)$$

where f_3 is a polynomial of degree 3 in $K(U)$ whose roots correspond to the places of $K(U)$ extending $\mathfrak{p}_a$ with odd ramification order.

Conversely we can begin with $K(X) \subset K(U)$ with $[K(U):K(X)] = p$ and the ramification pattern of $K(U)/K(X)$ as described above. Then $C : Z^2 = U(U-1)f_3(U)$ is a curve of genus 2 covering

$$E : Y^2 = X(X-1)(X-a)$$

of degree p.

Hence to determine all elliptic curves E/K such that there exists K-rational anti-isometric isomorphisms

$$\alpha_p : E_{0,p} \to E_p$$

is equivalent with the determination of all subfields $K(X)$ of $K(U)$ with the ramification pattern as above.

So assume that

$$X = \frac{P(U)}{Q(U)}$$

with relatively prime $P, Q \in K(U)$ and $\deg(X) = \max(\deg(P), \deg Q) = p$.

The ramification conditions at $\mathfrak{p}_0$ resp. $\mathfrak{p}_\infty$ imply:

$$X = \frac{P_1(U)^2 U}{P_2(U)^2}$$

with

$$\deg P_1 = \frac{p-1}{2},$$

$$\deg P_2 = \begin{cases} \frac{p-3}{2} & \text{if type II case i occurs} \\ \frac{p-1}{2} & \text{else} \end{cases}$$

(Note that $K^{*2} = K^*$ and so constants are absorbed by P_1.)

Now look at the place $\mathfrak{p}_1(X=1)$.

We get:

$$P_1(U)^2 U - P_2(U)^2 = S_1(U)^2(U-1)$$

with $S_1 \in K[U]$, $\deg S_1 = \frac{p-1}{2}$ and $S_1(1) \neq 0$.

So $Q_1 = \frac{P_1}{S_1}$ and $Q_2 = \frac{P_2}{S_1}$ are elements in $K(U)$ which solve

$$Q_2^2 - Q_1^2 U = (1-U)$$

a norm equation with respect to the extension $K(\sqrt{U})$ of $K(U)$ with $Q_i(1) \neq 0$.

We know one solution of this equation, namely

$$Q_1^0 = Q_2^0 = 1, \qquad \text{and hence} \qquad \frac{Q_2 + Q_1\sqrt{U}}{1+\sqrt{U}}$$

has norm 1. By Hilbert's theorem so it follows:

$$\frac{Q_2 + Q_1\sqrt{U}}{1+\sqrt{U}} = \frac{f_1(U) - f_2(U)\sqrt{U}}{f_1(U) + f_2(U)\sqrt{U}} \qquad \text{with} \qquad f_i \in K(U).$$

Hence

$$Q_2 + Q_1(\sqrt{U}) = \frac{f_1^2 + Uf_2^2 - 2f_1f_2U + \sqrt{U}(f_1^2 + Uf_2^2 - 2f_1f_2)}{f_1^2 - Uf_2^2}.$$

So

$$Q_2 = \frac{f_1^2 + Uf_2^2 - 2f_1f_2U}{f_1^2 - Uf_2^2}$$

and

$$Q_1 = \frac{f_1^2 + Uf_2^2 - 2f_1f_2}{f_1^2 - Uf_2^2}.$$

Because of homogeneity we can assume that $f_i \in K[U]$ and f_1 and f_2 are relatively prime with $f_1(1) \neq f_2(1)$. Hence $f_1^2 + Uf_2^2 - 2f_1f_2$ is prime to $f_1^2 - Uf_2^2$ and because of

$$\frac{P_1}{S_1} = \frac{f_1^2 + Uf_2^2 - 2f_1f_2}{f_1^2 - Uf_2^2}, \quad \frac{P_2}{S_1} = \frac{f_1^2 + Uf_2^2 - 2f_1f_2U}{f_1^2 - Uf_2^2}$$

we get

$$P_1 = f_1^2 + Uf_2^2 - 2f_1f_2, \quad P_2 = f_1^2 + Uf_2^2 - 2f_1f_2U, \quad S_1 = f_1^2 - Uf_2^2.$$

It remains to exploit the ramification pattern at $\mathfrak{p}_a$:
We must have:

$$P_1^2U - aP_2^2 = S_2^2 f_3(U)$$

where $f_3 \in K[U]$ is a polynomial with three distinct zeros $\lambda_1, \lambda_2, \lambda_3$ in $\bar{K}$ different from 0 and 1. Hence

$$\begin{aligned} U(f_1^4 + U^2f_2^4 + 4f_1^2f_2^2 + 2Uf_1^2f_2^2 - 4Uf_1f_2^3 - 4f_1^3f_2) \\ -a(f_1^4 + U^2f_2^4 + 4f_1^2f_2^2U^2 + 2Uf_1^2f_2^2 \\ -4U^2f_1f_2^3 - 4Uf_1^3f_2) = S_2^2f_3(U). \end{aligned}$$

Let us compute the degree of these polynomials:
Since $\deg(P_1) = \frac{p-1}{2}$ we get: If $p \equiv 1 \bmod 4$:

$\deg f_1 = \frac{p-1}{4}$ and $\deg f_2 < \deg f_1$ for type I-covers or for type II-covers, case ii
$\deg f_1 = \frac{p-1}{4}$ and $\deg f_2 = \frac{p-5}{4}$ else

If $p \equiv 3 \bmod 4$:

$\deg f_2 = \frac{p-3}{4}$ and $\deg f_1 \leq \deg f_2$ for type I-covers or type II-covers, case ii
$\deg f_2 = \frac{p-3}{4} = \deg f_1$ else.

Now put $w = \frac{f_1}{f_2}$:
Then

$$(U - a)w^4 + 4U(a-1)w^3 + (4U + 2U^2 - 4aU^2 - 2aU)w^2$$

$$+(4aU^2 - 4U^2)w + (U^3 - aU^2) = S_2^2 f_3(U).$$

Hence $(w, \frac{S_2}{f_2^2})$ is a solution of $E_{a,f_3(U)}$, the twist by $f_3(U)$ of the curve E_a of genus 1 defined over $K(U)$ and given by the equation

$$E_a : (U-a)W^2 + 4U(a-1)W^3 + 2U(U(1-2a)+2-a)W^2$$
$$+4U^2(a-1)W + U^2(U-a) = Z^2$$

This solution has the special property that the degree of its W-coordinate is equal to $\frac{p-1}{4}$ if $p \equiv 1 \bmod 4$, and equal to $\frac{p-3}{4}$, if $p \equiv 3 \bmod 4$.

Conversely if (w, z) is a solution of $E_{a,f_3(U)}$ where $f_3 \in K[U]$ is a separable polynomial of degree 3 prime to $U(U-1)$ and $w = \frac{f_1}{f_2}$; $f_i \in K[U]$ which satisfy the degree conditions from above with $f_i(0) \neq 0, f_1(0) \neq 2f_2(0)$ and $f_1(1) \neq f_2(1)$ then define,

$$P_1 := f_1^2 + f_2^2 U - 2f_1 f_2, \quad P_2 := f_1^2 + f_2^2 U - 2f_1 f_2 U$$

and

$$X := \frac{P_1^2 U}{P_2^2}.$$

Then $K(X) \subset K(U)$ has index p with ramification pattern as required.

So we get

Theorem 4.1. *For $a \in K^* \backslash \{0,1\}$ let E_a be the curve of genus 1 defined over $K(U)$ by the equation*

$$E_a : Z^2 = (U-a)W^4 + 4U(a-1)W^3 + 2U(U(1-2a)+2-a)W^2$$
$$+4U^2(a-1)W + U^2(U-a).$$

Let $f_3(U)$ be a separable polynomial in $K(U)$ with zeros different from 0 and 1.

Then the curve

$$C : V^2 = U(U-1)f_3(U)$$

is a curve of genus 2 covering the elliptic curve

$$Y^2 = X(X-1)(X-a)$$

of degree p if and only if the quadratic twist of E_a by f_3 has a $K(U)$-rational point $\left(\frac{f_1}{f_2}, z\right)$ with $f_i \in K[U]$, relatively prime, and

$$\begin{array}{ll} \deg f_1 \equiv \frac{p-1}{4}, \deg f_2 \leq \frac{p-5}{4} & \text{if } p \equiv 1 \bmod 4 \\ \deg f_2 = \frac{p-3}{4}, \deg f_1 \leq \deg f_2 & \text{if } p \equiv 3 \bmod 4 \end{array}$$

and $f_1(1) \neq f_2(1), 2f_2(0) \neq f_1(0)$.

Remark. 1) If E is not isogenous to $Y^2 = f_3(U)$ then $\left(\frac{f_1}{f_2}, z\right)$ is a solution of minimal degree of $E_{a,f_3(U)}$.

2) If K is a function field we know that Conjecture $\underline{(R_{E,K})}$ is true for all elliptic curves E/K. Hence there is a number M such that for $f_3(U) \in K[U]$ with height $(f_3) > M$ the equation E_{a,f_3} has no solution in $K[U]$.

We end by discussing the cases $p = 3$ and $p = 5$.

First assume that $p = 3$.

In this case the degrees of f_1 and f_2 have to be zero. So we have to determine all $f_3(U) \in K[U]$ of the form

$$f_3(U) = (U-a)w^4 + 4U(a-1)w^3 + 2U(U(1-2a)+2-a)w^2 \\ +4U^2(a-1)w + U^2(U-a)$$

with $w \in K^*\backslash\{1,2\}$.

For $f_3(U) = c(U^3 + \alpha U^2 + \beta U + \gamma)$ this implies:

$$\begin{aligned} c &= 1 \\ \gamma &= -aw^4 \\ \beta &= w^4 + 4(a-1)w^3 + (4-2a)w^2 \\ \alpha &= (2-4a)w^2 + 4(a-1)w - a \end{aligned}$$

Type II case i) occurs if $\deg P_2 = 0$ which is equivalent with $f_2^2 - 2f_1 f_2 = 0$, or: $w = \frac{1}{2}$.

The cases II ii) and II iii) cannot occur for $p = 3$.

We can interprete this result as follows:

Proposition 4.2. *Assume that $K^* = K^{*2}$ and that E is an elliptic curve given by the equation:*

$$Y^2 = X(X-1)(X-a).$$

Then the pairs {elliptic curves E' defined over K, anti-isometric isomorphisms

$$\alpha_3 : E_3 \to E_3'\}$$

are parametrized by the rational family of curves of genus 2 given by the equation

$$C_w : Y^2 = X(X-1)X^3 + [(4a-2)w^2 + (4a-4)w + a]X^2 \\ +[w^4 + (4a-4)w^3 + (4-2a)w^2]X \\ +[(2-4a)w^2 + 4(a-1)w - a]$$

where w runs over $K^*\backslash\{1,2\}$. The value $w = \frac{1}{2}$ gives rise to the unique covering of E_0 of degree 3 by a curve of genus 2 which is of type II.

REMARKS.

1) In **[10]** one finds a parametrization of elliptic curves E' with isometric isomorphisms from E to E_3'. We recall that E has a point P of order 2 by assumption, and if $\alpha : E_3 \to E_3'$ is isometric then $\alpha' : E/ <P> \to E \overset{\alpha}{\to} E'$ is anti-isometric.

Hence Proposition 4.2 as well as the equations of [R-S] can be used to describe all elliptic curves E' with isomorphisms from E_3 to E_3'.

2) Having C_w it is possible to determine the complementary elliptic curve E' of E in J_{C_w}. If we want to determine its j-invariant we can assume that $E_2' \subset E'(K)$ and so we have to find a rational function $Z = Z(U)$ of degree 3 in $K(U)$ with the following properties: Let $\lambda_1, \lambda_2, \lambda_3$ be the zeros of $U^3 + \alpha(w)U^2 + \beta(w)U + \gamma(w)$. Then the place $U = \lambda_1$ is the only unramified extension

of $Z = 0$, $U = \lambda_2$ is the only unramified extension of $Z = 1$, $U = \lambda_3$ is the only unramified extension of $\frac{1}{Z} = 0$, and the place $Z = \beta$ has the extensions $U = 0$, $U = 1$ and $U = \infty$. Then E' is given by $Y^2 = X(X-1)(X-\beta)$.
To determine Z and β one can proceed completely analogous as in the proof of proposition 4.2. Since Rubin and Silverberg give a very nice description of the curves E' (up to a 2-isogeny) we don't carry out this procedure explicitely here.

Next we discuss the case $p = 5$.

We get: $\deg f_1 = 1$ and $\deg f_2 = 0$.
Take $w = w_1 U + w_2$ with $w_i \in K^*$, $w_1 + w_2 \neq 1$, $w_2 \neq 2$
We have to solve:

$(U-a)(w_1U+w_2)^4 + (4U(a-1)(w_1U+w_2)^2 + 2U(U+2-2a-4aU)(w_1U+w_2)^2 + (4U^2a - 4U^2)(w_1U+w_2) + U(U-a) = (s_1U+s_2)^2 f_3(U)$
with $f_3(U) = U^3 + \alpha U^2 + \beta U + \gamma$.

By comparing coefficients one gets the following system of equations:

$$-s_1^2 + w_1^4 = 0 \tag{1}$$

$$-(\alpha s_1^2) - 2s_1s_2 + 2w_1^2 - 8aw_1^2 - aw_1^4 + 4w_1^3w_2 = 0 \tag{2}$$

$$\begin{aligned} -(\beta s_1^2) - 2\alpha s_1 s_2 - s_2^2 - 4w_1 + 4aw_1 \\ +4w_1w_2 - 16aw_1w_2 - 4aw_1^3w_2 + 6w_1^2w_2^2 = 0 \end{aligned} \tag{3}$$

$$\begin{aligned} 1 - \gamma s_1^2 - 2\beta s_1 s_2 - \alpha s_2^2 - 4w_2 + 4aw_2 \\ +2w_2^2 - 8aw_2^2 - 6aw_1^2w_2^2 + 4w_1w_2^3 = 0 \end{aligned} \tag{4}$$

$$-a - 2\gamma s_1 s_2 \beta s_2^2 - 4aw_1w_2^3 + w_2^4 = 0 \tag{5}$$

$$\gamma s_2^2 + aw_2^4 = 0 \tag{6}$$

Let us discuss type II. In case i) P_2 has to have degree 1, and so $w_1^2U^2 + U - 2w_1U^2$ has degree 1 which means that $w_1 = \frac{1}{2}$. Hence α_1 has to be $\pm\frac{1}{4}$, and the remaining 5 inequalities determine $f_3(U)$ uniquely.

In case ii) $s_1U - s_2$ has to divide $f_3(U)$, and so $\frac{s_2}{s_1}$ is one of the zeros of $f_3(U)$. Again we see that there are only finitely many possibilities to $f_3(U)$.

REMARKS.

1. It would be nice to "solve" this system of equations, i.e. to find a rational parametrization of the solutions.

2. These two examples illustrate the general situation, too. For each p we'll get a system of $p+1$ equations with $(p+2)$ variables which determine all possible polynomials $f_3(U)$. Since for fixed p and fixed E, the set of curves of genus 2 which cover E of degree p, is parametrized by a curve $\tilde{X}(p)$ which is a twist of the modular curve $X(p)$, $X(p)$ is birationally equivalent over $\bar{K}$ to the variety defined by these equations. The occurence of type II implies one more condition either for the leading coefficients of f_1 and f_2 or for the zeros of $f_3(U)$, hence for fixed p and E this occurs only finitely often.

References

[1] G. Frey, Links between solutions of $A - B = C$ and elliptic curves; in: Number Theory, Ulm 1987, Springer LNM 1380 (1989), (31-62).

[2] G. Frey, E. Kani, Curves of genus 2 covering elliptic curves and an arithmetical application; in: Arithmetic Algebraic Geometry; Progr. Math. 89, 1991, Boston- Basel-Berlin, (153-175).

[3] T. Hayashida, M. Nishi, Existence of curves of genus 2 on a product of two elliptic curves; J. Math. Soc. Japan (17) (1965) (1-16).

[4] E. Kani, Curves of genus 2 with elliptic differentials and the height conjecture for elliptic curves; Proc. Conf. Numb. Theory and Arithm. Geom.; Preprint 18 (1991); Inst. f. Exp. Math. Essen; (30-39).

[5] E. Kani, Curves with elliptic differentials; preprint.

[6] E. Kani, The number of curves of genus two with elliptic differentials; preprint.

[7] I. Kiming, On certain problems in the analytical arithmetic of quadratic forms arising from the theory of curves of genus 2 with elliptic differentials.

[8] A.N. Parshin, The Bogomolov-Miyaoka-Yau inequality for arithmetical surfaces and its applications; Sem. Théorie des Nombres Paris 1986/87; Progr. Math. 75, Boston-Basel-Berlin (1989), 299-312.

[9] K. Ribet, On modular representations of $G(\bar{\mathbb{Q}}/\mathbb{Q})$ arising from modular forms; Inv. Math. 100 (1990), 431-476.

[10] K. Rubin, A. Silverberg, Families of elliptic curves with constant mod p representations, this volume.

[11] J.P. Serre, Sur les représentations modulaires de degré 2 de $G(\bar{\mathbb{Q}}/\mathbb{Q})$; Duke Math. J. 54 (1987), 179-230.

[12] J.P. Serre, Propriétés galoisiennes des points d'ordre fini des courbes elliptiques; Inv. Math. 15 (1972), 259-331.

[13] H. Wielandt, Finite permutation groups, sec. ed.; New York-London, 1968.

Conference on Elliptic Curves and Modular Forms
Hong Kong, December 18-21, 1993

Complete intersections and Gorenstein rings

H.W. Lenstra, Jr.
Department of Mathematics # 3840
University of California, Berkeley, CA 94720–3840 USA

This paper is devoted to the proof of the following fact from commutative algebra, which is a slight sharpening of a result of Wiles. Let $\mathcal{O}$ be a complete discrete valuation ring, R a complete noetherian local $\mathcal{O}$-algebra, B a finite flat local $\mathcal{O}$-algebra, and $\varphi\colon R \to B$, $\pi\colon B \to \mathcal{O}$ surjective $\mathcal{O}$-algebra homomorphisms. Suppose that the length of the $\mathcal{O}$-module $(\ker \pi\varphi)/(\ker \pi\varphi)^2$ is finite and bounded by the length of $\mathcal{O}/\pi(\mathrm{Ann}_B \ker \pi)$. Then φ is an isomorphism and B is a complete intersection.

We prove the following fact from commutative algebra, due to Wiles in the case that B is a Gorenstein ring.

Theorem. *Let $\mathcal{O}$ be a complete discrete valuation ring, R a complete noetherian local $\mathcal{O}$-algebra, B a finite flat local $\mathcal{O}$-algebra, and $\varphi\colon R \to B$, $\pi\colon B \to \mathcal{O}$ surjective $\mathcal{O}$-algebra homomorphisms. Then the following are equivalent:*

(i) *the length of the $\mathcal{O}$-module $(\ker \pi\varphi)/(\ker \pi\varphi)^2$ is finite and less than or equal to the length of $\mathcal{O}/\pi(\mathrm{Ann}_B \ker \pi)$;*

(ii) *the length of the $\mathcal{O}$-module $(\ker \pi\varphi)/(\ker \pi\varphi)^2$ is finite and equal to the length of $\mathcal{O}/\pi(\mathrm{Ann}_B \ker \pi)$;*

(iii) *B is a complete intersection, $\pi(\mathrm{Ann}_B \ker \pi) \neq 0$, and φ is an isomorphism.*

The terms are explained below.

Rings are supposed to be commutative with 1. For the basic definitions from commutative algebra we refer to [**1**]. We write $\mathfrak{m}_R$ for the maximal ideal of a local ring R. By $\mathcal{O}$ we shall always denote a complete discrete valuation ring; the completeness assumption can be dropped, except where complete intersections are involved (this is mostly due to the naïve nature of our definition of a complete intersection below).

Finite flat. We shall call an $\mathcal{O}$-algebra *finite flat* if it is finitely generated and free as an $\mathcal{O}$-module. This is equivalent to it being finitely generated and *flat* as an $\mathcal{O}$-module, which is an easy fact that we shall not need (cf. [**1**], Exercise 7.16). For a finitely generated free $\mathcal{O}$-module M we shall put $M^\dagger = \mathrm{Hom}_{\mathcal{O}}(M, \mathcal{O})$; this is an autoduality of the category of finitely generated free $\mathcal{O}$-modules.

Key words: one-dimensional local ring, complete intersection, congruence ideal. 1991 Mathematics subject classification: 13H10. The author thanks B. de Smit, B. Mazur, and R. Pink for their assistance and comments. He was supported by NSF under grant No. DMS 92-24205. Part of the work reported in this paper was done while the author was on appointment as a Miller Research Professor in the Miller Institute for Basic Research in Science.

Local $\mathcal{O}$-algebras. A *local* $\mathcal{O}$-algebra is an $\mathcal{O}$-algebra B that is local as a ring and for which the structure map $\mathcal{O} \to B$ maps $\mathfrak{m}_{\mathcal{O}}$ inside $\mathfrak{m}_B$.

Gorenstein rings. A finite flat local $\mathcal{O}$-algebra B is called *Gorenstein* if $B^\dagger$ is free of rank 1 as a B-module. This is *not* a relative notion: there is an absolute notion of "Gorenstein ring" that is equivalent to the given one for finite flat local algebras over a discrete valuation ring (see [**3**], Section 18), and which we will not need.

Complete intersections. Let B be a finite flat local $\mathcal{O}$-algebra that has the same residue class field as $\mathcal{O}$. The latter condition means that the natural map $\mathcal{O}/\mathfrak{m}_{\mathcal{O}} \to B/\mathfrak{m}_B$ is an isomorphism; it is satisfied if there is an $\mathcal{O}$-algebra homomorphism $B \to \mathcal{O}$, which is the case in the Theorem. We call B a *complete intersection* if, for some non-negative integer n, there are elements $f_1, \ldots, f_n \in \mathcal{O}[[X_1, \ldots, X_n]]$ (the two n's are the same!) such that $B \cong \mathcal{O}[[X_1, \ldots, X_n]]/(f_1, \ldots, f_n)$ as $\mathcal{O}$-algebras. Again, this is *not* a relative notion: there is an absolute notion of "complete intersection" that is equivalent to the given one for finite flat local algebras over a complete discrete valuation ring with the same residue class field (see [**3**], Section 21). We will not need this fact. What we do need about complete intersections is summarized in the following lemma.

Lemma 1. *Let n be a non-negative integer and $f_1, \ldots, f_n \in \mathcal{O}[[X_1, \ldots, X_n]]$. Suppose that $B = \mathcal{O}[[X_1, \ldots, X_n]]/(f_1, \ldots, f_n)$ is finitely generated and non-zero as an $\mathcal{O}$-module. Then B is a finite flat local $\mathcal{O}$-algebra with the same residue class field as $\mathcal{O}$; it is Gorenstein; and $B^\dagger$ has a B-generator λ with the property that the trace map $\mathrm{Tr}_{B/\mathcal{O}}\colon B \to \mathcal{O}$ is given by $\mathrm{Tr}_{B/\mathcal{O}} = d \cdot \lambda$, where d is the image of $\det\big(\frac{\partial f_i}{\partial X_j}\big)_{i,j}$ in B.*

Proof. (*Sketch.*) Let $\pi_{\mathcal{O}}$ be a prime element of $\mathcal{O}$. For the first statement, it suffices to check that $f_1, \ldots, f_n, \pi_{\mathcal{O}}$ is a "regular" $\mathcal{O}[[X_1, \ldots, X_n]]$-sequence. One way to do this is by means of the "Koszul-complex" ([**3**], Theorem 16.8). Once one knows about regular sequences and the Koszul complex, one can prove the remaining statements by means of an argument due to Tate ([**4**], Appendix). □

A more general version of Lemma 1, in which $\mathcal{O}$ is allowed to be any noetherian ring, is proved in [**2**]. The main tool in the proof is again the Koszul complex.

The congruence ideal. Let B, C be rings and let $\pi\colon B \to C$ be a surjective ring homomorphism. Then the *congruence ideal* η_π of π is defined to be the C-ideal $\pi(\mathrm{Ann}_B \ker \pi)$, where $\mathrm{Ann}_B \ker \pi$ denotes the annihilator of $\ker \pi$ in B.

The terminology is explained by the following example. Let C, D be rings with ideals I, J, and suppose that an isomorphism $C/I \cong D/J$ is given. Put $B = C \times_{C/I} D = \{(x, y) \in C \times D : x \text{ and } y \text{ have the same image in } C/I\}$, and let $\pi\colon B \to C$ be the first projection. Then $\ker \pi = \{0\} \times J$, and if $\mathrm{Ann}_D J = 0$ then $\mathrm{Ann}_B \ker \pi = I \times \{0\}$ so that $\eta_\pi = I$. Since the elements of B are defined by means of a congruence mod I (more precisely, an equality in C/I), the ideal I may indeed be called a "congruence ideal".

The definition can be reformulated as follows. View C as a B-algebra via π. Then there is an isomorphism $\mathrm{Hom}_B(C, B) \cong \mathrm{Ann}_B \ker \pi$ sending f to $f(1)$, so η_π is just the image of the map $\mathrm{Hom}_B(C,B) \to \mathrm{Hom}_B(C,C) \cong C$ induced by π.

If $\eta_\pi = C$ then the sequence $0 \to \ker\pi \to B \to C \to 0$ of B-modules splits, so that B becomes a product of two rings. Hence if B and C are local, then one has $\eta_\pi = C$ if and only if π is an isomorphism. A "relative" version of this statement will, under additional hypotheses, be proved below (Lemma 3).

Suppose now that B is a flat $\mathcal{O}$-algebra and that $\pi\colon B \to \mathcal{O}$ is an $\mathcal{O}$-algebra homomorphism (necessarily surjective) with $\eta_\pi \neq 0$. We prove that

$$(\ker\pi) \cap (\mathrm{Ann}_B \ker\pi) = 0, \tag{2}$$

so that the surjective map $\mathrm{Ann}_B \ker\pi \to \eta_\pi$ given by π is actually an isomorphism. Let $x \in (\ker\pi) \cap (\mathrm{Ann}_B \ker\pi)$, choose $a \in \eta_\pi$, $a \neq 0$, and write $a = \pi(b)$ with $b \in \mathrm{Ann}_B \ker\pi$. Then we have $ax = (a-b)x = 0$, the first equality because $b \in \mathrm{Ann}_B \ker\pi$ and $x \in \ker\pi$, and the second because $a - b \in \ker\pi$ and $x \in \mathrm{Ann}_B \ker\pi$. Since B is flat, multiplication by a is injective, so $x = 0$, as required.

Gorenstein rings and the congruence ideal. Suppose that B and C are finite flat local $\mathcal{O}$-algebras that are Gorenstein, and let $\pi\colon B \to C$ be a surjective $\mathcal{O}$-algebra homomorphism. Choosing isomorphisms $B \cong B^\dagger$, $C \cong C^\dagger$ of B-modules we find that $\mathrm{Hom}_B(C,B) \cong_B \mathrm{Hom}_B(C^\dagger, B^\dagger)$. The latter module is, by duality, isomorphic to $\mathrm{Hom}_B(B,C)$, which is easily seen to be generated by the map π. Thus η_π is a principal C-ideal, generated by the image of π under the map $\mathrm{Hom}_B(B,C) \cong \mathrm{Hom}_B(C,B) \to C$. This can be used as an alternative definition of the congruence ideal in the Gorenstein situation.

Lemma 3. *Let A and B be finite flat local $\mathcal{O}$-algebras, and let $\varphi\colon A \to B$, $\pi\colon B \to \mathcal{O}$ be surjective $\mathcal{O}$-algebra homomorphisms. Suppose that A is Gorenstein and that $\eta_{\pi\varphi} = \eta_\pi \neq 0$. Then φ is an isomorphism.*

Proof. One easily checks that φ induces a map $\mathrm{Ann}_A \ker\pi\varphi \to \mathrm{Ann}_B \ker\pi$. Applying (2) to π and to $\pi\varphi$ we find that π and $\pi\varphi$ induce isomorphisms $\mathrm{Ann}_B \ker\pi \to \eta_\pi$ and $\mathrm{Ann}_A \ker\pi\varphi \to \eta_{\pi\varphi}$. Thus from $\eta_{\pi\varphi} = \eta_\pi$ it follows that $\mathrm{Ann}_A \ker\pi\varphi \to \mathrm{Ann}_B \ker\pi$ is an isomorphism as well, and that $\varphi\mathrm{Ann}_A \ker\pi\varphi = \mathrm{Ann}_B \ker\pi$. Therefore we have

$$A/(\ker\varphi + \mathrm{Ann}_A \ker\pi\varphi) \cong \varphi A/\varphi\mathrm{Ann}_A \ker\pi\varphi = B/\mathrm{Ann}_B \ker\pi,$$

which is free as an $\mathcal{O}$-module since $B/\mathrm{Ann}_B \ker\pi$ can be viewed as a submodule of $\mathrm{End}_{\mathcal{O}} \ker\pi$. Also, applying (2) to $\pi\varphi$ we obtain

$$\ker\varphi \cap \mathrm{Ann}_A \ker\pi\varphi \subset \ker\pi\varphi \cap \mathrm{Ann}_A \ker\pi\varphi = 0.$$

We conclude that there is an exact sequence of A-modules

$$0 \to \ker\varphi \oplus \mathrm{Ann}_A \ker\pi\varphi \to A \to B/\mathrm{Ann}_B \ker\pi \to 0$$

consisting of finitely generated free $\mathcal{O}$-modules. Dualizing, we obtain an exact sequence of A-modules

$$0 \to (B/\mathrm{Ann}_B \ker\pi)^\dagger \to A^\dagger \to (\ker\varphi)^\dagger \oplus (\mathrm{Ann}_A \ker\pi\varphi)^\dagger \to 0.$$

Since A is supposed to be Gorenstein, we have $A^\dagger \cong_A A$. Tensoring with the residue class field k of A we find that $\dim_k(A^\dagger \otimes_A k) = 1$. By the exact sequence, this implies that one of $\dim_k((\ker\varphi)^\dagger \otimes_A k)$ and $\dim_k((\mathrm{Ann}_A \ker\pi\varphi)^\dagger \otimes_A k)$ is 0. Hence by Nakayama's lemma and duality one of $\ker\varphi$ and $\mathrm{Ann}_A \ker\pi\varphi$ is 0. But$\mathrm{Ann}_A \ker\pi\varphi \cong \eta_{\pi\varphi} \neq 0$, so $\ker\varphi = 0$ and φ is an isomorphism. □

The condition that A be Gorenstein cannot be omitted in Lemma 3. This is shown by the example $A = \{(x,y,z) \in \mathcal{O}\times\mathcal{O}\times\mathcal{O} : x \equiv y \equiv z \bmod \mathfrak{m}_\mathcal{O}\}$, $B = \{(x,y) \in \mathcal{O}\times\mathcal{O} : x \equiv y \bmod \mathfrak{m}_\mathcal{O}\}$, $\varphi((x,y,z)) = (x,y)$, $\pi((x,y)) = x$, in which $\eta_{\pi\varphi} = \eta_\pi = \mathfrak{m}_\mathcal{O}$. The ring B in this example is Gorenstein (even a complete intersection).

Intermezzo on the Fitting ideal. Let B be a ring and let M be a finitely generated B-module, with generators $m_1, \ldots, m_r$. Let $f\colon B^r \to M$ map $(b_i)_{i=1}^r$ to $\sum_i b_i m_i$. Then the *Fitting ideal* $F_B M$ is the B-ideal generated by all elements of B of the form $\det(v_1, \ldots, v_r)$, with $v_1 \in \ker f, \ldots, v_r \in \ker f$ (viewed as column vectors); evidently, it suffices to let the v_i range over a set of *generators* for $\ker f$. The Fitting ideal is independent of the choice of the generators m_i. To see this, let $m_{r+1} = \sum_{i=1}^r c_i m_i$, with $c_i \in B$. One obtains generators for the kernel of $f'\colon B^{r+1} \to M$, $f'((b_i)_{i=1}^{r+1}) = \sum_{i=1}^{r+1} b_i m_i$, by taking generators for $\ker f$ (with a zero coordinate appended) together with the element $(-c_1, \ldots, -c_r, 1)$. The latter element will have to occur in any non-zero determinant built up from these generators of $\ker f'$. It follows that the Fitting ideal does not change if the system of generators $m_1, \ldots, m_r$ is changed into $m_1, \ldots, m_r, m_{r+1}$. Inductively, this implies that any two systems $m_1, \ldots, m_r$ and $m'_1, \ldots, m'_s$ of generators give rise to the same Fitting ideal as their union $m_1, \ldots, m_r, m'_1, \ldots, m'_s$.

We need three properties of the Fitting ideal. The first is

$$F_B M \subset \mathrm{Ann}_B M. \tag{4}$$

Namely, if $\sum_{j=1}^r v_{ij} m_j = 0$ for $1 \le i \le r$, then "multiplying by the adjoint" we see that $\det(v_{ij})$ annihilates each m_j and therefore M. Secondly, we have

$$F_C(M \otimes_B C) = \pi(F_B M) \tag{5}$$

when $\pi\colon B \to C$ is a surjective ring homomorphism. This is because $M \otimes_B C$ is, as a C-module, defined by the 'same' relations as those that define M as a B-module.

Thirdly, for $B = \mathcal{O}$ the Fitting ideal just measures the length: if M is a finitely generated $\mathcal{O}$-module, then $F_\mathcal{O} M = \mathfrak{m}_\mathcal{O}^{\text{length } M}$ (if M does not have finite length, interpret the right side to be 0). To prove this, one writes M as a direct sum of cyclic modules $\mathcal{O}/I_i$ and checks that $F_\mathcal{O} M$ is the product of the ideals I_i.

The congruence ideal and $(\ker\pi)/(\ker\pi)^2$. Let B and C be rings and let $\pi\colon B \to C$ be a surjective ring homomorphism for which $\ker\pi$ is finitely generated. Then $(\ker\pi)/(\ker\pi)^2$ is a C-module, and one has

$$F_C((\ker\pi)/(\ker\pi)^2) \subset \eta_\pi \subset \mathrm{Ann}_C((\ker\pi)/(\ker\pi)^2). \tag{6}$$

Namely, we have $F_B(\ker\pi) \subset \mathrm{Ann}_B \ker\pi \subset \mathrm{Ann}_B((\ker\pi)/(\ker\pi)^2)$, the first inclusion by (4) and the second one trivially. Now apply π. By (5) and $\mathrm{Ann}_B((\ker\pi)/(\ker\pi)^2) = \pi^{-1}\mathrm{Ann}_C((\ker\pi)/(\ker\pi)^2)$ this gives (6).

In the case that is of interest to us, the first inclusion of (6) can be found in [5], Proposition 6.2, with a rather more complicated proof.

Complete intersections and the congruence ideal. Let B be a finite flat local $\mathcal{O}$-algebra, let $\pi\colon B \to \mathcal{O}$ be an $\mathcal{O}$-algebra map, and suppose that B is a complete intersection. Then we have

$$F_{\mathcal{O}}((\ker\pi)/(\ker\pi)^2) = \eta_\pi. \tag{7}$$

This is proved by an explicit computation. Let $B = \mathcal{O}[[X_1,\dots,X_n]]/(f_1,\dots,f_n)$. The images b_j of X_j in B belong to $\mathfrak{m}_B$, and replacing X_j by $X_j - \pi(b_j)$ we may assume that $b_j \in \ker\pi$. Then $f_i(0) = 0$ for all i. To describe the $\mathcal{O}$-module $(\ker\pi)/(\ker\pi)^2$, one considers the ideal $\mathfrak{p}$ of $\mathcal{O}[[X_1,\dots,X_n]]$ generated by $X_1, \dots, X_n$; then $\mathfrak{p}/\mathfrak{p}^2$ is $\mathcal{O}$-free of rank n, and $(\ker\pi)/(\ker\pi)^2$ is $\mathfrak{p}/\mathfrak{p}^2$ modulo the submodule spanned by the images of $f_1, \dots, f_n$. The definition of the Fitting ideal now gives

$$F_{\mathcal{O}}((\ker\pi)/(\ker\pi)^2) = \mathcal{O}\det\Big(\frac{\partial f_i}{\partial X_j}(0)\Big)_{i,j} = \mathcal{O}\pi(d),$$

with d as in Lemma 1. To prove (7), it suffices, by (6), to prove the inclusion $\supset$. Let $x \in \eta_\pi$, and write $x = \pi(y)$ with $y \in \operatorname{Ann}_B \ker\pi$. By Lemma 1, we can choose $\lambda \in B^\dagger$ with $\operatorname{Tr}_{B/\mathcal{O}} = d\lambda$. From $\pi(d) - d \in \ker\pi$ we see that $(\pi(d) - d)y = 0$, so $\pi(d)\lambda(y) = (d\lambda)(y) = \operatorname{Tr}_{B/\mathcal{O}}(y)$. The trace of y can be computed from the action of y on the exact sequence $0 \to \ker\pi \to B \to \mathcal{O} \to 0$, and one finds that $\operatorname{Tr}_{B/\mathcal{O}}(y) = \pi(y) = x$. Therefore $x = \pi(d)\lambda(y) \in \mathcal{O}\pi(d) = F_{\mathcal{O}}((\ker\pi)/(\ker\pi)^2)$, as required.

The following result shows that one can recognize isomorphisms to complete intersections by looking at $(\ker\pi)/(\ker\pi)^2$.

Lemma 8. *Let R be a complete noetherian local $\mathcal{O}$-algebra, let B be a finite flat local $\mathcal{O}$-algebra, and let $\varphi\colon R \to B$, $\pi\colon B \to \mathcal{O}$ be surjective $\mathcal{O}$-algebra homomorphisms. Suppose that B is a complete intersection, that the map $(\ker\pi\varphi)/(\ker\pi\varphi)^2 \to (\ker\pi)/(\ker\pi)^2$ induced by φ is an isomorphism, and that these modules are of finite length over $\mathcal{O}$. Then φ is an isomorphism.*

Proof. Let n, f_i and the elements $b_j \in \ker\pi$ be as above, so that $\sum_j a_{ij}b_j \in (\ker\pi)^2$, where $a_{ij} = \frac{\partial f_i}{\partial X_j}(0)$. Since $(\ker\pi)/(\ker\pi)^2$ is of finite length we have $\det(a_{ij}) \neq 0$. Choose $r_j \in R$ with $\varphi(r_j) = b_j$. The hypothesis of the lemma implies that $r_1, \dots, r_n$ generate $\ker\pi\varphi$ modulo $(\ker\pi\varphi)^2$, so by Nakayama's lemma they generate $\ker\pi\varphi$. Hence together with $\mathfrak{m}_{\mathcal{O}}$ they generate $\mathfrak{m}_R$, so that there is a surjective $\mathcal{O}$-algebra map $\psi\colon \mathcal{O}[[X_1,\dots,X_n]] \to R$ sending X_j to r_j. The hypothesis of the lemma implies that $\sum_j a_{ij}r_j \in (\ker\pi\varphi)^2$, so there are $g_i \in \ker\psi$ with $\frac{\partial g_i}{\partial X_j}(0) = a_{ij}$. We have $g_i \in \ker\varphi\psi = (f_1,\dots,f_n)$ and therefore $g_i = \sum_l h_{il}f_l$, with $h_{il} \in \mathcal{O}[[X_1,\dots,X_n]]$. From $a_{ij} = \frac{\partial g_i}{\partial X_j}(0) = \sum_l h_{il}(0)\frac{\partial f_l}{\partial X_j}(0) = \sum_l h_{il}(0)a_{lj}$ and $\det(a_{ij}) \neq 0$ we see that $(h_{il}(0))$ is the identity matrix. This implies that the matrix (h_{il}) is invertible, so that $f_l \in \ker\psi$. Hence the map ψ factors through B and gives a map $B \to R$ that is inverse to φ. This proves the lemma. □

We need one more technical result before we can prove the Theorem.

Lemma 9. *Let B be a finite flat local $\mathcal{O}$-algebra, and let $\pi: B \to \mathcal{O}$ be an $\mathcal{O}$-algebra homomorphism. Then there is a finite flat local $\mathcal{O}$-algebra A, together with a surjective $\mathcal{O}$-algebra homomorphism $\varphi: A \to B$, such that A is a complete intersection and the map $(\ker \pi\varphi)/(\ker \pi\varphi)^2 \to (\ker \pi)/(\ker \pi)^2$ induced by φ is an isomorphism.*

Proof. Let $b_1, \dots, b_n$ generate $\ker \pi$. We first prove that $\mathcal{O}[b_1, \dots, b_n] = B$. Let $C = \mathcal{O}[b_1, \dots, b_n]$. Since B is finite over C the ring C is local, and its maximal ideal $\mathfrak{m}_C = \mathfrak{m}_B \cap C$ contains $b_1, \dots, b_n$. Clearly C is Noetherian. We have $B = \mathcal{O} + \ker \pi = \mathcal{O} + (\sum_j Bb_j) \subset C + \mathfrak{m}_C \cdot B$, so Nakayama's lemma implies that $C = B$.

The surjective B-linear map $B^n \to \ker \pi$ sending $(c_j)_{j=1}^n$ to $\sum_j c_j b_j$ gives upon tensoring with $\mathcal{O}$ a surjective map $\mathcal{O}^n \to (\ker \pi)/(\ker \pi)^2$. Choose generators $(a_{ij})_{j=1}^n$, $i = 1, \dots, n$, for the kernel of the latter map. This can be done, since every submodule of $\mathcal{O}^n$ is generated by n elements. For each i we have $\sum_j a_{ij} b_j \in (\ker \pi)^2$, so $g_i(b_1, \dots, b_n) = 0$ for some polynomial $g_i \in \mathcal{O}[X_1, \dots, X_n]$ of the form $g_i = (\sum_j a_{ij} X_j) + (\text{terms of degree} \geq 2)$; here, and below, "degree" means "total degree".

Since B is finite over $\mathcal{O}$, there is a non-negative integer m with the property that the expressions $\prod_j b_j^{m_j}$ of degree $\sum_j m_j \leq m$ span B as an $\mathcal{O}$-module. Enlarging m, if necessary, we can achieve that each g_i has degree at most $m+2$. Write $b_i^{m+1} = h_i(b_1, \dots, b_n)$, where $h_i \in \mathcal{O}[X_1, \dots, X_n]$ has degree at most m. We define

$$f_i = X_i^{m+3} - X_i^2 h_i + g_i \in \mathcal{O}[X_1, \dots, X_n] \qquad (1 \leq i \leq n).$$

Evidently, we have

$$f_i(b_1, \dots, b_n) = 0,$$
$$f_i = X_i^{m+3} + (\text{terms of degree} \leq m+2),$$
$$f_i = \sum_j a_{ij} X_j + (\text{terms of degree} \geq 2)$$

for $1 \leq i \leq n$.

Put $D = \mathcal{O}[X_1, \dots, X_n]/(f_1, \dots, f_n)$. Then there is a surjective $\mathcal{O}$-algebra map $\psi: D \to B$ sending the image of X_j to b_j. Each monomial of degree greater than $n(m+2)$ in $X_1, \dots, X_n$ is divisible by X_i^{m+3} for some i, so is modulo f_i congruent to an $\mathcal{O}$-linear combination of monomials of smaller degrees. This implies that the monomials of degree at most $n(m+2)$ span D as an $\mathcal{O}$-module, so that D is finite over $\mathcal{O}$. Since $\mathcal{O}$ is complete, it follows that D is a product of complete local rings: $D = \prod_{\mathfrak{n}} D_{\mathfrak{n}}$, where $\mathfrak{n}$ ranges over the maximal ideals of D; to see this, write for each positive integer t the Artin ring $D/\mathfrak{m}_{\mathcal{O}}^t D$ as a product of local rings (see [**1**], Theorem 8.7), and take the projective limit over t. One of these maximal ideals is the image of the maximal ideal $\mathfrak{m} = (\mathfrak{m}_{\mathcal{O}}, X_1, \dots, X_n)$ of $\mathcal{O}[X_1, \dots, X_n]$ in D; so we have $D = D' \times D_{\mathfrak{m}}$, where $\mathfrak{m}D' = D'$ and $D_{\mathfrak{m}}$ is complete. Thus, if we complete at $\mathfrak{m}$ then the equality $D = \mathcal{O}[X_1, \dots, X_n]/(f_1, \dots, f_n)$ turns into $D_{\mathfrak{m}} = \mathcal{O}[[X_1, \dots, X_n]]/(f_1, \dots, f_n)$, and the surjection $\psi: D \to B$ turns into a surjection $\varphi: D_{\mathfrak{m}} \to B$. By Lemma 1 it follows that $D_{\mathfrak{m}}$ is a complete intersection. From $f_i = \sum_j a_{ij} X_j + (\text{terms of degree} \geq 2)$ we see that the kernel of the surjective map $\mathcal{O}^n \to (\ker \pi\varphi)/(\ker \pi\varphi)^2$ that sends $(c_j)_{j=1}^n$ to the image of $\sum_j c_j X_j$ is generated by the elements $(a_{ij})_{j=1}^n$, $i = 1, \dots, n$. This implies that the map $(\ker \pi\varphi)/(\ker \pi\varphi)^2 \to (\ker \pi)/(\ker \pi)^2$ induced by φ is an isomorphism. The lemma follows, with $A = D_{\mathfrak{m}}$. □

Corollary 10. *Let B be a finite flat local $\mathcal{O}$-algebra, and let $\pi\colon B \to \mathcal{O}$ be an $\mathcal{O}$-algebra homomorphism with $\eta_\pi \neq 0$. Then B is a complete intersection if and only if $F_{\mathcal{O}}((\ker \pi)/(\ker \pi)^2) = \eta_\pi$.*

Proof. "Only if" we know from (7). To prove "if", we choose $\varphi\colon A \to B$ as in Lemma 9. Then we have

$$\eta_{\pi\varphi} = F_{\mathcal{O}}((\ker \pi\varphi)/(\ker \pi\varphi)^2) = F_{\mathcal{O}}((\ker \pi)/(\ker \pi)^2) = \eta_\pi,$$

the first equality by (7), the second from Lemma 9, and the last by hypothesis. Lemma 1 asserts that A is Gorenstein. Now apply Lemma 3. □

We prove the Theorem. The implication (iii)⇒(ii) is immediate from (7) and the second inclusion of (6), and (ii)⇒(i) is clear. To prove (i)⇒(iii), we note that

$$\eta_\pi \subset F_{\mathcal{O}}((\ker \pi\varphi)/(\ker \pi\varphi)^2) \subset F_{\mathcal{O}}((\ker \pi)/(\ker \pi)^2) \subset \eta_\pi,$$

the first inclusion by the hypothesis in (i), the second because there is a surjective map $(\ker \pi\varphi)/(\ker \pi\varphi)^2 \to (\ker \pi)/(\ker \pi)^2$, and the third by (6). We conclude that we have equality everywhere. The finite length hypothesis in (i) now implies that $\eta_\pi \neq 0$, so Corollary 10 shows that B is a complete intersection. Lemma 8, finally, shows that φ is an isomorphism. This completes the proof of the Theorem.

REMARK. R. Pink pointed out that Lemma 1, of which we only sketched the proof, can be bypassed entirely. To do this one verifies the conclusion of Lemma 1 and the equality (7) directly for the only ring to which it needs to be applied, namely the ring A constructed in the proof of Lemma 9. One proceeds as follows.

One starts by proving that the $\mathcal{O}$-algebra $D = \mathcal{O}[X_1, \dots, X_n]/(f_1, \dots, f_n)$ constructed in the proof of Lemma 9 has the two following properties: first, D is free of rank $(m+3)^n$ as an $\mathcal{O}$-module, the images of the monomials $\prod_{i=1}^n X_i^{k_i}$ with $0 \le k_i \le m+2$ forming a basis; and, second, the D-module $D^\dagger = \operatorname{Hom}_{\mathcal{O}}(D, \mathcal{O})$ is free of rank 1, a basis being formed by the linear map $\lambda\colon D \to \mathcal{O}$ that sends the monomial $\prod_{i=1}^n X_i^{m+2}$ to 1 and the other basis elements to 0. The proof of these properties is a straightforward verification that exploits the shape of the relations f_i. It follows that D is Gorenstein. From $D = D' \times A$ it follows that A is Gorenstein as well.

Next one studies $\eta_{\pi\psi}$, where $\psi\colon D \to B$ is as in the proof of Lemma 9. Since D is Gorenstein, one has $\operatorname{Ann}_D(\ker \pi\psi) = \operatorname{Hom}_D(\mathcal{O}, D) \cong \operatorname{Hom}_D(D, \mathcal{O}) = \mathcal{O} \cdot \pi\psi$, which shows that $\operatorname{Ann}_D(\ker \pi\psi)$ is free of rank 1 over $\mathcal{O}$. To exhibit a generator, one writes $f_i = \sum_{j=1}^n f_{ij} X_j$, where the polynomials f_{ij} are such that $f_{ii} - X_i^{m+2}$ and f_{ij} (for $i \neq j$) have degree at most $m+1$. From (4), with $M = \ker \pi\psi$, one sees that $\det(f_{ij})$ belongs to $\operatorname{Ann}_D(\ker \pi\psi)$, and it is in fact a generator of $\mathrm{Ann}_D(\ker \pi\psi)$ since $\lambda(\det(f_{ij})) = 1$. Applying $\pi\psi$ one finds that $\eta_{\pi\psi}$ is generated by $\det(a_{ij})$. This is the same as saying that $F_{\mathcal{O}}((\ker \pi\psi)/(\ker \pi\psi)^2) = \eta_{\pi\psi}$. Passing to A one concludes that $F_{\mathcal{O}}((\ker \pi\varphi)/(\ker \pi\varphi)^2) = \eta_{\pi\varphi}$.

Now that one knows the Gorenstein property and equality (7) for the complete intersections constructed in Lemma 9 one can pass to the more general case of Corollary 10. That is, if B and π are as in Corollary 10, then B is a complete intersection if and only if B is a complete intersection and Gorenstein, and if and only if $F_{\mathcal{O}}((\ker \pi)/(\ker \pi)^2) = \eta_\pi$. To prove this, suppose that B has one of these properties, and let $\varphi\colon A \to B$ be as in Lemma 9. Since A is known to have all three properties, it suffices to show that φ is an isomorphism. In the case that $F_{\mathcal{O}}((\ker \pi)/(\ker \pi)^2) = \eta_\pi$ this is done as in the proof of Corollary 10 given above. In the other cases B is a complete intersection, so φ is an isomorphism by Lemma 8; note that $(\ker \pi)/(\ker \pi)^2$ has finite length by the second inclusion of (6).

In all our results we assumed that the finite flat local $\mathcal{O}$-algebra B is provided with an $\mathcal{O}$-algebra homomorphism $\pi\colon B \to \mathcal{O}$. Similar results can be proved for more general finite flat $\mathcal{O}$-algebras B. The role of π can then be played by the multiplication map $\mu\colon B \otimes_{\mathcal{O}} B \to B$, which is defined by $\mu(b_1 \otimes b_2) = b_1 b_2$, and the role of the base ring $\mathcal{O}$ is taken over by B. As an example, we prove the following proposition, which was suggested by B. Mazur. Recall that the module $\Omega_{B/\mathcal{O}}$ of Kähler differentials is defined to be the B-module $(\ker \mu)/(\ker \mu)^2$ (see [**3**], Section 25).

Proposition. *Let B be a finite flat local $\mathcal{O}$-algebra that has the same residue class field as $\mathcal{O}$, and denote by μ the multiplication map $B \otimes_{\mathcal{O}} B \to B$. Suppose that the B-module $\Omega_{B/\mathcal{O}}$ has finite length. Then B is a complete intersection if and only if the congruence ideal η_μ is principal and equal to $F_B(\Omega_{B/\mathcal{O}})$.*

We note that the finite length condition for $\Omega_{B/\mathcal{O}}$ is equivalent to the K-algebra $B \otimes_{\mathcal{O}} K$ being étale, where K is the field of fractions of $\mathcal{O}$; if K has characteristic 0, then it equivalent to the nil-radical of B being zero.

The proof of the Proposition is analogous to the proof of Corollary 10. We go through the changes that need to be made.

For the "only if" part, suppose that $B \cong \mathcal{O}[[X_1, \ldots, X_n]]/(f_1, \ldots, f_n)$ as $\mathcal{O}$-algebras. Since B is finite flat over $\mathcal{O}$, there is a B-algebra isomorphism $\mathcal{O}[[X_1, \ldots, X_n]] \otimes_{\mathcal{O}} B \cong B[[X_1, \ldots, X_n]]$. This implies that we have $B \otimes_{\mathcal{O}} B \cong B[[X_1, \ldots, X_n]]/(f_1, \ldots, f_n)$ as B-algebras, where $B \otimes_{\mathcal{O}} B$ is viewed as a B-algebra via the second factor. As in the first half of the proof of (7) one now checks that $F_B(\Omega_{B/\mathcal{O}})$ equals the principal ideal $B \cdot d$, where d is as in Lemma 1. The equality $\mathrm{Tr}_{B/\mathcal{O}} = d\lambda$ from Lemma 1 implies that $\mathrm{Tr}_{B \otimes B/B} = (d \otimes 1)(\lambda \otimes 1)$. This is used to show that $\eta_\mu \subset B \cdot d$, as in the second half of the proof of (7). Hence the inclusion $\eta_\mu \subset F_B(\Omega_{B/\mathcal{O}})$ holds, and by (6) one has equality. This proves the "only if" part.

The proof of the "if" part depends on the following generalization of Lemma 3. Let a finite flat local algebra over a noetherian local ring C be defined in the same way as for $C = \mathcal{O}$.

Lemma 11. *Let C be a one-dimensional local noetherian ring, let A and B be finite flat local C-algebras, and let $\varphi\colon A \to B$, $\pi\colon B \to C$ be surjective*

C-algebra homomorphisms. Suppose that $\mathrm{Hom}_C(A, C)$ *is free of rank* 1 *as an* A*-module, that* $\eta_{\pi\varphi} = \eta_\pi$, *and that* η_π *is free of rank* 1 *as a* C*-module. Then* φ *is an isomorphism.*

Proof. If $\eta_\pi = C$ then $\pi\varphi$ and π are both isomorphisms, so φ is an isomorphism as well. For the rest of the proof we assume that $\eta_\pi \neq C$, so that $\eta_\pi \subset \mathfrak{m}_C$. Since η_π is C-free of rank 1, we have $\eta_\pi = Ca$, where a is a non-zero-divisor of C. The proof of (2) now carries through. Hence π induces an isomorphism $\mathrm{Ann}_B \ker \pi \to \eta_\pi$.

Let $\mathfrak{p}$ be a minimal prime ideal of C. Then $C_\mathfrak{p}$ is an Artin ring, and $(\eta_\pi)_\mathfrak{p}$ is a $C_\mathfrak{p}$-ideal that is free of rank 1. Since $C_\mathfrak{p}$ has finite length as a module over itself, this implies that $(\eta_\pi)_\mathfrak{p} = C_\mathfrak{p}$, so $\eta_\pi \not\subset \mathfrak{p}$. Because C is one-dimensional, this implies that the only prime ideal of the ring C/η_π is $\mathfrak{m}_C/\eta_\pi$. It follows that C/η_π is a local Artin ring. Therefore there exists $c \in C$, $c \notin \eta_\pi$, such that $c\mathfrak{m}_C \subset \eta_\pi$.

Next we prove that $B/\mathrm{Ann}_B \ker \pi$ is free as a C-module. Write $I = \mathrm{Ann}_B \ker \pi$, so that $I \cong \eta_\pi$. Both B and I are C-free, so it suffices to prove that a basis for I can be supplemented to a basis for B. By Nakayama's lemma, this can be done if the natural map $I/\mathfrak{m}_C I \to B/\mathfrak{m}_C B$ is injective, i.e., if $\mathfrak{m}_C I = I \cap \mathfrak{m}_C B$. Let $x \in I \cap \mathfrak{m}_C B$. Then $cx \in I \cap c\mathfrak{m}_C B \subset I \cap \eta_\pi B = I \cap aB$. Since a acts as a non-zero-divisor on the free C-module B, it follows from the definition of I that $I \cap aB = aI$, which equals $\eta_\pi I$. Hence x is an element of I with $cx \in \eta_\pi I$. Since I is C-free and $c \notin \eta_\pi$, this implies that $x \in \mathfrak{m}_C I$, as required.

Once (2) and the fact that $B/\mathrm{Ann}_B \ker \pi$ is C-free are known, the proof that we gave for Lemma 3 generalizes easily to a proof for Lemma 11. □

Let now B and μ be as in the Proposition, and suppose that the congruence ideal η_μ is principal and equal to $F_B(\Omega_{B/\mathcal{O}})$. We wish to prove that B is a complete intersection.

View $B \otimes_\mathcal{O} B$ as a B-algebra via the second factor. We start by constructing a finite flat local B-algebra A of the form $A = B[[X_1, \dots, X_n]]/(f_1, \dots, f_n)$ together with a surjective B-algebra homomorphism $\varphi\colon A \to B \otimes_\mathcal{O} B$ for which the induced map $(\ker \mu\varphi)/(\ker \mu\varphi)^2 \to (\ker \mu)/(\ker \mu)^2$ is an isomorphism. This is done as in the proof of Lemma 9, with $B \otimes_\mathcal{O} B$, μ, and B in the roles of B, π, and $\mathcal{O}$. There are two changes.

First, we need a new argument, in the second paragraph, to show that the kernel of any surjective B-linear map $f\colon B^n \to (\ker \mu)/(\ker \mu)^2$ is generated by n elements. This depends on the hypothesis that the ideal $F_B((\ker \mu)/(\ker \mu)^2)$ is *principal*, say with generator a. Since the module $(\ker \mu)/(\ker \mu)^2$ is supposed to be of finite length, its Fitting ideal contains a power of a prime element of $\mathcal{O}$, and therefore a is not a zero-divisor. This implies that $F_B((\ker \mu)/(\ker \mu)^2)$ is B-free of rank 1. By Nakayama's lemma, any set of generators for $F_B((\ker \mu)/(\ker \mu)^2)$ contains a basis, so one can choose the element a to be of the form $\det(v_1, \dots, v_n)$, where $v_1, \dots, v_n \in \ker f$. Let $v \in \ker f$, and replace, for some $1 \le i \le n$, the ith column of the matrix $(v_1, \dots, v_n)$ by v. The determinant of the resulting matrix belongs to $F_B((\ker \mu)/(\ker \mu)^2)$, and is therefore equal to $b_i a$ for some uniquely determined $b_i \in B$. One now verifies in a straightforward way that $v = \sum_{i=1}^n b_i v_i$ ("Cramer's rule"), so that $v_1, \dots, v_n$ span $\ker f$.

Second, we need a new proof that A is finite flat as a B-algebra. For this one can apply a version of Lemma 1 that is valid for general base rings (as in [**2**]), or one uses R. Pink's argument that we sketched above. In the same way one proves that $\mathrm{Hom}_B(A, B)$ is A-free of rank 1 and that the analogue of (7) is valid for A, that is, $F_B((\ker \mu\varphi)/(\ker \mu\varphi)^2) = \eta_{\mu\varphi}$.

Having constructed A, one shows that the map

$$\varphi\colon A = B[[X_1, \dots, X_n]]/(f_1, \dots, f_n) \to B \otimes_{\mathcal{O}} B$$

is an isomorphism of B-algebras. To do this one simply copies the proof of Corollary 10, replacing Lemma 3 by Lemma 11 (applied to $B \otimes_{\mathcal{O}} B$ and B in the roles of B and C).

We now know that B becomes a "relative complete intersection" after base extension with itself. To finish the proof of the Proposition we descend to $\mathcal{O}$.

Let, generally, C be a complete local noetherian ring, and R a complete local noetherian C-algebra with the same residue class field k as C. Then there exists m such that $C[[X_1, \dots, X_m]]$ has an ideal J for which $R \cong C[[X_1, \dots, X_m]]/J$ as C-algebras. The minimal number of generators of the ideal J equals $\dim_k J/\mathfrak{m}J$, where $\mathfrak{m}$ denotes the maximal ideal of $C[[X_1, \dots, X_m]]$. The number $m - \dim_k J/\mathfrak{m}J$ only depends on the C-algebra R, and not on the presentation $R \cong C[[X_1, \dots, X_m]]/J$; this is proved by a straightforward argument, which resembles the proof, given above, that the Fitting ideal is well-defined. Write $\epsilon(R, C) = m - \dim_k J/\mathfrak{m}J$. If D is a finite flat local C-algebra, then one readily verifies that $\epsilon(R \otimes_C D, D) = \epsilon(R, C)$.

With $C = \mathcal{O}$, $D = R = B$ we now find that $\epsilon(B, \mathcal{O}) = \epsilon(B \otimes_{\mathcal{O}} B, B) \geq 0$, the inequality coming from the isomorphism $B[[X_1, \dots, X_n]]/(f_1, \dots, f_n) \cong B \otimes_{\mathcal{O}} B$. It follows that there exist m and $g_1, \dots, g_m \in \mathcal{O}[[X_1, \dots, X_m]]$ such that $\mathcal{O}[[X_1, \dots, X_m]]/(g_1, \dots, g_m) \cong B$ as $\mathcal{O}$-algebras. Hence B is a complete intersection. (One actually has $\epsilon(B, \mathcal{O}) = 0$, by [**3**], Theorem 21.1.) This completes the proof of the Proposition. □

REMARK. Under the hypotheses of the Proposition, B is actually a complete intersection if and only if the Fitting ideal $F_B(\Omega_{B/\mathcal{O}})$ is principal. This can be deduced from Theorem 9.5 in E. Kunz, *Kähler differentials* (Vieweg, Braunschweig, 1986).

References

[1] M. F. Atiyah, I. G. Macdonald, *Introduction to commutative algebra*, Addison-Wesley, Reading, Mass., 1969.

[2] B. de Smit, H. W. Lenstra, Jr., *Finite complete intersection algebras*, Report 9453/B, Econometric Institute, Erasmus University Rotterdam, The Netherlands, 1994.

[3] H. Matsumura, *Commutative ring theory*, Cambridge University Press, Cambridge, 1986.

[4] B. Mazur, L. Roberts, *Local Euler characteristics*, Invent. Math. **9** (1970), 201–234.

[5] J. Tilouine, *Théorie d'Iwasawa classique et de l'algèbre de Hecke ordinaire*, Compositio Math. **65** (1988), 265–320.

Conference on Elliptic Curves and Modular Forms
Hong Kong, December 18-21, 1993

Homologie des courbes modulaires affines et paramétrisations modulaires

L. MEREL
INSTITUT DE MATHEMATIQUE,
UNIVERSITE PARIS 6, 4 PLACE TUSSIEU
75252 PARIS CEDEX, FRANCE

Introduction. Soit E une courbe elliptique modulaire définie sur $\mathbb{Q}$ de conducteur N. Supposons que E soit une courbe elliptique modulaire forte, et soit ϕ une paramétrisation modulaire forte de E : ϕ est un morphisme non constant défini sur $\mathbb{Q}$

$$X_0(N) \to E,$$

transformant la pointe $\Gamma_0(N)\infty$ en 0 et qui ne se factorise par aucune isogénie de courbes elliptiques $E' \to E$ définie sur $\mathbb{Q}$ de degré > 1.

Théorème 1 *a) Il existe un élément non nul* $\sum \mu_x x \in \mathbb{Q}^{(\mathbb{P}^1(\mathbb{Z}/N\mathbb{Z}))}$, *unique au signe près, satisfaisant les conditions* i$^-$), ii$^-$), iii$^-$) *et* iv$^-$) *ci-dessous. Il existe un élément non nul* $\sum \lambda_x x \in \mathbb{Z}^{(\mathbb{P}^1(\mathbb{Z}/N\mathbb{Z}))}$, *unique au signe près modulo* R_N *(où* R_N *est défini dans* i$^+$)*), satisfaisant les conditions* ii$^+$), iii$^+$) *et* iv$^+$) *énoncées dans la suite de l'introduction.*

b) Le degré deg ϕ *de* ϕ *et son nombre d'enroulement* n_ϕ *(voir [9] et section 3.2) sont donnés par les formules*

$$deg\,\phi = \frac{\epsilon}{2}|\sum_{x\in\mathbb{P}^1(\mathbb{Z}/N\mathbb{Z})} \lambda_x \mu_x|$$

et

$$n_\phi = \frac{\epsilon}{2}|\mu_{(0,1)}|,$$

où ϵ *est le nombre de composantes connexes de l'ensemble des points réels de* E *(c'est-à-dire* 2 *si le discriminant de* E *est positif,* 1 *s'il est négatif).*

Les conditions indiquées dans le théorème 1, a) suffisent à déterminer algorithmiquement les familles μ_x et λ_x à partir du conducteur de la courbe elliptique et du nombre de points de quelques réductions modulo p de la courbe. Le théorème 1 fournit donc un algorithme pour calculer deg ϕ et n_ϕ. Remarquons que l'expression de deg ϕ et n_ϕ ne dépend pas du choix des λ_x et des μ_x.

Posons $c = \begin{pmatrix} -1 & 0 \\ 0 & 1 \end{pmatrix}$, $\sigma = \begin{pmatrix} 0 & -1 \\ 1 & 0 \end{pmatrix}$ et $\tau = \begin{pmatrix} 0 & -1 \\ 1 & 1 \end{pmatrix}$.

On fait opérer à droite les matrices $M = \begin{pmatrix} a & b \\ c & d \end{pmatrix} \in M_2(\mathbb{Z})$ de déterminant premier à N sur $\mathbb{P}^1(\mathbb{Z}/N\mathbb{Z})$ en posant $(u,v)M = (au+cv, bu+dv)$.

Pour tout nombre premier p, notons $\mathcal{S}_p$ l'ensemble (fini) des matrices $\begin{pmatrix} a & b \\ c & d \end{pmatrix} \in M_2(\mathbb{Z})$ de déterminant p vérifiant $a > b \geq 0$ et $d > c \geq 0$. Posons $a_p = 1 + p - |E(\mathbb{F}_p)|$, où $|E(\mathbb{F}_p)|$ est le nombre de points de la réduction modulo p d'un modèle minimal de Weierstrass de E sur $\mathbb{Z}$.

Notons P_N la réunion disjointe des ensembles $(\mathbb{Z}/(\delta, N/\delta)\mathbb{Z})^*$, où δ parcourt les diviseurs positifs de N et où par convention $(\mathbb{Z}/1\mathbb{Z})^*$ est constitué d'un élément. Les éléments de P_N sont notés sous la forme $[\delta, x]$ où δ est un diviseur de N et x un élément de $(\mathbb{Z}/(\delta, N/\delta)\mathbb{Z})^*$ (par abus de notation si x est donné dans $\mathbb{Z}/N\mathbb{Z}$, il s'agit de sa réduction modulo $(\delta, N/\delta)$).

Les conditions mentionnées dans l'énoncé du théorème 1 s'énoncent ainsi :

i⁻) Pour tout $x \in \mathbb{P}^1(\mathbb{Z}/N\mathbb{Z})$ on a

$$\mu_x + \mu_{x\sigma} = \mu_x + \mu_{x\tau} + \mu_{x\tau^2} = 0.$$

ii⁻) Pour tout $x \in \mathbb{P}^1(\mathbb{Z}/N\mathbb{Z})$ on a

$$\mu_x = \mu_{xc}.$$

iii⁻) Pour tout nombre premier p ne divisant pas N et tout $x \in \mathbb{P}^1(\mathbb{Z}/N\mathbb{Z})$, on a

$$a_p\mu_x = \sum_{M\in\mathcal{S}_p} \mu_{xM}.$$

iv⁻) Pour tout élément $\sum_{(u,v)} \nu_{(u,v)}(u,v) \in \mathbb{Z}^{(\mathbb{P}^1(\mathbb{Z}/N\mathbb{Z}))}$ vérifiant

$$\sum_{(u,v)\in\mathbb{P}^1(\mathbb{Z}/N\mathbb{Z})} \nu_{(u,v)}([(u,N), v^{-1}] - [(v,N), -u^{-1}]) = 0$$

dans $\mathbb{Z}^{(P_N)}$ (c.f. notations ci dessus), on a

$$\sum_x \mu_x\nu_x \in \mathbb{Z}.$$

De plus $\sum \mu_x x$ est minimal pour cette propriété (i.e. n'est multiple par aucun entier $n > 1$ d'un élément de $\mathbb{Q}^{(\mathbb{P}^1(\mathbb{Z}/N\mathbb{Z}))}$ vérifiant cette propriété).

i⁺) Notons R_N le sous-groupe de $\mathbb{Z}^{(\mathbb{P}^1(\mathbb{Z}/N\mathbb{Z}))}$ engendré par les éléments de la forme $x + x\sigma$, $x + x\tau + x\tau^2$ ($x \in \mathbb{P}^1(\mathbb{Z}/N\mathbb{Z})$) et les éléments $x \in \mathbb{P}^1(\mathbb{Z}/N\mathbb{Z})$ vérifiant $x = x\sigma$ ou $x = x\tau$.

ii⁺) Pour tout $x \in \mathbb{P}^1(\mathbb{Z}/N\mathbb{Z})$ on a

$$\sum_x \lambda_x x \equiv \sum_x \lambda_x xc \pmod{R_N}.$$

iii⁺) Pour tout nombre premier p ne divisant pas N, on a

$$\sum_{M\in\mathcal{S}_p} \sum_x \lambda_x xM \equiv a_p \sum \lambda_x x \pmod{R_N}.$$

iv$^+$*)* On a dans $\mathbb{Z}^{(P_N)}$ l'égalité

$$\sum_{(u,v)\in\mathbf{P}^1(\mathbf{Z}/N\mathbf{Z})} \lambda_{(u,v)}([(u,N),v^{-1}] - [(v,N),-u^{-1}]) = 0.$$

Il existe $\sum \nu_x x \in \mathbb{Z}^{(\mathbf{P}^1(\mathbf{Z}/N\mathbf{Z}))}$ vérifiant pour tout $x \in \mathbb{P}^1(\mathbb{Z}/N\mathbb{Z})$ l'égalité suivante $\nu_x + \nu_{x\sigma} = \nu_x + \nu_{x\tau} + \nu_{x\tau^2} = 0$ et tel que

$$\sum_x \lambda_x \nu_x = 1.$$

Remarques.- Les éléments $\sum_x \mu_x x$ et $\sum_x \lambda_x x$ du théorème 1 satisfont pour $p|N$ les conditions *iii*$^-$*)* et *iii*$^+$*)* respectivement à condition de restreindre la sommation aux matrices $M \in S_p$ pour lesquelles xM est défini (cf. théorème 6).

Il existe un élément $\sum_x \mu'_x x \in \mathbb{Q}^{(\mathbf{P}^1(\mathbf{Z}/N\mathbf{Z}))}$ unique au signe près satisfaisant les conditions *i*$^-$*)*, *iii*$^-$*)*, *iv*$^-$*)* et $\mu'_x = -\mu'_{xc}$ pour tout $x \in \mathbb{P}^1(\mathbb{Z}/N\mathbb{Z})$. Il existe un élément $\sum_x \lambda'_x x$ de $\mathbb{Z}^{(\mathbf{P}^1(\mathbf{Z}/N\mathbf{Z}))}$ unique au signe près modulo R_N satisfaisant *iii*$^+$*)*, *iv*$^+$*)* et $\sum_x \lambda'_x x \equiv -\sum_x \lambda'_x xc \pmod{R_N}$. On a

$$\deg \phi = \frac{\epsilon}{2}|\sum_{x\in\mathbf{P}^1(\mathbf{Z}/N\mathbf{Z})} \lambda'_x \mu'_x|.$$

Cela se déduit de la partie 3 en échangeant les rôles de ω^+ et ω^-.

Le théorème 1 caractérise les courbes elliptiques modulaires en le sens suivant. Soit E une courbe elliptique définie sur $\mathbb{Q}$. Pour que E soit modulaire, il faut et il suffit qu'il existe $\sum_x \mu_x x \in \mathbb{Q}^{(\mathbf{P}^1(\mathbf{Z}/N\mathbf{Z}))}$ satisfaisant les conditions *i*$^-$*)* et *iii*$^-$*)*. Cela se déduit aisément des théorèmes 3 et 5.

La preuve du théorème 1 est fondée sur l'étude de l'homologie singulière des courbes modulaires $X_0(N)$. Nous étudierons l'homologie des courbes modulaires définies plus généralement par des sous-groupes d'indice fini de $SL_2(\mathbb{Z})$. Cet article est divisé en trois parties. La première partie donne une présentation (très semblable à celle obtenue dans [**14**]) de l'homologie des courbes modulaires privées de leurs pointes. Cette homologie est duale de l'homologie relative aux pointes. La dualité est fournie par les produits d'intersection. Une présentation par générateurs et relations de l'homologie relative a été obtenue par Manin [**6**]. Notre présentation de l'homologie des courbes modulaires affines s'interprète, grâce aux produits d'intersection, comme duale de la présentation de Manin.

Dans la deuxième partie nous nous restreignons aux sous-groupes de congruence pour étudier l'action des opérateurs de Hecke sur l'homologie des courbes modulaires affines. Les formules obtenues généralisent celles de [**12**].

La troisième partie est consacrée à la preuve du théorème 1. Nous illustrons ce théorème en donnant des coefficients μ_x et λ_x associés aux deux courbes elliptiques modulaires fortes de conducteur 37.

Je tiens à remercier Joseph Oesterlé, mon directeur de thèse, pour de nombreux commentaires très utiles à la rédaction de cet article.

1. Présentation des groupes d'homologie.

1.1. Homologie des courbes modulaires affines. Soit Γ un sous-groupe d'indice fini de $SL_2(\mathbb{Z})$ contenant $\begin{pmatrix} -1 & 0 \\ 0 & -1 \end{pmatrix}$. Il opère par homographies sur le demi-plan de Poincaré $\mathfrak{H}$. Notons Y_Γ la surface de Riemann quotient $\Gamma\backslash\mathfrak{H}$, X_Γ sa compactifiée et *ptes* l'ensemble des pointes de X_Γ. Notons π la surjection canonique

$$\mathfrak{H} \cup \mathbb{P}^1(\mathbb{Q}) \to X_\Gamma.$$

Si A et B sont deux parties finies et disjointes de X_Γ, notons H_A^B le groupe d'homologie singulière relative $\mathrm{H}_1(X_\Gamma - A, B; \mathbb{Z})$ (l'ensemble vide étant omis par convention).

Posons $\rho = e^{\frac{2\pi i}{3}}$. Notons δ le chemin géodésique reliant i à ρ. Posons $R = \pi(SL_2(\mathbb{Z})\rho)$ et $I = \pi(SL_2(\mathbb{Z})i)$. Ces ensembles sont disjoints. Pour $g \in SL_2(\mathbb{Z})$ notons $[g]_*$ la classe de $\pi(g\delta)$ dans $\mathrm{H}_{ptes}^{R\cup I}$. Cette classe ne dépend que de la classe Γg de g dans $\Gamma\backslash SL_2(\mathbb{Z})$. Nous la noterons encore $[\Gamma g]_*$.

Théorème 2 *L'homomorphisme de groupes*

$$\mathbb{Z}^{(\Gamma\backslash SL_2(\mathbb{Z}))} \to H_{ptes}^{R\cup I}$$

prolongeant l'application

$$\begin{aligned} \Gamma\backslash SL_2(\mathbb{Z}) &\to H_{ptes}^{R\cup I} \\ \Gamma g &\mapsto [g]_* \end{aligned}$$

est un isomorphisme de groupes.

Le groupe H_{ptes} s'identifie à un sous-groupe de $\mathrm{H}_{ptes}^{R\cup I}$. Ce sous-groupe est décrit par le théorème suivant.

Théorème 3 *Soit $\sum \mu_g g \in \mathbb{Z}^{(\Gamma\backslash SL_2(\mathbb{Z}))}$. L'élément $\sum \mu_g [g]_*$ de $H_{ptes}^{R\cup I}$ appartient à l'image canonique de H_{ptes} si et seulement si on a pour tout $g \in \Gamma\backslash SL_2(\mathbb{Z})$*

$$\mu_g + \mu_{g\sigma} = 0 \text{ et } \mu_g + \mu_{g\tau} + \mu_{g\tau^2} = 0.$$

Ce théorème est dual d'un théorème de Manin (**[6]**). Les théorèmes 2 et 3 sont très proches de ceux démontrés dans **[14]**. Nous emploierons les mêmes techniques pour les démontrer, c'est-à-dire notamment des formules de produits d'intersection.

1.2. Produits d'intersection entre les groupes $H_{ptes}^{R\cup I}$ et $H_{R\cup I}^{ptes}$. Soit $z_0 \in \mathfrak{H}$ tel que $|z_0| = 1$ et $-\frac{1}{2} < \Re(z_0) < 0$. Notons γ le chemin de $\bar{\mathfrak{H}}$ composé des chemins géodésiques reliant 0 à z_0 et z_0 à ∞. Pour $g \in SL_2(\mathbb{Z})$ notons $[g]^*$ la classe dans $\mathrm{H}_{R\cup I}^{ptes}$ de $\pi(g\gamma)$.

Les produits d'intersection fournissent un accouplement unimodulaire noté $\bullet$

$$\mathrm{H}_{R\cup I}^{ptes} \times \mathrm{H}_{ptes}^{R\cup I} \to \mathbb{Z}.$$

Proposition 1 *Soient g et h deux éléments de $SL_2(\mathbb{Z})$. Le nombre d'intersection de l'élément $[g]^*$ de $H_{R\cup I}^{ptes}$ et de l'élément $[h]_*$ de $H_{ptes}^{R\cup I}$ est donné par la formule*

$$\begin{aligned} &[g]^* \bullet [h]_* = 1 \qquad \text{si} \qquad \Gamma g = \Gamma h, \\ &[g]^* \bullet [h]_* = 0 \qquad \text{sinon.} \end{aligned}$$

Démonstration.- Le triangle hyperbolique T de sommets ∞, 0 et ρ est un domaine fondamental pour $SL_2(\mathbb{Z})$. Les points des chemins γ et δ distincts des extrémités sont intérieurs à T. On en déduit que $g\gamma$ et $h\delta$ n'ont pas de point d'intersection si $g \notin \{h, -h\}$. On a donc

$$[g]^* \bullet [h]_* = 0$$

si $\Gamma g \neq \Gamma h$.

Si $g \in \{h, -h\}$, on constate que les chemins $g\gamma$ et $h\delta$ ont un unique point d'intersection, en lequel ils se coupent transversalement avec pour nombre d'intersection +1. On a donc

$$[g]^* \bullet [h]_* = 1$$

si $\Gamma g = \Gamma h$. Cela achève la preuve de la proposition.

Proposition 2 *L'ensemble* $\{[g]_*/g \in SL_2(\mathbb{Z})\}$ *engendre* $H^{R\cup I}_{ptes}$.

Démonstration.- Le groupe $\mathrm{H}^{R\cup I}_{ptes}$ est l'image par l'homomorphisme de groupes déduit fonctoriellement de π de $\mathrm{H}_1(\mathfrak{H}, SL_2(\mathbb{Z})\{\rho, i\}; \mathbb{Z})$. Les translatés du triangle hyperbolique T par les éléments de $SL_2(\mathbb{Z})$ constituent un pavage connexe par bords de $\mathfrak{H}$. Pout tout $g \in SL_2(\mathbb{Z})$, le chemin $g\delta$ relie les points gi et $g\rho$, et ces points sont les seuls de gT appartenant à $SL_2(\mathbb{Z})\{\rho, i\}$. Comme $\mathfrak{H}$ est simplement connexe on en déduit que $\mathrm{H}_1(\mathfrak{H}, SL_2(\mathbb{Z})\{\rho, i\}; \mathbb{Z})$ est engendré par les classes de $g\delta$ ($g \in SL_2(\mathbb{Z})$). On en déduit la proposition.

Prouvons maintenant le théorème 2. D'après la proposition 2 l'homomorphisme

$$\mathbb{Z}^{(\Gamma\backslash SL_2(\mathbb{Z}))} \rightarrow \mathrm{H}^{R\cup I}_{ptes}$$

est surjectif. D'après la proposition 1 il est injectif.

Corollaire 1 *L'homomorphisme de groupes*

$$\mathbb{Z}^{(\Gamma\backslash SL_2(\mathbb{Z}))} \rightarrow H^{ptes}_{R\cup I}$$

prolongeant l'application

$$\begin{aligned} \Gamma\backslash SL_2(\mathbb{Z}) &\rightarrow H^{R\cup I}_{ptes} \\ \Gamma g &\mapsto [g]^* \end{aligned}$$

est un isomorphisme de groupes.

Démonstration.- Cela résulte immédiatement du fait que l'accouplement

$$\mathrm{H}^{ptes}_{R\cup I} \times \mathrm{H}^{R\cup I}_{ptes} \rightarrow \mathbb{Z}$$

défini par les produits d'intersection est unimodulaire et de la proposition 1.

1.3. Application aux groupes H_{ptes} *et* H^{ptes}. Notons $<\sigma>$ et $<\tau>$ les sous-groupes finis de $SL_2(\mathbb{Z})$ engendrés par σ et τ respectivement. Les applications

$$\Gamma\backslash SL_2(\mathbb{Z})/<\sigma> \rightarrow I \quad \text{et} \quad \Gamma\backslash SL_2(\mathbb{Z})/<\tau> \rightarrow R$$

déduites par passage aux quotients de $g \mapsto \Gamma g i$ et $g \mapsto \Gamma g \rho$ sont des bijections. On en déduit la série d'isomorphismes canoniques suivante

$$\begin{aligned}\mathrm{H}_0(R \cup I; \mathbb{Z}) &\simeq \mathbb{Z}^{(R\cup I)} \simeq \mathbb{Z}^{(R)} \times \mathbb{Z}^{(I)} \\ &\simeq \mathbb{Z}^{(\Gamma\backslash SL_2(\mathbb{Z})/<\tau>)} \times \mathbb{Z}^{(\Gamma\backslash SL_2(\mathbb{Z})/<\sigma>)}.\end{aligned}$$

Cela permet d'écrire la suite exacte longue d'homologie relative

$$0 \leftarrow \mathrm{H}_0(X_\Gamma - ptes; \mathbb{Z}) \leftarrow \mathrm{H}_0(R \cup I; \mathbb{Z}) \leftarrow \mathrm{H}_{ptes}^{R\cup I} \leftarrow \mathrm{H}_{ptes} \leftarrow 0$$

sous la forme

$$\begin{aligned}0 \leftarrow \mathbb{Z} \leftarrow \mathbb{Z}^{(\Gamma\backslash SL_2(\mathbb{Z})/<\tau>)} &\times \mathbb{Z}^{(\Gamma\backslash SL_2(\mathbb{Z})/<\sigma>)} \\ \leftarrow \mathbb{Z}^{(\Gamma\backslash SL_2(\mathbb{Z}))} &\leftarrow \mathrm{H}_{ptes} \leftarrow 0.\end{aligned}$$

Les produits d'intersection permettent d'en déduire par dualité une suite exacte

$$\begin{aligned}0 \rightarrow \mathbb{Z} \rightarrow \mathbb{Z}^{(\Gamma\backslash SL_2(\mathbb{Z})/<\tau>)} &\times \mathbb{Z}^{(\Gamma\backslash SL_2(\mathbb{Z})/<\sigma>)} \\ \rightarrow \mathbb{Z}^{(\Gamma\backslash SL_2(\mathbb{Z}))} &\rightarrow \mathrm{H}^{ptes} \rightarrow 0,\end{aligned}$$

dans laquelle la surjection $\mathbb{Z}^{(\Gamma\backslash SL_2(\mathbb{Z}))} \rightarrow \mathrm{H}^{ptes}$ s'identifie à la surjection canonique $\mathrm{H}_{R\cup I}^{ptes} \rightarrow \mathrm{H}^{ptes}$.

Prouvons maintenant le théorème 3. Soit $\sum \mu_g g \in \mathbb{Z}^{(\Gamma\backslash SL_2(\mathbb{Z}))}$. L'élément $\sum \mu_g [g]_*$ de $\mathrm{H}_{ptes}^{R\cup I}$ est l'image d'un élément de H_{ptes} par l'injection canonique si et seulement si son bord est nul, c'est-à-dire si on a dans $\mathbb{Z}^{(R\cup I)}$

$$\sum \mu_g (g\rho - gi) = 0.$$

On obtient la condition énoncée dans le théorème 3 lorsqu'on identifie $\mathbb{Z}^{(R\cup I)}$ à $\mathbb{Z}^{(\Gamma\backslash SL_2(\mathbb{Z})/<\tau>)} \times \mathbb{Z}^{(\Gamma\backslash SL_2(\mathbb{Z})/<\sigma>)}$ comme ci-dessus.

Pour $\sum_x \mu_x x \in \mathbb{Z}(\Gamma\backslash SL_2(\mathbb{Z}))$ vérifiant la condition du théorème 3, notons $\xi_0(\sum_x \mu_x x)$ l'élément correspondant de H_{ptes}. Pour $g \in SL_2(\mathbb{Z})$ ou $g \in \Gamma\backslash SL_2(\mathbb{Z})$, notons $\xi^0(g)$ l'image dans H^{ptes} de $[g]^*$ par la surjection canonique

$$\mathrm{H}_{R\cup I}^{ptes} \rightarrow \mathrm{H}^{ptes}.$$

On déduit immédiatement des suites exactes longues figurant plus haut le corollaire suivant.

Corollaire 2 *Les classes $\xi^0(g)$ engendrent H^{ptes} lorsque g décrit $SL_2(\mathbb{Z})$ et on a les relations*

$$\xi^0(g) + \xi^0(g\sigma) = 0,$$

$$\xi^0(g) + \xi^0(g\tau) + \xi^0(g\tau^2) = 0$$

pour tout $g \in SL_2(\mathbb{Z})$.

Remarque.- Pour $(\alpha,\beta) \in \mathbb{P}^1(\mathbb{Q})$ notons $\{\alpha,\beta\}$ la classe dans H^{ptes} de l'image par π du chemin géodésique reliant α à β dans $\bar{\mathfrak{H}}$. On a $\xi^0(g) = \{g0, g\infty\}$. Signalons que le corollaire 2 est un théorème de Manin ([**6**]).

On a la formule suivante relative à l'accouplement

$$\mathrm{H}^{ptes} \times \mathrm{H}_{ptes} \to \mathbb{Z}.$$

Corollaire 3 *Soit $g \in \Gamma\backslash SL_2(\mathbb{Z})$. Soit $\sum_h \mu_h h \in \mathbb{Z}^{(\Gamma\backslash SL_2(\mathbb{Z}))}$ tel que $\sum_h \mu_h [h]_*$ soit dans l'image de H_{ptes} par l'injection canonique. On a*

$$\xi^0(g) \bullet \xi_0(\sum_h \mu_h h) = \mu_g.$$

Démonstration.- Cela résulte de l'égalité

$$\xi^0(g) \bullet \xi_0(\sum_h \mu_h h) = [g]^* \bullet (\sum_h \mu_h [h]_*) = \mu_g.$$

1.4. L'homologie de X_Γ. Remarquons qu'on a une surjection canonique

$$\alpha \,:\, \mathrm{H}_{ptes} \to H$$

où suivant nos conventions de la section 1.1 H désigne le groupe $\mathrm{H}_1(X_\Gamma;\mathbb{Z})$.

Proposition 3 *Soit $\sum_x \mu_x x \in \mathbb{Q}^{(\Gamma\backslash SL_2(\mathbb{Z}))}$ vérifiant pour tout $x \in \Gamma\backslash SL_2(\mathbb{Z})$*

$$\mu_x + \mu_{x\sigma} = \mu_x + \mu_{x\tau} + \mu_{x\tau^2} = 0.$$

Alors l'image de $\xi_0(\sum_x \mu_x x) \in H_{ptes} \otimes \mathbb{Q}$ dans $H \otimes \mathbb{Q}$ par $\alpha \otimes 1_\mathbb{Q}$ appartient à H si et seulement si pour tout élément $\sum_x \nu_x x \in \mathbb{Z}^{(\Gamma\backslash SL_2(\mathbb{Z}))}$ vérifiant

$$\sum_x \nu_x((x\infty) - (x0)) = 0$$

dans $\mathbb{Z}^{(\Gamma\backslash \mathbf{P}^1(\mathbf{Q}))}$ on a

$$\sum_x \nu_x \mu_x \in \mathbb{Z}.$$

Démonstration.- On a une injection canonique

$$\beta \,:\, H \to \mathrm{H}^{ptes}.$$

adjointe de α pour les produits d'intersection. Observons d'abord que si $\sum_x \nu_x x \in \mathbb{Z}^{(\Gamma\backslash SL_2(\mathbf{Z}))}$, alors $\sum_x \nu_x \xi^0(x)$ appartient à β(H) si et seulement si son bord est nul, c'est-à-dire si et seulement si on a

$$\sum_x \nu_x((x\infty) - (x0)) = 0$$

dans $\mathbb{Z}^{(\Gamma\backslash \mathbf{P}^1(\mathbf{Q}))}$.

Soit alors $\sum_x \mu_x x \in \mathbb{Q}^{(\Gamma\backslash SL_2(\mathbf{Z}))}$ tel que $\mu_x + \mu_{x\sigma} = \mu_x + \mu_{x\tau} + \mu_{x\tau^2} = 0$ pour tout $x \in \Gamma\backslash SL_2(\mathbb{Z})$. Notons y l'image de $\xi_0(\sum_x \mu_x x)$ par $\alpha \otimes 1_\mathbf{Q}$. Elle appartient à H si et seulement si on a $y' \bullet y \in \mathbb{Z}$ pour tout $y' \in H$, c'est-à-dire $\beta(y') \bullet \xi_0(\sum_x \mu_x x) \in \mathbb{Z}$ pour tout $y' \in H$, ou encore, d'après ce qui précède, si et seulement si, pour tout élément $\sum_x \nu_x x$ de $\mathbb{Z}^{(\Gamma\backslash SL_2(\mathbf{Z}))}$ vérifiant $\sum_x \nu_x((x\infty) - (x0)) = 0$ dans $\mathbb{Z}^{(\Gamma\backslash \mathbf{P}^1(\mathbf{Q}))}$, le produit d'intersection $\xi^0(\sum_x \nu_x x) \bullet \xi_0(\sum_x \mu_x x)$, qui est égal à $\sum_x \mu_x \nu_x$, appartient à $\mathbb{Z}$.

2. Opérateurs de Hecke.

2.1. L'action des opérateurs de Hecke sur H^{ptes}. Soient N un entier > 0 et K un sous-groupe de $GL_2(\mathbb{Z}/N\mathbb{Z})$ tel que l'application déterminant $GL_2(\mathbb{Z}/N\mathbb{Z}) \to (\mathbb{Z}/N\mathbb{Z})^*$ soit surjective. Notons ϕ_N la surjection canonique $M_2(\mathbb{Z}) \to M_2(\mathbb{Z}/N\mathbb{Z})$. Supposons qu'on ait

$$\Gamma = \phi_N^{-1}(K) \cap SL_2(\mathbb{Z}).$$

Soit m un entier > 0 premier à N. Posons

$$A_m = \{g \in M_2(\mathbb{Z})/\det(g) = m\} \text{ et } A_{m,K} = \phi_N^{-1}(K) \cap A_m.$$

Le groupe Γ opère à gauche et à droite sur $A_{m,K}$. Soit R un système de représentants de $\Gamma\backslash A_{m,K}$. Notons $T_{m,K}$ la correspondance de Hecke

$$\Gamma z \mapsto \sum_{r \in R} \Gamma r z$$

sur la courbe modulaire Y_Γ. Elle définit un endomorphisme encore noté $T_{m,K}$ du groupe H^{ptes} qui à l'image par π de la classe d'un chemin c de $\bar{\mathfrak{H}}$ d'extrémités dans $\mathbb{P}^1(\mathbb{Q})$ associe la somme des images par π des classes des chemins rc, pour $r \in R$.

Soit $\sum_M u_M M \in \mathbb{Z}^{(A_m)}$ vérifiant la condition (C) suivante : pour tout $\alpha \in A_m/SL_2(\mathbb{Z})$ on a dans $\mathbb{Z}^{(\mathbb{P}^1(\mathbb{Q}))}$

$$\sum_{M \in \alpha} u_M((M\infty) - (M0)) = (\infty) - (0).$$

De tels éléments existent ; des exemples en sont donnés dans [**12**], [**10**] et [**11**].

Pour $M = \begin{pmatrix} a & b \\ c & d \end{pmatrix}$ on pose $\tilde{M} = \begin{pmatrix} d & -b \\ -c & a \end{pmatrix}$ (si M est inversible on a $\tilde{M} = M^{-1}\det M$). Notons $\tilde{K}$ l'image de K par $M \mapsto \tilde{M}$.

L'application canonique $\Gamma\backslash SL_2(\mathbb{Z}) \to \tilde{K}\backslash GL_2(\mathbb{Z}/N\mathbb{Z})$ est bijective. Cela permet d'associer à tout élément $x \in \tilde{K}\backslash GL_2(\mathbb{Z}/N\mathbb{Z})$ un élément $[x]_*^{\tilde{K}}$ de $H_{ptes}^{R\cup I}$. Pour $\sum_x \mu_x x \in \mathbb{Z}^{(\tilde{K}\backslash GL_2(\mathbb{Z}/N\mathbb{Z}))}$ tel que $\sum_x \mu_x [x]_*^{\tilde{K}}$ soit dans l'image de H_{ptes} par l'injection canonique, notons $\xi_0^{\tilde{K}}(\sum_x \mu_x x)$ l'élément correspondant de H_{ptes}.

Pour $x \in \tilde{K}\backslash GL_2(\mathbb{Z}/N\mathbb{Z})$, notons encore $\xi_{\tilde{K}}^0(x)$ l'image de de x dans H^{ptes} par l'application composée

$$\mathbb{Z}^{(\tilde{K}\backslash GL_2(\mathbb{Z}/N\mathbb{Z}))} \to \mathbb{Z}^{(\Gamma\backslash SL_2(\mathbb{Z}))} \to H^{ptes}.$$

Théorème 4 *Soit $x \in \tilde{K}\backslash GL_2(\mathbb{Z}/N\mathbb{Z})$. On a*

$$T_{m,K}(\xi_{\tilde{K}}^0(x)) = \sum_M u_M \xi_{\tilde{K}}^0(xM).$$

Démonstration.- Soit $\lambda_{\tilde{K}} : \tilde{K}\backslash GL_2(\mathbb{Z}/N\mathbb{Z}) \to SL_2(\mathbb{Z})$ une section de la surjection canonique $SL_2(\mathbb{Z}) \to \tilde{K}\backslash GL_2(\mathbb{Z}/N\mathbb{Z})$. Soit R_m un système de représentants de $A_m/SL_2(\mathbb{Z})$. Il existe $g \in SL_2(\mathbb{Z})$ tel que $x = \tilde{K}g$. On a

$$\sum_M u_M \xi_{\tilde{K}}^0(xM) = \sum_{r \in R_m} \sum_{M \in rSL_2(\mathbb{Z})} u_M \xi_{\tilde{K}}^0(xM)$$

$$= \sum_{r\in R_m} \sum_{M\in rSL_2(\mathbb{Z})} u_M\{\lambda_{\tilde{K}}(xM)0, \lambda_{\tilde{K}}(xM)\infty\}$$
$$= \sum_{r\in R_m} \sum_{M\in rSL_2(\mathbb{Z})} u_M\{\lambda_{\tilde{K}}(xr)r^{-1}M0, \lambda_{\tilde{K}}(xr)r^{-1}M\infty\}.$$

La dernière égalité provient du fait que $\lambda_{\tilde{K}}(xM)$ et $\lambda_{\tilde{K}}(xr)r^{-1}M$ diffèrent par multiplication à gauche par un élément de Γ. Utilisons maintenant la condition (C) et les propriétés des symboles modulaires. On a

$$\sum_M u_M \xi^0_{\tilde{K}}(xM) = \sum_{r\in R_m} \{\lambda_{\tilde{K}}(xr)r^{-1}0, \lambda_{\tilde{K}}(xr)r^{-1}\infty\}$$
$$= \sum_{r\in R_m} \{\lambda_{\tilde{K}}(xr)r^{-1}g^{-1}\begin{pmatrix} m & 0 \\ 0 & m\end{pmatrix} g0,$$
$$\lambda_{\tilde{K}}(xr)r^{-1}g^{-1}\begin{pmatrix} m & 0 \\ 0 & m\end{pmatrix} g\infty\}.$$

Posons $s = gr$. Quand r parcourt R_m, s parcourt encore un système de représentants S_m de $A_m/SL_2(\mathbb{Z})$. Il nous faut donc vérifier que

$$\lambda_{\tilde{K}}(\tilde{K}s)s^{-1}\begin{pmatrix} m & 0 \\ 0 & m\end{pmatrix}$$

parcourt un système de représentants de $\Gamma\backslash A_{m,K}$. On a

$$\lambda_{\tilde{K}}(\tilde{K}s)s^{-1}\begin{pmatrix} m & 0 \\ 0 & m\end{pmatrix} \in A_{m,K}$$

car $\phi_N(\lambda_{\tilde{K}}(\tilde{K}s)s^{-1}\begin{pmatrix} m & 0 \\ 0 & m\end{pmatrix})$ s'écrit $\tilde{k}s\tilde{s}$ avec $k \in K$ de déterminant m^{-1} modulo N ; on a donc $\tilde{k}s\tilde{s} = \tilde{k}\begin{pmatrix} m & 0 \\ 0 & m\end{pmatrix} = k^{-1} \in K$.

Deux éléments s et s' de S_m définissent deux classes $\Gamma\lambda_{\tilde{K}}(\tilde{K}s)s^{-1}\begin{pmatrix} m & 0 \\ 0 & m\end{pmatrix}$ et $\Gamma\lambda_{\tilde{K}}(\tilde{K}s')s'^{-1}\begin{pmatrix} m & 0 \\ 0 & m\end{pmatrix}$ distinctes. Sinon il existerait $\gamma \in \Gamma$ tel que

$$s\lambda_{\tilde{K}}(\tilde{K}s)^{-1} = s'\lambda_{\tilde{K}}(\tilde{K}s')^{-1}\gamma$$

et les classes $sSL_2(\mathbb{Z})$ et $s'SL_2(\mathbb{Z})$ seraient égales.

Prouvons maintenant que $\lambda_{\tilde{K}}(\tilde{K}s)s^{-1}\begin{pmatrix} m & 0 \\ 0 & m\end{pmatrix}$ parcourt toutes les classes de $\Gamma\backslash A_{m,K}$ quand s parcourt S_m. Il suffit de prouver que $\Gamma\backslash A_{m,K}$ et S_m ont même cardinal. Cela résulte de la bijectivité de l'application canonique

$$\Gamma\backslash A_{m,K} \to SL_2(\mathbb{Z})\backslash A_m$$

et de l'égalité des cardinaux de $SL_2(\mathbb{Z})\backslash A_m$ et $A_m/SL_2(\mathbb{Z})$. Cela prouve le théorème.

Remarque.- Dans **[12]** et dans **[10]** nous avons prouvé des théorèmes très voisins relatifs dans le premier cas à $\Gamma = \Gamma_0(N)$ et dans le deuxième cas aux sous-groupes de congruence de $\Gamma(2)$.

Exemple-. Notons $\mathcal{S}_m$ l'ensemble des matrices $\begin{pmatrix} a & b \\ c & d \end{pmatrix} \in A_m$ vérifiant $a > b \geq 0$ et $d > c \geq 0$. Alors (voir [**11**]) $\sum_{M \in \mathcal{S}_m} M$ vérifie la condition (C) . Lorsque $m = p$ est un nombre premier, nous utilisons ces formules dans les conditions *iii*$^+$*)* et *iii*$^-$*)* du théorème 1. A titre d'exemple, lorsque $m = 2$, l'élément

$$\begin{pmatrix} 2 & 0 \\ 0 & 1 \end{pmatrix} + \begin{pmatrix} 1 & 0 \\ 0 & 2 \end{pmatrix} + \begin{pmatrix} 2 & 1 \\ 0 & 1 \end{pmatrix} + \begin{pmatrix} 1 & 0 \\ 1 & 2 \end{pmatrix}$$

vérifie la condition (C).

2.2. Action des opérateurs de Hecke sur H_{ptes}. Rappelons que m est un entier > 0 premier à N et que $\sum_M u_M M$ est un élément de $\mathbb{Z}^{(A_m)}$ satisfaisant la condition (C) introduite dans la section 2.1.

Théorème 5 *Soit*

$$\sum_x \mu_x x \in \mathbb{Z}^{(\tilde{K}\backslash GL_2(\mathbb{Z}/N\mathbb{Z}))}$$

tel que pour tout $x \in \tilde{K}\backslash GL_2(\mathbb{Z}/N\mathbb{Z})$ on ait $\mu_x + \mu_{x\sigma} = \mu_x + \mu_{x\tau} + \mu_{x\tau^2} = 0$. Alors $\sum_M \sum_x u_M \mu_x x\tilde{M}$ vérifie une condition analogue et on a

$$T_{m,K}(\xi_0^{\tilde{K}}(\sum_x \mu_x x)) = \xi_0^{\tilde{K}}(\sum_M \sum_x u_M \mu_x x\tilde{M}).$$

Démonstration.- Reprenons la définition des opérateurs de Hecke donnée plus haut. L'ensemble $\tilde{K}$ déduit de K est également un sous-groupe de $GL_2(\mathbb{Z}/N\mathbb{Z})$ avec déterminant surjectif. On a

$$\Gamma = \phi_N^{-1}(\tilde{K}) \cap SL_2(\mathbb{Z}).$$

Le groupe $\tilde{K}$ permet donc de définir une correspondance $T_{m,\tilde{K}}$ sur la courbe modulaire X_Γ par un procédé analogue à celui qui définit $T_{m,K}$. On constate que les correspondances $T_{m,K}$ et $T_{m,\tilde{K}}$ sont adjointes. Elles définissent donc deux endomorphismes $T_{m,K}$ et $T_{m,\tilde{K}}$ sur H_{ptes} et H^{ptes} qui sont adjoints pour les accouplements définis par les produits d'intersection.

Soit $h \in SL_2(\mathbb{Z})$. Posons $y = Kh \in K\backslash GL_2(\mathbb{Z}/N\mathbb{Z})$. On a d'après les relations d'adjonction mentionnées ci-dessus les égalités

$$\begin{aligned}
\xi_K^0(y) \bullet T_{m,K}(\xi_0^{\tilde{K}}(\sum_x \mu_x x)) &= T_{m,\tilde{K}}(\xi_K^0(y)) \bullet \xi_0^{\tilde{K}}(\sum_x \mu_x x) \\
&= (\sum_M u_M \xi_K^0(yM)) \bullet \xi_0^{\tilde{K}}(\sum_x \mu_x x) \\
&= \sum_M u_M \xi_K^0(yM) \bullet \xi_0^{\tilde{K}}(\sum_x \mu_x x) \\
&= \sum_M u_M \sum_x \mu_x [yM]_K^* \bullet [x]_*^{\tilde{K}}.
\end{aligned}$$

Démontrons que l'on a

$$[yM]_K^* \bullet [x]_*^{\tilde{K}} = [y)]_K^* \bullet [x\tilde{M}]_*^{\tilde{K}}.$$

Soit $g \in SL_2(\mathbb{Z})$ tel que $\tilde{K}g = x$. D'après la proposition 1, le membre de gauche (resp. de droite) est nul, sauf lorsque $\phi_N(g) \in KhM$ (resp. $\phi_N(h) \in \tilde{K}g\tilde{M}$), et dans ce cas il est égal à 1. Or on a $\phi_N(g) \in KhM$ si et seulement si $\phi_N(hMg^{-1}) \in K$, i.e. si et seulement si $\phi_N(g\tilde{M}h^{-1}) \in \tilde{K}$, i.e. si et seulement si $\phi_N(h) \in \tilde{K}g\tilde{M}$.

On en déduit l'égalité

$$\begin{aligned}\xi_K^0(y) \bullet T_{m,K}(\xi_0^{\tilde{K}}(\sum_x \mu_x x)) &= \sum_M u_M \sum_x \mu_x [y]_K^* \bullet [x\tilde{M}]_*^{\tilde{K}} \\ &= [y]_K^* \bullet (\sum_M \sum_x u_M \mu_x [x\tilde{M}]_*^{\tilde{K}}).\end{aligned}$$

Si on écrit l'élément $\sum_M \sum_x u_M \mu_x x\tilde{M}$ de $\mathbb{Z}^{(\tilde{K}\backslash GL_2(\mathbb{Z}/N\mathbb{Z}))}$ sous la forme $\sum \nu_t \tilde{K} t$, où t parcourt $\Gamma\backslash SL_2(\mathbb{Z})$, on a d'après ce qui précède

$$\begin{aligned}\nu_t &= [t]^* \bullet \sum_M u_M \sum_x \mu_x [x\tilde{M}]_*^{\tilde{K}} \\ &= \xi^0(t) \bullet T_{m,K}(\xi_0^{\tilde{K}}(\sum_x \mu_x x)).\end{aligned}$$

Etant données les relations $\xi^0(t) + \xi^0(t\sigma) = \xi^0(t) + \xi^0(t\tau) + \xi^0(t\tau^2) = 0$, on a

$$\nu_t + \nu_{t\sigma} = \nu_t + \nu_{t\tau} + \nu_{t\tau^2} = 0.$$

L'élément $\sum_M \sum_x u_M \mu_x [x\tilde{M}]_*^{\tilde{K}}$ de $\mathrm{H}_{ptes}^{R\cup I}$ est donc l'image canonique d'un élément de H_{ptes}, noté $\xi_0^{\tilde{K}}(\sum_M u_M \mu_x x\tilde{M})$. Par conséquent on a, pour tout $y \in K\backslash GL_2(\mathbb{Z}/N\mathbb{Z})$, l'égalité

$$\xi_K^0(y) \bullet T_{m,K}(\xi_0^{\tilde{K}}(\sum_x \mu_x x)) = \xi_K^0(y) \bullet \xi_0^{\tilde{K}}(\sum_M \sum_x u_M \mu_x x\tilde{M}).$$

Comme les $\xi_K^0(y)$ engendrent H^{ptes} et comme l'accouplement $\mathrm{H}^{ptes} \times \mathrm{H}_{ptes} \to \mathbb{Z}$ est non dégénéré cette égalité prouve le théorème 5.

2.3. Application à $\Gamma = \Gamma_0(N)$. Soit N un entier > 0. Posons

$$K = \{\begin{pmatrix} a & b \\ 0 & d \end{pmatrix} \in GL_2(\mathbb{Z}/N\mathbb{Z})\}.$$

Le groupe $\Gamma = SL_2(\mathbb{Z}) \cap \phi_N^{-1}(K)$ est alors

$$\Gamma_0(N) = \{\begin{pmatrix} a & b \\ c & d \end{pmatrix} \in SL_2(\mathbb{Z})/N|c\}.$$

L'application $\begin{pmatrix} a & b \\ c & d \end{pmatrix} \mapsto (c,d)$ définit par passage aux quotients une bijection canonique

$$K\backslash GL_2(\mathbb{Z}/N\mathbb{Z}) \simeq \mathbb{P}^1(\mathbb{Z}/N\mathbb{Z})$$

compatible à la multiplication à droite par les matrices de $M_2(\mathbb{Z})$ de déterminant premier à N.

Remarquons qu'on a $K = \tilde{K}$. Pout tout entier $m > 0$ premier à N, les correspondances $T_{m,K}$ et $T_{m,\tilde{K}}$ sont donc égales.

Dans ces conditions le théorème 3 peut être reformulé ainsi. Soit $\sum_x \mu_x x \in \mathbb{Z}^{(\mathbb{P}^1(\mathbb{Z}/N\mathbb{Z}))}$. Alors l'élément $\sum_x \mu_x [x]_*$ de $\mathrm{H}_{ptes}^{R\cup I}$ est l'image canonique d'un élément de H_{ptes} si et seulement si pour tout $x \in \mathbb{P}^1(\mathbb{Z}/N\mathbb{Z})$ on a

$$\mu_x + \mu_{x\sigma} = 0 \text{ et } \mu_x + \mu_{x\tau} + \mu_{x\tau^2} = 0.$$

Dans ce cas on note $\xi_0(\sum_x \mu_x x)$ l'élément de H_{ptes} dont il est l'image.

On note $\xi^0(x)$ l'image d'un élément x de $\mathbb{Z}^{(\mathbb{P}^1(\mathbb{Z}/N\mathbb{Z}))}$ par la surjection canonique

$$\mathbb{Z}^{(\mathbb{P}^1(\mathbb{Z}/N\mathbb{Z}))} \to \mathbb{Z}^{(K\backslash GL_2(\mathbb{Z}/N\mathbb{Z}))} \to \mathrm{H}^{ptes}.$$

Ajoutons un mot sur la conjugaison complexe. L'involution $z \mapsto -\bar{z}$ de $\mathfrak{H}$ définit une involution, la conjugaison complexe, sur la courbe modulaire $X_0(N) = X_{\Gamma_0(N)}$. Cette involution laisse stable l'ensemble des pointes de $X_0(N)$; de plus elle laisse stable l'image de $SL_2(\mathbb{Z})\{\rho, i\}$ dans $X_0(N)$. Elle définit des involutions de H_{ptes}, H^{ptes}, $\mathrm{H}_{ptes}^{R\cup I}$ et $\mathrm{H}_{R\cup I}^{ptes}$. On note ces involutions par une barre horizontale. Rappelons qu'on a posé dans l'introduction $c = \begin{pmatrix} -1 & 0 \\ 0 & 1 \end{pmatrix}$.

Proposition 4 *Soit $x \in \mathbb{P}^1(\mathbb{Z}/N\mathbb{Z})$. On a*

$$\overline{[x]^*} = -[xc\sigma]^* \quad et \quad \overline{[x]_*} = [xc\sigma]_*.$$

De plus on a

$$\overline{\xi^0(x)} = \xi^0(xc).$$

Soit $\sum_y \mu_y y \in \mathbb{Z}^{(\mathbb{P}^1(\mathbb{Z}/N\mathbb{Z}))}$ tel que $\mu_y + \mu_{y\sigma} = \mu_y + \mu_{y\tau} + \mu_{y\tau^2} = 0$. Alors $-\sum_y \mu_y yc$ satisfait une condition analogue et on a

$$\overline{\xi_0(\sum_y \mu_y y)} = -\xi_0(\sum_y \mu_y yc).$$

Démonstration.- Remarquons d'abord que les chemins $\bar{\delta}$ et $\bar{\gamma}$ sont égaux à $\sigma\delta$ et au chemin opposé à $\sigma\gamma$ respectivement. Soit $g \in SL_2(\mathbb{Z})$ tel que Kg s'identifie à x. On a

$$\overline{g\delta} = cgc\bar{\delta} \quad \text{et} \quad \overline{g\gamma} = cgc\bar{\gamma}.$$

Comme $\phi_N(c) \in K$, $Kcgc$ s'identifie à xc. On en déduit les deux premières égalités de la proposition.

Comme $-\xi^0(xc\sigma) = \xi^0(xc)$, on a $\overline{\xi^0(x)} = \xi^0(xc)$.

L'élément $\sum_y \mu_y [yc\sigma]_*$ de $\mathrm{H}_{ptes}^{R\cup I}$ est l'image canonique de l'élément $\overline{\xi_0(\sum_y \mu_y y)}$ de H_{ptes}. Comme on a $yc\sigma = y\sigma c$ et $\mu_{y\sigma} = -\mu_y$ pour tout $y \in \mathbb{P}^1(\mathbb{Z}/N\mathbb{Z})$, cela prouve la dernière assertion.

Décrivons plus complètement l'action des opérateurs de Hecke pour $\Gamma_0(N)$. Pour le reste de cette section considérons un entier $m \geq 1$ non nécessairement premier à N. Posons

$$A_{m,N} = \{\begin{pmatrix} a & b \\ c & d \end{pmatrix} \in A_m/N|c,\ (N,a) = 1\}.$$

Cet ensemble coïncide avec $A_{m,K}$ si m est premier à N. Le groupe $\Gamma_0(N)$ opère à gauche et à droite sur $A_{m,N}$. Soit $R_{m,N}$ un système de représentants de $\Gamma_0(N)\backslash A_{m,N}$. Notons T_m la correspondance de Hecke définie sur $X_0(N)$ par

$$\Gamma_0(N)z \mapsto \sum_{r\in R_{m,N}} \Gamma_0(N)rz.$$

Elle définit comme dans la section 2.1 un endomorphisme encore noté T_m de H^{ptes}. Notons $\tilde{T}_m$ l'endomorphisme de H_{ptes} adjoint de T_m ; c'est l'opérateur déduit de la correspondance T_m comme dans la section 2.1 si m et N sont premiers entre eux.

Soient $M = \begin{pmatrix} a & b \\ c & d \end{pmatrix} \in M_2(\mathbb{Z})$ et $x = (u,v) \in \mathbb{P}^1(\mathbb{Z}/N\mathbb{Z})$. Nous dirons que xM est défini si $au+cv$ et $bu+dv$ engendrent $\mathbb{Z}/N\mathbb{Z}$ et nous noterons dans ce cas xM le point de $\mathbb{P}^1(\mathbb{Z}/N\mathbb{Z})$ de coordonnées homogènes $(au+cv, bu+dv)$. Lorsque le déterminant de M est premier à N, xM est toujours défini et l'on retrouve les conventions considérées dans l'introduction.

L'action des opérateurs de Hecke est alors donnée par le théorème suivant, qui est un corollaire du théorème 4 et du théorème 5 si m et N sont premiers entre eux.

Théorème 6 *Soit $x \in \mathbb{P}^1(\mathbb{Z}/N\mathbb{Z})$. Soit $\sum_M u_M M$ satisfaisant la condition (C). On a*

$$T_m(\xi^0(x)) = \sum_M{}' u_M \xi^0(xM),$$

où $\sum'$ signifie que l'on somme sur tous les M tels que xM soit défini. Soit $\sum_x \mu_x x \in \mathbb{Z}^{(\mathbb{P}^1(\mathbb{Z}/N\mathbb{Z}))}$ vérifiant pour tout $x \in \mathbb{P}^1(\mathbb{Z}/N\mathbb{Z})$ les égalités $\mu_x + \mu_{x\sigma} = \mu_x + \mu_{x\tau} + \mu_{x\tau^2} = 0$, alors $\sum_M \sum_x \mu_x u_M x\tilde{M}$ vérifie également cette condition et on a

$$\tilde{T}_m(\xi_0(\sum_x \mu_x x)) = \xi_0(\sum_{M,x}{}' \mu_{xM} u_M x),$$

avec les mêmes conventions pour $\sum'$ que ci-dessus.

Démonstration.- Notons $\tilde{A}_{m,N}$ l'image de $A_{m.N}$ par $M \mapsto \tilde{M}$. L'application canonique

$$\Gamma_0(N)\backslash A_{m,N} \to SL_2(\mathbb{Z})\backslash SL_2(\mathbb{Z})A_{m,N}$$

est surjective ; prouvons qu'elle est injective. Soient $\gamma = \begin{pmatrix} a & b \\ c & d \end{pmatrix} \in A_{m,N}$ et $t = \begin{pmatrix} a' & b' \\ c' & d' \end{pmatrix} \in SL_2(\mathbb{Z})$ tels que $t\gamma \in A_{m,N}$. Le coefficient inférieur gauche de $t\gamma$ est égal à $ac' + cd'$. Comme il est divisible par N, comme $N|c$ et comme a est premier à N on en déduit $N|c'$, i.e. $t \in \Gamma_0(N)$.

Cela permet de définir une application

$$\begin{aligned} \eta \ : \ \tilde{A}_{m,N}SL_2(\mathbb{Z}) &\to \Gamma_0(N)\backslash SL_2(\mathbb{Z}) \\ \gamma g &\mapsto \Gamma_0(N)g. \end{aligned}$$

Notons que l'ensemble $\tilde{A}_{m,N}SL_2(\mathbb{Z})$ est l'ensemble des matrices $\begin{pmatrix} a & b \\ c & d \end{pmatrix} \in A_m$ telles que c et d engendrent $\mathbb{Z}/N\mathbb{Z}$. L'image de $\eta \begin{pmatrix} a & b \\ c & d \end{pmatrix}$ dans $\mathbb{P}^1(\mathbb{Z}/N\mathbb{Z})$ est égale à (c, d). Soit $g \in SL_2(\mathbb{Z})$ d'image x dans $\mathbb{P}^1(\mathbb{Z}/N\mathbb{Z})$ et soit $M \in A_m$. On a $gM \in \tilde{A}_{m,N}SL_2(\mathbb{Z})$ si et seulement si xM est défini, et alors l'image de $\eta(gM)$ dans $\mathbb{P}^1(\mathbb{Z}/N\mathbb{Z})$ est égale à xM. Soit $\bar{\eta} : \tilde{A}_{m,N}SL_2(\mathbb{Z}) \to SL_2(\mathbb{Z})$ un relèvement de η.

Démontrons la première égalité du théorème. Soit $g \in SL_2(\mathbb{Z})$ d'image x dans $\mathbb{P}^1(\mathbb{Z}/N\mathbb{Z})$. Fixons un système de représentants R de $g^{-1}\tilde{A}_{m,N}SL_2(\mathbb{Z})/SL_2(\mathbb{Z})$. Partons du terme de droite de l'égalité figurant dans le théorème. On a (comme dans la démonstration du théorème 4

$$\begin{aligned}\sum_M {}'u_M \xi^0(xM) &= \sum_{M \in g^{-1}\tilde{A}_{m,N}SL_2(\mathbb{Z})} u_M \{\bar{\eta}(gM)0, \bar{\eta}(gM)\infty\} \\ &= \sum_{r \in R} \sum_{M \in rSL_2(\mathbb{Z})} u_M \{\bar{\eta}(gr)r^{-1}M0, \\ &\qquad\qquad \bar{\eta}(gr)r^{-1}M\infty\} \\ &= \sum_{r \in R} \{\bar{\eta}(gr)r^{-1}g^{-1} \begin{pmatrix} m & 0 \\ 0 & m \end{pmatrix} g0, \bar{\eta}(gr)r^{-1}g^{-1} \begin{pmatrix} m & 0 \\ 0 & m \end{pmatrix} g\infty\}.\end{aligned}$$

Il reste à vérifier que $\bar{\eta}(gr)r^{-1}g^{-1} \begin{pmatrix} m & 0 \\ 0 & m \end{pmatrix}$ parcourt un système de représentants de $\Gamma_0(N)\backslash A_{m,N}$ lorsque r parcourt R. Posons $s = gr$. Les matrices s parcourent un système de représentants de $\tilde{A}_{m,N}SL_2(\mathbb{Z})/SL_2(\mathbb{Z})$.

Vérifions que $\bar{\eta}(s)s^{-1} \begin{pmatrix} m & 0 \\ 0 & m \end{pmatrix} \in A_{m,N}$. La matrice s s'écrit $\gamma\bar{\eta}(s)$ avec $\gamma \in \tilde{A}_{m,N}$. On a donc $\bar{\eta}(s)s^{-1} \begin{pmatrix} m & 0 \\ 0 & m \end{pmatrix} = \gamma^{-1} \begin{pmatrix} m & 0 \\ 0 & m \end{pmatrix} = \tilde{\gamma}$. Cela prouve que $\bar{\eta}(s)s^{-1} \begin{pmatrix} m & 0 \\ 0 & m \end{pmatrix}$ appartient à $A_{m,N}$.

Deux éléments s et s' de $\tilde{A}_{m,N}SL_2(\mathbb{Z})$ non congrus à droite modulo $SL_2(\mathbb{Z})$ définissent deux classes $\Gamma_0(N)\bar{\eta}(s)s^{-1} \begin{pmatrix} m & 0 \\ 0 & m \end{pmatrix}$ et $\Gamma_0(N)\bar{\eta}(s')s'^{-1} \begin{pmatrix} m & 0 \\ 0 & m \end{pmatrix}$ distinctes. En effet, dans le cas contraire il existerait $\gamma_0 \in \Gamma_0(N)$ tel que $s\bar{\eta}(s)^{-1}\gamma_0 = s'\bar{\eta}s'^{-1}$. On aurait alors $sSL_2(\mathbb{Z}) = s'SL_2(\mathbb{Z})$, ce qui est exclu par hypothèse

On déduit le fait $\bar{\eta}(s)s^{-1} \begin{pmatrix} m & 0 \\ 0 & m \end{pmatrix}$ parcourt un sytème de représentants de $\Gamma_0(N)\backslash A_{m,N}$ de la bijectivité de l'application composée

$$\begin{array}{ccccc} \Gamma_0(N)\backslash A_{m,N} & \to & SL_2(\mathbb{Z})\backslash SL_2(\mathbb{Z})A_{m,N} & \to & \tilde{A}_{m,N}SL_2(\mathbb{Z})/SL_2(\mathbb{Z}) \\ \Gamma_0(N)\gamma & \mapsto & SL_2(\mathbb{Z})\gamma & \mapsto & \tilde{\gamma}SL_2(\mathbb{Z}) \end{array}$$

Cela prouve la première égalité du théorème.

La seconde égalité est une conséquence du fait que $\tilde{T}_m$ est par définition adjoint de T_m, de la première égalité du théorème et des formules de produits d'intersection établis par le corollaire 3.

Indiquons maintenant comment se comprennent les conditions évoquées sur les pointes dans la proposition 3 dans le cadre du groupe $\Gamma_0(N)$. Voir l'introduction pour la définition de l'ensemble P_N.

Proposition 5 *Soit* $\sum_x \lambda_x x \in \mathbb{Z}^{(\mathbb{P}^1(\mathbb{Z}/N\mathbb{Z}))}$. *Le bord de l'élément* $\sum_x \lambda_x \xi^0(x)$ *de* $H_1(X_0(N), ptes; \mathbb{Z})$ *est nul si et seulement si on a dans* $\mathbb{Z}^{(P_N)}$ *l'égalité suivante :*

$$\sum_{(u,v)\in\mathbb{P}^1(\mathbb{Z}/N\mathbb{Z})} \lambda_{(u,v)}([(u,N),v^{-1}] - [(v,N),-u^{-1}] = 0.$$

Démonstration.- On a une bijection

$$\begin{aligned} \Gamma_0(N)\backslash\mathbb{P}^1(\mathbb{Q}) &\to P_N \\ \Gamma_0(N)\tfrac{x}{y} &\mapsto [(y,N),x], \end{aligned}$$

où par abus de notation $[(y,N),x]$ désigne l'élément $x \pmod{(y,N,N/(y,N))}$ dans $(\mathbb{Z}/(y,N,N/(y,N))\mathbb{Z})^* \subset P_N$ (voir l'introduction). Par conséquent $\mathbb{Z}^{(\Gamma_0(N)\backslash\mathbb{P}^1(\mathbb{Q}))}$ s'identifie à $\mathbb{Z}^{(P_N)}$. Soit $(u,v) \in \mathbb{P}^1(\mathbb{Z}/N\mathbb{Z})$. Soit $\begin{pmatrix} a & b \\ c & d \end{pmatrix} \in SL_2(\mathbb{Z})$ tel que la classe de (c,d) dans $\mathbb{P}^1(\mathbb{Z}/N\mathbb{Z})$ soit égale à (u,v). Le bord de l'élément $\xi^0((u,v)) = \{\frac{b}{d}, \frac{a}{c}\}$ dans $\mathbb{Z}^{(\Gamma_0(N)\backslash\mathbb{P}^1(\mathbb{Q}))}$ est égal à $(\frac{a}{c}) - (\frac{b}{d})$, ou encore dans $\mathbb{Z}^{(P_N)}$ à $[(c,N),a] - [(d,N),b]$, ou encore en utilisant l'égalité $ad - bc = 1$ et le fait que la classe de (c,d) dans $\mathbb{P}^1(\mathbb{Z}/N\mathbb{Z})$ est égale à (u,v) :

$$[(u,N),v^{-1}] - [(v,N),-u^{-1}].$$

On en déduit la proposition.

3. Paramétrisations modulaires.

3.1. Degré des paramétrisations modulaires. Soit N un entier > 0. Dans cette partie on aura toujours $\Gamma = \Gamma_0(N)$. Notons $S_2(N)$ l'espace des formes modulaires paraboliques holomorphes de poids 2 pour $\Gamma_0(N)$. Soit E une courbe elliptique modulaire définie sur $\mathbb{Q}$ de conducteur N. Soit ϕ :

$$X_0(N) \to E$$

une paramétrisation modulaire de E. L'image réciproque d'une forme différentielle de E par ce morphisme est une forme différentielle sur $X_0(N)$. Elle définit une forme modulaire nouvelle normalisée $f \in S_2(N)$, propre pour tous les opérateurs de Hecke. La valeur propre de T_p correspondant au nombre premier p est $1 + p - |E(\mathbb{F}_p)|$.

Les parties invariantes et antiinvariantes de $\mathrm{H}_1(E;\mathbb{Z})$ par la conjugaison complexe sont des $\mathbb{Z}$-modules libres de rang 1. Soient ω^+ et ω^- deux générateurs respectivement de ces modules. On a un homomorphisme de groupes ϕ^* :

$$\mathrm{H}_1(E;\mathbb{Z}) \to \mathrm{H}_1(X_0(N);\mathbb{Z})$$

déduit fonctoriellement de ϕ . Posons

$$c_0^+ = \phi^*(\omega^+) \text{ et } c_0^- = \phi^*(\omega^-).$$

Notons $H^0(X_0(N),\Omega^1)$ l'espace vectoriel complexe des formes différentielles holomorphes sur $X_0(N)$. On a un isomorphisme de $\mathbb{R}$-espaces vectoriels :

$$H_1(X_0(N);\mathbb{Z})\otimes\mathbb{R}\to \mathrm{Hom}_{\mathbb{C}}(H^0(X_0(N),\Omega^1),\mathbb{C})$$

$$c\otimes t\mapsto (\omega\mapsto t\int_c\omega).$$

L'intégration de formes différentielles le long de cycles relatifs permet de définir un homomorphisme canonique de groupes (compatible aux actions des opérateurs de Hecke):

$$\begin{aligned}\mathrm{H}_1(X_0(N),ptes;\mathbb{Z})&\to \mathrm{Hom}_{\mathbb{C}}(\mathrm{H}^0(X_0(N),\Omega^1),\mathbb{C})\\ &\simeq \mathrm{H}_1(X_0(N);\mathbb{Z})\otimes\mathbb{R}.\end{aligned}$$

Le théorème de Manin-Drinfeld nous assure que cet homomorphisme est à valeurs dans $\mathrm{H}_1(X_0(N);\mathbb{Z})\otimes\mathbb{Q}$. Cet homomorphisme a été étudié pour N premier dans **[13]** et fera l'objet d'une autre étude dans le cas général dans un article à paraître. Grâce aux produits d'intersection on a un homomorphisme canonique de groupes dual du précédent

$$\mathrm{H}_1(X_0(N);\mathbb{Z})\to \mathrm{H}_1(Y_0(N);\mathbb{Z})\otimes\mathbb{Q}.$$

Notons c^- l'image de c_0^- par ce dernier homomorphisme. L'image de c^- par l'application canonique $\mathrm{H}_1(Y_0(N);\mathbb{Z})\otimes\mathbb{Q}\to \mathrm{H}_1(X_0(N);\mathbb{Z})\otimes\mathbb{Q}$ est $c_0^-\otimes 1$. Notons c^+ l'image de c_0^+ par l'injection canonique

$$\mathrm{H}_1(X_0(N);\mathbb{Z})\to \mathrm{H}_1(X_0(N),ptes;\mathbb{Z}).$$

Les éléments c^+ et c^- sont propres pour l'action des opérateurs de Hecke d'indice premier à N avec les mêmes valeurs propres que f. Ils sont respectivement invariant et antiinvariant par la conjugaison complexe. D'après ce qui précède, il existe $\sum_x \lambda_x x\in\mathbb{Z}^{(\mathbb{P}^1(\mathbb{Z}/N\mathbb{Z}))}$ (bien défini modulo R_N, voir l'introduction) et $\sum_x \mu_x x\in\mathbb{Q}^{(\mathbb{P}^1(\mathbb{Z}/N\mathbb{Z}))}$ tels qu'on ait

$$\mu_x+\mu_{x\sigma}=\mu_x+\mu_{x\tau}+\mu_{x\tau^2}=0$$

pout tout $x\in\mathbb{P}^1(\mathbb{Z}/N\mathbb{Z})$ et

$$\xi^0(\sum_x\lambda_x x)=c^+\quad\text{et}\quad \xi_0(\sum_x\mu_x x)=c^-.$$

Notons $\deg\phi$ le degré de ϕ et ϵ le nombre de composantes connexes de l'ensemble des points réels de E.

Proposition 6 *On a*

$$deg\,\phi=\frac{\epsilon}{2}|\sum_x\lambda_x\mu_x|.$$

Démonstration.- Les produits d'intersection sur $X_0(N)$ et E sont liés par la formule suivante

$$\phi^*(\omega^+)\bullet\phi^*(\omega^-)=\deg\phi\,\omega^+\bullet\omega^-.$$

En examinant les cas où un parallélogramme fondamental d'un réseau complexe déterminant E est un losange ou un rectangle on constate que l'on a

$$2 = \epsilon|\omega^+ \bullet \omega^-|.$$

On en déduit les égalités

$$\begin{aligned}\frac{2}{\epsilon}\deg\phi &= |c_0^+ \bullet c_0^-| \\ &= |c^+ \bullet c^-| \\ &= |\xi^0(\sum_x \lambda_x x) \bullet \xi_0(\sum_x \mu_x x)| \\ &= |\sum_x \lambda_x \mu_x|\end{aligned}$$

en utilisant la première partie. Cela prouve la proposition.

Signalons qu'une majoration polynômiale ne dépendant que de N de $\deg\phi$ aurait d'intéressantes conséquences arithmétiques (voir [**15**], [**3**] et [**17**]).

Supposons maintenant que E soit une courbe elliptique modulaire forte et que ϕ soit une paramétrisation modulaire forte. L'homomorphisme de groupes ϕ^* a pour image un facteur direct de $\mathrm{H}_1(X_0(N);\mathbb{Z})$. Cela implique que c_0^+ et c_0^- ne sont multiples entiers d'aucun élément de $\mathrm{H}_1(X_0(N);\mathbb{Z})$ par un entier > 1.

Prouvons le théorème 1. Démontrons que les éléments $\sum_x \lambda_x x$ de $\mathbb{Z}^{(\mathbb{P}^1(\mathbb{Z}/N\mathbb{Z}))}$ et $\sum_x \mu_x x$ de $\mathbb{Q}^{(\mathbb{P}^1(\mathbb{Z}/N\mathbb{Z}))}$ dont c^+ et c^- sont les images par ξ^0 et ξ_0 respectivement satisfont les conditions du théorème 1. La condition *i^-)* et l'expression de R_N proviennent des présentations de $\mathrm{H}_1(X_0(N), ptes;\mathbb{Z})$ et $\mathrm{H}_1(Y_0(N);\mathbb{Z})$ respectivement (voir partie 1). Les conditions *ii^+)* et *ii^-)* expriment l'invariance et l'antiinvariance respectivement de c^+ et c^- par la conjugaison complexe (voir section 2.3). Les conditions *iii^+)* et *iii^-)* traduisent que c^+ et c^- sont vecteurs propres pour les opérateurs de Hecke d'indice premier à N avec les valeurs propres associées à f. Comme $\xi^0(\sum_x \lambda_x x)$ provient de $\mathrm{H}_1(X_0(N);\mathbb{Z})$, son bord est nul. On a donc la première partie de la condition *iv^+)*. La première partie de la condition *iv^-)* exprime que l'image c_0^- de c^- dans $\mathrm{H}_1(X_0(N);\mathbb{Z})\otimes\mathbb{Q}$ est en réalité dans $\mathrm{H}_1(X_0(N);\mathbb{Z})$, i.e. son intersection avec tout élément de $\mathrm{H}_1(X_0(N);\mathbb{Z})$ est un entier (voir section 1.4). Les secondes parties des conditions *iv^+)* et *iv^-)* traduisent le fait que c_0^+ et c_0^- ne sont multiples d'aucun élément de $\mathrm{H}_1(X_0(N);\mathbb{Z})$ par un entier > 1.

Il reste à prouver que $\sum_x \lambda_x x + R_N$ et $\sum_x \mu_x x$ sont uniques au signe près. Soient $\sum_x \lambda'_x x \in \mathbb{Z}^{(\mathbb{P}^1(\mathbb{Z}/N\mathbb{Z}))}$ et $\sum_x \mu'_x x \in \mathbb{Q}^{(\mathbb{P}^1(\mathbb{Z}/N\mathbb{Z}))}$ vérifiant les conditions *i^+), ii^+), iii^+), iv^+)* et *i^-), ii^-), iii^-), iv^-)* respectivement. Considérons leurs images respectives c'^+ et c'^- par ξ^0 et ξ_0 respectivement. D'après la condition *iv^-)*, l'image $c_0'^-$ de c'^- dans $\mathrm{H}_1(X_0(N);\mathbb{Z})\otimes\mathbb{Q}$ est un élément de $\mathrm{H}_1(X_0(N);\mathbb{Z})$. D'après la condition *$iv^+$)*, c'^+ est l'image $c_0'^+$ d'un élément de $\mathrm{H}_1(X_0(N);\mathbb{Z})$. D'après les conditions *ii^+), iii^+), ii^-), iii^-)*, $c_0'^+$ et $c_0'^-$ sont vecteurs propres pour les opérateurs de Hecke avec les valeurs propres correspondant à f et sont respectivement invariant et antiinvariant par la conjugaison complexe. Les parties invariantes et antiinvariantes par la conjugaison complexe de $\mathrm{H}_1(X_0(N);\mathbb{Z})$

sont isomorphes, après extension des scalaires à $\mathbb{C}$, à $S_2(N)$. Cet isomorphisme est compatible aux actions des opérateurs de Hecke d'indice premier à N ([**16**]). Les systèmes de valeurs propres pour les opérateurs de Hecke associés à ${c'_0}^+$ et ${c'_0}^-$ correspondent à celui de f. Or f est une forme modulaire nouvelle pour $\Gamma_0(N)$. On a donc un théorème de multiplicité 1 pour ce système de valeurs propres ([**1**]). On en déduit que ${c'_0}^+$ et ${c'_0}^-$ sont des multiples de c_0^+ et c_0^-. Les conditions de minimalité et d'intégralité *iv*$^+$*)* et *iv*$^-$*)* nous assurent que ces multiples sont égaux à $+1$ ou -1. Cela achève la preuve du théorème 1.

Remarque.- Manin a introduit dans [**6**] (de façon transcendante) les coefficients μ_x. Si $g \in SL_2(\mathbb{Z})$ a pour image x dans $\Gamma_0(N)\backslash SL_2(\mathbb{Z}) \simeq \mathbb{P}^1(\mathbb{Z}/N\mathbb{Z})$, μ_x est égal à la demi-somme des intégrales de la forme différentielle $2\pi i f(z)\,dz$ sur les chemins reliant $g0$ à $g\infty$ et $cgc0$ à $cgc\infty$ ($g \in SL_2(\mathbb{Z})$) dans $\bar{\mathfrak{H}}$ divisée par la période réelle de la courbe elliptique et multipliée par la constante de Manin. Grâce à ce procédé, Manin dresse des tables de valeurs des coefficients μ_x dans le cas des courbes modulaires $X_0(N)$ de genre 1.

3.2. Nombre d'enroulement. Reprenons toutes les notations de la section précédente. Notons ω_e l'image dans $\mathrm{H}_1(E;\mathbb{Z})\otimes\mathbb{Q}$ du symbole modulaire $\{0,\infty\}$ à travers les applications canoniques

$$\mathrm{H}_1(X_0(N), ptes;\mathbb{Z}) \to \mathrm{H}_1(X_0(N);\mathbb{Z})\otimes\mathbb{Q} \to \mathrm{H}_1(E;\mathbb{Z})\otimes\mathbb{Q}.$$

Comme $\{0,\infty\}$ est invariant par la conjugaison complexe il en est de même de ω_e. On a donc

$$\omega_e = n\omega^+ \quad (n \in \mathbb{Q}).$$

Le nombre rationnel $n_\phi = |n|$ est par définition le *nombre d'enroulement* de ϕ. Ce nombre d'enroulement est égal à celui de [**9**] multiplié par ϵ.

Proposition 7 *On a*

$$n_\phi = \frac{\epsilon}{2}|\mu_{(0,1)}|.$$

Démonstration.- Notons que le symbole modulaire $\{0,\infty\}$ est égal à $\xi^0((0,1))$. Notons e l'image de $\xi^0((0,1))$ dans $\mathrm{H}_1(X_0(N);\mathbb{Z})\otimes\mathbb{Q}$ (e est le *winding element* de [**8**]). On a les égalités

$$\begin{aligned}\frac{2}{\epsilon}n_\phi &= |n\omega^+ \bullet \omega^-| \\ &= |\omega_e \bullet \omega^-| \\ &= |e \bullet c_0^-| \\ &= |\xi^0((0,1)) \bullet c^-| \\ &= |\xi^0((0,1)) \bullet \xi_0(\sum_x \mu_x x)| \\ &= |\mu_{(0,1)}|.\end{aligned}$$

Cela prouve la proposition.

Supposons maintenant que E soit une courbe elliptique modulaire forte. On dispose alors d'une formule conjecturale pour n_ϕ provenant des conjectures de

Birch et Swinnerton-Dyer. Lorsque le groupe de Mordell-Weil $E(\mathbb{Q})$ de E est infini on a

$$n_\phi = 0.$$

Lorsque $E(\mathbb{Q})$ est fini, notons $|E(\mathbb{Q})|$ son ordre, $|\text{Ш}|$ l'ordre (fini dans ce cas, voir [4]) du groupe de Shafarevich-Tate de E, n_p le nombre de composantes rationnelles de la fibre d'un modèle de Néron en p de E (p premier divisant N) et c_ϕ la constante de Manin de la paramétrisation (conjecturalement c'est toujours 1, voir [5]). On a alors ([9])

$$n_\phi = c_\phi \frac{\epsilon |\text{Ш}| \prod_p n_p}{|E(\mathbb{Q})|^2}.$$

Si on oublie la constante de Manin, tous les facteurs figurant dans le membre de droite de l'égalité ne font intervenir que les invariants intrinsèques de la courbe elliptique, indépendamment de toute paramétrisation modulaire. Essentiellement on a donc une formule donnant le coefficient $\mu_{(0,1)}$ donnée directement par la courbe elliptique.

3.3. Deux exemples. Considérons les deux courbes elliptiques modulaires fortes de conducteur 37. Donnons une liste de coefficients μ_x et λ_x associés à ces courbes. Des coefficients μ_x associés à des courbes modulaires $X_0(N)$ de genre 1, qui sont nécessairement des courbes elliptiques modulaires fortes, sont donnés dans [6].

La courbe (37C). La courbe notée (37C) dans les tables d'Anvers [2] a pour équation minimale

$$y^2 + y = x^3 + x^2 - 23x - 50.$$

Son conducteur est 37 et son discriminant est > 0. On a donc $\epsilon = 2$. Des coefficients μ_x et λ_x associés à la courbe elliptique par le théorème 1 sont donnés par la table suivante.

x	$(0,1)$	$(1,1)$	$(2,1)$	$(3,1)$	$(4,1)$	$(5,1)$	$(6,1)$	$(7,1)$
μ_x	$\frac{2}{3}$	0	2	1	1	1	0	1
λ_x	0	0	1	0	0	0	0	0

x	$(8,1)$	$(9,1)$	$(10,1)$	$(11,1)$	$(12,1)$	$(13,1)$	$(14,1)$	$(15,1)$
μ_x	0	-1	0	0	-1	1	0	-1
λ_x	0	0	0	0	0	0	0	0

x	$(16,1)$	$(17,1)$	$(18,1)$	$(19,1)$	$(20,1)$	$(21,1)$	$(22,1)$	$(23,1)$
μ_x	-1	-1	-2	-2	-1	-1	-1	0
λ_x	0	0	0	0	0	0	0	0

x	$(24,1)$	$(25,1)$	$(26,1)$	$(27,1)$	$(28,1)$	$(29,1)$	$(30,1)$	$(31,1)$
μ_x	1	-1	0	0	-1	0	1	0
λ_x	0	0	0	0	0	0	0	0

x	$(32,1)$	$(33,1)$	$(34,1)$	$(35,1)$	$(36,1)$	$(1,0)$		
μ_x	1	1	1	1	0	$-\frac{2}{3}$		
λ_x	0	0	0	0	0	0		

On en déduit la valeur de n_ϕ et de d_ϕ

$$n_\phi = \frac{2}{3} \quad \text{et} \quad d_\phi = 2.$$

La courbe (37A). La courbe notée (37A) dans les tables de [2] a pour équation minimale

$$y^2 + y = x^3 - x.$$

Son conducteur est 37 et son discriminant est > 0. On a donc $\epsilon = 2$. Le rang du groupe de Mordell-Weil de la courbe (37A) est égal à 1. Dressons une liste de coefficients μ_x et λ_x.

x	(0,1)	(1,1)	(2,1)	(3,1)	(4,1)	(5,1)	(6,1)	(7,1)
μ_x	0	0	0	0	0	1	0	1
λ_x	0	0	0	1	0	−1	0	0

x	(8,1)	(9,1)	(10,1)	(11,1)	(12,1)	(13,1)	(14,1)	(15,1)
μ_x	1	0	0	0	0	0	−1	−1
λ_x	0	0	0	0	0	0	1	0

x	(16,1)	(17,1)	(18,1)	(19,1)	(20,1)	(21,1)	(22,1)	(23,1)
μ_x	−1	0	0	0	0	−1	−1	−1
λ_x	0	0	0	0	0	0	0	0

x	(24,1)	(25,1)	(26,1)	(27,1)	(28,1)	(29,1)	(30,1)	(31,1)
μ_x	0	0	0	0	0	1	1	0
λ_x	0	0	0	0	0	0	0	0

x	(32,1)	(33,1)	(34,1)	(35,1)	(36,1)	(1,0)		
μ_x	1	0	0	0	0	0		
λ_x	0	0	0	0	0	0		

Signalons que Manin a déjà donné cette liste de coefficients μ_x dans [7]. On déduit de cette table

$$n_\phi = 0 \quad \text{et} \quad d_\phi = 2.$$

References

[1] Atkin A.O.L., Lehner J., *Hecke operators on* $\Gamma_0(m)$, Math. Ann., 185:134–160, 1970.

[2] Atkin A.O.L. et al. *Numerical tables for elliptic curves. Modular functions of one variable IV*, number 476 in Lecture notes in mathematics. Springer -Verlag, 1972.

[3] Frey G. *Links between solutions of* $a - b = c$ *and elliptic curves*, In E. Wirsing H.P. Schlickewei, editor, *Number theory, Ulm 1987*, number 1380 in Lecture notes in mathematics, pages 31–62. Springer -Verlag, 1989.

[4] Kolyvagin V. A., *Finiteness of* $E(\mathbb{Q})$ *and* $Ш(E,\mathbb{Q})$ *for a subclass of Weil curves*, Izvestija AN SSSR, Ser. Math., 52(3):522–540, 1988.

[5] Manin Y., *Cyclotomic fields and modular curves*, Russian Math. Survey, 26(6):7–71, 1971.

[6] Manin Y., *Parabolic points and zeta function of modular curves*, Math. USSR Izvestija, 6(1):19–64, 1972.

[7] Manin Y., *Modular forms and number theory*, In Ohli Lehto, editor, *Proceedings of the international congress of mathematicians 1978*, pages 177–186. 1980.

[8] Mazur B., *Modular curves and the Eisenstein ideal*, Pub. math. de l'IHES, 47:33–186, 1977.

[9] Mazur B., Swinnerton-Dyer H.P.F., *The arithmetic of Weil curves*, Inv. math., 25:1–61, 1974.

[10] Merel L., *Opérateurs de Hecke et sous-groupes de* $\Gamma(2)$, Journal of Number theory, A paraître. (= Thèse, chapitre 5).

[11] Merel L., *Universal fourier expansions of modular forms*, In Gerhard Frey, editor, *On Artin's conjecture for 2-dimensional, odd Galois representations*, Lecture Notes in Mathematics, 1585, Springer Verlag.

[12] Merel L., *Opérateurs de Hecke pour* $\Gamma_0(N)$ *et fractions continues*, Ann. Inst. Fourier, 41(3), 1991. (= Thèse, chapitre 2).

[13] Merel L., *L'accouplement de Weil entre le sous-groupe cuspidal et le sous-groupe de Shimura de* $J_0(p)$, 1992. A paraître. (= Thèse, chapitre 4).

[14] Merel L., *Intersections sur des courbes modulaires*, Manuscripta mathematica, 80:283–289, 1993.(= Thèse, chapitre 1).

[15] Oesterlé J., *Nouvelles approches du "théorème" de Fermat, séminaire Bourbaki* n^0 *694*, In *Séminaire Bourbaki 1987-88*, volume 161-162, pages 165–186. Astérisque, 1988.

[16] Stevens G., *Arithmetic on modular curves*, Number 20 in Progress in mathematics. Birkhaüser, 1982.

[17] Zagier D., *Modular parametrization of elliptic curves*, Canad. Math. Bull., 28(3):377–389, 1985.

Conference on Elliptic Curves and Modular Forms
Hong Kong, December 18-21, 1993

Irreducible Galois representations arising from component groups of Jacobians

KENNETH A. RIBET
UC MATHEMATICS DEPARTMENT
BERKELEY, CA 94720-3840 USA

1. Introduction. Much has been written about component groups of Néron models of Jacobians of modular curves. In a variety of contexts, these groups have been shown to be "Eisenstein," which implies that they can be neglected in the study of irreducible two-dimensional representations of $\mathrm{Gal}(\overline{\mathbf{Q}}/\mathbf{Q})$.

The first theorem to this effect may be extracted from Mazur's landmark article [**11**], which concerns the Jacobian $J_0(N)$ when N is a prime. In this article, Mazur studies the action of Hecke operators on the "cuspidal subgroup" of the Jacobian, and obtains information about the relevant component group as an application [**11**], p. 98. More precisely, consider the fiber in characteristic N of the Néron model of $J_0(N)$. This commutative group scheme is an extension of a finite étale group scheme Φ by an algebraic torus; one is interested in the functorial action of Hecke operators on Φ. Mazur proves for all prime numbers $p \neq N$ that the pth Hecke operator T_p acts on Φ by multiplication by $1+p$, and also that T_N is the identity on Φ. The "Eisenstein" terminology arsies from the fact that $1+p$ is the pth coefficient of the standard Eisenstein series of weight two on $\Gamma_0(N)$.

In his first article on Serre's conjectures [**13**], the author generalized Mazur's result to the Jacobian $J_0(N)$ where N is the product of a positive integer M and a prime number q prime to M. According to Theorem 3.12 of [**13**], one again has the identity $T_p = 1+p$ for all prime numbers p prime to N on the group of connected components of the Néron reduction $J_0(N)_{/\mathbf{F}_q}$. (A slightly more refined statement appears as Theorem 3.22 of [**13**]; for the proof, see [**14**].) Subsequently, B. Edixhoven showed in [**6**] that an analogous result holds for every reduction $J_0(N)_{/\mathbf{F}_q}$ with $q > 3$; here, the new element is that N is allowed to be divisible by an arbitrary power of q.

In another article on Serre's conjectures [**15**], the author discusses the component group Φ attached to the mod q reduction of the Jacobian of the modular curve derived from $\Gamma_1(N) \cap \Gamma_0(q)$; here N is a positive integer and q is a prime number not dividing N. The analysis of [**15**], pp. 672–673 proves the identity $T_p = 1+p$ on Φ for all prime numbers $p \nmid qN$ and establishes at the same time that the "diamond bracket" operators $\langle d \rangle$ act as the identity on Φ.

This work was partially supported by NSF Grant #DMS 93-06898. The author wishes to thank B. Edixhoven, H. W. Lenstra, Jr., D. Lorenzini, N. Skoruppa, D. Ulmer, and A. Wilkinson for helpful correspondence and suggestions. We also thank the International Scientific Committee and the local organizers of the Hong Kong conference on Fermat's Last Theorem for their generous invitation.

In this article, we consider the situation in which $\Gamma_1(N)$ is replaced by a subgroup Γ intermediate between $\Gamma_0(N)$ and $\Gamma_1(N)$. In other words, we study the component group Φ_Γ associated to the mod q reduction coming from $\Gamma \cap \Gamma_0(q)$. (We continue to assume that q is prime to N.) Since Φ_Γ is Eisenstein in the two extreme situations $\Gamma = \Gamma_0(N)$ and $\Gamma = \Gamma_1(N)$, it is natural to ask whether Φ_Γ is Eisenstein for intermediate groups.

In order to rule out trivial counterexamples, we will generalize slightly our definition of "Eisenstein." Let S be the complex vector space of weight-two cusp forms on $\Gamma \cap \Gamma_0(q)$, and let $\widetilde{\mathbb{T}}$ be the subring of End S generated by the Hecke operators T_n and $\langle d\rangle$ for n a prime number prime to qN and d prime to N. As we shall recall below, there is a natural action of these operators on Φ_Γ. We shall be interested in the set of maximal ideals $\mathfrak{m}$ of $\widetilde{\mathbb{T}}$ which lie in the support of Φ_Γ as a $\widetilde{\mathbb{T}}$-module, i.e., the set of maximal ideals in the image of $\widetilde{\mathbb{T}}$ in End Φ_Γ.

To each $\mathfrak{m}$, one associates as usual a continuous semisimple representation

$$\rho_\mathfrak{m} : \mathrm{Gal}(\overline{\mathbf{Q}}/\mathbf{Q}) \to \mathbf{GL}(2, \widetilde{\mathbb{T}}/\mathfrak{m}).$$

This representation is characterized up to isomorphism by the fact that the trace of $\rho_\mathfrak{m}(\mathrm{Frob}_p)$ is T_p mod $\mathfrak{m}$ and the determinant of this matrix is $p\langle p\rangle$ mod $\mathfrak{m}$, for all but finitely many prime numbers p. (Here, Frob_p is an arithmetic Frobenius element for p in $\mathrm{Gal}(\overline{\mathbf{Q}}/\mathbf{Q})$. See, e.g., **[15]**, Section 7 for some of the relevant background.) It is worth stressing that Φ_Γ gives rise through this construction to a collection of two-dimensional representations of $\mathrm{Gal}(\overline{\mathbf{Q}}/\mathbf{Q})$, despite the fact there is no natural action of $\mathrm{Gal}(\overline{\mathbf{Q}}/\mathbf{Q})$ on the group Φ_Γ.

We shall say that Φ_Γ is *Eisenstein* if all $\mathfrak{m}$ in the support of Φ_Γ give rise to representations $\rho_\mathfrak{m}$ which are *reducible*. We say that Φ_Γ is *strongly Eisenstein* if the operators $\langle d\rangle$ are trivial on Φ_Γ and one has $T_p = 1 + p$ on Φ_Γ for all but finitely many prime numbers p. It is easy to show that "strongly Eisenstein" implies "Eisenstein" by using the Cebotarev Density Theorem and the Brauer-Nesbitt Theorem **[13]**, 5.2c.

To be sure, one's guess that Φ_Γ is Eisenstein turns out to be not very far off the mark. Indeed, Φ_Γ is strongly Eisenstein in most cases, and the prime-to-6 part of Φ_Γ is strongly Eisenstein in *all* cases. Nonetheless, the blanket assertion that Φ_Γ is Eisenstein is definitely false. To convince the reader of this fact, we exhibit in the next section some non-Eisenstein component groups; our construction is a digression which is intended to motivate the more systematic study which follows.

This article's main contribution is an analysis of Φ_Γ in the case of general Γ and q. As we indicated above, the ℓ-primary parts of Φ_Γ are strongly Eisenstein (and cyclic as abelian groups) for all prime numbers $\ell \geq 5$. After recalling this simple result, we investigate the 2-primary and the 3-primary components of Φ_Γ. In case Φ_Γ is non-Eisenstein, we identify the non-Eisenstein "pieces" and describe on these pieces the action of the Hecke operators T_n and $\langle d\rangle$.

From the point of view of Galois representations, our main result is the identification of all irreducible representations with values in $\mathbf{GL}(2, \overline{\mathbf{F}}_2)$ and $\mathbf{GL}(2, \overline{\mathbf{F}}_3)$ that can be associated with component groups. These are induced representations, coming from certain characters $\mathrm{Gal}(\overline{\mathbf{Q}}/\mathbf{Q}(\sqrt{-1})) \to \overline{\mathbf{F}}_2^*$ and $\mathrm{Gal}(\overline{\mathbf{Q}}/\mathbf{Q}(\sqrt{-3})) \to \overline{\mathbf{F}}_3^*$, respectively. For a character θ to intervene in some component group, it is necessary and sufficient that all residue classes modulo its conductor be represented by rational integers.

This article is an outgrowth of the author's talks at the Hong Kong conference on Fermat's Last Theorem. These talks outlined portions of A. Wiles's manuscript [**20**]. This manuscript provides ample motivation for a detailed analysis of mod 3 representations of $\mathrm{Gal}(\overline{\mathbf{Q}}/\mathbf{Q})$ which arise from modular forms. Indeed, Wiles's attack on the Taniyama-Shimura conjecture begins with a theorem of J. Tunnell [**19**] which implies that the mod 3 representation of $\mathrm{Gal}(\overline{\mathbf{Q}}/\mathbf{Q})$ arising from an elliptic curve over $\mathbf{Q}$ is associated to some Hecke eigenform. Assuming that this mod 3 representation is irreducible, one shows that the eigenform may be chosen in accordance with the conjectures of [**16**]. For this, attention must be directed to the special problems posed by mod 3 representations; see F. Diamond's article in this volume [**5**], which provides an update on Serre's conjectures. In reflecting on mod 3 representations, the author was forced to abandon his "axiom" that component groups are Eisenstein; this led to the study which is presented below.

2. Some non-Eisenstein component groups. Consider the irreducible continuous representation $\rho : \mathrm{Gal}(\overline{\mathbf{Q}}/\mathbf{Q}) \to \mathbf{GL}(2, \mathbf{F}_3)$ arising from the space of forms of weight two on $\Gamma_1(13)$. This space has dimension two, and the two Hecke eigenforms in the space are complex conjugates of each other. Their coefficients lie in the ring of integers of $\mathbf{Q}(\sqrt{-3})$; ρ is the mod $(\sqrt{-3})$ representation of $\mathrm{Gal}(\overline{\mathbf{Q}}/\mathbf{Q})$ associated with either of them. The determinant of ρ is the product of the mod 3 cyclotomic character of $\mathrm{Gal}(\overline{\mathbf{Q}}/\mathbf{Q})$ and the character of order two which corresponds to the quadratic extension $\mathbf{Q}(\sqrt{13})$ of $\mathbf{Q}$. Let H be the group of squares in $(\mathbf{Z}/13\mathbf{Z})^*$. Then conjecture (3.2.4?) of Serre's article [**16**] predicts that ρ should arise from a Hecke eigenform in the space of weight-two cusp forms on the group

$$\Gamma = \Gamma_H(13) := \left\{ \begin{pmatrix} a & b \\ c & d \end{pmatrix} \in \mathbf{SL}(2, \mathbf{Z}) \;\middle|\; c \equiv 0 \bmod 13,\ d \in H \right\}.$$

However, as Serre pointed out in a letter to the author [**17**], this space of forms is zero—the conjecture is false as stated.

One way to deal with this apparent difficulty is to reformulate Serre's conjecture as a relation between mod p Galois representations and mod p modular forms in the sense of Katz [**9**]. Such a reformulation is presented as Conjecture 4.2 of Edixhoven's article [**7**], which attributes the reformulation to Serre. The characters which appear in the conjecture are then naturally $\overline{\mathbf{F}}_p^*$-valued, and the difficulty becomes invisible. On the other hand, this solution hides a problem that may be genuinely of interest, since one wants to characterize those spaces of forms which give rise to a given Galois representation.

One certainly knows that problems of the sort exemplified by ρ occur only for mod 2 and mod 3 representations of $\mathrm{Gal}(\overline{\mathbf{Q}}/\mathbf{Q})$. Further, among all mod 3 representations, only those which become abelian on $\mathrm{Gal}(\overline{\mathbf{Q}}/\mathbf{Q}(\sqrt{-3}))$ can give trouble, cf. [**2**], p. 796. Finally, let $q > 2$ be a prime number which is congruent modulo 3. Then the difficulty in the case of ρ "disappears" when the level is augmented by an auxiliary $\Gamma_0(q)$-structure: Lemme 1 of [**2**] implies that ρ arises from the space of weight-two cusp forms on $\Gamma \cap \Gamma_0(q)$. (The author is grateful to N. Skoruppa for undertaking a numerical verification of this assertion when $q = 5$.)

Let J be the Jacobian of the modular curve over $\mathbf{Q}$ associated with $\Gamma \cap \Gamma_0(q)$. (We again write J for the Néron model of this abelian variety.) As above, we let Φ_Γ be the component group associated with the reduction of J mod q.

Theorem 1. *The group Φ_Γ is non-Eisenstein.*

Proof. Let V be a two-dimensional $\mathbf{F}_3$-vector space affording the representation ρ, and for each $r \not| 13q$ let $a_r \in \mathbf{F}_3$ be the trace of $\rho(\mathrm{Frob}_r)$, where Frob_r is a Frobenius element for r in $\mathrm{Gal}(\overline{\mathbf{Q}}/\mathbf{Q})$. The fact that ρ arises from $\Gamma \cap \Gamma_0(q)$ implies that there is a $\mathrm{Gal}(\overline{\mathbf{Q}}/\mathbf{Q})$-equivariant embedding $V \hookrightarrow J(\overline{\mathbf{Q}})$ with the following property: For each prime number $r \not| 13q$, the endomorphism T_r of J acts on V as the homothety a_r. Since q is prime to 39, V is unramified at q. Thus (after choosing a place of $\overline{\mathbf{Q}}$ lying over q), we may view V as embedded in $J_{/\mathbf{F}_q}(\overline{\mathbf{F}}_q)$.

We claim now that V does not land entirely in the connected component $J^o_{/\mathbf{F}_q}$ of $J_{/\mathbf{F}_q}$. In other words, we assert that the image W of V in the group of connected components of $J_{/\mathbf{F}_q}$ is non-trivial. Since the group of connected components in question is none other than Φ_Γ, our claim implies that Φ_Γ contains a non-zero subgroup of exponent 3 on which each Hecke operator T_r acts as a_r. That the a_r are traces of an *irreducible* representation indicates that this subgroup is non-Eisenstein, thereby proving the Theorem.

The claim concerning V is proved by an appeal to results of Deligne-Rapoport [**4**] and general theorems of Grothendieck and Raynaud. This body of work is summarized in [**12**], [**13**], Section 2, [**15**], Section 8, and the discussion which occurs in Section 3 below. The main point is that by using [**4**], Th. 6.9, p. 286, one sees that $J^o_{/\mathbf{F}_q}$ is a certain extension of an abelian variety by a torus. The abelian variety in question appears as the product of two copies of the Jacobian of the modular curve associated with the group Γ. However, this modular curve has genus zero, so its Jacobian is zero. Thus $J^o_{/\mathbf{F}_q}$ coincides with its "toric part" T. Therefore, the statement that V falls entirely in $J^o_{/\mathbf{F}_q}(\overline{\mathbf{F}}_q)$ is the statement that V is a subgroup of $T(\overline{\mathbf{F}}_q)$. The argument given at the conclusion of [**15**], Section 8 shows that the inclusion $V \hookrightarrow T(\overline{\mathbf{F}}_q)$ is possible only when $q \equiv 1$ mod 3; the choice of q thus guarantees that W is non-zero. This proves the claim. □

3. A concrete description of Φ_Γ. In this section, we present a nuts-and-bolts description of Φ_Γ as an abelian group furnished with a family of Hecke operators T_r and $\langle d \rangle$. This material is now quite well known, at least in the case where $\Gamma = \Gamma_0(N)$ [**1, 12**].

Let N be a positive integer. Subgroups Γ between $\Gamma_1(N)$ and $\Gamma_0(N)$ are in 1-1 correspondence with subgroups of $(\mathbf{Z}/N\mathbf{Z})^*$. Let H be such a subgroup, and set

$$\Gamma = \Gamma_H(N) = \left\{ \begin{pmatrix} a & b \\ c & d \end{pmatrix} \in \mathbf{SL}(2, \mathbf{Z}) \;\middle|\; c \equiv 0 \bmod N,\ d \in H \right\}.$$

Thus $\Gamma = \Gamma_1(N)$ in case H is the trivial subgroup of $(\mathbf{Z}/N\mathbf{Z})^*$, while $\Gamma = \Gamma_0(N)$ if $H = (\mathbf{Z}/N\mathbf{Z})^*$. The modular curve corresponding to Γ will be called $X_H(N)$; thus, $X_H(N)$ is the quotient of $X_1(N)$ by the image of H in the Galois group $(\mathbf{Z}/N\mathbf{Z})^*/\{\pm 1\}$ of the covering $X_1(N) \to X_0(N)$. Postponing the assumption that q is prime, we let q be a positive integer prime to N, and put

$$\Gamma(H,q) = \Gamma_H(N) \cap \Gamma_0(q).$$

(A more precise, but less compact, name for this group would have been $\Gamma_H(N,q)$.)

The modular curve $X(H,q)$ corresponding to $\Gamma(H,q)$ is the quotient of $X_1(Nq)$ by the image of $H \times (\mathbf{Z}/q\mathbf{Z})^*$ in $(\mathbf{Z}/Nq\mathbf{Z})^*/\{\pm 1\}$. This image is unchanged if we augment H by $\{\pm 1\}$; thus we can, and will, assume that H contains -1. (If $N \le 2$, we have $-1 = +1$ in $(\mathbf{Z}/N\mathbf{Z})^*$, and we are imposing no condition on H.) We view $X(H,q)$ as classifying elliptic curves E which are furnished with a subgroup C of order q and a point P of order N. In the classification, the point P is considered "mod H" in the sense that the triples (E,C,P) and (E,C,hP) are identified for all $h \in H$. (Compare [**4**], IV, Section 3.) The orbit of P mod H will be called α_P; we will say that $X(H,q)$ classifies triples (E,C,α) where α is a point of order N on E which is taken "mod H."

Define $J_H(N)$ and $J(H,q)$ to be the Jacobians of the modular curves $X_H(N)$ and $X(H,q)$, respectively.

From now on, we assume that q is a *prime* number. For convenience, we impose the assumption $q \ge 5$ at this point. (The cases $q = 2$ and $q = 3$ presumably could be included with little modification in what follows.) Consider the reduction of $J(H,q)$ modulo q. The group of components associated with this reduction is the group Φ_Γ in the discussion above; of course, this group depends on q as well as on Γ.

A description of Φ_Γ can be deduced in a standard way from the theorem of Deligne and Rapoport which we cited above [**4**], Th. 6.9, p. 286. The theorem provides a model $\mathcal{C}$ of $X(H,q)$ over $\mathbf{Z}_q$ which is an "admissible curve" in the sense of [**8**], Section 3 and [**13**], Section 2. The special fiber of $\mathcal{C}$ has two irreducible components, each isomorphic to the modular curve $X_H(N)$.

The set of singular points of $\mathcal{C}_{\overline{\mathbf{F}}_q}$ is in bijection with the set S of supersingular $\overline{\mathbf{F}}_q$-valued points of $X_H(N)$, i.e., those points which arise from pairs (E,P) where E is a supersingular elliptic curve over $\overline{\mathbf{F}}_q$ and P is a point of order N on E. To each $s \in S$ we associate an integer $e(s) \in \{1,2,3\}$ in the following way: If s is represented by the pair (E,P), then $2e(s)$ is the number of automorphisms of E which map P to some point hP with $h \in H$; in other words, $2e(s)$ is the number of automorphisms of (E,α_P). (Because -1 belongs to H, the automorphism "-1" of E induces an automorphism of (E,α_P). The number of such automorphisms is then 2, 4, or 6 because we have assumed $q \ge 5$.) The singular point of $\mathcal{C}$ corresponding to s is then described in $\mathcal{C}$ by the local equation $XY = q^{e(s)}$.

Consider now the diagonal pairing on $\mathbf{Z}^S$ for which $\langle s,s\rangle$ is the positive integer $e(s)$. Let L be the group of degree 0 elements in $\mathbf{Z}^S$, and let $\iota : L \hookrightarrow \operatorname{Hom}(L,\mathbf{Z})$ be the linear map which describes the restriction of this pairing to L. The following result is a consequence of the theorem of Deligne and Rapoport and work of Grothendieck and Raynaud. (See [**12**], and perhaps the discussion in [13], p. 438.)

Theorem 2. *The component group Φ_Γ may be identified with the cokernel of ι.*

One understands that the identification between Φ_Γ and $\operatorname{coker}\iota$ is compatible with the functorial actions of Hecke operators and of the Galois group $\operatorname{Gal}(\overline{\mathbf{F}}_q/\mathbf{F}_q)$. We shall discuss the Hecke operators.

As usual, Hecke operators act on $J(H,q)$ in two different ways, because the Jacobian construction is both covariant and contravariant. In what follows we adopt the contravariant point of view in which $J(H,q)$ is regarded as $\operatorname{Pic}^0(X(H,q))$. We describe the action of Hecke operators on Φ_Γ without supplying any substantial justification. However, the reader may consult [12], which discusses in detail the special case case where $H = (\mathbf{Z}/N\mathbf{Z})^*$.

First, let d be an integer prime to N. The association $(E,C,P) \mapsto (E,C,dP)$ defines a "diamond bracket" automorphism $\langle d\rangle$ on $X(H,q)$. Using the contravariant functoriality, we obtain automorphisms of $J(H,q)$ and then Φ_Γ, both of which we shall call simply $\langle d\rangle$. In the description of Φ_Γ as the cokernel of an injection $L \hookrightarrow \operatorname{Hom}(L,\mathbf{Z})$, the map $\langle d\rangle$ on Φ_Γ is induced from two automorphisms of L. Namely, consider the permutation of S which takes the class of a pair (E,P) to the class of (E,dP). This permutation induces an automorphism $\langle d\rangle_L$ of L, which we regard as a subgroup of $\mathbf{Z}^S$. Let $\langle d\rangle'_L$ be the *inverse* of this permutation. Then we have a commutative diagram

$$\begin{array}{ccccccccc} 0 & \to & L & \stackrel{\iota}{\hookrightarrow} & \operatorname{Hom}(L,\mathbf{Z}) & \to & \Phi_\Gamma & \to & 0 \\ & & \downarrow & & \downarrow & & \downarrow & & \\ 0 & \to & L & \stackrel{\iota}{\hookrightarrow} & \operatorname{Hom}(L,\mathbf{Z}) & \to & \Phi_\Gamma & \to & 0 \end{array}$$

in which the three vertical arrows are respectively $\langle d\rangle'_L, \operatorname{Hom}(\langle d\rangle_L,\mathbf{Z})$ and $\langle d\rangle$.

Similarly, suppose that p is a prime number which does not divide qN. The symbol T_p then denotes a host of objects: the standard Hecke correspondence T_p on $X(H,q)$, the endomorphism of $J(H,q)$ which this correspondence induces by contravariant functoriality, and finally the endomorphism T_p of Φ_Γ which then arises from the functoriality of the Néron model. We seek to identify this latter endomorphism.

Fix p for the moment and let T be the correspondence on S defined by the rule $(E,\alpha) \mapsto \sum(E/D, \alpha \bmod D)$, where the sum is taken over the subgroups D of E having order p. In analogy with the situation above, we write T_L for the induced endomorphism of $L \subset \mathbf{Z}^S$. The analogue of $\langle d\rangle'_L$ is then the endomorphism $T'_L := \langle p\rangle_L^{-1} \circ T_L$. We have a commutative diagram like the one above in which the three vertical arrows are respectively T'_L, $\operatorname{Hom}(T_L,\mathbf{Z})$ and the endomorphism T_p of Φ_Γ. Since p may vary in what follows, we restore p to the notation, referring to the first of these operators as T'_p and the second as $\operatorname{Hom}(T_p,\mathbf{Z})$. In other words, we will permit T_p to denote the endomorphism of L induced from the correspondence $(E,\alpha) \mapsto \sum(E/D, \alpha \bmod D)$ on S by $\mathbf{Z}$-linearity.

4. Extra automorphisms.

We will now determine those points $s \in S$ with $e(s) > 1$. A point s in S is defined by a supersingular elliptic curve E

over $\overline{\mathbf{F}}_q$, together with a point P on E of order N. Write α for the orbit of P under H and C for the subgroup of E generated by P. Let $\overline{s}$ be the point on $X_0(N)$ defined by (E, C).

Let R be the subring $\mathbf{Z}[\operatorname{Aut} E]$ of $\operatorname{End} E$. If $e(s)$ is different from 1, then $\operatorname{Aut} E$ is different from $\{\pm 1\}$. Since $\operatorname{End}(E)$ is a definite quaternion algebra over $\mathbf{Q}$, $\operatorname{Aut} E$ is different from $\{\pm 1\}$ if and only if E has an automorphism of order 4 or 6. In fact, the condition $q > 3$ implies that the automorphism group of E can only be cyclic of order 2, order 4, or order 6. Hence R is either $\mathbf{Z}$, or else the integer ring in an imaginary quadratic field of discriminant -3 or -4. As is well known, E has an automorphism of order 4 if and only if its j-invariant is 1728. Since E is supersingular, the prime q must be congruent to 3 mod 4. Similarly, E has an automorphism of order 3 if and only if its j-invariant is zero; if the curve with this j-invariant is supersingular, we have $q \equiv 2 \bmod 3$. (Compare, for example, **[18]**, page 103, pp. 143–144.)

Suppose that $\operatorname{Aut} E \neq \{\pm 1\}$. Then (E, C) has a non-trivial automorphism if and only if $\operatorname{Aut} E$ coincides with its subgroup $\operatorname{Aut}(E, C)$, i.e., if and only if C is stable under R. If this condition is satisfied, then the group C is free of rank 1 over R/I, where I is the annihilator of C in R. Since C is cyclic, this puts a numerical constraint on N: No prime factor of N can remain inert in the ring R, and the prime which ramifies in R can occur in N only to the first power if it occurs at all. Further, the inclusion of $\mathbf{Z}$ in R induces an identification $R/I \xrightarrow{\sim} \mathbf{Z}/N\mathbf{Z}$. Thus the pair (E, C) has a non-trivial automorphism if and only if C is the kernel on E of an ideal I of R such that R/I is isomorphic to $\mathbf{Z}/N\mathbf{Z}$. Thus there is a 1-1 correspondence between cyclic subgroups C of order N which are stable by R and ideals I of R such that the additive group of R/I is cyclic of order N. One sees that distinct subgroups C lead to distinct point of $X_0(N)$. Indeed, if (E, C) and (E, C') are isomorphic, then the isomorphism between them is induced by an automorphism of E, which preserves C (and C') by the definition of R.

The conjugation map $r \mapsto \overline{r}$ of R induces an an involution on the set of I. This Atkin-Lehner style involution has no fixed points if $N > 3$. Indeed, suppose that $I = \overline{I}$. For each $r \in R$, there is an integer n such that $r - n$ lies in I. Since $\overline{n} = n$, $r - \overline{r}$ belongs to $I + \overline{I} = I$. Hence I contains $\sqrt{-3}$ or 2, according as the discriminant of R is -3 or -4. Hence N divides 3 in the former case and N divides 2 in the latter.

Suppose now that (E, C) has a non-trivial automorphism. Then $e(s) > 1$ if and only if $\operatorname{Aut}(E, \alpha) = \operatorname{Aut}(E, C)$. This equality translates into the statement that

$$(*) \qquad \mu \overset{?}{\subseteq} H \subseteq (\mathbf{Z}/N\mathbf{Z})^*,$$

where μ is the image in $(R/I)^*$ of the unit group of R. This condition involves only (E, C) and H; hence it applies simultaneously to all points on $X_H(N)$ which lie above the point $\overline{s}$ on $X_0(N)$. One checks that when $(*)$ is satisfied, the number of points on $X_H(N)$ which lie above $\overline{s}$ is precisely $((\mathbf{Z}/N\mathbf{Z})^* : H)$.

In summary, to find those points $s \in S$ with $e(s) > 1$, we first look for the supersingular points $\overline{s}$ in $X_0(N)(\overline{\mathbf{F}}_q)$ which satisfy the analogous condition $e(\overline{s}) > 1$. Mark off those points (if any) which satisfy the supplementary condition $(*)$. Above each marked point, we find $((\mathbf{Z}/N\mathbf{Z})^* : H)$ points in S with "extra automorphisms."

The condition $(*)$ may be true for some points $\overline{s}$ with $e(\overline{s}) > 1$ that arise from a given supersingular elliptic curve and false for others. For example, suppose that E has six automorphisms and that $N = 7 \cdot 13 = 91$. There are four ideals I of R with $R/I \approx \mathbf{Z}/91\mathbf{Z}$. For two of these ideals, the image of R^* in $(\mathbf{Z}/91\mathbf{Z})^*$ is the cyclic group generated by 10; for the other two, the image is generated by 17. If H is one of these two groups, then $(*)$ will be satisfied for two of the I, but not for all. Notice, however, that the set of I for which $(*)$ is satisfied is stable under the natural conjugation map on R. Indeed, R^* is stable under this conjugation. Furthermore, we have observed that the inequality $N > 3$ implies that an ideal I with $R/I \approx \mathbf{Z}/N\mathbf{Z}$ can never be its own conjugate. This gives

Lemma 1. *Suppose that $N \geq 4$. Then the set of points $\overline{s}$ for which $e(\overline{s}) > 1$ and for which $(*)$ is satisfied has an even number of elements. In particular, the number of such points is different from 1.*

The case $N \leq 3$ does not play an important role in the analysis which follows. In fact, one has

Lemma 2. *If $N \leq 12$ and there is at least one point s with $e(s) > 1$, then $H = (\mathbf{Z}/N\mathbf{Z})^*$.*

Proof. Assume that $N < 13$. If $e(s) = 3$, then there is an ideal I of $R = \mathbf{Z}[\frac{1+\sqrt{-3}}{2}]$ with $R/I \approx \mathbf{Z}/N\mathbf{Z}$. This implies that N is 3 or 7. In both cases, R^* maps onto $(\mathbf{Z}/N\mathbf{Z})^*$. If $e(s) = 2$, there is an ideal I of $R = \mathbf{Z}[\sqrt{-1}]$ with $R/I \approx \mathbf{Z}/N\mathbf{Z}$. The possibilities for N are 2, 5, and 10. Again, R^* maps onto $(\mathbf{Z}/N\mathbf{Z})^*$ in all cases. □

The prime 13 splits both in $\mathbf{Z}[\frac{1+\sqrt{-3}}{2}]$ and in $\mathbf{Z}[\sqrt{-1}]$. If R is one of these two rings, and if I is a prime lying over 13 in R, then the image of R^* in $(\mathbf{Z}/13\mathbf{Z})^*$ is strictly smaller than $(\mathbf{Z}/13\mathbf{Z})^*$. This circumstance (if not the discussion in Section 2) shows that 12 cannot be replaced by a larger integer in the statement of the Lemma.

5. A canonical subgroup of Φ_Γ. Let $\Phi = \Phi_\Gamma$. We shall discuss a cyclic subgroup Φ_0 of Φ which has already been studied in case $\Gamma = \Gamma_0(N)$ (see **[13]** and **[12]**).

For each element s of S, let $\varphi_s : L \to \mathbf{Z}$ be the linear form $\ell \mapsto \langle \ell, s \rangle$ and let ω_s be the linear form on L

$$\sum_t n_t t \mapsto n_s.$$

Clearly, $\operatorname{Hom}(L, \mathbf{Z})$ is generated by the ω_s, so that Φ is generated by their images $\overline{\omega}_s$ in $\Phi = \operatorname{Hom}(L, \mathbf{Z})/L$. One has $e(s)\omega(s) = \varphi_s$.

The class $\overline{\varphi}$ of φ_s in $\Phi = \operatorname{Hom}(L, \mathbf{Z})/L$ is independent of s, since $\varphi_s - \varphi_t$ is the image in $\operatorname{Hom}(L, \mathbf{Z})$ of $s - t \in L$, if $s, t \in S$. Let Φ_0 be the subgroup of Φ generated by the canonically given element $\overline{\varphi}$. We set

$$Q = \Phi/\Phi_0.$$

In view of the formula $e(s)\omega(s) = \varphi_s$, we have $e(s)\overline{\omega}_s = \overline{\varphi}$ for each s. Hence, as noted in [**13**], Section 2, we have a surjection

$$\bigoplus_{s\in S} (\mathbf{Z}/e(s)\mathbf{Z}) \longrightarrow Q, \qquad (a_s) \mapsto \sum_s a_s\overline{\omega}_s.$$

One may view $\mathbf{Z}^S$ as embedded in $\mathrm{Hom}(\mathbf{Z}^S, \mathbf{Z})$ via the symmetric bilinear pairing $\langle\,,\,\rangle$ on $\mathbf{Z}^S$. Let $L^\perp \subset \mathrm{Hom}(\mathbf{Z}^S, \mathbf{Z})$ be the group of linear forms which vanish on L, and let $U = L^\perp \cap \mathbf{Z}^S$ be group of vectors in $\mathbf{Z}^S$ which are orthogonal to L under the pairing. Since $\mathrm{Hom}(L, \mathbf{Z}) = \mathrm{Hom}(\mathbf{Z}^S, \mathbf{Z})/L^\perp$, we have $\Phi = \mathrm{Hom}(\mathbf{Z}^S, \mathbf{Z})/(L \oplus L^\perp)$. The image of $\mathbf{Z}^S/U$ in Φ, i.e., the group $\mathbf{Z}^S/(L\oplus U)$, is clearly Φ_0. (Indeed, the image of $s \in \mathbf{Z}^S$ in $\mathrm{Hom}(\mathbf{Z}^S, \mathbf{Z})$ is φ_s.)

Now U is free of rank one over $\mathbf{Z}$; it consists of those multiples of $\sum \frac{1}{e(s)} s \in \mathbf{Q}^S$ which lie in $\mathbf{Z}^S$. Hence U is generated by $\sum \frac{m}{e(s)} s$, where m is the least common multiple of the $e(s)$. (Thus, m divides 6.) It follows that the order of $\Phi_0 = \mathbf{Z}^S/(L\oplus U)$ is $m \cdot \sum e(s)^{-1}$, cf. [**12**], p. 16. Also, $L^\perp/U$ has order m, since $L^\perp$ is generated by the element $\sum \omega_s$ and since $m \sum \omega_s = \sum \frac{m}{e(s)} s$ in $\mathrm{Hom}(\mathbf{Z}^S, \mathbf{Z})$.

Consider the commutative diagram with exact rows

$$\begin{array}{ccccccc} 0 \to & L^\perp & \to & \mathrm{Hom}(\mathbf{Z}^S, \mathbf{Z}) & \to & \mathrm{Hom}(\mathbf{Z}^S, \mathbf{Z})/L^\perp & \to 0 \\ & \uparrow & & \uparrow & & \uparrow & \\ 0 \to & U & \to & \mathbf{Z}^S & \to & \mathbf{Z}^S/U & \to 0 \end{array}$$

in which the central vertical map is the inclusion given by $\langle\,,\,\rangle$; the two other vertical maps are the evident, related inclusions. The cokernels of the vertical maps form the exact sequence $0 \to \mathbf{Z}/m\mathbf{Z} \to \bigoplus_{s\in S}(\mathbf{Z}/e(s)\mathbf{Z}) \to Q \to 0$, where the map from $m\mathbf{Z}$ to the direct sum takes $1 \in \mathbf{Z}/m\mathbf{Z}$ to the vector $(1, \ldots, 1)$. Note that Q is now seen to have order $m^{-1}\prod e(s)$. Therefore, we recover the formula $\left(\prod_s e(s)\right) \cdot \left(\sum_s e(s)^{-1}\right)$ for the order of Φ, cf. [**1**], Ch. 9, Prop. 10.

Lemma 3. *The group Φ_0 is strongly Eisenstein. The identities $\langle d\rangle = 1$ and $T_p = 1 + p$ hold on Φ_0 for $d \in (\mathbf{Z}/N\mathbf{Z})^*$ and p prime to qN.*

Proof. We first consider the action of T_p on $\overline{\varphi}$. Choose $s \in S$. From the perspective introduced at the end of Section 3, we see that the action of T_p on this element of Φ is derived from the action of $\mathrm{Hom}(T_L, \mathbf{Z})$ on $\varphi_s \in \mathrm{Hom}(L, \mathbf{Z})$. The map $\varphi_s \circ T_L$ is the linear form mapping $\ell \in L$ to $\langle T_L\ell, s\rangle = \langle \ell, T_L' s\rangle = \langle \ell, \langle p\rangle^{-1} T_L s\rangle$. Since $\langle p\rangle^{-1}T_L s$ is an element of $\mathbf{Z}^S$ of degree $p+1$, the image in Φ of $\ell \mapsto \langle \ell, \langle p\rangle^{-1}T_L s\rangle$ is $(p+1)\overline{\varphi}$. This establishes the first identity. The second is proved by an analogous computation. □

Corollary. *The group Φ is Eisenstein in each of the following situations:*

(i) $e(s) = 1$ *for all* $s \in S$;

(ii) $H = (\mathbf{Z}/N\mathbf{Z})^*$;

(iii) $N < 13$;

(iv) $H = \{\pm 1\}$.

Proof. In the first case, we have $\Phi_0 = \Phi$, so that the assertion to be proved is that given by the lemma. The assertion in the second case is given as Theorem 3.12 in **[13]**. In the third case, we are either in case (i) or case (ii) by Lemma 2. Suppose now that we are in the fourth case, which corresponds to the equality $X_H(N) = X_1(N)$. If N is at least 4, we are in case (i) by **[10]**, Corollary 2.7.4. If $N \leq 4$, then we are of course in case (iii). □

6. The structure of Q. In view of the Corollary above, we will now impose the condition $N \geq 13$.

Lemma 4. *Suppose that (E, C) represents a supersingular point of $X_0(N)_{/\overline{\mathbf{F}}_q}$. Let D be a cyclic subgroup of $E(\overline{\mathbf{F}}_q)$ of order prime to qN. Assume that the automorphism group of the triple (E, C, D) is larger than $\{\pm 1\}$. Then the pair $(E/D, C \bmod D)$ is isomorphic to (E, C).*

Proof. This statement is proved as Proposition 2 in **[14]**; we recall the proof for the convenience of the reader. Let ϵ be an automorphism of (E, C, D) which is different from ± 1, and let R be the subring of $\mathrm{End}(E, C, D)$ which is generated by ϵ. Thus R is isomorphic either to the ring of Gaussian integers or to the ring of integers of $\mathbf{Q}(\sqrt{-3})$. If $J = \mathrm{Ann}_R(D)$, then J is a principal ideal (r) of R, and D is the kernel of J on E. Thus the map "multiplication by r" on E induces an isomorphism $(E/D, (C \oplus D)/D) \stackrel{\sim}{\to} (E, C)$ as required. □

Recall now that Q is generated as an abelian group by the elements $\overline{\omega}_s$ with $s \in S$. Since $e(s)\overline{\omega}_s = 0$, it suffices to consider only those $\overline{\omega}_s$ with $e(s) > 1$. As we have seen, the inequality $e(s) > 1$ means that s lies over a supersingular point $\overline{s}$ of $X_0(N)(\overline{\mathbf{F}}_q)$ which satisfies the numerical condition $e(\overline{s}) > 1$ and the supplementary condition $(*)$.

Let $\overline{s}$ be such a point. Define $Q_{\overline{s}}$ to be the subgroup of Q generated by the $\overline{\omega}_s$ with s lying over $\overline{s}$. Thus $Q_{\overline{s}}$ is a quotient of the elementary abelian group $\bigoplus(\mathbf{Z}/e(\overline{s})\mathbf{Z})$, where the sum is extended over the set of points s lying over $\overline{s}$. As was mentioned above, the number of such points is the index $((\mathbf{Z}/N\mathbf{Z})^* : H)$. More precisely, the set of points s is a principal homogeneous space over $(\mathbf{Z}/N\mathbf{Z})^*/H$, the action of this group being given by the "diamond bracket" operators.

Proposition 1. *The group $Q_{\overline{s}}$ is free of rank $((\mathbf{Z}/N\mathbf{Z})^* : H)$ over $\mathbf{Z}/e(\overline{s})\mathbf{Z}$.*

Proof. Let S_0 be the inverse image of $\overline{s}$ in S. We must show that the direct sum $\bigoplus_{s \in S_0}(\mathbf{Z}/e(s)\mathbf{Z})$ has trivial intersection with the kernel of the map

$$\bigoplus_{s \in S}(\mathbf{Z}/e(s)\mathbf{Z}) \to Q.$$

As we have seen, this kernel is the cyclic group generated by the vector $(1, \ldots, 1)$ in the full direct sum. Hence it suffices to check that there is a point $s \in S \setminus S_0$ with $e(s) > 1$. This follows from Lemma 1, since $N \geq 13$. □

Proposition 2. *The subgroup $Q_{\overline{s}}$ of Q is stable under the Hecke operators T_p and $\langle d \rangle$.*

Proof. Let p be a prime number not dividing qN. As at the end of Section 3, we write T for the correspondence on S defined by the formula $(E, \alpha) \mapsto \sum(E/D, \alpha \bmod D)$. This correspondence is summarized by the matrix of natural numbers (a_{tu}) which one constructs by writing $Tu = \sum_{t \in S} a_{tu} \cdot t$ for $u \in S$. Recall that the action of T_p on Φ arises from the action of a map labeled $\mathrm{Hom}(T_L, \mathbf{Z})$ on $\mathrm{Hom}(L, \mathbf{Z})$. This group is generated by the elements ω_s for $s \in S$, and one finds that $\mathrm{Hom}(T_L, \mathbf{Z})$ maps ω_s to the sum $\sum_{t \in S} a_{st} \omega_t$.

To prove that $Q_{\overline{s}}$ is stable under T_p, it suffices to show that a_{st} is divisible by $e(t)$ whenever s maps to $\overline{s}$ and t does *not*. In other words, we wish to show that a_{tu} is divisible by $e(u)$ whenever t and u have distinct images on $X_0(N)$.

This divisibility can be established by the method used to prove [**13**], Th. 3.12. Indeed, suppose that u is represented by (E, α). Let C be the cyclic subgroup of order N on E which is associated with α. The group $\mathrm{Aut}(E, \alpha)/\{\pm 1\}$ operates on the set of cyclic subgroups of E of order p. Clearly, if D and D' are such subgroups which are equivalent under this action, then $(E/D, \alpha \bmod D)$ and $(E/D', \alpha \bmod D')$ are isomorphic. Moreover, suppose that the cyclic subgroup D has a non-trivial stabilizer under this action. Then Lemma 4 shows that (E, C) and $(E/D, (C \oplus D)/D)$ are isomorphic, i.e., that (E, α) and $(E/D, \alpha \bmod D)$ map down to the same point on $X_0(N)$. Now let t be a point of S whose image on $X_0(N)$ is distinct from that of s. Then the set of D for which $(E/D, \alpha \bmod D)$ represents t is a union of copies of the group $\mathrm{Aut}(E, \alpha)/\{\pm 1\}$, whose cardinality is $e(u)$. Hence a_{tu} is divisible by $e(u)$, as was claimed.

Now let d be an integer prime to N. The action of $\langle d \rangle$ on Φ is deduced from the automorphism $\mathrm{Hom}(\langle d \rangle_L, \mathbf{Z})$ introduced at the end of Section 3. This automorphism maps a given linear form ω_s to the linear form $\omega_{\langle d \rangle^{-1} s}$. Here, the automorphism labeled $\langle d \rangle^{-1}$ sends the class of (E, P) to the class of $(E, d^{-1}P)$; the quantity d^{-1} is computed mod N. Since P and $d^{-1}P$ generated the same subgroup of E, it is clear that $\langle d \rangle$ permutes the $\overline{\omega}_s$ with s having a given image on $X_0(N)$. Therefore, $\langle d \rangle$ preserves $Q_{\overline{s}}$. □

Let S_0 again be the inverse image of $\overline{s}$ in S. We observed above that the diamond bracket operation of $(\mathbf{Z}/N\mathbf{Z})^*$ on S_0 makes S_0 a principal homogeneous space over the group $\Delta := (\mathbf{Z}/N\mathbf{Z})^*/H$. Let us identify $Q_{\overline{s}}$ with the direct sum $\bigoplus_{s \in S_0} (\mathbf{Z}/e(\overline{s})\mathbf{Z})$, i.e., with the space of functions from S_0 to $\mathbf{Z}/e(\overline{s})\mathbf{Z}$. Then the diamond bracket operation discussed in Proposition 2, which involves an inverse, may be viewed as the natural action coming from the diamond bracket action of Δ on S_0.

Let $\ell = e(\overline{s})$; thus $\ell = 2$ or $\ell = 3$. Let $\mathbb{T} = \mathbf{F}_\ell[\Delta]$ be the group algebra consisting of sums $\sum_{\delta \in \Delta} n_\delta[\delta]$ where the n_δ are integers mod ℓ. Since $Q_{\overline{s}}$ is an $\mathbf{F}_\ell$-vector space with an action of Δ, $Q_{\overline{s}}$ may be viewed as a $\mathbb{T}$-module in a natural way.

Lemma 5. *The $\mathbb{T}$-module $Q_{\overline{s}}$ is free of rank one.*

Proof. Proposition 1 shows that $Q_{\overline{s}}$ is an $\mathbf{F}_\ell$ vector space of dimension $\#(\Delta) = \dim_{\mathbf{F}_\ell} \mathbb{T}$. Further, it is clear that $Q_{\overline{s}}$ is a cyclic $\mathbb{T}$-module, since Δ permutes the generators ω_s of $Q_{\bar{s}}$. □

Now choose a supersingular elliptic curve E over $\overline{\mathbf{F}}_q$ and a cyclic subgroup C on E so that the pair (E,C) defines the point $\bar{s}$ of $X_0(N)$. As usual, let R be the ring $\mathbf{Z}[\operatorname{Aut} E]$, and let I be the annihilator of C in R. The isomorphism $R/I \approx (\mathbf{Z}/N\mathbf{Z})^*$ gives a meaning to $[r]$ whenever $r \in R$ is prime to I. The element $[r]$ depends only on the ideal generated by r in R, since H contains the image of R^* in $(\mathbf{Z}/N\mathbf{Z})^*$.

In fact, this might be a good time to relate Δ to the Galois group $\operatorname{Gal}(\overline{\mathbf{Q}}/\mathbf{Q})$. Let K be the imaginary quadratic field $\mathbf{Q}[\operatorname{Aut} E] = R \otimes \mathbf{Q}$. Since all fractional ideals of K are principal, the quotient of $(R/I)^*$ by R^* is the ray class group of K with conductor I. Via class field theory, this class group corresponds with an abelian extension L of K. It will be useful to fix an embedding $K \hookrightarrow \overline{\mathbf{Q}}$ and to view L as a subfield of $\overline{\mathbf{Q}}$. Since Δ is then a quotient of $\operatorname{Gal}(L/K)$, Δ becomes the Galois group of an abelian extension of K in $\overline{\mathbf{Q}}$. The conductor of this extension divides I.

For each prime $p \nmid qN$, we define an element τ_p of $\mathbb{T}$ by the formula:

$$\tau_p = \begin{cases} 0 & \text{if } p \text{ is inert in } R, \\ [\pi] + [\pi'] & \text{if } p = (\pi)(\pi') \text{ is split in } R, \\ [\pi] & \text{if } p = (\pi)^2 \text{ is ramified in } R. \end{cases}$$

The third case occurs only when $p = \ell$.

Theorem 3. *For each $p \nmid qN$, the Hecke operator T_p acts on $Q_{\bar{s}}$ by multiplication by τ_p.*

Proof. Let (a_{st}) be the Brandt matrix which was introduced in the course of the proof of Proposition 2. Recall that if t is the isomorphism class of the pair (E,P), where P is a point of order N lying in C, then a_{st} is the number of subgroups D of order p in E such that $(E/D, P \bmod D)$ defines s. We have seen that $a_{st} \equiv 0 \bmod \ell$ if s and t do not have the same image on $X_0(N)$. We must now calculate $a_{st} \bmod \ell$ in the case where s and t have the same image, namely $\bar{s}$.

Suppose first that p is inert in R. Then there is no subgroup of order p in E which is stable under $\operatorname{Aut} E$. Accordingly, the proof of Proposition 2 shows that a_{st} is divisible by ℓ, since the group $(\operatorname{Aut} E)/\{\pm 1\}$ acts freely on the set of D for which $(E/D, P \bmod D)$ defines s.

Next, let $p = (\pi)^2$ be ramified in R. Then $E[\pi]$ is the unique subgroup of order p on E which is stable under R. The pair $(E/D, P \bmod D)$ is isomorphic to $(E, \pi\cdot P)$ via π. With the obvious meaning of $\langle\pi\rangle$, we may write $a_{st} \equiv 0 \bmod \ell$ for $s \neq \langle\pi\rangle t$, while $a_{st} \equiv 1 \bmod \ell$ for $s = \langle\pi\rangle t$. Tracing through the definitions, one emerges with the desired assertion that T_p operates on $Q_{\bar{s}}$ as $[\pi]$.

In the final case where $p = (\pi)(\pi')$, there are two distinct subgroups D which are stable under $\operatorname{Aut} E$. The corresponding quotients of (E,α) are represented by $(E, \pi\cdot P)$ and $(E, \pi'\cdot P)$. Let $s_1 = \langle\pi\rangle t$ and $s_2 = \langle\pi'\rangle t$. If the two s_i are distinct, the numbers $a_{st} \bmod \ell$ are 1 for $s = s_1, s_2$ and zero otherwise. If $s_1 = s_2$, then $a_{st} = 2$ when $s = s_1 = s_2$ and $a_{st} = 0$ otherwise. This leads to the required formula $T_p = [\pi] + [\pi']$. □

Let $\widetilde{\mathbb{T}}$ be the subring of $\operatorname{End} J(H,q)$ generated by the endomorphisms T_p and $\langle d\rangle$ of $J(H,q)$. (There is a functorial, faithful action of $\widetilde{\mathbb{T}}$ on the space of weight-two cusp forms for $\Gamma \cap \Gamma_0(q)$; thus $\widetilde{\mathbb{T}}$ may be defined alternatively as in the Introduction.) Theorem 3 states that the action of $\widetilde{\mathbb{T}}$ on $Q_{\bar{s}}$ factors through the action of $\mathbb{T}$ on $Q_{\bar{s}}$, in such a way that $\langle d\rangle \in \widetilde{\mathbb{T}}$ acts as $[d] \in \mathbb{T}$ for all $d \in (\mathbf{Z}/N\mathbf{Z})^*$ that $T_p \in \widetilde{\mathbb{T}}$ acts as τ_p. Equivalently, there is a commutative triangle

$$\begin{array}{ccc} \widetilde{\mathbb{T}} & \longrightarrow & \mathbb{T} \\ & \searrow \quad \swarrow & \\ & \operatorname{End} Q_{\bar{s}} & \end{array}$$

in which the left-hand diagonal arrow describes the action of Hecke operators on $Q_{\bar{s}}$ and the right-hand diagonal arrow is the structural map making $Q_{\bar{s}}$ into a $\mathbb{T}$-module. The horizontal arrow is visibly surjective, since $\mathbb{T} = \mathbf{F}_\ell[\Delta]$ is generated by the various $[d]$. By Lemma 5, $Q_{\bar{s}}$ is a free rank-one $\mathbb{T}$-module. Therefore, the map $\mathbb{T} \to \operatorname{End} Q_{\bar{s}}$ is injective. Accordingly, the image of $\widetilde{\mathbb{T}}$ in $\operatorname{End} Q_{\bar{s}}$ may be identified with $\mathbb{T}$. Via this identification, the set of maximal ideals of $\widetilde{\mathbb{T}}$ in the support of $Q_{\bar{s}}$ becomes the set of maximal ideals of $\mathbb{T} = \mathbf{F}_\ell[\Delta]$.

This latter set may be viewed as a set of conjugacy classes of characters. Indeed, if $\mathfrak{m}$ is a maximal ideal of $\mathbb{T}$, then $\mathbb{T}/\mathfrak{m}$ is a finite field of characteristic ℓ. Hence $\mathfrak{m}$ may be obtained as the kernel of some ring homomorphism $\mathbb{T} \to \overline{\mathbf{F}}_\ell$. On the other hand, restriction to Δ yields a 1-1 correspondence between ring homomorphisms $\mathbb{T} \to \overline{\mathbf{F}}_\ell$ and group homomorphisms $\theta : \Delta \to \overline{\mathbf{F}}_\ell^*$. By Galois theory, two ring homomorphisms $\mathbb{T} \rightrightarrows \overline{\mathbf{F}}_\ell$ have the same kernel if and only if they are conjugate under $\operatorname{Gal}(\overline{\mathbf{F}}_\ell/\mathbf{F}_\ell)$. It follows that the set of maximal ideals of $\mathbb{T}$ is in correspondence with the set of $\operatorname{Gal}(\overline{\mathbf{F}}_\ell/\mathbf{F}_\ell)$-conjugacy classes of characters θ.

As was noted above, Δ may be viewed as the Galois group of an abelian extension of K in $\overline{\mathbf{Q}}$. Accordingly, the characters θ become homomorphisms $\operatorname{Gal}(\overline{\mathbf{Q}}/K) \to \overline{\mathbf{F}}_\ell^*$ which factor through the Galois group we called $\operatorname{Gal}(L/K)$. For each such homomorphism, we let θ' be the map $\operatorname{Gal}(\overline{\mathbf{Q}}/K) \to \overline{\mathbf{F}}_\ell^*$ defined by

$$\theta'(x) = \theta(gxg^{-1}),$$

where g is an element of $\operatorname{Gal}(\overline{\mathbf{Q}}/\mathbf{Q})$ which does not belong to $\operatorname{Gal}(\overline{\mathbf{Q}}/K)$.

Lemma 6. *We have $\theta = \theta'$ if and only if θ is identically 1.*

Proof. If θ is identically 1, it certainly coincides with θ'. Conversely, suppose that $\theta = \theta'$. Let $\mathfrak{f}$ be the conductor of θ. As is well known, the conductor of θ' is the image of $\mathfrak{f}$ under the non-trivial involution $K \to K$. Because $\theta = \theta'$, $\mathfrak{f}$ is invariant under this involution. It is in any case a divisor of I. It follows that $\mathfrak{f}$ divides $(\sqrt{-3})$ in the case where $K = \mathbf{Q}(\sqrt{-3})$ and that $\mathfrak{f}$ divides $(1+\sqrt{-1})$ in case $K = \mathbf{Q}(\sqrt{-1})$. In the former case, θ is a map $\mathbf{F}_3^* \to \overline{\mathbf{F}}_3^*$ which is trivial on $\{\pm 1\}$. In the latter case, θ is a map $\mathbf{F}_2^* \to \overline{\mathbf{F}}_2^*$. Hence θ is the trivial map in either case. □

Now let $\operatorname{Ind}\theta$ be the induced representation

$$\operatorname{Ind}_{\operatorname{Gal}(\overline{\mathbf{Q}}/K)}^{\operatorname{Gal}(\overline{\mathbf{Q}}/\mathbf{Q})}\theta : \operatorname{Gal}(\overline{\mathbf{Q}}/\mathbf{Q}) \to \mathbf{GL}(2, \overline{\mathbf{F}}_\ell).$$

We can make an explicit model for $\operatorname{Ind}\theta$ by choosing $g \notin \operatorname{Gal}(\overline{\mathbf{Q}}/K)$ as above and defining

$$\operatorname{Ind}\theta(x) = \begin{cases} \begin{pmatrix} \theta(x) & 0 \\ 0 & \theta'(x) \end{pmatrix} & \text{for } x \in \operatorname{Gal}(\overline{\mathbf{Q}}/K), \\ \begin{pmatrix} 0 & \theta(xg) \\ \theta(g^{-1}x) & 0 \end{pmatrix} & \text{for } x \notin \operatorname{Gal}(\overline{\mathbf{Q}}/K). \end{cases}$$

In particular, one reads from this model the well known formula

$$\operatorname{tr}(\operatorname{Ind}\theta(x)) = \begin{cases} \theta(x) + \theta'(x) & \text{if } x \in \operatorname{Gal}(\overline{\mathbf{Q}}/K), \\ 0 & \text{if } x \notin \operatorname{Gal}(\overline{\mathbf{Q}}/K) \end{cases}$$

for the trace of $\operatorname{Ind}\theta$. The determinant of $\operatorname{Ind}\theta$ is the product of two characters, the first of which is the homomorphism $\operatorname{Gal}(\overline{\mathbf{Q}}/\mathbf{Q}) \to \{\pm 1\}$ which corresponds to the quadratic subfield K of $\overline{\mathbf{Q}}$. This homomorphism may be described alternatively as the mod ℓ cyclotomic character χ_ℓ which gives the action of $\operatorname{Gal}(\overline{\mathbf{Q}}/\mathbf{Q})$ on the group of ℓth roots of 1 in $\overline{\mathbf{Q}}$. It has order two if $\ell = 3$, while it is trivial if $\ell = 2$. The second of the two characters is $\theta \circ \operatorname{Ver}$, where Ver is the Verlagerung map from $\operatorname{Gal}(\overline{\mathbf{Q}}/\mathbf{Q})$ to the abelianization of $\operatorname{Gal}(\overline{\mathbf{Q}}/K)$. It is obtained by composing: the mod N cyclotomic character $\operatorname{Gal}(\overline{\mathbf{Q}}/\mathbf{Q}) \to (\mathbf{Z}/N\mathbf{Z})^*$, the identification $(\mathbf{Z}/N\mathbf{Z})^* \xrightarrow{\sim} (R/I)^*$, and the character $\theta : (R/I)^* \to \overline{\mathbf{F}}_\ell^*$.

Lemma 7. *If θ is non-trivial, the representation* $\operatorname{Ind}\theta$ *is irreducible. Suppose instead that θ is trivial. Then the image of* $\operatorname{Ind}\theta$ *has order two; the kernel of* $\operatorname{Ind}\theta$ *is* $\operatorname{Gal}(\overline{\mathbf{Q}}/K)$. *When* $\ell = 3$, $\operatorname{Ind}\theta$ *is the direct sum of the trivial representation and the one-dimensional representation with character* χ_3. *When* $\ell = 2$, $\operatorname{Ind}\theta$ *is indecomposable and its semisimplification is the direct sum of two copies of the trivial representation.*

Proof. Suppose that θ is non-trivial. Then by Lemma 6, θ and θ' are distinct. The restriction of $\operatorname{Ind}\theta$ to $\operatorname{Gal}(\overline{\mathbf{Q}}/K)$ is thus the direct sum of two *distinct* characters. It follows that there are precisely two lines in the representation space of $\operatorname{Ind}\theta$ which are invariant under $\operatorname{Gal}(\overline{\mathbf{Q}}/K)$. These lines are permuted by the element g; in particular, no line is stable under the full Galois group $\operatorname{Gal}(\overline{\mathbf{Q}}/\mathbf{Q})$. If θ is trivial, then the required assertions are clear from the model of $\operatorname{Ind}\theta$ presented above. □

Note that, in the case where $\ell = 2$ and $\theta = 1$, the representation $\operatorname{Ind}\theta$ is that given by the unique non-trivial extension of $\mathbf{Z}/2\mathbf{Z}$ by μ_2 in the category of commutative group schemes of type $(2, \ldots, 2)$ over $\operatorname{Spec}\mathbf{Z}$. (See **[11]**, Ch. II, Section 12.)

Let $\mathfrak{m}$ be the maximal ideal of $\widetilde{\mathbb{T}}$ arising from θ. The quotient $\widetilde{\mathbb{T}}/\mathfrak{m}$ becomes a subfield of $\overline{\mathbf{F}}_\ell$ via the embedding $\widetilde{\mathbb{T}}/\mathfrak{m} \hookrightarrow \overline{\mathbf{F}}_\ell$ induced by θ. To θ we associate the representation

$$\rho_\theta = \rho_\mathfrak{m} \otimes_{\widetilde{\mathbb{T}}/\mathfrak{m}} \overline{\mathbf{F}}_\ell \ : \ \mathrm{Gal}(\overline{\mathbf{Q}}/\mathbf{Q}) \to \mathbf{GL}(2, \overline{\mathbf{F}}_\ell)$$

obtained by composing $\rho_\mathfrak{m}$ with the inclusion $\mathbf{GL}(2, \widetilde{\mathbb{T}}/\mathfrak{m}) \hookrightarrow \mathbf{GL}(2, \overline{\mathbf{F}}_\ell)$. By varying θ over the set of characters $\Delta \to \overline{\mathbf{F}}_\ell^*$, we obtain from $Q_{\overline{s}}$ a family of representations $\mathrm{Gal}(\overline{\mathbf{Q}}/\mathbf{Q}) \to \overline{\mathbf{F}}_\ell^*$. We seek to describe this family.

Proposition 3. *Let p be a prime number which does not divide $qN\ell$. Then the trace of $\rho_\theta(\mathrm{Frob}_p)$ is*

$$\begin{cases} 0 & \text{if } p \text{ is inert in } K, \\ \theta(\pi) + \theta(\pi') & \text{if } p = (\pi)(\pi') \text{ is split in } K, \end{cases}$$

while the determinant of $\rho_\theta(\mathrm{Frob}_p)$ is $p\theta(p) = \pm\theta(p)$.

Proof. The matrix $\rho_\mathfrak{m}(\mathrm{Frob}_p)$ has trace T_p mod $\mathfrak{m}$ and determinant $p\langle p\rangle$. Viewing $\widetilde{\mathbb{T}}/\mathfrak{m}$ as a quotient of $\mathbb{T}$, we can replace $\langle p\rangle$ by $[p]$ and T_p by τ_p. The map θ sends each $[\delta] \in \mathbb{T}$ to $\theta(\delta)$. The desired formulas now follow. □

Theorem 4. *For each θ, ρ_θ is the semisimplification of the induced representation* $\mathrm{Ind}\,\theta$.

Proof. This follows easily from Proposition 3, the Cebotarev Density Theorem, the Brauer-Nesbitt Theorem and the formulas given above for the trace and determinant of $\mathrm{Ind}\,\theta$. □

It is time now to recapitulate. The group $Q = \Phi/\Phi_0$ is the sum of subgroups $Q_{\overline{s}}$ belonging to the supersingular points $\overline{s}$ of $X_0(N)(\overline{\mathbf{F}}_q)$ which have extra automorphisms and which satisfy the condition $(*)$ of Section 4. To each such point is associated one of the quadratic fields $\mathbf{Q}(\sqrt{-3})$, $\mathbf{Q}(\sqrt{-1})$; call this field K. The point $\overline{s}$ and the group H determine an abelian extension L/K. We induce those characters $\theta : \mathrm{Gal}(\overline{\mathbf{Q}}/K) \to \overline{\mathbf{F}}_\ell^*$ which factor through $\mathrm{Gal}(L/K)$ to obtain two-dimensional ρ representations of $\mathrm{Gal}(\overline{\mathbf{Q}}/\mathbf{Q})$. According to Lemma 7, the representations ρ are irreducible when θ is non-trivial. The semisimplifications of the induced representations ρ are the $\overline{\mathbf{F}}_\ell$-linear two-dimensional representations of $\mathrm{Gal}(\overline{\mathbf{Q}}/\mathbf{Q})$ which arise from $Q_{\overline{s}}$.

From another point of view, let K be one of the two fields $\mathbf{Q}(\sqrt{-3})$, $\mathbf{Q}(\sqrt{-1})$. To fix ideas, take the former field and embed it in $\overline{\mathbf{Q}}$. Let R be the ring of integers of K, and let $\mathfrak{f}$ be a non-zero integral ideal of R for which $R/\mathfrak{f}$ is cyclic as an abelian group. Consider the group $C_\mathfrak{f}$ of ideal classes of K modulo $\mathfrak{f}$, and let θ be a non-trivial homomorphism $C_\mathfrak{f} \to \overline{\mathbf{F}}_3^*$. Using class field theory, regard θ as being defined on $\mathrm{Gal}(\overline{\mathbf{Q}}/K)$ and induce θ to obtain a two-dimensional representation ρ of $\mathrm{Gal}(\overline{\mathbf{Q}}/\mathbf{Q})$. This representation is irreducible. Let N be the norm of $\mathfrak{f}$, and let H be the image of R^* in $(R/\mathfrak{f})^* = (\mathbf{Z}/N\mathbf{Z})^*$.

Theorem 5. *The representation ρ arises from the component group in the mod q reduction of $J(H, q)$ whenever q is a prime number which does not divide $2N$ and which is congruent to 2 modulo 3.*

Proof. Since q is 2 modulo 3, we may choose a supersingular elliptic curve E over $\overline{\mathbf{F}}_q$ whose automorphism group is cyclic of order six. Identify $\mathbf{Z}[\mathrm{Aut}\,E]$

with R and let C be the kernel of $\mathfrak{f}$ on E. The pair (E, C) then represents a point $\bar{s}$ on $X_0(N)$ which has extra automorphisms and satisfies condition $(*)$. We have seen that ρ arises from the subquotient $Q_{\bar{s}}$ of the component group associated with $J(H, q)$. □

As an application of the Theorem, we obtain the following statement, which could presumably be proved much more directly.

Corollary. *The representation ρ arises from the space of weight-two cusp forms on $\Gamma_1(N)$.*

Proof. Let q be a prime number as in the statement of Theorem 5. The conclusion of the Theorem implies that ρ arises from the space of weight-two cusp forms on $\Gamma_H(N) \cap \Gamma_0(q)$, and hence from the space of cusp forms on $\Gamma_1(N) \cap \Gamma_0(q)$. Since ρ is unramified at q, the desired conclusion now follows from "Mazur's Principle" [**15**], Theorem 8.1. □

In the situation of Theorem 5, one can ask whether or not ρ arises from the space of weight-two cusp forms on $\Gamma_H(N)$. In the example treated in Section 2, this space is zero, so the response is negative. It might be interesting to answer this question in general.

References

[1] S. Bosch, W. Lütkebohmert and M. Raynaud, *Néron models*, Ergebnisse der Mathematik und ihrer Grenzgebiete, Springer-Verlag, Berlin and New York, 1990.

[2] H. Carayol, *Sur les représentations galoisiennes modulo ℓ attachées aux formes modulaires*, Duke Math. J., vol. 59 (1989), p. 785–801.

[3] H. Cohen, N. Skoruppa and D. Zagier, *Tables of coefficients of modular forms*, to appear.

[4] P. Deligne and M. Rapoport, *Les schémas de modules de courbes elliptiques*, Lecture Notes in Math., vol. 349, Springer-Verlag, Berlin and New York, 1973, p. 143–316.

[5] F. Diamond, *The refined conjecture of Serre*, This volume.

[6] B. Edixhoven, *L'action de l'algèbre de Hecke sur les groupes de composantes des jacobiennes des courbes modulaires est "Eisenstein"*, Astérisque, vol. 196–197 (1991), p. 159–170.

[7] B. Edixhoven, *The weight in Serre's conjectures on modular forms*, Invent. Math., vol. 109 (1992), p. 563–594.

[8] B. Jordan, R. Livné, *Local diophantine properties of Shimura curves*, Math. Ann., vol. 270 (1985), p. 235–248.

[9] N. M. Katz, *p-adic properties of modular schemes and modular forms*, Lecture Notes in Math., vol. 350, Springer-Verlag, Berlin and New York, 1973, p. 69–190.

[10] N. M. Katz, B. Mazur, *Arithmetic moduli of elliptic curves*, Ann. of Math. Studies **108**, Princeton Univ. Press, Princeton, 1985.

[11] B. Mazur, *Modular curves and the Eisenstein ideal*, Publ. Math. IHES, vol. 47 (1977), p. 33–186.

[12] M. Raynaud, *Jacobienne des courbes modulaires et opérateurs de Hecke*, Astérisque, vol. 196–197 (1991), p. 9–25.

[13] K. A. Ribet, *On modular representations of* Gal($\overline{\mathbf{Q}}/\mathbf{Q}$) *arising from modular forms*, Invent. Math., vol. 100 (1990), p. 431–476.

[14] K. A. Ribet, *On the Component Groups and the Shimura Subgroup of* $J_o(N)$, Sém. Th. des Nombres de l'Université de Bordeaux, 1987–88.

[15] K. A. Ribet, *Report on mod* ℓ *representations of* Gal($\overline{\mathbf{Q}}/\mathbf{Q}$), Proc. of Symposia in Pure Mathematics, vol. 55 (2) (1994), p. 639–676.

[16] J-P. Serre, *Sur les représentations modulaires de degré 2 de* Gal($\overline{\mathbf{Q}}/\mathbf{Q}$), Duke Math. J., vol. 54 (1987), p. 179–230.

[17] J-P. Serre, Letter to K. Ribet (15 April 1987), unpublished.

[18] J. H. Silverman, *The arithmetic of elliptic curves*, Graduate Texts in Math., vol. 106, Springer-Verlag, Berlin and New York, 1986.

[19] J. Tunnell, *Artin's conjecture for representations of octahedral type*, Bull. AMS, (new series), vol. 5 (1981), p. 173–175.

[20] A. Wiles, *Modular elliptic curves and Fermat's Last Theorem*, Annals of Math., to appear.

Conference on Elliptic Curves and Modular Forms
Hong Kong, December 18-21, 1993

Families of elliptic curves with constant mod p representations

K. RUBIN
DEPARTMENT OF MATHEMATICS
OHIO STATE UNIVERSITY, COLUMBUS, OHIO 43210, USA

A. SILVERBERG
DEPARTMENT OF MATHEMATICS
OHIO STATE UNIVERSITY, COLUMBUS, OHIO 43210, USA

Introduction. Suppose E is an elliptic curve over $\mathbf{Q}$ and p is 3 or 5. Then the collection of elliptic curves over $\mathbf{Q}$ having the same mod p representation as E forms an infinite family. In this paper we show how to construct these families explicitly. One reason for the interest in these families is the result announced by Wiles, that if E and E' are elliptic curves over $\mathbf{Q}$ with good reduction at an odd prime p and with the same mod p representation, and E has complex multiplication, then E' is modular.

The proofs of the results in Section 4 and Section 5 rely on symbolic computer computations, which were done using the programs Pari and Mathematica.

Note that all the results of this paper remain true with $\mathbf{Q}$ replaced by any number field.

The authors thank Tricia Pacelli and Glenn Stevens for asking a question which led us to simplify the polynomials in Theorems 4.1 and 5.1.

Notation. If N is a positive integer and E is an elliptic curve over a field k with algebraic closure $\bar{k}$, let $E[N]$ denote the kernel of multiplication by N on $E(\bar{k})$. Let $\mathfrak{H}$ denote the complex upper half plane,

$$\Gamma(N) = \{\left(\begin{smallmatrix} a & b \\ c & d \end{smallmatrix}\right) \in SL_2(\mathbf{Z}) : \left(\begin{smallmatrix} a & b \\ c & d \end{smallmatrix}\right) \equiv \left(\begin{smallmatrix} 1 & 0 \\ 0 & 1 \end{smallmatrix}\right) \pmod{N}\},$$

and $G_{\mathbf{Q}} = \mathrm{Gal}(\bar{\mathbf{Q}}/\mathbf{Q})$.

If E is an elliptic curve over $\mathbf{Q}$ and p is a prime, write

$$\rho_{E,p} : G_{\mathbf{Q}} \to \mathrm{Aut}(E[p]) \cong GL_2(\mathbf{F}_p)$$

for the (isomorphism class of the) mod p representation of E.

1. Modular curves.

Suppose p is an odd rational prime. Denote by Y_p the (non-compact) modular curve over $\mathbf{Q}$ which parametrizes isomorphism classes of pairs (E, ϕ) where E is an elliptic curve and

$$\phi : \mathbf{Z}/p\mathbf{Z} \times \mu_p \to E[p]$$

The authors thank the NSF for financial support.

is an isomorphism with the property that

$$\langle \phi(a_1, \zeta_1), \phi(a_2, \zeta_2) \rangle = \zeta_2^{a_1}/\zeta_1^{a_2}$$

where $\langle\, ,\, \rangle$ denotes the Weil pairing on $E[p]$. Equivalently, Y_p parametrizes triples (E, P, C) where E is an elliptic curve, P is a point of exact order p on E, and C is a subgroup of order p on E, not containing P. (Given (E, P, C), define ϕ by $\phi(a, \zeta) = aP + Q$ for the unique $Q \in C$ such that $\langle P, Q \rangle = \zeta$.) Let $Y(N)$ denote the modular curve which parametrizes elliptic curves with full level N structure (see [**5**]). Fixing a p-th root of unity ζ, the map

$$(E, \phi) \mapsto (E, \phi(1, 1), \phi(0, \zeta)),$$

induces an isomorphism (defined over $\mathbf{Q}(\zeta)$) from Y_p onto one connected component of $Y(p)$. Thus $Y_p(\mathbf{C})$ is isomorphic to $\mathfrak{H}/\Gamma(p)$. Let X_p and $X(N)$ denote the compactifications of Y_p and $Y(N)$, respectively. Then X_p has genus 0 when $p \leq 5$ and genus at least 3 when $p \geq 7$ (see p. 23 of [**5**]).

From now on we will identify $X(1)$ with $\mathbf{P}^1$ by the map which sends an elliptic curve E to its j-invariant $j(E)$. From the description over $\mathbf{C}$ we see that the forgetful morphism $X_p \to X(1)$ induced by $(E, \phi) \mapsto E$ has degree

$$\frac{1}{2}[SL_2(\mathbf{Z}) : \Gamma(p)] = \frac{1}{2}\#(SL_2(\mathbf{F}_p)) = \frac{p^3 - p}{2}.$$

Let W_p be a compactification of the elliptic surface associated to the universal elliptic curve over Y_p. Then W_p is a variety over $\mathbf{Q}$, and W_p is an elliptic surface over X_p equipped with a morphism $W_p \to X_p$ and a zero section $\iota_p : X_p \to W_p$, both defined over $\mathbf{Q}$, and $p^2 - 1$ additional sections of order p defined over $\bar{\mathbf{Q}}$. We will denote the $G_{\mathbf{Q}}$-module of these p^2 sections by $W_p[p]$. The definitions of X_p and W_p imply that $W_p[p]$ is isomorphic to $\mathbf{Z}/p\mathbf{Z} \times \mu_p$ as a $G_{\mathbf{Q}}$-module. Note that W_p is not uniquely determined, because there is a choice of compactification, but the choice is not important for our purposes. See [4] for the arithmetic of elliptic modular surfaces.

1.1. Level three. Let A'_u denote the elliptic curve (the Hesse cubic)

$$X^3 + Y^3 + Z^3 = 3uXYZ$$

over $\mathbf{Q}(u)$, with origin $(0, 1, -1)$. This curve has a Weierstrass model

$$A_u : y^2 = x^3 - 27u(u^3 + 8)x + 54(u^6 - 20u^3 - 8) \tag{1}$$

so in particular

$$j(A'_u) = j(A_u) = 27\frac{u^3(u^3 + 8)^3}{(u^3 - 1)^3}. \tag{2}$$

Let $P_u = (-1, 1, 0) \in A'_u[3]$. If ζ is a primitive cube root of unity, then

$$C_u = \{(0, 1, -1), (0, \zeta, -1), (0, \zeta^2, -1)\}$$

is a $\mathrm{Gal}(\overline{\mathbf{Q}(u)}/\mathbf{Q}(u))$-invariant subgroup of $A'_u[3]$. Thus the map

$$u \mapsto (A'_u, P_u, C_u)$$

induces a morphism $f : \mathbf{P}^1 \to X_3$ defined over $\mathbf{Q}$. To see that f is an isomorphism it suffices to observe that by (2), the degree of the composition $\mathbf{P}^1 \xrightarrow{f} X_3 \to X(1)$ is 12 which is the same as the degree of X_3 over $X(1)$. From now on we will identify X_3 with $\mathbf{P}^1$ using the isomorphism f. With this identification we can take W_3 to be given by (1).

1.2. Level five. Following Klein (see [1] and p. 130 of [2]), let B_u denote the elliptic curve

$$(3)\quad y^2 = x^3 - \frac{u^{20} - 228u^{15} + 494u^{10} + 228u^5 + 1}{48}x + \frac{u^{30} + 522u^{25} - 10005u^{20} - 10005u^{10} - 522u^5 + 1}{864}$$

over $\mathbf{Q}(u)$. The j-invariant of B_u is

$$(4)\qquad j(B_u) = \frac{-(u^{20} - 228u^{15} + 494u^{10} + 228u^5 + 1)^3}{u^5(u^{10} + 11u^5 - 1)^5}.$$

Let

$$\begin{aligned} x_0(u) =& (u^{10} + 12u^8 - 12u^7 + 24u^6 + 30u^5 + \\ &+ 60u^4 + 36u^3 + 24u^2 + 12u + 1)/12, \\ y_0(u) =& (u^{13} + u^{12} + 4u^{11} + 5u^9 + 6u^8 + 21u^7 + \\ &+ 29u^6 + 25u^5 + 15u^4 + 9u^3 + 4u^2 + u)/2, \end{aligned}$$

and let $Q_u = (x_0(u), y_0(u))$. One can verify that $Q_u \in B_u[5]$. Let ζ denote a primitive fifth root of unity. Since the equation defining B_u is invariant under $u \mapsto \zeta u$, we have $R_u = (x_0(\zeta u), y_0(\zeta u)) \in B_u[5]$ as well. Since $Q_u \in B_u(\mathbf{Q}(u))$ and $R_u \notin B_u(\mathbf{Q}(u))$, these are independent points of order 5. Further, one can verify that for $\sigma \in G_{\mathbf{Q}}$,

$$\sigma R_u = \epsilon(\sigma)R_u + (1 - \epsilon(\sigma))Q_u$$

where $\epsilon : G_{\mathbf{Q}} \to (\mathbf{Z}/5\mathbf{Z})^\times$ is the cyclotomic character. Thus the subgroup C_u of $B_u[5]$ generated by $Q_u - R_u$ is stable under $\mathrm{Gal}(\overline{\mathbf{Q}(u)}/\mathbf{Q}(u))$. Therefore the map

$$u \mapsto (B_u, Q_u, C_u)$$

induces a morphism $g : \mathbf{P}^1 \to X_5$ defined over $\mathbf{Q}$. To see that g is an isomorphism it is enough to observe that by (4), the degree of the composition $\mathbf{P}^1 \xrightarrow{g} X_5 \to X(1)$ is 60 which is the same as the degree of X_5 over $X(1)$. From now on we will identify X_5 with $\mathbf{P}^1$ using this isomorphism. With this identification we can take W_5 to be given by (3).

2. Twists of modular curves.

Proposition 2.1. *Suppose p is a prime and there are commutative diagrams*

$$(5)\qquad \begin{array}{ccc} W_p & \xleftarrow{\psi} & W' \\ \downarrow & & \downarrow \\ X_p & \xleftarrow{\psi_0} & X' \\ & \searrow \quad \swarrow & \\ & X(1) & \end{array} \qquad\qquad \begin{array}{ccc} W_p & \xleftarrow{\psi} & W' \\ \iota_p\uparrow & & \uparrow\iota' \\ X_p & \xleftarrow{\psi_0} & X' \end{array}$$

where X' is a curve and W' is an elliptic surface over X' with zero-section ι', all defined over $\mathbf{Q}$, ψ and ψ_0 are isomorphisms defined over $\bar{\mathbf{Q}}$, the maps $W_p \to X_p \to X(1)$ are the natural ones (see §1), and the maps $W' \to X' \to X(1)$ are defined over $\mathbf{Q}$. Let $S \subset X'(\bar{\mathbf{Q}})$ denote the inverse image under ψ_0 of the (finite) set of cusps of X_p. Then there is a $G_{\mathbf{Q}}$-module V such that for every point $t \in X'(\mathbf{C}) - S$, the fiber E_t of W' over t is an elliptic curve defined over $\mathbf{Q}(t)$ and $E_t[p] \cong V$ as Gal($\overline{\mathbf{Q}(t)}/\mathbf{Q}(t)$)-modules.

Proof. Let V be the group of p^2 sections from X' to W' of order dividing p,

$$V = \{\psi^{-1} \circ s \circ \psi_0 : s \in W_p[p]\}.$$

These sections are defined over $\bar{\mathbf{Q}}$, so V is a $G_{\mathbf{Q}}$-module and the map from V to $E_t[p]$ obtained by restricting the sections in V to t is a Gal($\overline{\mathbf{Q}(t)}/\mathbf{Q}(t)$)-isomorphism. □

Corollary 2.2. *With notation as in Proposition 2.1, if E_1, E_2 are the fibers of W' over $t_1, t_2 \in X'(\mathbf{Q}) - S$, then $E_1[p] \cong E_2[p]$ as $G_{\mathbf{Q}}$-modules.*

Proof. Immediate from Proposition 2.1. □

Remark 2.3. Under the hypotheses of Proposition 2.1, the resulting module V is endowed with a pairing coming from the Weil pairing on $W_p[p]$, and X' is the moduli space for isomorphism classes of pairs (E, ϕ) where E is an elliptic curve and $\phi : V \xrightarrow{\sim} E[p]$ is an isomorphism which takes the pairing on V to the Weil pairing on $E[p]$.

Remark 2.4. Fix an odd prime p, let $V_0 = \mathbf{Z}/p\mathbf{Z} \times \mu_p$, and let η_0 be the $G_{\mathbf{Q}}$-equivariant pairing on V_0

$$\eta_0((a_1, \zeta_1), (a_2, \zeta_2)) = \zeta_2^{a_1}/\zeta_1^{a_2}.$$

There are natural $G_{\mathbf{Q}}$-equivariant maps

$$\mathrm{Aut}(V_0, \eta_0) \to \mathrm{Aut}(W_p) \to \mathrm{Aut}(X_p),$$

where $\mathrm{Aut}(V_0, \eta_0)$ is the group of automorphisms of V_0 which preserve η_0. Suppose V is a two-dimensional $\mathbf{F}_p$-vector space with an action of $G_{\mathbf{Q}}$ and a non-degenerate skew-symmetic μ_p-valued $G_{\mathbf{Q}}$-equivariant pairing η. Let $\rho : G_{\mathbf{Q}} \to$

$\mathrm{Aut}(V)$ and $\rho_0 : G_{\mathbf{Q}} \to \mathrm{Aut}(V_0)$ denote the representations induced by the $G_{\mathbf{Q}}$-actions, and fix a group isomorphism $\varphi : V_0 \to V$ which takes the pairing η_0 to the pairing η. Then $\sigma \mapsto \varphi^{-1}\rho(\sigma)\varphi\rho_0(\sigma)^{-1}$ defines a cocycle on $G_{\mathbf{Q}}$ with values in $\mathrm{Aut}(V_0, \eta_0)$, which induces cohomology classes in $H^1(G_{\mathbf{Q}}, \mathrm{Aut}(W_p))$ and $H^1(G_{\mathbf{Q}}, \mathrm{Aut}(X_p))$ that are independent of the choice of φ. The twists of W_p and X_p by these cohomology classes (see [**3**]) give the varieties W' and X' satisfying Proposition 2.1 for the given $G_{\mathbf{Q}}$-module V.

Suppose E is an elliptic curve defined over $\mathbf{Q}$. Our aim is to explicitly construct W', X', and corresponding morphisms satisfying the hypotheses of Proposition 2.1, so that E is a fiber of W' over some rational point of X'. Then, by Corollary 2.2, the points of $X'(\mathbf{Q}) - \mathcal{S}$ will correspond to elliptic curves E' over $\mathbf{Q}$ with $\rho_{E',p} \cong \rho_{E,p}$.

From now on suppose that $p = 3$ or 5, so that X_p has genus 0. Recall that we have identified both X_p and $X(1)$ with $\mathbf{P}^1$, and with these identifications the morphism from X_p to $X(1)$ is given by the rational function $\mathcal{J}(u) \in \mathbf{Q}(u)$ defined by

$$\mathcal{J}(u) = \begin{cases} j(A_u) & \text{if } p = 3, \\ j(B_u) & \text{if } p = 5 \end{cases}$$

(see (2) and (4)). Let $X' = \mathbf{P}^1$. An isomorphism $\psi_0 : X' \to X_p$ defined over a number field K is given by a linear fractional transformation

$$u = A(t) = \frac{at + b}{ct + d} \qquad \text{where} \qquad A = \begin{pmatrix} a & b \\ c & d \end{pmatrix} \in GL_2(K).$$

The map from X' to $X(1)$ in (5) will be defined over $\mathbf{Q}$ if and only if

$$\mathcal{J}(A(t)) \in \mathbf{Q}(t). \tag{6}$$

3. Algorithm. Fix $p = 3$ or 5 and suppose E is an elliptic curve over $\mathbf{Q}$. We will construct the family of all elliptic curves over $\mathbf{Q}$ with the same mod p representation as E. When applying Wiles' result one is interested in curves E with complex multiplication, but the method works in general.

The construction proceeds as follows. Define polynomials $a_4(u), a_6(u) \in \mathbf{Q}[u]$ so that

$$y^2 = x^3 + a_4(u)x + a_6(u)$$

is the elliptic curve A_u of (1) if $p = 3$, or B_u of (3) if $p = 5$. Also let $n = 1$ if $p = 3$ and $n = 5$ if $p = 5$, so that a_4 has degree $4n$ and a_6 has degree $6n$.

Step 1. Find $u_0 \in \bar{\mathbf{Q}}$ such that

$$\mathcal{J}(u_0) = j(E).$$

Step 2. Find $\mu \in \bar{\mathbf{Q}}^\times$ such that

$$\mu^{-2}a_4(u_0), \mu^{-3}a_6(u_0) \in \mathbf{Q}$$

and

$$y^2 = x^3 + \mu^{-2}a_4(u_0)x + \mu^{-3}a_6(u_0) \quad \text{is isomorphic over } \mathbf{Q} \text{ to } E.$$

Step 3. Find $\alpha, \gamma \in \bar{\mathbf{Q}}$ such that

$$\mu^{-2}(1+\gamma t)^{4n}a_4(A(t)), \quad \mu^{-3}(1+\gamma t)^{6n}a_6(A(t)) \in \mathbf{Q}[t], \tag{7}$$

where

$$A = \begin{pmatrix} \alpha & u_0 \\ \gamma & 1 \end{pmatrix}.$$

REMARK 3.1. For Step 1, let u_0 be any root of the numerator of $\mathcal{J}(u) - j(E) \in \mathbf{Q}(t)$. For Step 2, after putting E in Weierstrass form one can solve for μ directly. Step 2 ensures that the constant terms of the polynomials in (7) are rational numbers, and choosing rational values for the coefficients of t gives values for α and γ.

REMARK 3.2. The condition in Step 3 implies (6). Conversely, if $\mathrm{Aut}(E) = \{\pm 1\}$, and we have $u_0, \mu, \alpha, \gamma \in \bar{\mathbf{Q}}$ and $A = \left(\begin{smallmatrix} \alpha & u_0 \\ \gamma & 1 \end{smallmatrix}\right)$ which satisfy (6) and the conditions in Steps 1 and 2, then it can be shown that the condition in Step 3 is satisfied.

Theorem 3.3. *Suppose E is an elliptic curve defined over $\mathbf{Q}$, and we have elements $u_0, \mu, \alpha, \gamma \in \bar{\mathbf{Q}}$ and a matrix $A \in GL_2(\bar{\mathbf{Q}})$ satisfying the conditions in Steps 1, 2, and 3. Let $\mathcal{E}_t$ be the curve*

$$y^2 = x^3 + \mu^{-2}(1+\gamma t)^{4n}a_4(A(t))x + \mu^{-3}(1+\gamma t)^{6n}a_6(A(t)) \tag{8}$$

over $\mathbf{Q}(t)$. Then for every rational number t which is not a pole of $\mathcal{J} \circ A$, $\mathcal{E}_t$ is an elliptic curve over $\mathbf{Q}$ and $\mathcal{E}_t[p] \cong E[p]$ as $G_{\mathbf{Q}}$-modules.

Proof. Let $X' = \mathbf{P}^1$ and let W' be the elliptic surface over X' defined by (8), with W' mapping to X' by $(x, y, t) \mapsto t$. Let $\psi_0 : X' \xrightarrow{\sim} X_p$ be given by A, and $\psi : W' \xrightarrow{\sim} W_p$ by

$$(x, y, t) \mapsto (\mu(1+\gamma t)^{-2n}x, \mu^{3/2}(1+\gamma t)^{-3n}y, A(t)).$$

Then all the hypotheses of Proposition 2.1 are satisfied. By Step 2, the fiber of W' over $t = 0$ is isomorphic over $\mathbf{Q}$ to E. The theorem now follows from Corollary 2.2. $\square$

4. Examples with $p = 3$.

4.1. Generic example.

Theorem 4.1. *Fix an elliptic curve E over $\mathbf{Q}$,*

$$y^2 = x^3 + ax + b,$$

and let $J = j(E)/1728 = 4a^3/(4a^3 + 27b^2)$. Define

$$\begin{aligned} a(t) &= (-3(J-1)^2t^4 - 8(J-1)^2t^3 + 6(J-1)t^2 + 1)a, \\ b(t) &= (-(J-1)^3(8J+1)t^6 + 6(J-1)^2(2J+1)t^5 + 15(J-1)^2t^4 \\ &\quad + 20(J-1)^2t^3 - 15(J-1)t^2 + 6t + 1)b \end{aligned}$$

and define $\mathcal{E}_t$ by

$$y^2 = x^3 + a(t)x + b(t).$$

Then for every rational number t such that $\mathcal{E}_t$ is nonsingular, $\mathcal{E}_t[3]$ is isomorphic as a $G_{\mathbf{Q}}$-module to $E[3]$.

Proof. If $a = 0$ then $\mathcal{E}_t$ is $y^2 = x^3 + (t+1)^6b$ and if $b = 0$ then $\mathcal{E}_t$ is $y^2 = x^3 + ax$. In both cases the elliptic surface is isotrivial and the theorem holds. Now assume a and b are both nonzero and let $j = j(E)$. Using (2), a computation shows that

$$\mathcal{J}(u) - j = \frac{27u^{12} + (648 - j)u^9 + 3(1728 + j)u^6 + 3(4608 - j)u^3 + j}{(u^3 - 1)^3}.$$

Let u_0 be a root of the numerator of $\mathcal{J}(u) - j$, so that Step 1 of §3 is satisfied. Let

$$\begin{aligned} \mu &= \frac{a_6(u_0)a}{a_4(u_0)b} = \frac{-2(u_0^6 - 20u_0^3 - 8)a}{(u_0^3 + 8)u_0 b}, \\ \alpha &= \frac{\mu^3 b u_0}{144(u_0^3 - 1)^2}, \\ \gamma &= \frac{-\mu^3 b(u_0^3 + 2)}{864(u_0^3 - 1)^2}. \end{aligned}$$

Since a and b are nonzero, we have $\mu \in \bar{\mathbf{Q}}^\times$. A symbolic computer calculation shows that with these choices, Steps 2 and 3 are satisfied, and the curve $\mathcal{E}_t$ is the same as the curve $\mathcal{E}_t$ of Theorem 3.3. Thus Theorem 3.3 completes the proof. □

REMARK 4.2. If a and b are both nonzero, then the elliptic surface $\mathcal{E}_t$ in Theorem 4.1 is not isotrivial. Further, if E' is an elliptic curve over $\mathbf{Q}$ such that there is an isomorphism between the $G_{\mathbf{Q}}$-modules $E[3]$ and $E'[3]$ which preserves the Weil pairings, then E' is isomorphic over $\mathbf{Q}$ to $\mathcal{E}_t$ for some $t \in \mathbf{P}^1(\mathbf{Q})$ (see Remark 2.3).

REMARK 4.3. The values of α and γ in the proof of Theorem 4.1 were obtained by choosing 0 and $6b$ for the linear coefficients of the polynomials in (7) and then solving for α and γ (see Remark 3.1). This choice gives rise to relatively simple polynomials $a(t)$ and $b(t)$. Choosing other values (not both zero) for the linear coefficients leads to other α, γ, $a(t)$, and $b(t)$, giving rise to an isomorphic elliptic surface.

4.2. $y^2 = x^3 - Dx$: supersingular at 3, CM by $\sqrt{-1}$.

Theorem 4.4. *Fix a nonzero integer D and define E_t by*

$$y^2 = x^3 + D(27D^2t^4 - 18Dt^2 - 1)x + 4D^2t(27D^2t^4 + 1).$$

(i) *For every rational number t, E_t is an elliptic curve over $\mathbf{Q}$ and $E_t[3]$ is isomorphic as a $G_{\mathbf{Q}}$-module to $E[3]$, where E is the elliptic curve $y^2 = x^3 - Dx$.*

(ii) *If D is prime to 3 and $t \in \mathbf{Q}$ is integral at 3 then E_t has good reduction at 3.*

Proof. Using (2), a computation shows that

$$\mathcal{J}(u) - j(E) = \frac{27(u^2 - 2u - 2)^2(u^4 + 2u^3 + 6u^2 - 4u + 4)^2}{(u^3 - 1)^3}.$$

Let u_0 be a root of $u^2 - 2u - 2$ and let ν be a fixed square root of $(1 + 2u_0)/D$. Let

$$\mu = 18\nu,$$
$$\alpha = (4 - u_0)\nu D,$$
$$\gamma = (1 - u_0)\nu D.$$

(We can take $u_0 = 1 + \sqrt{3}$, $\mu = \pm 18\sqrt{(3 + 2\sqrt{3})/D}$, $\alpha = \pm\sqrt{6D\sqrt{3}}$, and $\gamma = \pm\sqrt{(9 + 6\sqrt{3})D}$, where the signs of the square roots are taken appropriately and compatibly.) Then Steps 1, 2, and 3 are satisfied, and with these choices E_t is the curve $\mathcal{E}_t$ of Theorem 3.3. The discriminant of E_t is

$$\Delta(E_t) = -2^6D^3(27D^2t^4 + 18Dt^2 - 1)^3,$$

which has no rational roots, so E_t is an elliptic curve for every $t \in \mathbf{P}^1(\mathbf{Q})$. Theorem 3.3 implies (i). Statement (ii) follows immediately from the formula for the discriminant. □

REMARK 4.5. The j-invariant of the curve E_t of Theorem 4.4 is

$$j(E_t) = \frac{1728(27D^2t^4 - 18Dt^2 - 1)^3}{(27D^2t^4 + 18Dt^2 - 1)^3}.$$

The result of Wiles mentioned in the introduction, together with Theorem 4.4, implies that if E' is an elliptic curve over $\mathbf{Q}$ whose j-invariant belongs to the set

$$\left\{\frac{1728(27D^2t^4 - 18Dt^2 - 1)^3}{(27D^2t^4 + 18Dt^2 - 1)^3} : \right.$$
$$\left. D \in \mathbf{Z} \text{ is prime to 3 and } t \in \mathbf{Q} \text{ is integral at } 3\right\}$$

then E' is modular.

4.3. $y^2 = x^3 - 432D$*: additive at* 3*, CM by* $\sqrt{-3}$.

Theorem 4.6. *Fix a nonzero integer* D *and define* E_t *by*

$$y^2 = x^3 - 27Dt(Dt^3+8)x + 54D(D^2t^6 - 20Dt^3 - 8).$$

If $t \in \mathbf{Q}$ *and* $Dt^3 \neq 1$, *then* E_t *is an elliptic curve over* $\mathbf{Q}$ *and* $E_t[3]$ *is isomorphic as a* $G_{\mathbf{Q}}$*-module to* $E[3]$, *where* E *is the elliptic curve* $y^2 = x^3 - 432D$.

Note that when $D = 1$, this is the Weierstrass model (1) of the Hesse family. The proof of Theorem 4.6 is the same as that of Theorem 4.4, after letting

$$u_0 = 0, \quad \mu = D^{-1/3}, \quad \alpha = D^{1/3}, \quad \text{and} \quad \gamma = 0.$$

The discriminant and j-invariant of E_t are given by

$$\Delta(E_t) = 2^{12}3^9D^2(Dt^3-1)^3, \qquad j(E_t) = \frac{27Dt^3(Dt^3+8)^3}{(Dt^3-1)^3}.$$

4.4. $y^2 + y = x^3 - x^2 - 7x + 10$*: ordinary at* 3*, CM by* $\sqrt{-11}$.

Theorem 4.7. *Define polynomials*

$$a(t) = -5346t^4 + 2079t^3 - 99t^2 + 77t - 7,$$
$$b(t) = -154366t^6 + 73507t^5 - 26805t^4 - 694t^3 + 1091t^2 - 127t + 10$$

and for each $t \in \mathbf{Q}$ *let* E_t *be the elliptic curve*

$$y^2 + (t^3+t^2+t+1)y = x^3 - x^2 + a(t)x + b(t).$$

For every rational number t, E_t *is an elliptic curve over* $\mathbf{Q}$ *and* $E_t[3]$ *is isomorphic as a* $G_{\mathbf{Q}}$*-module to* $E[3]$, *where* E *is the elliptic curve* $y^2 + y = x^3 - x^2 - 7x + 10$. *If* $t \in \mathbf{Q}$ *is integral at* 3 *and* $t \equiv 0$ *or* 1 (mod 3) *then* E_t *has good reduction at* 3.

Proof. The elliptic curve E has complex multiplication by $\mathbf{Q}(\sqrt{-11})$ and good reduction at 3. We apply Theorem 4.1 with the Weierstrass model

$$y^2 = x^3 - \frac{22}{3}x + \frac{847}{108}$$

of E. The change of variables

$$(x,y,t) \mapsto \left(\frac{56^2(3x-1)}{3(56-147t)^2}, \frac{56^3(2y+t^3+t^2+t+1)}{2(56-147t)^3}, \frac{27t}{56-147t}\right),$$

transforms the curve $\mathcal{E}_t$ of Theorem 4.1 (with $a = -22/3$ and $b = 847/108$) to the curve E_t. The assertion about good reduction follows by computing that the discriminant of E_t is

$$-11^3(27t^2-8t+1)^3(27t^2+36t+1)^3.$$

□

REMARK 4.8. The curve E_t of Theorem 4.7 has been chosen so that it is an integral model which is minimal when $t = 0$. The j-invariant of E_t is

$$\left(\frac{-16(54t^2 - 27t + 2)(27t^2 + 3t + 1)}{(27t^2 - 8t + 1)(27t^2 + 36t + 1)}\right)^3.$$

5. Examples with $p = 5$.

5.1. Generic example.

Theorem 5.1. *Fix an elliptic curve E over $\mathbf{Q}$,*

$$y^2 = x^3 + ax + b.$$

Define

$$a(t) = a\sum_{k=0}^{20} \alpha_k t^k, \quad b(t) = b\sum_{k=0}^{30} \beta_k t^k,$$

where $\alpha_k, \beta_k \in \mathbf{Q}[J]$, with $J = j(E)/1728 = 4a^3/(4a^3 + 27b^2)$, are the polynomials given in the appendix. Define $\mathcal{E}_t$ by

$$y^2 = x^3 + a(t)x + b(t).$$

Then for every rational number t such that $\mathcal{E}_t$ is nonsingular, $\mathcal{E}_t[5]$ is isomorphic as a $G_{\mathbf{Q}}$-module to $E[5]$.

Proof. If $a = 0$ then $\mathcal{E}_t$ is $y^2 = x^3 + (t+1)^{30}b$ and if $b = 0$ then $\mathcal{E}_t$ is $y^2 = x^3 + ax$. In both cases the elliptic surface is isotrivial and the theorem holds. Now assume a and b are both nonzero and let $j = j(E)$. Using (4), a computation shows that the numerator of $\mathcal{J}(u) - j$ is

$$\begin{aligned} u^{60} &+ (j - 684)u^{55} + (55j + 157434)u^{50} + 5(241j - 2505492)u^{45} \\ &+ 35(374j + 2213157)u^{40} + 3(23195j - 43563048)u^{35} \\ &+ (134761j - 33211924)u^{30} - 3(23195j - 43563048)u^{25} \\ &+ 35(374j + 2213157)u^{20} - 5(241j - 2505492)u^{15} \\ &+ (55j + 157434)u^{10} - (j - 684)u^5 + 1. \end{aligned}$$

Let u_0 be a root of this polynomial and let

$$\begin{aligned} \mu &= \frac{a_6(u_0)a}{a_4(u_0)b} = \frac{-(u_0^{30} + 522u_0^{25} - 10005u_0^{20} - 10005u_0^{10} - 522u_0^5 + 1)a}{18(u_0^{20} - 228u_0^{15} + 494u_0^{10} + 228u_0^5 + 1)b}, \\ \alpha &= \frac{6\mu^3 b(57u_0^{15} - 247u_0^{10} - 171u_0^5 - 1)}{u_0^4(u_0^{10} + 11u_0^5 - 1)^4}, \\ \gamma &= \frac{6\mu^3 b(u_0^{15} - 171u_0^{10} + 247u_0^5 + 57)}{(u_0^{10} + 11u_0^5 - 1)^4}. \end{aligned}$$

A symbolic computer calculation shows that with these choices, the polynomials of (7) are $a(t)$ and $b(t)$. Thus the theorem follows from Theorem 3.3. □

REMARK 5.2. If $ab \neq 0$, then the elliptic surface $\mathcal{E}_t$ in Theorem 5.1 is not isotrivial. Further, if E' is an elliptic curve over $\mathbf{Q}$ such that there is an isomorphism between the $G_{\mathbf{Q}}$-modules $E[5]$ and $E'[5]$ which preserves the Weil pairings, then E' is isomorphic over $\mathbf{Q}$ to $\mathcal{E}_t$ for some $t \in \mathbf{P}^1(\mathbf{Q})$ (see Remark 2.3).

5.2. $y^2 = x^3 - Dx$: ordinary at 5, CM by $\sqrt{-1}$.

Theorem 5.3. *Fix a nonzero integer D and define polynomials*

$$\begin{aligned} a(t) &= D^5t^9 - Dt, \\ b(t) &= -3125D^{11}t^{20} - 39583D^{10}t^{18} + 11875D^9t^{16} - 95000D^8t^{14} \\ &\quad + 61750D^7t^{12} + 41166D^6t^{10} + 12350D^5t^8 - 3800D^4t^6 \\ &\quad + 95D^3t^4 - 63D^2t^2 - D, \\ c(t) &= -521875D^{16}t^{29} - 1355787D^{15}t^{27} - 7366875D^{14}t^{25} \\ &\quad + 9635000D^{13}t^{23} - 8315875D^{12}t^{21} - 3678639D^{11}t^{19} \\ &\quad - 10560675D^{10}t^{17} + 30400D^9t^{15} + 2091615D^8t^{13} \\ &\quad + 134479D^7t^{11} + 62583D^6t^9 - 14200D^5t^7 \\ &\quad + 2327D^4t^5 + 107D^3t^3 + 7D^2t, \end{aligned}$$

and for each $t \in \mathbf{Q}$ let E_t be the elliptic curve

$$y^2 = x^3 + a(t)x^2 + b(t)x + c(t).$$

For every rational number t, E_t is an elliptic curve over $\mathbf{Q}$ and $E_t[5]$ is isomorphic as a $G_{\mathbf{Q}}$-module to $E[5]$, where E is the elliptic curve $y^2 = x^3 - Dx$. If D is prime to 5, $t \in \mathbf{Q}$ is integral at 5, and $Dt^2 \not\equiv 3 \pmod 5$ then E_t has good reduction at 5.

The elliptic curve E has complex multiplication by $\mathbf{Q}(i)$ and good reduction at 5. Let ν be a fixed square root of $-(1+2i)/D$. The proof of Theorem 5.3 is the same as the proof of Theorem 4.4, using the values

$$u_0 = i, \quad \mu = (1+2i)^2\nu/2, \quad \alpha = -\nu D, \quad \gamma = -i\nu D,$$

and the change of variables $x \mapsto x + (D^5t^9 - Dt)/3$. The discriminant of E_t is

$$4^3D^3(5D^2t^4 - 2Dt^2 + 1)^5(25D^4t^8 - 100D^3t^6 - 210D^2t^4 - 20Dt^2 + 1)^5$$

and the j-invariant is

$$\frac{4^3(15D^2t^4 + 10Dt^2 + 3)^3(25D^4t^8 + 70D^2t^4 + 1)^3}{(5D^2t^4 - 2Dt^2 + 1)^5} \times$$

$$\times \frac{(25D^4t^8 + 300D^3t^6 - 370D^2t^4 + 60Dt^2 + 1)^3}{(25D^4t^8 - 100D^3t^6 - 210D^2t^4 - 20Dt^2 + 1)^5}.$$

5.3. $y^2 = x^3 + 16D$: *supersingular at* 5, *CM by* $\sqrt{-3}$.

Theorem 5.4. *Fix a nonzero integer D and define polynomials*

$$\begin{aligned} a(t) &= -15000D^7t^{19} - 106875D^6t^{16} + 85500D^5t^{13} - 185250D^4t^{10} \\ &\quad - 17100D^3t^7 - 4275D^2t^4 + 120Dt, \\ b(t) &= 50000D^{11}t^{30} + 3625000D^{10}t^{27} + 4893750D^9t^{24} \\ &\quad + 25012500D^8t^{21} - 47523750D^7t^{18} - 9504750D^5t^{12} \\ &\quad - 1000500D^4t^9 + 39150D^3t^6 - 5800D^2t^3 + 16D, \end{aligned}$$

and for each $t \in \mathbf{Q}$ let E_t be the elliptic curve

$$y^2 = x^3 + a(t)x + b(t).$$

For every rational number t, E_t is an elliptic curve over $\mathbf{Q}$ and $E_t[5]$ is isomorphic as a $G_{\mathbf{Q}}$-module to $E[5]$, where E is the elliptic curve $y^2 = x^3 + 16D$. If D is prime to 5 and $t \in \mathbf{Q}$ is integral at 5, then E_t has good reduction at 5.

The elliptic curve E has j-invariant 0, complex multiplication by $\mathbf{Q}(\sqrt{-3})$, and good reduction at 5. The proof of Theorem 5.4 is the same as the proof of Theorem 4.4, after letting

$$\begin{aligned} &u_0 \text{ be a root of the polynomial } u^4 - 3u^3 - u^2 + 3u + 1, \\ &\mu = 25(151711644u_0^3 - 6630528u_0^2 - 171313440u_0 \\ &\qquad - 51318165)^{1/3}/(12D^{1/3}), \\ &\alpha = -((2u_0^3 - 6u_0^2 + 3)D)^{1/3}, \\ &\gamma = ((3u_0^3 - 6u_0 - 2)D)^{1/3}, \end{aligned}$$

using the real cube roots. The discriminant of E_t is

$$-2^{12}3^3D^2(25D^4t^{12} - 275D^3t^9 - 165D^2t^6 + 55Dt^3 + 1)^5$$

and the j-invariant is

$$\frac{-15^3Dt^3(5D^2t^6 + 40Dt^3 - 1)^3(5D^2t^6 - 5Dt^3 + 8)^3}{(25D^4t^{12} - 275D^3t^9 - 165D^2t^6 + 55Dt^3 + 1)^5} \times$$
$$\times (40D^2t^6 + 5Dt^3 + 1)^3.$$

Appendix. Since the polynomials $\alpha_k, \beta_k \in \mathbf{Q}[J]$ which occur in Theorem 5.1 are of marginal interest, we include them in this appendix.

$$\begin{aligned} \alpha_0 &= 1, \\ \alpha_1 &= 0, \\ \alpha_2 &= 190(J-1), \\ \alpha_3 &= -2280(J-1)^2, \\ \alpha_4 &= 855(J-1)^2(-17+16J), \end{aligned}$$

$$
\begin{aligned}
\alpha_5 &= 3648(J-1)^3(17-9J),\\
\alpha_6 &= 11400(J-1)^3(17-8J),\\
\alpha_7 &= -27360(J-1)^4(17+26J),\\
\alpha_8 &= 7410(J-1)^4(-119-448J+432J^2),\\
\alpha_9 &= 79040(J-1)^5(17+145J-108J^2),\\
\alpha_{10} &= 8892(J-1)^5(187+2640J-5104J^2+1152J^3),\\
\alpha_{11} &= 98800(J-1)^6(-17-388J+864J^2),\\
\alpha_{12} &= 7410(J-1)^6(-187-6160J+24464J^2-24192J^3),\\
\alpha_{13} &= 54720(J-1)^7(17+795J-3944J^2+9072J^3),\\
\alpha_{14} &= 2280(J-1)^7(221+13832J-103792J^2+554112J^3-373248J^4),\\
\alpha_{15} &= 1824(J-1)^8(-119-9842J+92608J^2-911520J^3+373248J^4),\\
\alpha_{16} &= 4275(J-1)^8(-17-1792J+23264J^2-378368J^3+338688J^4),\\
\alpha_{17} &= 18240(J-1)^9(1+133J-2132J^2+54000J^3-15552J^4),\\
\alpha_{18} &= 190(J-1)^9(17+2784J-58080J^2+2116864J^3-946944J^4+2985984J^5),\\
\alpha_{19} &= 360(J-1)^{10}(-1+28J-1152J^2)(1+228J+176J^2+1728J^3),\\
\alpha_{20} &= (J-1)^{10}(-19-4560J+144096J^2-9859328J^3-8798976J^4-226934784J^5+\\
&\quad 429981696J^6),\\
\beta_0 &= 1,\\
\beta_1 &= 30,\\
\beta_2 &= -435(J-1),\\
\beta_3 &= 580(J-1)(-7+9J),\\
\beta_4 &= 3915(J-1)^2(7-8J),\\
\beta_5 &= 1566(J-1)^2(91-78J+48J^2),\\
\beta_6 &= -84825(J-1)^3(7+16J),\\
\beta_7 &= 156600(J-1)^3(-13-91J+92J^2),\\
\beta_8 &= 450225(J-1)^4(13+208J-144J^2),\\
\beta_9 &= 100050(J-1)^4(143+4004J-5632J^2+1728J^3),\\
\beta_{10} &= 30015(J-1)^5(-1001-45760J+44880J^2-6912J^3),\\
\beta_{11} &= 600300(J-1)^5(-91-6175J+9272J^2-2736J^3),\\
\beta_{12} &= 950475(J-1)^6(91+8840J-7824J^2),\\
\beta_{13} &= 17108550(J-1)^6(7+926J-1072J^2+544J^3),\\
\beta_{14} &= 145422675(J-1)^7(-1-176J+48J^2-384J^3),\\
\beta_{15} &= 155117520(J-1)^8(1+228J+176J^2+1728J^3),\\
\beta_{16} &= 145422675(J-1)^8(1+288J+288J^2+5120J^3-6912J^4),\\
\beta_{17} &= 17108550(J-1)^8(7+2504J+3584J^2+93184J^3-283392J^4+165888J^5),
\end{aligned}
$$

$$\beta_{18} = 950475(J-1)^9(-91-39936J-122976J^2-2960384J^3+11577600J^4-5971968J^5),$$

$$\beta_{19} = 600300(J-1)^9(-91-48243J-191568J^2-6310304J^3+40515072J^4-46455552J^5+11943936J^6),$$

$$\beta_{20} = 30015(J-1)^{10}(1001+634920J+3880800J^2+142879744J^3-1168475904J^4+1188919296J^5-143327232J^6),$$

$$\beta_{21} = 100050(J-1)^{10}(143+107250J+808368J^2+38518336J^3-451953408J^4+757651968J^5-367276032J^6),$$

$$\beta_{22} = 450225(J-1)^{11}(-13-11440J-117216J^2-6444800J^3+94192384J^4-142000128J^5+95551488J^6),$$

$$\beta_{23} = 156600(J-1)^{11}(-13-13299J-163284J^2-11171552J^3+217203840J^4-474406656J^5+747740160J^6-429981696J^7),$$

$$\beta_{24} = 6525(J-1)^{12}(91+107536J+1680624J^2+132912128J^3-3147511552J^4+6260502528J^5-21054173184J^6+10319560704J^7),$$

$$\beta_{25} = 1566(J-1)^{12}(91+123292J+2261248J^2+216211904J^3-6487793920J^4+17369596928J^5-97854234624J^6+96136740864J^7-20639121408J^8),$$

$$\beta_{26} = 3915(J-1)^{13}(-7-10816J-242352J^2-26620160J^3+953885440J^4-2350596096J^5+26796552192J^6-13329432576J^7),$$

$$\beta_{27} = 580(J-1)^{13}(-7-12259J-317176J^2-41205008J^3+1808220160J^4-5714806016J^5+93590857728J^6-70131806208J^7-36118462464J^8),$$

$$\beta_{28} = 435(J-1)^{14}(1+1976J+60720J^2+8987648J^3-463120640J^4+1359157248J^5-40644882432J^6-5016453120J^7+61917364224J^8),$$

$$\beta_{29} = 30(J-1)^{14}(1+2218J+77680J^2+13365152J^3-822366976J^4+2990693888J^5-118286217216J^6-24514928640J^7+509958291456J^8-743008370688J^9),$$

$$\beta_{30} = (J-1)^{15}(-1-2480J-101040J^2-19642496J^3+1399023872J^4-4759216128J^5+315623485440J^6+471904911360J^7-2600529297408J^8+8916100448256J^9).$$

References

[1] F. Klein, *Vorlesungen über das Ikosaeder und die Auflösung der Gleichungen fünften Grades*, Leipzig (1884) (= Lectures on the icosahedron and the solution of equations of the fifth degree, London (1913)).

[2] F. Klein, *Elementary mathematics from an advanced standpoint: Arithmetic, Algebra, Analysis*, Dover Publications, New York (1945).

[3] J-P. Serre, *Cohomologie galoisienne*, Lecture Notes in Mathematics **5**, Springer-Verlag, Berlin-New York (1965).

[4] G. Shimura, *Moduli and fibre systems of abelian varieties*, Ann. Math. **83** (1966) 294–338.

[5] G. Shimura, *Introduction to the arithmetic theory of automorphic functions*, Princeton Univ. Press, Princeton (1971).

Conference on Elliptic Curves and Modular Forms
Hong Kong, December 18-21, 1993

A review of non-archimedean elliptic functions

JOHN TATE
DEPT. OF MATHEMATICS, UNIVERSITY OF TEXAS,
AUSTIN, TX 78712, USA

This expository article consists of two parts. The first is an old manuscript dating from 1959, entitled "Rational points on elliptic curves over complete fields" containing my first proof of the isomorphism $k^*/t^{\mathbb{Z}} \simeq E_t(k)$. The second part is a discussion of some further aspects of the theory. It begins with a sketch of some topics I had hoped to add to the old manuscript before publishing it, namely, the description of the rational functions on E_t as "rigid analytic" meromorphic functions on k^* with multiplicative period t, the construction of these via theta-functions, and the classification of isogenies between the E_t's. Then, after a discussion of some consequences of the isogeny classification, there is a description of the kernel $E_t[m]$ of multiplication by m on E_t as finite flat group scheme, and an indication of its relevance to the main theme of this conference. Finally, curves E_t over more general base rings than local fields are discussed, in particular, the "universal curve" E_q over $\mathbb{Z}[[q]][q^{-1}]$, and the connection with moduli. First, here is the old manuscript.

Rational Points on Elliptic Curves Over Complete Fields.

Let k be a field complete with respect to a non-trivial real valued valuation of the type satisfying

$$|x| \geq 0 , \quad |xy| = |x|\,|y| , \quad |x+y| \leq |x| + |y| .$$

According to a theorem of Ostrowski, k is either the real or complex field, with a valuation equivalent to the ordinary one, or else k is 'non-archimedean' in the sense that the valuation satisfies the stronger condition

$$|x+y| \leq \max\{|x|, |y|\} .$$

In the 'classical' case when k is the complex field, the group of points on an elliptic curve defined over k can be represented as the quotient of the additive group of k by a discrete subgroup generated by two independent 'periods' ω_1 and ω_2. Passing from the additive group to the multiplicative group by means of the exponential function one can absorb one of these periods and obtain a representation of the group of points as the quotient of the multiplicative group of k by a discrete subgroup generated by one multiplicative period $t = e^{2\pi i\tau}$, where $\tau = \omega_2/\omega_1$. The explicit formulas giving this multiplicative representation are the well-known Fourier expansions of the Weierstrass functions $\wp, g_2, g_3$, etc. These Fourier expansions, suitably normalized, yield 'universal' identities among power series with rational integral coefficients. Our aim in this paper is to show that these identities can be used to obtain an exactly parallel 'multiplicative' representation (Theorem 1) for the group of rational points on certain elliptic curves over an arbitrary complete field k. Unfortunately when k is non-archimedean this method works only for curves whose absolute invariant j satisfies $|j| > 1$. In a second paper we will consider the case $|j| \leq 1$, which is of quite a different nature.[1]

Let t be an element of k such that $0 < |t| < 1$ and consider the series

$$(1) \qquad x(w) = \sum_{m=-\infty}^{\infty} \frac{t^m w}{(1-t^m w)^2} - 2\sum_{m=1}^{\infty} \frac{t^m}{(1-t^m)^2},$$

where w is a non-zero variable in k. Using the identity

$$\frac{w}{(1-w)^2} = \frac{1}{w+w^{-1}-2} = \frac{w^{-1}}{(1-w^{-1})^2}$$

we can rewrite our series in the form

$$(2) \quad x(w) = \frac{w}{(1-w)^2} + \sum_{m=1}^{\infty} \left(\frac{t^m w}{(1-t^m w)^2} + \frac{t^m w^{-1}}{(1-t^m w^{-1})^2} - 2\frac{t^m}{(1-t^m)^2} \right)$$

which shows, by comparison with the geometric series $\sum_1^\infty t^m$, that the convergence is absolute for all $w \in k^*$ and is uniform for w in a subset of the form $r_1 \leq |w| \leq r_2$ and $|w - t^m| \geq \varepsilon$ for all $m \in \mathbb{Z}$, with $0 < r_1 < r_2$ and $\varepsilon > 0$. The functional equations

$$(3) \qquad x(tw) = x(w) = x(w^{-1})$$

are now obvious from (1) and (2) respectively. In the restricted range $|t| < |w| < |t|^{-1}$ we have $|t^m w| < 1$ and $|t^m w^{-1}| < 1$ for all positive integers m and can therefore expand the fractions under the summation sign in (2), obtaining (4)

$$\begin{aligned} x(w) &= \frac{w}{(1-w)^2} + \sum_{m=1}^{\infty}\sum_{n=1}^{\infty}(nt^{mn}w^n + nt^{mn}w^{-n} - 2nt^{mn}) \\ &= \frac{1}{w+w^{-1}-2} + \sum_{n=1}^{\infty} \frac{nt^n}{1-t^n}(w^n + w^{-n} - 2), \quad \text{for } |t| < |w| < |t|^{-1}. \end{aligned}$$

In case k is the complex field the classical Fourier expansions to which we alluded in the first paragraph can be written in the form

$$(5) \qquad \wp(u) = x(w) + \frac{1}{12} \qquad (w = e^u)$$

[1]This "second paper," never written, was to have discussed kernel of the reduction map via the formal group, especially in the ordinary case when that group is of height 1, so twisted multiplicative.

and

$$
\tag{6}
\begin{cases}
g_2 = \dfrac{1}{12} + 20\displaystyle\sum_{n=1}^{\infty} \frac{n^3 t^n}{1-t^n} \\
g_3 = -\dfrac{1}{216} + \dfrac{7}{3}\displaystyle\sum_{n=1}^{\infty} \frac{n^5 t^n}{1-t^n}
\end{cases} ,
$$

where $\wp(u), g_2$ and g_3 are the Weierstrass functions belonging to the periods $\omega_1 = 2\pi i$ and $\omega_2 = \log t$ (any value). A good reference for this classical theory is [**4**], Abschnitt II; formulas (5) and (6) are proved there in 2, §12. However, we shall pause here to give a direct proof for the sake of completeness. The function $x(w)$ is obviously meromorphic in the domain $k^* = k - \{0\}$, with multiplicative period t, its only singularities being double poles at the points $w = t^m$, $m \in \mathbb{Z}$. Therefore the function $x(e^u)$ is meromorphic in the whole u plane with additive periods $\omega_1 = 2\pi i$ and $\omega_2 = \log t$, its only singularities being double poles at the period points $m_1\omega_1 + m_2\omega_2$; $m_1, m_2 \in \mathbb{Z}$. The same is true of $\wp(u)$. The expansion of $\wp(u)$ in powers of u is

$$
\tag{7}
\wp(u) = \frac{1}{u^2} + \frac{1}{20} g_2 u^2 + \frac{1}{28} g_3 u^4 + \cdots .
$$

The corresponding expansion of $x(e^u)$ is readily obtained by substituting $w = e^u = 1 + u + \cdots$ in (4), namely

$$
\tag{8}
\begin{aligned}
x(w) &= \frac{1}{u^2 + \frac{1}{12}u^4 + \frac{1}{360}u^6 + \frac{1}{20160}u^8 + \cdots} \\
&\quad + \sum_{n=1}^{\infty} \frac{nt^n}{(1-t^n)} \left(n^2 u^2 + \frac{n^4}{12} u^4 + \cdots \right) \\
&= \frac{1}{u^2} - \frac{1}{12} + \left(\frac{1}{240} + \sum_{n=1}^{\infty} \frac{n^3 t^n}{1-t^n} \right) u^2 \\
&\quad + \left(-\frac{1}{6048} + \frac{1}{12} \sum_{n=1}^{\infty} \frac{n^5 t^n}{1-t^n} \right) u^4 + \cdots .
\end{aligned}
$$

Since the pole terms cancel, the difference $\wp(u) - x(w)$ is an entire meromorphic function, hence constant. Formulas (5) and (6) now follow by comparison of coefficients in the two expansions (7) and (8).

Differentiating (5) we obtain

$$
\tag{9}
\begin{aligned}
\wp'(u) = w\tfrac{d}{dw}x(w) &= \sum_{m=-\infty}^{\infty} \left(\frac{t^m w}{(1-t^m w)^2} + 2\frac{(t^m w)^2}{(1-t^m w)^3} \right) \\
&= x(w) + 2y(w) ,
\end{aligned}
$$

where

$$
\tag{10}
y(w) = \sum_{m=-\infty}^{\infty} \frac{(t^m w)^2}{(1-t^m w)^3} + \sum_{m=1}^{\infty} \frac{t^m}{(1-t^m)^2}
$$

$$
\tag{11}
= \frac{w^2}{(1-w)^3} + \sum_{m=1}^{\infty} \left(\frac{t^{2m} w^2}{(1-t^m w)^3} - \frac{t^m w^{-1}}{(1-t^m w^{-1})^3} + \frac{t^m}{(1-t^m)^2} \right) .
$$

Substituting (5) and (9) in the identity

$$\wp'^2 = 4\wp^3 - g_2\wp - g_3 \tag{12}$$

we find

$$y^2 + xy = x^3 - b_2 x - b_3 \ , \tag{13}$$

where

$$\begin{cases} b_2 &= \dfrac{1}{4}\left(g_2 - \dfrac{1}{12}\right) = 5\displaystyle\sum_{n=1}^{\infty} \dfrac{n^3 t^n}{1-t^n} = 5t + 45t^2 + 140t^3 + \cdots \\ b_3 &= \dfrac{1}{4}\left(g_3 + \dfrac{g_2}{12} - \dfrac{1}{432}\right) = \displaystyle\sum_{n=1}^{\infty}\left(\dfrac{7n^5 + 5n^3}{12}\right)\dfrac{t^n}{1-t^n} \\ &= t + 23t^2 + 154t^3 + \cdots \ . \end{cases} \tag{14}$$

The coefficients of these power series for b_2 and b_3 are integers because

$$7n^5 + 5n^3 = 7n^3(n^2-1) + 12n^3 \equiv 12n^3 \pmod{24} \ .$$

We now abandon the assumption that k is the complex field and see how much of the preceding theory carries over to an arbitrary complete field k, as described in the opening paragraph of this paper. Certainly the series (10) has the same convergence properties as (1) and can therefore be used to define a function $y(w)$ for non-zero $w \in k$. Trivial rearrangements of the defining series show that y satisfies the functional equations

$$y(tw) = y(w) \quad \text{and} \quad y(w^{-1}) + y(w) = -x(w) \ . \tag{15}$$

Moreover we can use the power series (14) to define elements b_2 and b_3 in k, no matter what its characteristic, because the coefficients are rational integers; the convergence for $|t| < 1$ in the non-archimedean case is obvious because the coefficients have absolute value ≤ 1. Thus it is natural to conjecture that for any complete field k and any element $t \in k$ with $0 < |t| < 1$ the equation (13) defines an elliptic curve, and the map $w \to (x(w), y(w))$ gives a parametrization of the points on this curve by the transcendental variable $w \in k^*$.

Let us denote by A the plane cubic curve which is defined over k by equation (13). To show that A is non-singular of genus 1 (rather than singular of genus 0) we must show that the discriminant Δ does not vanish. The expression for Δ as polynomial in b_2 and b_3, and hence as power series in t can be obtained by way of the classical formula in terms of g_2 and g_3:

$$\Delta = g_2^3 - 27g_3^2 = \left(4b_2 + \frac{1}{12}\right)^3 - 27\left(4b_3 - \frac{1}{3}b_2 - \frac{1}{216}\right)^2$$

$$= b_3 + b_2^2 + 72b_2b_3 - 432b_3^2 + 64b_2^3 \tag{16}$$

$$= t - 24t^2 + 252t^3 + \cdots \ . \tag{17}$$

In the classical case, hence also in the real case, we know $\Delta \neq 0$. In the non-archimedean case we see that $\Delta \neq 0$ because $\Delta \equiv t \pmod{t^2}$. Although we shall not require it, we mention here that the classical expansion of Δ as infinite product,

$$\Delta = t \prod_{n=1}^{\infty} (1 - t^n)^{24} \tag{17'}$$

holds for all t with $|t| < 1$ in any complete field, because the validity of this formula in the classical case ensures that it is a formal identity in the power series ring $\mathbb{Z}[\![t]\!]$, resulting from the substitution of (14) in (16). From (17′) the non-vanishing of Δ follows without consideration of the different cases. At any rate, our curve A is elliptic, with the absolute invariant

$$\begin{aligned} j &= \frac{(12g_2)^3}{\Delta} = \frac{(1+48b_2)^3}{\Delta} = \frac{1 + 240t + 2160t^2 + \cdots}{t - 24t^2 + 252t^3 + \cdots} \\ &= \frac{1}{t}(1 + 744t + 196884t^2 + \cdots)\ , \end{aligned} \tag{18}$$

just as in the classical case.

In the projective plane our curve A is complete and non-singular with just one point at infinity, where $x/y = 0$. We shall designate this infinite point by 0 and shall from now on view A as an abelian variety in the canonical way, with 0 as origin. Thus the addition of points on A is determined uniquely by the fact that the map which attaches to each point $P \in A$ the linear equivalence class of the divisor $(P) - (0)$ is an isomorphism between the group of points on A and the group of divisor classes of degree 0 on A; that this map is a bijection follows from the Riemann-Roch theorem which assures us that each divisor class of degree 1 on A contains one and only one point. For three points P_i on A we have $P_1 + P_2 + P_3 = 0$ if and only if the divisors $(P_1) + (P_2) + (P_3)$ and $3(0)$ are linearly equivalent. But $3(0)$ is the intersection divisor of A with the infinite line; hence our condition is simply that $(P_1) + (P_2) + (P_3)$ be the intersection divisor of A with some line in the projective plane.

Let $P_1 = (x_1, y_1)$ and $P_2 = (x_2, y_2)$ be points $\neq 0$ on A and consider the line L joining P_1 and P_2 (or tangent to A at $P_1 = P_2$ if the points coincide). L is parallel to the y-axis in the (x, y) plane if and only if

$$x_1 = x_2 \quad \textit{and} \quad y_1 + y_2 = -x_1\ . \tag{19}$$

Thus (19) is necessary and sufficient for L to pass through 0, or, equivalently, for $P_1 + P_2 = 0$. If $P_1 + P_2 \neq 0$, then L has an equation of the form

$$y = \lambda x + \nu\ , \tag{20}$$

where

$$\begin{aligned} \lambda &= \frac{y_1 - y_2}{x_1 - x_2} = \frac{x_1^2 + x_1 x_2 + x_2^2 - b_2 - y_2}{y_1 + y_2 + x_1} \\ \nu &= y_1 - \lambda x_1 = y_2 - \lambda x_2\ . \end{aligned} \tag{21}$$

(Notice that the two expressions for λ are identically equal when both are defined, and that at least one of them is defined since we are assuming (19) is false; in fact the second expression is needed only when $P_1 = P_2$, in which case it reduces to dy/dx at $P = P_1$.) Let $(x_3, \lambda x_3 + \nu)$ be the third intersection of L with A. Then x_3 is the third root of the cubic equation

$$x^3 - (\lambda^2 + \lambda)x^2 - \big(\nu(2\lambda + 1) + b_2\big)x - (b_3 + \nu^2) = 0 \tag{22}$$

which results from substituting (20) in (13). Hence $x_1 + x_2 + x_3 = \lambda^2 + \lambda$. Taking the negative of the third intersection using (19) we find that the sum $P_1 + P_2 = P_3$ has the coordinates

$$\begin{aligned} x_3 &= \lambda^2 + \lambda - x_1 - x_2 \\ y_3 &= -x_3 - \lambda x_3 - \nu , \end{aligned} \tag{23}$$

where $\lambda = \lambda(x_1, x_2, y_1, y_2)$ and $\nu = \nu(x_1, x_2, y_1, y_2)$ are given by (21).

Let $t^{\mathbb{Z}} = \{t^m \mid m \in \mathbb{Z}\}$ denote the infinite cyclic discrete subgroup of k^* which is generated by our element t, and let φ be the map of k^* into the projective plane which is defined by

$$\begin{aligned} \varphi(w) &= \big(x(w), y(w)\big) , \qquad && \text{if } w \notin t^{\mathbb{Z}} \\ \varphi(w) &= 0 , && \text{if } w \in t^{\mathbb{Z}} . \end{aligned} \tag{24}$$

Let A_k denote the group of points on A which are rational over k.

Theorem 1. *The map φ is a homomorphism of k^* onto A_k with kernel $t^{\mathbb{Z}}$.*

We prove first that φ maps k^* *into* A. Since $0 \in A$, this amounts to proving that equation (13) is an identity between the functions $x(w)$ and $y(w)$, holding for all values of the argument $w \in k^*$, $w \notin t^{\mathbb{Z}}$, for which the functions are defined. Since these functions have multiplicative period t, it is enough to consider values of w such that $|t| < |w| \leq 1$ and $w \neq 1$. In this range we can use formula (4) which expresses x as a power series in t with coefficients which are rational functions of w. There is an analogous expression for y which is obtained by expanding the fractions under the summation sign in (11) and which we don't bother to write out. Thus our contention will be proved if we can show that (13) is a *formal* identity when we interpret x and y (and b_2 and b_3) as *formal* power series in t with coefficients which are rational functions of an *indeterminate* w. Now in fact the coefficients of the formal power series in question are expressed as elements of the ring $\mathbb{Z}[w, w^{-1}, (1-w)^{-1}]$, *i.e.*, the subring of $\mathbb{Q}(w)$ generated by $1, w, w^{-1}$ and $(1-w)^{-1}$. The canonical homomorphism $\mathbb{Z} \to k$ extends to a homomorphism $\mathbb{Z}[w, w^{-1}, (1-w)^{-1}] \to k(w)$. Hence the formal identity we are trying to establish is a 'universal' one, and will hold in any characteristic provided it holds in characteristic 0. From the classical theory we know that our equation (13) holds true *numerically* if we substitute any pair of complex numbers w and t in the domain of convergence $|t| < |w| < |t|^{-1}$, $w \neq 1$. Fixing first w such that $|w| < 1$ and letting t vary we conclude that the resulting power series in t with complex coefficients are equal coefficient-wise; then letting w vary, we conclude that the coefficients are formally equal as rational functions of an indeterminate, as was to be shown.

Next we prove that φ is a homomorphism. Given $w_1, w_2 \in k^*$, we put $w_3 = w_1 w_2$ and must prove

$$P_3 = P_1 + P_2 , \qquad P_i = \varphi(w_i) , \quad i = 1, 2, 3 . \tag{25}$$

In view of the periodicity $\varphi(tw) = \varphi(w)$ we can restrict our consideration to values of w_1 and w_2 in the ranges $|t| < |w_1| \leq 1$ and $1 \leq |w_2| < |t|^{-1}$. Then $|t| < |w_3| < |t|^{-1}$, so that all three w_i are within the domain of convergence of the power series expressions for x and y considered in the previous paragraph. Since $\varphi(1) = 0$ by definition, (25) holds trivially if $w_1 = 1$ or $w_2 = 1$. From (3), (15), and (19) we see that it also holds if $w_1 w_2 = 1$. Thus we may assume all three points P_i are different from 0 and write $P_i = (x_i, y_i)$, *i.e.*, put $x_i = x(w_i)$, $y_i = y(w_i)$ for $i = 1, 2, 3$. Suppose $x_1 \neq x_2$. Then, using the first expression for λ in (21), substituting (21) in (23), and clearing denominators we see that (25) is equivalent to the simultaneous identities

$$(25') \quad \begin{aligned} (x_1 - x_2)^2 x_3 &= (y_1 - y_2)^2 + (y_1 - y_2)(x_1 - x_2) - (x_1 - x_2)^2(x_1 + x_2) \\ (x_1 - x_2) y_3 &= -(x_1 - x_2)(y_1 + x_3) + (y_1 - y_2)(x_1 - x_3) \,. \end{aligned}$$

Now we can argue just as in the preceding paragraph: (25′) holds in the classical case (being in fact just the addition formulas for $\wp(u)$ and $\wp'(u)$). Hence (25′) is an identity in the ring of formal power series in t with coefficients in the ring

$$\mathbb{Z}\left[w_1, w_1^{-1}, w_2, w_2^{-1}, (1 - w_1)^{-1}, (1 - w_2)^{-1}, (1 - w_1 w_2)^{-1}\right] ,$$

and is therefore a functional identity in any complete field k. We could take care of the remaining case $x_1 = x_2$ by other explicit formulas, or by a continuity argument, but perhaps the simplest way out is to observe that $x_1 = x_2$ if and only if $P_1 = \pm P_2$ and to appeal to

Lemma 1. *Let φ be a map of a (multiplicative) group into an (additive) group which takes on an infinite number of distinct values and satisfies the identity $\varphi(w_1 w_2) = \varphi(w_1) + \varphi(w_2)$ whenever $\varphi(w_1) \neq \pm\varphi(w_2)$. Then φ is a homomorphism.*

Indeed, given w_1, w_2 we can select w so that

$$\varphi(w) \neq \pm\varphi(w_1) \quad , \quad \varphi(w) + \varphi(w_1) \neq \pm\varphi(w_2) \quad , \quad \varphi(w) \neq \pm\varphi(w_1 w_2)$$

and have then $\varphi(w w_1) = \varphi(w) + \varphi(w_1)$. Then

$$\varphi(w) + \varphi(w_1) + \varphi(w_2) = \varphi(w w_1) + \varphi(w_2) = \varphi(w w_1 w_2) = \varphi(w) + \varphi(w_1 w_2)$$

and the lemma follows upon cancelling $\varphi(w)$. It is obvious that our φ takes an infinity of values; for example (2) shows that $|x(1 + t^r)| = |t|^{-2r}$ in the non-archimedean case.

Thus φ is a homomorphism. That its kernel is $t^{\mathbb{Z}}$ is apparent from its very definition (24), and to complete the proof of Theorem 1 we have only to show that φ is surjective. This is true in the classical case, and the case in which k is the real field can be obtained as a corollary by means of the following lemma, which will also be of assistance in the non-archimedean case.

Lemma 2. *If $w \neq 0$ is separably algebraic over k and $\varphi(w) \in A_k$, then $w \in k$.*

Let K be a finite Galois extension of k containing w. Since k is complete, there is a unique extension of its valuation to K, and K is complete in the extended valuation. Thus we can consider the resulting map $\varphi : K^* \to A_K$. According to what we have already proved (applied to K instead of k), φ is a homomorphism with kernel $t^{\mathbb{Z}}$. Each automorphism σ of K over k preserves the extended valuation and therefore commutes with φ. Thus if $\varphi(w) \in A_k$, then $\varphi(w^\sigma) = \varphi(w)$, and w^σ/w is a power of t. But w^σ and w have the same absolute value so this power of t must be 1. Thus $w^\sigma = w$ for all σ and $w \in k$ as contended.

Before treating the case of non-archimedean k we first recall some facts about power series in such a field. Any non-zero power series can be written in the canonical form

$$f(z) = \alpha z^\nu(1 + a_1 z + a_2 z^2 + \cdots) \quad \text{with} \quad \alpha \neq 0\,. \tag{26}$$

We define the *norm* of f by

$$\|f\| = \sup_{1 \le n < \infty} \{|a_n|^{1/n}\}\,. \tag{27}$$

Notice that this is a sup rather than a lim sup; $\|f\|$ is the smallest real number such that

$$|a_n| \le \|f\|^n \quad \text{for } all \quad n = 1, 2, 3, \ldots\,, \tag{28}$$

or is $+\infty$ if no such number exists. We call the open circle $|z| < \|f\|^{-1}$ the *inner* circle of convergence of f. Of course f may converge in some larger circle, but if so it does so only hesitantly. In the inner circle the convergence is direct and business-like right from the start, for we have $|a_n z^n| < (\|f\|\,|z|)^n$ for $n = 1, 2, 3, \ldots$ there, and $\|f\|\,|z| < 1$. This shows also that we have $|f(z) - \alpha z^\nu| < |\alpha z^\nu|$ in the inner circle, because the valuation is non-archimedean.

Proposition 1. *The non-zero power series with norm less than or equal to a fixed real number $\rho \ge 0$ form a group under multiplication.*

Let $f(z)$ be as in (26) and $g(z) = \beta z^\mu(1 + b_1 z + b_2 z^2 + \cdots)$. Their quotient is

$$\frac{f(z)}{g(z)} = \frac{\alpha}{\beta} z^{\nu-\mu}(1 + c_1 z + c_2 z^2 + \cdots)$$

where the c_n are determined recursively from the equation

$$a_n = b_n + b_{n-1}c_1 + b_{n-2}c_2 + \cdots + b_1 c_{n-1} + c_n\,.$$

From this equation we can prove inductively that $|c_n| \le \rho^n$ if we assume $|a_n| \le \rho^n$ and $|b_n| \le \rho^n$ for $n = 1, 2, 3, \ldots$. Thus $\|f\| \le \rho$ and $\|g\| \le \rho$ implies $\|f/g\| \le \rho$. This proves Proposition 1.

Proposition 2. *The power series of the form*

$$f(z) = z + a_1 z^2 + a_2 z^3 + \cdots \tag{29}$$

with norm less than or equal to a fixed real number $\rho \ge 0$ form a group with respect to the operation of composition $(f \circ g)(z) = f((g(z))$.

Let $f(z)$ be given by (29), let $g(z) = z + b_1 z^2 + b_2 z^3 + \cdots$ be another series of the same form, and let $h = f \circ \overset{-1}{g}$ be the series such that $f(z) = h(g(z))$. The coefficients c_i of $h(z) = z + c_1z^2 + c_2z^3 + \cdots$ are determined uniquely in terms of the a's and b's by comparing coefficients in the identity

$$z + a_1z^2 + a_2z^3 + \cdots = g(z) + c_1\left(g(z)\right)^2 + c_2\left(g(z)\right)^3 + \cdots .$$

Putting $g(z)^i = z^i + b_1^{(i)}z^{i+1} + b_2^{(i)}z^{i+2} + \cdots$ we can write the recursive equations for the c's in the form

$$a_n = b_n + c_1b_{n-1}^{(2)} + c_2b_{n-2}^{(3)} + \cdots + c_{n-1}b^{(n)} + c_n$$

for $n = 1, 2, \ldots$. Assume now $||f|| \le \rho$ and $||g|| \le \rho$. By Proposition 1 we have $||g^i|| \le \rho$ for all i. Thus $|a_n| \le \rho^n$, $|b_n| \le \rho^n$ *and* $|b_n^{(i)}| \le \rho^n$ for all $n \ge 1$, $i \ge 2$, and it follows by induction that $|c_n| \le \rho^n$ for all n. Hence $||h|| = ||f \circ \overset{-1}{g}|| \le \rho$, and the proposition is proved.

In particular if our power series (29) has norm $\le \rho$ then so does its inverse series

$$\overset{-1}{f}(z) = z - a_1z^2 - (a_2 - 2a_1^2)z^3 - (a_3 - 5a_2a_1 + 5a_1^3)z^4 + \cdots , \tag{30}$$

and since f is the inverse of its inverse it follows that f and $\overset{-1}{f}$ have the same norm and hence the same inner circle of convergence. Each maps that circle into itself, and the identities $z = \overset{-1}{f}(f(z)) = f(\overset{-1}{f}(z))$ are satisfied for all z in the circle because the absolute convergence ensures the validity of the rearrangements of the series necessary to prove them. Hence:

Corollary 1. *A power series of the form* (29) *maps its inner circle of convergence* $|z| < ||f||^{-1}$ *bijectively onto itself, the inverse mapping being given by the formal inverse series* (30).

Consider a power series of the form

$$g(z) = \frac{1}{z} + a_1 + a_2z + a_3z^2 + \cdots . \tag{31}$$

Its reciprocal $1/g(z)$ is of the form (29) and has the same norm. Applying Corollary 1 to this reciprocal we obtain

Corollary 2. *A power series of the form* (31) *maps its inner circle of convergence* $|z| < ||g||^{-1}$ *($z = 0$ excluded) bijectively onto the domain* $|z| > ||g||$.

These general facts about inversion of power series in a complete non-archimedean field will enable us to prove

Lemma 3. *If k is non-archimedean then for each $x \in k$ there exists an element w in a quadratic extension of k such that $x = x(w)$. Moreover w is separable over k except possibly if $x = 0$ and the characteristic of k is* 2.

The surjectivity of $\varphi : k^* \to A_k$ follows directly from Lemmas 2 and 3. Given a point $P = (x, y)$ in A_k we choose w as in Lemma 3 such that $x = x(w)$.

Then $\varphi(w)$ and $\varphi(w^{-1}) = -\varphi(w)$ are points on A with the same x-coordinate as P. Since they are opposites, they are the only such points, so one of them is our given point P. By Lemma 2 we have $w \in k$ if w is separable. Thus every point of A_k is the image of some $w \in k^*$, except possibly the point $(0, \sqrt{b_3}\,)$ in characteristic 2. But the image of k^* under φ is a subgroup of A_k consisting of more than one element and its complement, which is a union of cosets, cannot possibly consist of one point. Thus the exception does not occur.

To prove Lemma 3 we treat the cases $|x| > |t|^{1/2}$ and $|x| < 1$ by different but analogous methods. In the first case we use the new variable

$$r = w + w^{-1} - 2\ . \tag{32}$$

Each fixed value $r \in k$ determines two reciprocal values of w, namely the roots of the quadratic equation

$$w^2 - (r+2)w + 1 = 0 \tag{33}$$

and these roots are separable over k because $w = w^{-1}$ implies $w = \pm 1 \in k$. For all integers $n \geq 1$ we have

$$w^n + w^{-n} - 2 = F_n(r) = r^n + c_{n,1}r^{n-1} + \cdots + c_{n,n-1}r \tag{34}$$

where $F_n(r)$ is a monic polynomial of degree n with rational integral coefficients $c_{n,i}$ and without constant term. (For example, this follows from the recurrence relation $F_{n+1} = (z+2)F_n - F_{n-1} + 2r$.) Substituting (34) in (4) and rearranging the resulting series in powers of r we obtain

$$x(w) = f(r) = \frac{1}{r} + a_1 r + a_2 r^2 + \cdots\ , \tag{35}$$

where the coefficients

$$a_n = \frac{nt^n}{1-t^n} + c_{n+1,1}\frac{(n+1)t^{n+1}}{1-t^{n+1}} + c_{n+2,2}\frac{(n+2)t^{n+2}}{1-t^{n+2}} + \cdots$$

are elements of k satisfying $|a_1| = |t|$, and $|a_n| \leq |t|^n$ for all $n \geq 1$. Notice that the numbering of the coefficients in (35) is shifted by one from that in the canonical form (31), so that the norm of f, being the supremum of the numbers $|a_n|^{1/(n+1)}$, is $|t|^{1/2}$ rather than $|t|$. Thus, although $f(r)$ converges for $|r| < |t|^{-1}$, the inner circle of convergence is only $|r| < |t|^{-1/2}$. Corollary 2 shows now that given $x \in k$, $|x| > |t|^{1/2}$, there exists $r \in k$, $|r| < |t|^{-1/2}$ such that $x = f(r)$ and consequently $x = x(w)$, where w is a root of (33). For values of x such that $|x| < 1$ we shall expand $x(w)$ in terms of another variable, namely

$$s = w + tw^{-1}\ . \tag{36}$$

Each value of $s \in k$ determines two values of w, namely the roots of the quadratic equation

$$w^2 - sw + t = 0\ . \tag{37}$$

These roots are separable over k except in the case $s = 0$, characteristic of $k = 2$, and $t \notin (k^*)^2$. For all integers $n \geq 1$ we have

$$w^n + (tw^{-1})^n = G_n(s,t) = s^n + d_{n,2}ts^{n-2} + d_{n,4}t^2s^{n-4} + \cdots\ , \tag{38}$$

where $G_n(s,t)$ is a polynomial in s and t with rational integral coefficients $d_{n,i}$ of the form indicated. This follows for example from the recursion relation $G_{n+1} = sG_n - tG_{n-1}$ and the initial conditions $G_0 = 2$, $G_1 = s$. From (2) we have

(39)

$$\begin{aligned} x(w) + 2\sum_{m=1}^{\infty} \frac{t^m}{(1-t^m)^2} &= \sum_{m=0}^{\infty}\left(\frac{t^m}{(1-t^m w)^2} + \frac{t^m(tw^{-1})}{(1-t^m(tw^{-1}))^2}\right) \\ &= \sum_{m=0}^{\infty}\sum_{n=1}^{\infty} nt^{mn}w^n + nt^{mn}(tw^{-1})^n \\ &= \sum_{n=1}^{\infty} \frac{n}{1-t^n}(s^n + d_{n,2}ts^{n-2} + d_{n,4}t^2s^{n-4} + \cdots) \\ &= a_0 + a_1 s + a_2 s^2 + \cdots , \end{aligned}$$

where

$$a_m = \frac{m}{1-t^m} + d_{m+2,2}\frac{m+2}{1-t^{m+2}}t + d_{m+4,4}\frac{m+4}{1-t^{m+4}}t^2 + \cdots . \tag{40}$$

In particular,

$$a_0 = d_{2,2}\frac{2t}{1-t^2} + d_{4,4}\frac{4t^2}{1-t^4} + \cdots$$

is divisible by $2t$, *i.e.*, $|a_0| \le |2t|$, and

$$a_1 = \frac{1}{1-t} + d_{3,2}\frac{3t}{1-t^3} + d_{5,4}\frac{5t^2}{1-t^5} + \cdots$$

is a unit, *i.e.*, $|a_1| = 1$, and all a_n are integers, *i.e.*, $|a_n| \le 1$ for all $n \ge 1$. Thus we have

$$x(w) = 2t\alpha + \beta f(s) , \tag{41}$$

where $|\alpha| \le 1$, $|\beta| = 1$, and where $f(s) = s + (a_2/a_1)s^2 + \cdots$ is a power series of the form (29) with norm 1. Since the map $z \to 2t\alpha + \beta z$ is a bijection of the circle $|z| < 1$ with itself we conclude from Corollary 1 that, given any $x \in k$ with $|x| < 1$, there exists $s \in k$ with $x = 2t\alpha + \beta f(s)$ and consequently $x = x(w)$ where w is a root of (37). This concludes the proof of Lemma 3 and Theorem 1.

Let us have a closer look at our homomorphism $\varphi : k^* \to A_k$ in the non-archimedean case. Because of the periodicity $\varphi(tw) = \varphi(w)$ we can restrict our attention to values of w such that $|t| < w \le 1$. For these we have by formulas (2) and (8′)

$$\left|x(w) - \frac{w}{(1-w)^2}\right| < 1 \quad \text{and} \quad \left|y(w) - \frac{w^2}{(1-w)^3}\right| < 1 . \tag{42}$$

Hence

$$\begin{aligned} &|t| < |w| < 1 \Longrightarrow |x(w)| < 1 \quad \text{and} \quad |y(w)| < 1 \\ &|w| = 1 \Longrightarrow |x(w)| = \frac{1}{|1-w|^2} \ge 1 \quad \text{and} \quad |y(w)| = \frac{1}{|1-w|^3} \ge 1 . \end{aligned} \tag{43}$$

Let us denote the ring of integers in k by $R_k = \{z \in k; |z| \leq 1\}$, and the group of units of R_k by $U_k = \{z \in k; |z| = 1\}$. Corresponding to each real number ρ, $0 < \rho \leq 1$, we have subgroups of U_k defined and denoted by

$$\begin{aligned} U_k[\rho] &= \{z \in U_k \,;\, |z-1| \leq \rho\} \\ U_k(\rho) &= \{z \in U_k \,;\, |z-1| < \rho\} \,. \end{aligned} \tag{44}$$

According to (43) the image of U_k under φ is the subset of A_k consisting of 0 and the points $P = (x, y)$ such that $|x| \geq 1$ and $|y| \geq 1$. This subset, which we shall denote by B_k, is therefore a subgroup of A_k, isomorphic to U_k. It also follows from (43) that the images under φ of the subgroups $U_k[\rho]$ and $U_k(\rho)$ are, respectively,

$$\begin{aligned} B_k[\rho] &= \{P = (x,y) \in A_k \,;\, |x| \geq \rho^{-2}\} \\ B_k(\rho) &= \{P = (x,y) \in A_k \,;\, |x| > \rho^{-2}\} \,. \end{aligned} \tag{45}$$

The fact that these subsets of A_k are subgroups can of course be proved directly from the algebraic addition formulas, without resort to our transcendental parameter w; the necessary argument becomes quite transparent if one uses the new variables $\xi = x/y$ and $\eta = 1/y$ which put 0 at the origin of the (ξ, η)-plane. Notice that ξ is a uniformizing parameter at 0. B_k is the subset of the (ξ, η)-plane where $|\xi| \leq 1$ and $|\eta| \leq 1$, and $B_k[\rho]$ is the subset of B_k where $|\xi| \leq \rho$. We don't go into details here.

The quotient group A_k/B_k is isomorphic to $k^*/\varphi^{-1}(B_k) = k^*/\langle t \rangle U_k$, and therefore to the quotient group of all absolute values of elements of k^* (the "value group" of k) by the group of powers of $|t|$. Thus, A_k/B_k is finite if and only if the valuation is discrete, in which case A_k/B_k is a cyclic group of order e, where e is the ordinal number of t in k. In particular, we have $A_k = B_k$ if and only if t is a prime element in R_k.

The quotient group $B_k/B_k(1)$ is isomorphic to $U_k/U_k(1)$ and therefore to the multiplicative group of the residue class field $\bar{k} = R_k/P_k$. (Here $P_k = \{z; |z| < 1\}$ is the maximal ideal of R_k.) Now the coefficients of the defining equation (9) of our curve A lie in R_k; reducing them mod y_k we obtain

$$\bar{y}^2 + \bar{x}\,\bar{y} = \bar{x}^3 \tag{46}$$

as the equation for the "reduced curve" $\bar{A}$, defined over the residue class field. $\bar{A}$ is singular, having an ordinary double point at the origin $\bar{x} = 0$, $\bar{y} = 0$. There is one point at infinity on $\bar{A}$, which we shall denote by $\bar{0}$. The set of *non-singular* points on $\bar{A}$ forms a group with $\bar{0}$ as neutral element, the addition of non-singular points on $\bar{A}$ being defined geometrically in the same way as the addition of points on A, by the rule $\bar{P}_1 + \bar{P}_2 + \bar{P}_3 = \bar{0}$ if and only if $(\bar{P}_1) + (\bar{P}_2) + (\bar{P}_3)$ is the intersection of $\bar{A}$ with a line in the projective plane not passing through the singular point $(0,0)$. Being singular and cubic, $\bar{A}$ must be of genus 0. In fact, the map

$$\bar{\varphi}(\bar{w}) = (\bar{w}, \bar{y}) \,, \quad \text{where } \bar{x} = \frac{\bar{w}}{(1-\bar{w})^2} \quad , \quad \bar{y} = \frac{\bar{w}^2}{(1-\bar{w})^3} \,, \tag{47}$$

gives a birational transformation of the $\bar{w}$-line onto $\bar{A}$, the inverse transform being $\bar{w} = \bar{y}^2/\bar{x}^3$. This transformation $\bar{\varphi}$ carries the points $\bar{w} = 0$ and $\bar{w} = \infty$

onto the singular point of $\bar{A}$; for the remaining points it is an isomorphism of the multiplicative group onto the group of non-singular points of $\bar{A}$. Applying the place $k \to \bar{k}$ to coordinates of points we obtain a *reduction map* $\theta : A \to \bar{A}$. Specifically, we have $\theta(P) = \bar{0}$ if and only if $P \in B(1)$, and for the remaining points $P = (x, y)$, $\theta(P) = (\bar{x}, \bar{y})$ where bar denotes the residue class of whatever is under it. Thus B is just the set of points P on A such that $\theta(P)$ is a simple point of $\bar{A}$, and we shall denote the group formed by these simple points by $\bar{B}$. The diagram

$$(48)\qquad \begin{array}{ccc} U_k & \xrightarrow{\ \varphi\ } & B_k \\ \downarrow & & \downarrow{\scriptstyle\theta} \\ \bar{k}^* & \xrightarrow{\ \bar{\varphi}\ } & \bar{B}_{\bar{k}} \end{array}$$

is commutative, as one sees from (47) and (42). Thus our homomorphism φ can be viewed as a transcendental "lifting" of the birational transformation $\bar{\varphi}$, and from this point of view the success of our method becomes more understandable.

In the following discussion we will denote the curve (13) of the above manuscript by E or E_t instead of A, and its group of k-rational points by $E(k)$ instead of A_k. We also assume that k is non-archimedean, and let C denote the completion of the algebraic closure of k.

The field of rational functions on E via theta functions.

We will denote by $K = k(x, y)$ the field of rational functions on E which are defined over k. In the classical case $k = \mathbb{C}$, K can be identified with the field of meromorphic functions f on $\mathbb{C}^*$ which have multiplicative period t, that is, satisfy $f(tw) = f(w)$ for all w in $\mathbb{C}^*$. The same is true for our non-archimedean k if we add "defined over k" and define "meromorphic function on C^* defined over k" to be an element of the field of fractions of the ring of holomorphic ones, where by "holomorphic function on C^* defined over k" we mean a function $f : C^* \to C^*$ which is representable by an everywhere convergent Laurent series with coefficients in k. For a non-zero such series, the number of zeros in an annulus $r \leq |w| \leq R$ is finite, and in fact, the exact number of such zeros (multiplicities counted) can be read off the Newton Polygon of the series in the usual way; see e.g., [**2**], Ch.II. The divisor of such a function is a collection of points c_i in C^* with multiplicities, such that each annulus contains only a finite number of c_i, and which is "defined over k" in the sense that the c_i are algebraic over k, and points conjugate over k have the same multiplicity, which is a multiple of the degree of inseparability of the point over k. Every such divisor is the divisor of a holomorphic function. In contrast to the classical case, the only holomorphic functions on k^* with no zeros in $\bar{k}$ are the monomials aw^n, $a \in k^*$, and in the Weierstrass product

$$\prod_{i,|c_i|\geq 1} \left(1 - \frac{w}{c_i}\right) \prod_{i,|c_i|<1} \left(1 - \frac{c_i}{w}\right)$$

showing the existence of a function with a given divisor $\{c_i\}$, no convergence factors are needed.

By an elliptic function (with period t) defined over k we mean a function meromorphic on C^* which is defined over k and invariant under $w \mapsto tw$. The above considerations show that the divisor of such a function $f(w)$ is the difference of two disjoint divisors with multiplicities > 0, each of which is invariant by $w \mapsto tw$ and defined over k. Consequently, $f = g/h$, where g and h are holomorphic functions on C^* defined over k, which satisfy a functional equation of the form $g(w) = aw^n g(tw)$. The monomial aw^n is the same for g and for h, since $f(w) = f(tw)$. Such functions are called theta functions of type aw^n, and n is called their degree. The dimension of the space of theta functions of a given type aw^n is n for $n > 0$, and is 0 for $n < 0$ (for $n = 0$ it is 1 if $a = t^m$ for some m in $\mathbb{Z}$, in which case the thetas are the monomials of degree $-m$, and is 0 otherwise, as is easily checked). Indeed, for $n \neq 0$, the space of *formal* Laurent series f satisfying $f(w) = aw^n f(tw)$ is of dimension $|n|$, because $|n|$ successive coefficients of the series can be arbitrarily prescribed, and the remaining ones are then determined by the relation $f(w) = aw^n f(tw)$. The resulting series converge everywhere if $n > 0$, but diverge if $n < 0$. The Newton Polygon of such a series is invariant under the map $(x, y) \mapsto (x + n, y - \log|a| - x \log|t|)$. From this it follows easily that a theta function $g(w)$ of degree $n > 0$ has exactly n zeros in an annulus of the form $r|t| < |w| \leq r$. Hence, an elliptic function f has the same number of zeros as poles in such a "period annulus", and if that number is n, then f is the quotient of two theta functions of degree n of the same type. For $n = 0$ or 1, the ratio of two such thetas is constant. Hence, an elliptic function with no pole is constant, and there is no elliptic function with just one simple pole in a period annulus.

Now one can proceed as in the classical Weierstrass theory. The role of the Weierstrass sigma function is played by a theta function of type $-w$, for example, the function

$$\theta(w) = \sum_{n \in \mathbb{Z}} (-1)^n t^{\frac{n^2-n}{2}} w^n .$$

The zeros of $\theta(w)$ are the points $w = t^n$, $n \in \mathbb{Z}$, each with multiplicity one. This follows from consideration of the Newton Polygon as above, together with the fact that $\theta(1) = 0$. It also follows from the classical identity

$$\theta(w) = (1 - w) \prod_{m=1}^{\infty} [(1 - t^m)(1 - t^m w)(1 - t^m w^{-1})]$$

Using that identity it is easy to show that $\theta(w)^2 x(w)$ and $\theta(w)^3 y(w)$ are holomorphic, and consequently, $x(w)$ and $y(w)$ are elliptic functions in the above sense. In fact, the differential operators $D := w d/dw$ and $D_2 := \frac{1}{2}(D^2 - D)$ act on the ring of holomorphic functions, and, putting

$$\zeta(w) = -\frac{D\theta(w)}{\theta(w)}$$

we have $\zeta(tw) = 1 + \zeta(w)$, and

$$x(w) = D\zeta(w) \quad , \quad y(w) = D_2\zeta(w) .$$

Thus $Dx = 2y + x$, or, in terms of differentials,

$$dx/(2y+x) = dw/w\ .$$

Further details (e.g., the fact that elliptic functions are rational functions of x and y, the expression of elliptic functions in terms of $\theta(w)$, etc.) are left to the reader; references are **[9]** and **[13]**. Everything is as in the classical case, except perhaps for some questions of rationality, like the surjectivity of the homomorphism $\varphi : k^* \to E(k)$. A shorter proof of that than the one in the manuscript above is as follows. Let $P = (x_0, y_0)$ be a point on $E(k)$. The function $x - x_0$ has zeros only at P and $-P$. If P is not in the image of φ, then the function $w \mapsto x(w) - x_0$ has no zero in k^*. That function has a pole at $w = 1$, hence it has a zero at some point w_0 in C^*. Then $\varphi(w_0) = P$ or $-P$, and, replacing w_0 by $1/w_0$ if necessary, we can assume $\varphi(w_0) = P$, i.e., that $w = w_0$ is a common zero of $x(w) - x_0$ and of $y(w) - y_0$. These two functions can't have two common zeros in a period annulus — otherwise the function $y(w)/x(w)$ would have exactly one pole there, which is impossible. Thus the divisor consisting of the points $w_0 t^n$, $n \in \mathbb{Z}$, with multiplicity one, is rational over k, i.e., w_0 is in k^*, as was to be shown.

At one point later on it will be convenient to consider the group of points of our curve E_t not only with coordinates in k or C, but also in a finite commutative k-algebra A. Any such algebra is a product of local ones, so it suffices to consider that case. We leave to the reader the exercise of showing that for a local A, our functions x and y, or perhaps better, the theta functions of type $-w^3$, give a surjective homomorphism

$$\varphi : A^* \to E_t(A)$$

with kernel $t^{\mathbb{Z}}$, so induce an isomorphism

$$A^*/t^{\mathbb{Z}} \simeq E(A)\ ,$$

which is of course functorial in A. One can argue by induction on the length of A. If I is an ideal of square zero and $B = A/I$, the induction step from B to A can be made with the diagram

$$\begin{array}{ccccccccc} 0 & \longrightarrow & \mathrm{Lie}(\mathbb{G}_m)\otimes I & \longrightarrow & A^*/t^{\mathbb{Z}} & \longrightarrow & B^*/t^{\mathbb{Z}} & \longrightarrow & 0 \\ & & \big\downarrow{\scriptstyle \mathrm{Lie}(\varphi)\otimes\mathrm{id}} & & \big\downarrow{\scriptstyle\varphi} & & \big\downarrow{\scriptstyle\varphi} & & \\ 0 & \longrightarrow & \mathrm{Lie}(E_t)\otimes I & \longrightarrow & E_t(A) & \longrightarrow & E_t(B) & \longrightarrow & 0 \end{array}$$

For non local A it follows that $E(A) = A^*/t^{\mathbb{Z}(A)}$, where $\mathbb{Z}(A)$ is the group of continuous maps $i : \mathrm{Spec}(A) \to \mathbb{Z}$.

Isogenies.

Suppose $t_1, t_2 \in k$, with $0 < |t_1|, |t_2| < 1$. We will say t_1 and t_2 are "commensurable" if there exist non-zero integers m, n such that $t_2^m = t_1^n$. When this is the case the left hand square of the following diagram is commutative

$$\begin{array}{ccccccccc} 0 & \longrightarrow & \mathbb{Z} & \xrightarrow{1\mapsto t_1} & C^* & \xrightarrow{\varphi} & E_{t_1}(C) & \longrightarrow & 0 \\ & & \big\downarrow{\scriptstyle m} & & \big\downarrow{\scriptstyle n} & & \big\downarrow{\scriptstyle \alpha_{m,n}} & & \\ 0 & \longrightarrow & \mathbb{Z} & \xrightarrow[1\mapsto t_2]{} & C^* & \xrightarrow{\varphi} & E_{t_2}(C) & \longrightarrow & 0 \end{array}$$

and consequently there is a unique homomorphism

$$\alpha_{m,n} : E_{t_1}(C) \to E_{t_2}(C)$$

making the right hand square commutative. This map $\alpha_{m,n}$ is in fact an isogeny defined over k, because composition with $w \mapsto w^n$ takes meromorphic functions on C^* with period t_2 into those with period t_1 and preserves the field of definition of such functions, and therefore composition with $\alpha_{m,n}$ carries rational functions on E_{t_2} defined over k into those on E_{t_1}.

Theorem. *The degree of $\alpha_{m,n}$ is mn. Every isogeny between curves of type E_t is of the form $\alpha_{m,n}$; the map $(m,n) \mapsto \alpha_{m,n}$ is an isomorphism*

$$\{(m,n) \in \mathbb{Z} \times \mathbb{Z} \mid t_2^m = t_1^n\} \simeq \mathrm{Hom}_k(E_{t_1}, E_{t_2}) = \mathrm{Hom}_C(E_{t_1}, E_{t_2}) .$$

In particular, E_{t_1} is isogenous to E_{t_2} if and only if t_1 and t_2 are commensurable.

Proof. If $t = t_2^m = t_1^n$, with $m > 0$, hence $n > 0$, then we have a commutative triangle.

$$\begin{array}{ccc} & E_t & \\ {}^{\alpha_{1,n}}\nearrow & & \searrow^{\alpha_{m,1}} \\ E_{t_1} & \xrightarrow[\alpha_{m,n}]{} & E_{t_2} \end{array}$$

The degree of $\alpha_{m,1}$ is m, because the corresponding function field extension is cyclic of degree m, with Galois group generated by the automorphism translation by t_2 on the field of meromorphic functions with period $t = t_2^m$. In case $m = n$ and $t_2 = t_1$ the map $\alpha_{n,n}$ is multiplication by n on E_{t_1}, which shows that $\alpha_{n,1}$ is dual to $\alpha_{1,n}$ and has therefore degree n. Hence the degree of $\alpha_{m,n}$ is mn, as claimed. (Note that $\alpha_{-m,-n} = -\alpha_{m,n}$.)

To prove the rest we can assume the ground field is C, since we have already noted that $\alpha_{m,n}$ is defined over k if t_1 and t_2 are in k. Since C is algebraically closed, every isogeny over C is a product of isogenies of prime degree. Since it is obvious that the composition of maps of the form $\alpha_{m,n}$ is of that same form, we are reduced to proving that an isogeny $\beta : E_t \to E_{t'}$ of prime degree p is of the form $\alpha_{p,1}$, or $\alpha_{1,p}$. The j-invariants of our curves are not 0 or 1728 because they satisfy $|j| > 1$. Consequently the only automorphisms of the curves (preserving the group structure) are ± 1, and it follows that an isogeny is determined up to sign by its kernel. Since $-\alpha_{m,n} = \alpha_{-m,-n}$, it suffices to show that the only finite subgroup-schemes of order p of E_t are the kernel of $\alpha_{1,p}$ and the kernels of the various $\alpha_{p,1}$'s. If p is not the characteristic of C, then this amounts to the fact that a subgroup of order p of $C^*/t^{\mathbb{Z}}$ is either the group of pth roots of 1, or is generated by one of the pth roots of t. Suppose $p = \mathrm{char}(C)$. Then $\alpha_{1,p} : E_t \to E_{t^p}$ is the Frobenius, and $\alpha_{p,1} : E_t \to E_{t^{1/p}}$ is the Verschiebung. The kernels of these two maps are the only subgroupschemes of order p, because their product is the kernel of multiplication by p and they are not isomorphic, being isomorphic to μ_p and $\mathbb{Z}/p\mathbb{Z}$, respectively. This concludes the proof of the Theorem.

Applications.

An immediate corollary is the fact that $\mathrm{Hom}(E_t, E_t) = \mathbb{Z}$, i.e., our curves do not have nontrivial endomorphisms. When I first told Serre about the existence of the curves E_t, he noted that if the curve is defined over a finite extension of $\mathbb{Q}_p$ the non-existence of non-trivial endomorphisms can be seen from the action of Galois on points of finite order, and that this gives a nice proof that the j-invariant of an elliptic curve with complex multiplication is an algebraic integer. Indeed, if j were not an integer at a place above a prime p, then there would be a finite extension of $\mathbb{Q}_p$ over which the curve is isomorphic to E_t for $t = j^{-1} + 744j^{-2} + 750420j^{-3} + \cdots$, contradicting the fact that E_t does not have non trivial endomorphisms. Incidentally, a generalization to abelian varieties of CM type is the fact that such a variety has potential good reduction at every place [**12**], Thm. 6(a).

Another corollary of the theorem is that if $\ell : k^* \to V$ is a homomorphism of k^* into a $\mathbb{Q}$-vector space V, and if the valuation v satisfies $v(k^*) \subset \mathbb{Q}$, then $\ell(t)/v(t)$ is an isogeny invariant of the curves of type E_t. The special case $k = \mathbb{Q}_p$, $V = k$, $\ell : k^* \to k$ is the p-adic logarithm $\log_p$, and $v = \mathrm{ord}_p$ is of interest. In that case, if E_t arises by base change from an elliptic curve E over $\mathbb{Q}$, Greenberg and Stevens have proved [**3**], at least for $p > 3$, that

$$L_p'(1) = \frac{\log_p(t)}{\mathrm{ord}_p(t)} \frac{L(1)}{\Omega}$$

where L_p is the p-adic and L the ordinary L-function of E, and Ω is the real period of E. This relation is predicted by the p-adic Birch and Swinnerton-Dyer conjecture proposed in [**7**]. The usual Birch and Swinnerton-Dyer conjecture involves the discriminant of a real-valued height pairing on the Mordell-Weil group $E(\mathbb{Q})$. The p-adic conjecture involves the discriminant of a p-adic height pairing on the so-called extended Mordell-Weil group which is the extension of $E(\mathbb{Q})$ by $\mathbb{Z}$ obtained by taking the inverse image of $E(\mathbb{Q})$ under the map $\varphi : \mathbb{Q}_p^* \to E(\mathbb{Q}_p)$. Call this inverse image $E'(\mathbb{Q})$, and let $E'(\mathbb{Q})_0$ denote its intersection with $\mathbb{Z}_p^*$. Then for a, b $E_0'(\mathbb{Q})$ and $c \in E'(\mathbb{Q})$, the p-adic height pairing satisfies

$$\langle a, b\rangle = \text{usual } p\text{-adic } \langle (a), (b)\rangle$$
$$\langle c, t\rangle = \log_p(c)/\mathrm{ord}_p(t)$$

In particular, $\langle t, t\rangle$ is the isogeny invariant $\log_p(t)/\mathrm{ord}_p(t)$ occurring in the result of Greenberg and Stevens mentioned above. In case the rank of $E(\mathbb{Q})$ is 0, their result is the p-adic Birch and Swinnerton-Dyer conjecture.

For each integer $n > 0$, let K_n be the field of elliptic functions on k^* with period t^n, isomorphic to the field of rational functions on E_{t^n} defined over k. Let K_∞ be the union of the K_n, and let s denote the automorphism of K_∞ given by translation by t. This automorphism s is of infinite order and its fixed field is $K_1 = K$. The extension K_∞/K is Galois with group $\widehat{\mathbb{Z}}$ and the powers of s form a canonically determined dense subgroup. (K_n/k, as abstract function field, determines the elliptic curve E_{t^n} over k, its j-invariant, hence its period t^n, and then, up to sign, the parametrization $\varphi : k^* \to E_{t^n}$, and the automorphism s.) The finite subextensions of K_∞/K are the K_n/K, and they have the remarkable property to be not only unramified, but split completely at every place of K, even though k may be far from algebraically closed.

Points of finite order.

For an integer $m > 0$ let $E_t[m]$ denote the kernel of multiplication by m on E_t. This is a finite flat group scheme over k whose structure is easy to describe. Let R be an arbitrary ground ring, and let u be an invertible element of R. Consider the functor from commutative R-algebras A to abelian groups which associates to A the quotient group of the group of pairs $(a, i) \in A^* \times \mathbb{Z}(A)$ such that $a^m = u^i$, modulo the subgroup of pairs of the form (u^i, im). (Here $\mathbb{Z}(A)$ denotes the group of locally constant functions $i : \operatorname{Spec} A \to \mathbb{Z}$.) An element of that quotient is represented by a unique pair (a, i) such that $0 \leq i(x) < m$ for $x \in \operatorname{Spec} A$, and consequently the functor is represented by the scheme

$$\coprod_{i=0}^{m-1} \operatorname{Spec}(B_i) = \operatorname{Spec}\left(\prod_{i=0}^{m-1} B_i\right)$$

where $B_i = R[X]/(X^m - u^i)$. The group law on the functor makes this scheme a groupscheme which we denote by $G = G_{R,u,m}$. There is an exact sequence

$$0 \to \mu_m \to G_{R,u,m} \to \mathbb{Z}/m\mathbb{Z} \to 0 \ ,$$

the maps being $z \mapsto (z, 0)$ and $(a, i) \mapsto i \pmod{m}$. Since $H^1(R, G_m) = 0$ in the flat topology ("Hilbert theorem 90"), every $(\mathbb{Z}/m\mathbb{Z})$-module scheme over R which is an extension of $\mathbb{Z}/m\mathbb{Z}$ by μ_m is of the form $G_{R,u,m}$ for some $u \in R^*$. The class of the extension is determined by the image of u in $R^*/(R^*)^m = H^1(R, \mu_m)$. It is easy to see that an isomorphism $G_{R,u,m} \to G_{R,v,m}$ which induces identity on μ_m and on $\mathbb{Z}/m\mathbb{Z}$ is of the form $(a, i) \to (ar^i, i)$, where $r \in R$ and $r^m = v/u$. More generally, for integers e, f, a homomorphism $G_{r,u,m} \to G_{r,v,m}$ which induces multiplication by e on μ_m and by f on $\mathbb{Z}/m\mathbb{Z}$ is of the form

$$(a, i) \mapsto (a^e r^i, fi) \ , \quad \text{where } r \in R \text{ satisfies } u^e r^m = v^f \ .$$

If u is of infinite order in A^* in the sense that $i \mapsto u^i$ is an isomorphism $\mathbb{Z}(A) \simeq u^{\mathbb{Z}(A)}$, then a pair (a, i) is determined by a alone and

$$G_{R,u,m}(A) = \text{ points of order dividing } m \text{ in the group } A^*/u^{\mathbb{Z}(A)} \ .$$

Now take $R = k$, $u = t$. The functorial isomorphism $\varphi : A^*/t^{\mathbb{Z}(A)} \simeq E_t(A)$ for finite dimensional k-algebras A induces an isomorphism of functors, and hence of finite flat group schemes,

$$G_{k,t,m} \simeq E_t[m] \ .$$

Of course, if m is prime to the characteristic of k, then we do not need the isomorphism $A^*/t^{\mathbb{Z}(A)} \simeq E_t(A)$ for all finite A, but only for A's which are (products of) fields. Indeed in that case, the group scheme $E_t[m]$ is etale and is determined by the Galois module $E_t[m](\bar{k})$. In any case, there is a canonical exact sequence of finite flat k-group schemes, or Galois modules,

$$0 \to \mu_m \to E_t[m] \xrightarrow{a} \mathbb{Z}/m\mathbb{Z} \to 0 \ .$$

The Weil pairing

$$E_t[m] \times E_t[m] \to \mu_m$$

is given by $(P,Q) \mapsto a(P)Q - a(Q)P$, or by the negative of that expression, according to one's convention.

The case $E_t[m]$ has good reduction.

Let R be the ring of integers in k. The elliptic curve E_t over k does not have good reduction, that is, there is no elliptic curve over R whose general fiber is E_t. However, it may happen for some m that $E_t[m]$ has good reduction, i.e., is the general fiber of a finite flat group scheme over R. From the above discussion it is clear that this happens if $|t| \in |k^*|^m$; then $t = ua^m$ with $u \in R^*$, $a \in k^*$, and $E_t[m]$ is isomorphic to the general fiber of $G_{R,u,m}$. This would be the situation at each place ℓ of bad reduction of the Frey curve corresponding to a counterexample to Fermat for a prime exponent m. That fact plays an essential role in the relation between the modularity of curves over $\mathbb{Q}$ and Fermat's last theorem suggested by Frey, made more precise by Serre [**11**], proved by Ribet [**8**] and recently exploited by Wiles.

Serre's isogeny theorem.

Suppose p is a prime and k is locally compact with residue characteristic p. Let E_{t_1} and E_{t_2} be two curves over k of our type, and let $E_{t_i}[p^\infty]$ denote the p-divisible group of E_{t_i} over k for $i = 1, 2$. Then the natural map

$$\mathbb{Z}_p \otimes \operatorname{Hom}(E_{t_1}, E_{t_2}) \to \operatorname{Hom}(E_{t_1}[p^\infty], E_{t_2}[p^\infty]) \ ,$$

is an isomorphism. This was proved by Serre [**10**], A1.4 in case k is of characteristic 0 and used by him to prove [**10**], 2.3 the global isogeny theorem for elliptic curves over a number field with non-integral j-invariant. The global theorem was later proved for all abelian varieties by Faltings, but Serre's result is all that is needed to show for semistable elliptic curves over $\mathbb{Q}$ that the modularity of their Galois representation implies that they are modular in the stronger sense of being covered by a modular curve.

We will prove the local isogeny theorem above by giving an explicit description of the two groups involved as indicated by the following diagram

$$\begin{array}{cccccc}
 & \operatorname{Hom}(E_{t_1}, E_{t_2}) & \longrightarrow & \operatorname{Hom}(E_{t_1}[p^\infty], E_{t_2}[p^\infty]) & \\
\alpha_{m,n} & & & & h \\
\uparrow & \wr\uparrow & & \downarrow\wr & \downarrow \\
(m,n) & & & & (e,f) \\
 & \{(m,n) \in \mathbb{Z} \times \mathbb{Z} \mid t_2^m = t_1^n\} & \longrightarrow & \{(e,f) \in \mathbb{Z}_p \times \mathbb{Z}_p \mid t_2^e = t_1^f \in \widehat{k^*}\} & \\
 & & (m,n) \longmapsto (m,n) & &
\end{array}$$

Here,

$$\widehat{k^*} := \varprojlim_r (k^*/(k^*)^{p^r}) .$$

The right hand vertical map is explained by the diagram

$$\begin{array}{ccccccccc} 0 & \longrightarrow & \mu_{p^\infty} & \longrightarrow & E_{t_1}[p^\infty] & \longrightarrow & \mathbb{Q}_p/\mathbb{Z}_p & \longrightarrow & 0 \\ & & \downarrow f & & \downarrow h & & \downarrow e & & \\ 0 & \longrightarrow & \mu_{p^\infty} & \longrightarrow & E_{t_2}[p^\infty] & \longrightarrow & \mathbb{Q}_p/\mathbb{Z}_p & \longrightarrow & 0 \end{array}$$

Here h is an arbitrary homomorphism of the p-divisible groups. It must respect the filtrations because there is no non-trivial homomorphism $\mu_{p^\infty} \to \mathbb{Q}_p/\mathbb{Z}_p$. (In characteristic p, μ_{p^∞} is connected and the connected component of $\mathbb{Q}_p/\mathbb{Z}_p$ is trivial; in characteristic 0 it is clear from the fully faithful representation of the situation by Galois modules, since the Galois action on μ_{p^∞} is non-trivial.) The endomorphisms induced by h on the submodule and quotient module are multiplications by p-adic integers f and e, respectively, because those groups have no other endomorphisms than such multiplications. Thus the restriction of h to the finite group schemes $E_t[p^r] \simeq G_{k,t}, p^r$ is of the type considered above and consequently $t_2^f = t_1^e$ in $k^*/(k^*)^{p^r}$ for all r. In other words, the equality $t_2^f = t_1^e$ holds in the multiplicatively written $\mathbb{Z}_p$-module

$$\widehat{k^*} := \varprojlim_r (k^*/(k^*)^{p^r})$$

The map $h \mapsto (e, f)$ is injective because there is no non-trivial homomorphism $\mathbb{Q}_p/\mathbb{Z}_p \to \mu_{p^\infty}$. It is surjective because by the above discussion, for given (e, f) satisfying $t_2^e = t_1^f$ in $\widehat{k^*}$, there is a homomorphism

$$h_j : E_{t_1}[p^j] \longrightarrow E_{t_2}[p^j]$$

for each j inducing e on the quotient module and f on the submodule, and h_j is unique up to the addition of homomorphisms $\mathbb{Z}/p^j\mathbb{Z} \to \mu_{p^i}$. Therefore if $p^c = \#(\mu_{p^\infty}(k))$ is the number of p-power roots of 1 in k, then all choices of h_j (for a given pair (e, f)) induce the same homomorphism on $E_{t_1}[p^{j-c}]$. Hence these latter homomorphisms are coherent and give the desired h which maps to (e, f).

Now the key idea in Serre's proof is that the valuation gives a homomorphism $v : \widehat{k^*} \to \mathbb{Z}_p$ which applied to the condition on e and f gives

$$ev(t_2) = fv(t_1)$$

Since $v(t_1)$ and $v(t_2)$ are non-zero rational integers, it follows that we can multiply the pair (e, f) by a p-adic unit so that it becomes a pair of ordinary integers. Then $t_2^e t_1^{-f}$ is in the kernel of the map $k^* \to \widehat{k^*}$ so is a root of unity of order prime to p in k. Multiplying the pair (e, f) by that order, we get a pair (m, n) such that and our original pair is a p-adic unit times (m, n), which is what we needed to show.

More general ground rings.

Suppose R is a commutative ring and I an ideal in R such that R is I-adically complete and separated. Suppose $t \in I$ is a non-zero-divisor in R, and put $k = R[1/t]$. Formulas (14) define elements b_2 and b_3 in I, and equation (13) defines an elliptic curve over k which we will denote by E_t. Putting

$$u = \frac{y+x}{y} = 1 + \frac{x}{y} \quad , \quad v = \frac{1}{y}$$

the equation for E_t becomes

$$F(u,v) := uv - (u-1)^3 + b_2(u-1)v^2 + b_3v^3 = 0 \ .$$

Let $E_t^0(k) = \{P = (u,v) \in E_t(k) \mid u \in R^* \text{ and } v \in R\}$. Then the map $(u,v) \mapsto u$ is a bijection from $E_t^0(k) \to R^*$. The inverse map can be defined by $v = v(u) := \lim(v_n(u))$, where the functions $v_n : R^* \to R$ are defined inductively by

$$v_1(u) = \frac{(u-1)^3}{u} \ , \quad \text{and } v_{n+1}(u) = v_n(u) - F(u, v_n(u))/u \ , \quad \text{for } n > 0 \ .$$

By induction, one proves that $F(u, v_n(u))$ is in I^n.

In fact, $E_t^0(k)$ is a subgroup of $E_t(k)$, "analytically" isomorphic to R^* in the following sense. There is a Laurent series $f(u)$ with coefficients in R, convergent for all $u \in R^*$, such that $f(u) \equiv u \pmod I$, and such that the map $P = (u,v) \mapsto f(u)$ is an isomorphism $E_t^0(k) \to R^*$. In case R is a discrete valuation ring this map is just the inverse of the negative of the map $\varphi : R^* \to E_t^0(k)$, and we can use the same formulas in the present more general setting. Formulas (2) and (11) yield

$$\begin{aligned} x + y &= \frac{w}{(1-w)^3} + g(w) \\ y &= \frac{w^2}{(1-w)^3} + h(w) \end{aligned}$$

where $g(w)$ and $h(w)$ are Laurent series with coefficients in I which are Laurent polynomials modulo I^n for every n. Hence

$$\begin{aligned} u = (x+y)/y &= (w^{-1} + g(w)(1-w)^3/w^2)/(1 + h(w)(1-w)^3/w^2) \\ &= F(w^{-1}) \ , \text{ say,} \end{aligned}$$

where $F(z)$ is a Laurent series in z with coefficients in R, $\equiv z \pmod I$ and polynomial $(\text{mod } I^n)$ for every n. Such an F maps R^* bijectively to R^*, and the inverse function is of the same type. To see this, note that all positive and negative powers of $F(z)$ are functions of a similar type, the monomial z being replaced by z^n in the nth power, so that we can define functions $f_n(Z)$ inductively by

$$f_1(z) = 1 \text{ and } f_{n+1}(z) = f_n(z) + H_n(z) \ , \quad \text{where } H_n(z) := z - f_n(F(z)) \ ,$$

for $n > 0$. Then $H_{n+1}(z) = H_n(z) - H_n(F(z))$, hence $H_n(z) \equiv 0 \pmod{I^n}$, and the f_n's converge to an f such that $f(F(z)) = z$.

We leave to the reader the task of showing that the map : $w \mapsto P = (u, v(u))$, with $u = f(w)$ is a group homomorphism. I do not know to what extent this isomorphism $R^* \simeq E_t^0(k)$ can be extended to $k^*/q^{\mathbb{Z}} \to E_t(k)$ in this more general situation, as it can in the case k is a local field. Presumably it can at least be extended to points $w \in k^*$, some power of which is in $R^* q^{\mathbb{Z}}$, with the image being the points in $E_t(k)$, some multiple of which is in $E_t^0(k)$. That would enable us to show that the isomorphism $G_{k,t,m} \simeq E_t[m]$ still holds in this more general situation. But that is known anyway, as we indicate in the next section.

The modular point of view.

All of the elliptic curves which we have discussed so far are obtainable by base change from one single curve, E_q, defined over the ring $\mathbb{Z}[[q]][1/q]$ of finite-tailed Laurent series in an indeterminate q with coefficients in $\mathbb{Z}$. Indeed, given any R, t as above, there is a unique homomorphism from $\mathbb{Z}[[q]][1/q]$ to $k = R[1/t]$ taking $q \to t$, and continuous from the q-adic topology in $\mathbb{Z}[[q]]$ to the I-adic topology in R, and thus E_t can be obtained as a base change of E_q. A more sophisticated and algebraic construction of E_q than the one we give here, due to Raynaud, is explained in detail in [**1**], VII. That construction, which involves no specific power series at all, uses the notion of "generalized elliptic curve" and Grothendieck's existence theorem. It produces E_q as the generalized elliptic curve over $\mathbb{Z}[[q]]$ which algebrizes a formal such curve which is obtained as the limit of schemes over $\mathbb{Z}[q]/(q^n)$ which are quotients by the action of $\mathbb{Z}$ on schemes whose geometric fibers are infinite strings of copies of P^1 indexed by $\mathbb{Z}$, the point ∞ of the ith copy being identified with the point 0 on the $(i+1)$th. This approach yields an isomorphism $G_{k,q,m} \to E_q[m]$ for each m, where here $k = \mathbb{Z}[[q]][1/q]$.

The curve E_q is used to study the modular curves in the neighborhood of cusps. For example, it identifies $\mathbb{Z}[[q]]$ with the formal completion of $\mathcal{M}_1$ at $j = \infty$, and if a modular form f of level 1 defined over a ring R is viewed as a function which attaches to every pair (E, ω) consisting of an elliptic curve E over an R-algebra A, together with a nowhere vanishing differential ω on E, an element $f(E, \omega) \in A$, then the Fourier expansion of f is the series $f(E_q \otimes R, dq/q) \in \mathbb{Z}[[q]][1/q] \otimes R$. Thus the coefficients of the expansion are contained in a finitely generated $\mathbb{Z}$-submodule of R. Working over $\mathbb{Z}[[q^{1/n}]][1/q][z]/(z^n - 1)$ one can similarly define the expansions of forms of higher level. But the story of the arithmetic of modular curves and modular forms is very long and all we can do here is to refer the reader to such accounts as [**1**], [**5**] and [**6**], where it is extensively explained.

References

[1] Deligne, P. and Rapoport, M., *Les schémas de modules de courbes elliptiques, in Modular Functions of One Variable II*, Springer Lecture Notes in Math. **349** (1973), 143–316.

[2] Dwork, B., Gerotto, G. and Sullivan, F.J., *An Introduction to G-functions*, Annals of Math. Studies **133**, Princeton U. Press, 1994.

[3] Greenberg, R. and Stevens, G., *p-adic L-functions and p-adic periods of modular forms*, Invent. Math. **111** (1993), 407–447.

[4] A. Hurwitz and R. Courant, *Funktionentheorie*, Springer, Berlin, 1929.

[5] Katz, N., *p-adic properties of modular schemes and modular forms, in Modular Functions of One Variable III*, Springer Lecture Notes in Math. **350** (1973), 69–190.

[6] Katz, N. and Mazur, B., *Arithmetic Moduli of Elliptic Curves*, Annals of Math. Studies **108**, Princeton U. Press, 1985.

[7] Mazur, B., Tate, J., Teitelbaum, J., *On p-adic analogs of the conjectures of Birch and Swinnerton-Dyer*, Invent. Math. **84** (1986), 1–48.

[8] Ribet, K. A., *On modular representations of* $Gal(\overline{\mathbb{Q}}/\mathbb{Q})$ *arising from modular forms*, Invent. Math. **100** (1989), 359–407.

[9] Roquette, P., *Analytic theory of elliptic functions over local fields*, Hamburger Math. Einzelschriften, Neue Folge - Heft 1, Vandenhoeck & Ruprecht, Gottingen.

[10] Serre, J-P., *Abelian l-adic representations and elliptic curves*, W. A. Benjamin, 1968.

[11] Serre, J-P., *Sur les représentations modulaires de degré 2 de* $Gal(\overline{\mathbb{Q}}/\mathbb{Q})$, Duke Math J. **54** (1987), 179–230.

[12] Serre, J-P. and Tate, J., *Good Reduction of Abelian Varieties*, Annals of Math. **88** (1968), 492–517.

[13] Silverman, J., *Advanced Topics in the Arithmetic of Elliptic Curves*, Springer, 1994.

Conference on Elliptic Curves and Modular Forms
Hong Kong, December 18-21, 1993

On Galois representations associated to Hilbert modular forms II

RICHARD TAYLOR
DEPARTAMENT OF MATHEMATICS, HARVARD UNIVERSITY
1 OXFORD STREET, CAMBRIDGE, MA 02138, USA

Introduction. Let K be a totally real field. In our previous paper we [**12**] we attached a system of λ-adic representations of Gal $(\overline{K}/K)$ to a cuspidal automorphic representation π of $GL_2(\mathbb{A}_K)$ for which π_∞ is regular algebraic. Based on the work of Carayol ([**2**]) one can determine the restriction of these representations to the decomposition group at all primes $\wp \not| l$. One can ask what happens at primes $\wp | l$. One would expect these representations always to be de Rham and to be crystalline if $\wp$ does not divide the conductor of π. When there is a direct geometric construction of these λ-adic representations (see [**2**] and [**1**]) results of this sort follow from the work of Faltings ([**4**]). Tony Scholl raised the question of what can be said in the remaining cases where the only construction is via a congruence argument (see [**12**]). In this paper we give a partial answer to this question. We show that for all but finitely many λ the λ-adic representations are crystalline. In fact the argument is rather easy given a recent, important, but entirely algebraic result of Carayol (proposition 1 of [**3**]). In fact we shall show that if π_∞ is the lowest discrete series representation, $\wp$ does not divide the conductor of π, and the mod λ representation associated to π is irreducible then the λ-adic representation associated to π is crystalline. This suffices as the mod λ representation associated to π can only be irreducible for finitely many λ. As this latter result is not explicitly in the literature we sketch a proof, but we stress that the result is well known to experts and the arguments are immediate generalisations of arguments of Ribet ([**9**]) and Serre ([**10**]).

1. Notation and Results. We shall let K denote a totally real field; $\mathbb{A}_K$ the ring of adeles of K; I the set of embeddings of $K \hookrightarrow \mathbb{R}$; and for $\tau \in I$ we shall let v_τ denote the corresponding place of K. We shall also let $\pi = \otimes_v \pi_v$ denote a cuspidal automorphic representation of $GL_2(\mathbb{A}_K)$ and $N(\pi)$ its conductor of π.

We shall let c denote complex conjugation. If $k \in \mathbb{Z}_{\geq 2}$ we shall let σ_k denote the irreducible admissible representation of $GL_2(\mathbb{R})$ with Langlands parameter

$$W_{\mathbf{R}} = \langle \mathbb{C}^\times, j \mid j^2 = -1;\ jzj^{-1} = {}^c z \rangle \longrightarrow GL_2(\mathbb{C})$$

$$z \longmapsto \begin{pmatrix} z^{1-k} & 0 \\ 0 & {}^c z^{1-k} \end{pmatrix} |z|$$

$$j \longmapsto \begin{pmatrix} 0 & 1 \\ (-1)^{1-k} & 0 \end{pmatrix}.$$

We shall say that π has weight $\mathbf{k} \in \mathbb{Z}^I_{\geq 2}$ if for all $\tau \in I$ the component $\pi_{v_\tau} = \sigma_{k_\tau}$. For the rest of this paper we shall assume that π has some such weight $\mathbf{k}$.

If N is an ideal of $\mathcal{O}_K$ then we shall let $\mathbb{T}_N$ denote the polynomial algebra over $\mathbb{Z}$ in variables $T_\wp$ and $S_\wp$ for primes $\wp$ of $\mathcal{O}_K$ not dividing N. If $N(\pi)|N$ then we get a character $\theta_\pi : \mathbb{T}_N \to \mathbb{C}$ which sends $T_\wp$ (resp. $S_\wp$) onto the eigenvalue of the Hecke operator $[GL_2(\mathcal{O}_{K,\wp}) \begin{pmatrix} 1 & 0 \\ 0 & \varpi_\wp \end{pmatrix} GL_2(\mathcal{O}_{K,\wp})]$ (resp. $[GL_2(\mathcal{O}_{K,\wp})\varpi_\wp GL_2(\mathcal{O}_{K,\wp})]$) on $\pi_\wp^{GL_2(\mathcal{O}_{K,\wp})}$. Here $\varpi_\wp$ denotes a uniformiser in $\mathcal{O}_{K,\wp}$. It is well known that the field generated by the image of θ_π is a number field, which we denote E_π, and that in fact θ_π is valued in the ring of integers of E_π. In the rest of this paper λ will denote a prime of E_π and l will denote its residue characteristic.

Then the main theorem of **[12]** is as follows.

Theorem 1.1. *There is a continuous representation*

$$\rho_{\pi,\lambda} : \mathrm{Gal}\,(\overline{K}/K) \to GL_2(E_{\pi,\lambda})$$

such that if $\wp \not| N(\pi)l$ then $\rho_{\pi,\lambda}$ is unramified at $\wp$ and $\rho_{\pi,\lambda}(\mathrm{Frob}_\wp)$ has characteristic polynomial

$$X^2 - \theta_\pi(T_\wp)X + (\mathbf{N}\wp)\theta_\pi(S_\wp).$$

In fact if $\wp$ is any prime of F not dividing l then the representation $\sigma_{\pi,\wp,\lambda}$ of the Weil-Deligne group of $F_\wp$ corresponding to the restriction of $\rho_{\pi,\lambda}$ to the decomposition group at $\wp$ (see [11]) is rational over E_π and has Frobenius semi-simplification $\sigma_{\pi,\wp}$, the representation of the Weil-Deligne group of $F_\wp$ associated to $\pi_\wp$ by the local Langlands correpondence (see [2]).

The behaviour of $\rho_{\pi,\lambda}$ at primes dividing l would be at least partly described by the following conjecture.

Conjecture 1 *If $\wp$ is a prime of $\mathcal{O}_K$ above l then the restriction of $\rho_{\pi,\lambda}$ to the decomposition group at $\wp$ is de Rham. If further $\wp \not| N(\pi)$ then it is crystalline.*

Combining the work of Carayol **[2]** with that of Faltings **[4]** it is not hard to see that the conjecture is true if either $[K : \mathbb{Q}]$ is odd or if π_v is discrete series at some finite place v. That is we have the following theorem.

Theorem 1.2. (Carayol, Faltings.) *Suppose that either $[F : \mathbb{Q}]$ is odd or that π_v is discrete series for some finite place v. Suppose also that $\wp$ is a prime of $\mathcal{O}_K$ above l. Then the restriction of $\rho_{\pi,\lambda}$ to the decomposition group at $\wp$ is de Rham and if $\wp \not| N(\pi)$ then it is crystalline.*

In the case that $k_\tau > 2$ for some $\tau \in I$, Blasius and Rogawski **[1]** gave a second construction of the representations $\rho_{\pi,\lambda}$ and hence, using Faltings' results **[4]**, can show the following result.

Theorem 1.3. **(Faltings, Blasius, Rogawski.)** *Suppose that for some $\tau \in I$ $k_\tau > 2$.*

1. *If $\wp | l$ then the restriction of $\rho_{\pi,\lambda}$ to the decomposition group at $\wp$ is de Rham.*
2. *For all but finitely many primes λ of E_π and for all primes $\wp$ of $\mathcal{O}_K$ dividing l (which depends on λ) the restriction of $\rho_{\pi,\lambda}$ to the decomposition group at $\wp$ is crystalline.*

It is not clear to us whether their method gives the rest of conjecture 1 in the case that $k_\tau > 2$ for some $\tau \in I$. In this paper we shall prove the following result.

Theorem 1.4. *For all but finitely many primes λ of E_π and for all primes $\wp$ of $\mathcal{O}_K$ dividing l (which depends on λ) the restriction of $\rho_{\pi,\lambda}$ to the decomposition group at $\wp$ is crystalline.*

From theorems 1 and 2 we see that we can restrict our attention to the case $[F : \mathbb{Q}]$ even and $k_\tau = 2$ for all τ. We shall assume this for the rest of this paper.

In fact, in this case, we shall prove a slightly more precise result. Recall that $\rho_{\pi,\lambda}$ is conjugate to a representation valued in $GL_2(\mathcal{O}_{E_\pi,\lambda})$ and hence we can form its reduction modulo λ, which we shall denote $\overline{\rho_{\pi,\lambda}}$. This may depend on the choice of $GL_2(E_{\pi,\lambda})$-conjugate we take, but its semi-simplification $\overline{\rho_{\pi,\lambda}}^{ss}$ depends only on π and λ. In section 3 we shall prove the following result.

Proposition 1.5. *$\overline{\rho_{\pi,\lambda}}^{ss}$ is irreducible for all but finitely many primes λ.*

Our main theorem will then be the following.

Theorem 1.6. *Suppose that $\wp$ is a prime of $\mathcal{O}_K$ and that λ is a prime of E_π, both lying above l. Suppose moreover that $\wp \nmid N(\pi)$ and that $\overline{\rho_{\pi,\lambda}}^{ss}$ is irreducible. Finally suppose that $k_\tau = 2$ for all $\tau \in I$. Then there is an l-divisible group $A/\mathcal{O}_{K,\wp}$ such that $\rho_{\pi,\lambda}$ is isomorphic to VA as $\mathbb{Q}_l[\mathrm{Gal}(\overline{K_\wp}/K_\wp)]$ modules.*

Here VA denotes the Tate module $(\varprojlim A[l^n](\overline{K})) \otimes_{\mathbb{Z}_l} \mathbb{Q}_l$. We note that the theorem and the proposition together imply theorem 1.4.

2. The Main Theorem.

We now turn to the proof of theorem 1.6. We first recall a result about finite flat group schemes.

Lemma 2.1. *Suppose that $L/\mathbb{Q}_l$ is a finite extension.*

1. *Suppose that A/L is an l-divisible group. Then A extends to an l-divisible group $\tilde{A}/\mathcal{O}_L$ if and only if for all n, $A[l^n]$ extends to a finite flat group scheme $\widetilde{A[l^n]}/\mathcal{O}_L$.*
2. *Suppose that $\tilde{G}/\mathcal{O}_L$ is a finite flat group scheme. If H is a subquotient of the generic fibre of $\tilde{G}$ then H extends to a finite flat group scheme $\tilde{H}/\mathcal{O}_L$.*
3. *Suppose that A and B are isogenous l-divisible groups over L. Then A extends to an l-divisible group $\tilde{A}/\mathcal{O}_L$ if and only if B extends to an l-divisible group $\tilde{B}/\mathcal{O}_L$.*

4. *Suppose that $T/\mathbb{Z}_l$ is a finite algebra and that $\rho : \mathrm{Gal}(\overline{L}/L) \to GL_d(T)$ is a continuous homomorphism. Suppose that there is an l-divisible group $\widetilde{A}/\mathcal{O}_L$ such that $V\widetilde{A} \cong (\rho \otimes_{\mathbb{Z}_l} \mathbb{Q}_l)$ as $\mathbb{Q}_l[\mathrm{Gal}(\overline{L}/L)]$-modules. Suppose finally that I is an ideal of T of finite index. Then there is a finite flat group scheme $\widetilde{G}/\mathcal{O}_L$ such that $\widetilde{G}(\overline{L}) \cong (\rho \bmod I)$ as $\mathrm{Gal}(\overline{L}/L)$-modules.*

The first part of this proposition is a theorem of Raynaud (see [**8**]). We thank Fontaine for pointing this out to us. For the second part there are sub-finite flat group schemes $G_1 \subset G_2$ of the special fibre of $\widetilde{G}$. Let $\widetilde{G_i}$ denote the scheme theoretic closure of G_i in $\widetilde{G}$. Then $\widetilde{G_1}/\widetilde{G_2}$ is a finite flat group scheme over $\mathcal{O}_L$ with generic fibre H. For the third part suppose that A extends to an l-divisible group scheme over $\mathcal{O}_L$ and that there is a surjection $A \twoheadrightarrow B$. Then for all n, $B[l^n]$ is a sub-quotient of $A[l^{n+r}]$ for suitable r. Hence $B[l^n]$ extends to a finite flat group scheme over $\mathcal{O}_L$ for all n, and so by the first part B extends to a l-divisible over $\mathcal{O}_L$.

For the final part note that $(T \otimes_{\mathbb{Z}_l} \mathbb{Q}_l)^d/I^d$ with the action of $\mathrm{Gal}(\overline{L}/L)$ defines an l-divisible group over L isogenous to the special fibre of $\widetilde{A}$. Hence it extends to an l-divisible group $\widetilde{B}/\mathcal{O}_L$. Then for sufficiently large n, the finite flat group scheme over L defined by $\rho \bmod I$ is a sub-group scheme of the generic fibre of $\widetilde{B}[l^n]$, and hence extends to a finite flat group scheme over $\mathcal{O}_L$, as desired.

We now return to theorem 1.6. Because $\overline{\rho_{\pi,\lambda}}^{ss}$ is irreducible any two $GL_2(E_{\pi,\lambda})$-conjugates of $\rho_{\pi,\lambda}$ which are valued in $GL_2(\mathcal{O}_{E_\pi,\lambda})$ are $GL_2(\mathcal{O}_{E_\pi,\lambda})$-conjugate. Hence $\rho_{\pi,\lambda}$ is well defined as a representation into $GL_2(\mathcal{O}_{E_\pi,\lambda})$ as is its reduction modulo λ^n, which we shall denote ρ_π/λ^n. It suffices to prove that for infinitely many n there is a finite flat group scheme $A_n/\mathcal{O}_{K,\wp}$ such that $A_n(\overline{K_\wp}) \cong \rho_{\pi/\lambda^n}$ as $\mathrm{Gal}(\overline{K_\wp}/K_\wp)$-modules. We now fix n for the rest of this section.

We must recall something of our construction of $\rho_{\pi,\lambda}$ (see [**12**]). We shall let $\widehat{\mathcal{O}_K}$ denote the product of all completions of $\mathcal{O}_K$ at finite places. If N is an ideal of $\mathcal{O}_K$ we let $U_1(N)$ denote the subgroup of $GL_2(\widehat{\mathcal{O}_K})$ consisting of matrices $\begin{pmatrix} a & b \\ c & d \end{pmatrix}$ with c and $d-1 \in N\widehat{\mathcal{O}_K}$. We let $\mathbb{T}(N)$ denote the image of $\mathbb{T}_N$ in the ring of endomorphisms of $\bigoplus_\pi (\pi^\infty)^{U_1(N)}$, where π runs over cuspidal automorphic representations of $GL_2(\mathbb{A}_F)$ of weight **2** and π^∞ denotes the product over all finite places v of π_v. If a prime $\mathfrak{r}$ divides N exactly once then we let $\mathbb{T}(N)^{\mathfrak{r}-new}$ denote the quotient of $\mathbb{T}(N)$ which acts faithfully on $\bigoplus_\pi (\pi^\infty)^{U_1(N)}$, where π runs over cuspidal automorphic representations of $GL_2(\mathbb{A}_F)$ of weight **2** and with $\pi_\mathfrak{r}$ special.

If $N(\pi)|N$ then θ_π factors through $\mathbb{T}(N)$ and we get an embedding

$$\prod_\pi \theta_\pi : \mathbb{T}(N) \hookrightarrow \prod_\pi \mathcal{O}_{E_\pi},$$

where the product is over a set of representatives of $\mathrm{Gal}(\overline{\mathbb{Q}}/\mathbb{Q})$ orbits of cuspidal automorphic representations π of weight two and conductor dividing N. This map becomes an isomorphism after tensoring with $\mathbb{Q}$. Similarly if $\mathfrak{r}$ divides N

exactly once one get a similar map from $\mathbb{T}(N)^{\mathfrak{r}-new}$ to a similar product but with the added restriction that only those π for which $\pi_{\mathfrak{r}}$ is special occur.

In [**12**] we prove that there is a prime $\mathfrak{r}\nmid N(\pi)l$ and a homomorphism θ : $\mathbb{T}(N(\pi)\mathfrak{r})^{\mathfrak{r}-new} \longrightarrow \mathcal{O}_{E_\pi}/\lambda^n$ such that

$$\begin{array}{ccc} \mathbb{T}(N(\pi)\mathfrak{r}) & \longrightarrow & \mathbb{T}(N(\pi)\mathfrak{r})^{\mathfrak{r}-new} \\ \downarrow & & \downarrow \\ \mathcal{O}_{E_\pi} & \longrightarrow & \mathcal{O}_{E_\pi}/\lambda^n \end{array}$$

commutes. Here the horizontal arrows are the natural surjections and the vertical arrows are the maps θ_π and θ respectively.

Let $\mathfrak{m}$ denote the kernel of the composite map

$$\mathbb{T}(N(\pi)\mathfrak{r})^{\mathfrak{r}-new} \xrightarrow{\theta} \mathcal{O}_{E_\pi}/\lambda^n \longrightarrow \mathcal{O}_{E_\pi}/\lambda,$$

and let $\mathbb{T}$ denote the completion of $\mathbb{T}(N(\pi)\mathfrak{r})^{\mathfrak{r}-new}$ at $\mathfrak{m}$. Then θ factors through $\mathbb{T}$. Let I denote the kernel of $\mathbb{T} \xrightarrow{\theta} \mathcal{O}_{E_\pi}/\lambda^n$. Now $\mathbb{T} \otimes_{\mathbb{Z}_l} \mathbb{Q}_l \cong \prod_j E_{\pi_j,\lambda_j}$, where j runs over some finite set; π_j is a cuspidal automorphic representation of $GL_2(\mathbb{A}_F)$ of weight **2**, of conductor dividing $N(\pi)\mathfrak{r}$ and with $\pi_{j,\mathfrak{r}}$ special; and λ_j is a prime of E_{π_j} above l. For each j there is a continuous representation ρ_j : $\mathrm{Gal}(\overline{K}/K) \to GL_2(E_{\pi_j,\lambda_j})$ such that if $\mathfrak{q}$ is a prime of K not dividing $N(\pi)\mathfrak{r}l$ then ρ_j is unramified at $\mathfrak{q}$ and $\rho_j(\mathrm{Frob}_{\mathfrak{q}})$ has characteristic polynomial

$$X^2 - \theta_{\pi_j}(T_{\mathfrak{q}})X + (\mathbf{N}\mathfrak{q})\theta_{\pi_j}(S_{\mathfrak{q}}).$$

Moreover Carayol ([**2**]) has shown that ρ_j occurs in the Tate module of a Jacobian of a Shimura curve defined over K and which has good reduction at $\wp$. Hence there is an l-divisible group $A_j/\mathcal{O}_{K,\wp}$ such that $VA_j \cong \rho_j$ as $\mathbb{Q}_l[\mathrm{Gal}(\overline{K_\wp}/K_\wp)]$-modules. Thus we get a continuous representation ρ : $\mathrm{Gal}(\overline{K}/K) \to GL_2(\mathbb{T} \otimes_{\mathbb{Z}_l} \mathbb{Q}_l)$ and an l-divisible group $A/\mathcal{O}_{K,\wp}$ such that $\rho \cong VA$ as $\mathbb{Q}_l[\mathrm{Gal}(\overline{K_\wp}/K_\wp)]$-modules and such that if $\mathfrak{q}$ is a prime of K not dividing $N(\pi)\mathfrak{r}l$ then ρ is unramified at $\mathfrak{q}$ and $\rho(\mathrm{Frob}_{\mathfrak{q}})$ has characteristic polynomial

$$X^2 - T_{\mathfrak{q}}X + (\mathbf{N}\mathfrak{q})S_{\mathfrak{q}}.$$

The representations $\overline{\rho_j}^{ss}$ have the same trace as $\overline{\rho_{\pi,\lambda}}^{ss}$ on Frobenius elements and hence are isomorphic. Thus they are all irreducible by assumption. Crucially, we can now apply proposition 1 of [**3**] to conclude that the representation ρ is conjugate to one valued in $GL_2(\mathbb{T})$. Now applying part 4 of our proposition we see that there is a finite flat group scheme $G/\mathcal{O}_{K,\wp}$ such that

$$G(\overline{K_\wp}) \cong (\theta \bmod I) \cong \theta \circ \rho \cong \rho_\pi/\lambda^n$$

as $\mathrm{Gal}(\overline{K_\wp}/K_\wp)$-modules. This completes the proof of theorem 1.6.

3. Irreducibility Results. In this section we remind the reader how to prove proposition 1. First we record the following result.

Proposition 3.1. *If π is a cuspidal automorphic representation of $GL_2(\mathbb{A}_K)$ of some weight $\mathbf{k}$ then $\rho_{\pi,\lambda}$ is irreducible.*

The proof is just Ribet's argument from **[9]**, we repeat it as we know no reference in this context. Suppose that $\rho_{\pi,\lambda}$ were reducible. Then $\rho_{\pi,\lambda}^{ss}$ is a continuous, semi-simple abelian representation of $\mathrm{Gal}(\overline{K}/K)$ such that all but finitely many Frobenius elements have characteristic polynomials which are rational over the number field E_π. By the results of Serre and Henniart (see [**5**]) we see that there are two grossencharacters of type A_0, $\chi_1, \chi_2 : \mathbb{A}_K^\times \to \mathbb{C}^\times$ such that for $\wp \not| N(\pi)l$ the characteristic polynomial of $\rho_{\pi,\lambda}(\mathrm{Frob}_\wp)$ is

$$(X - \chi_1(\varpi_\wp))(X - \chi_2(\varpi_\wp)) \in E_\pi[X].$$

Because K has a real place and the χ_i are of type A_0 we know that there is an integer $w_i \in \mathbb{Z}$ such that for all $\tau \in I$,

$$\chi_i|_{(K_{v_\tau}^\times)_{>0}} : t \longmapsto t^{w_i};$$

and hence that $|\chi_i(\varpi_\wp)| = |\mathbf{N}\wp|^{w_i}$. By corollary 2.5 of [**7**] we know that $(k-2)/2 < w_i < k/2$ and hence that $w_i = (k-1)/2$ for $i = 1, 2$. In particular χ_2/χ_1 is a unitary (in fact finite order) character. Let S denote the set of primes dividing $lN(\pi)$. Then we have an equality of partial L-functions

$$L_S(\pi \otimes \chi_1^{-1}, s) = \zeta_{K,S}(s) L_S(\chi_2/\chi_1, s).$$

As π is a cusp form $L_S(\pi \otimes \chi_1^{-1}, s)$ is entire (see [**6**]), while $\zeta_{K,S}(s)$ has a pole at $s = 1$. Thus $L_S(\chi_2/\chi_1, s)$ must vanish at $s = 1$, but this is impossible as χ_2/χ_1 is unitary. This contradiction proves the proposition.

We now turn to proposition 1. First we have the following lemma.

Lemma 3.2. *Suppose that π is a cuspidal automorphic representation of $GL_2(\mathbb{A}_K)$ of weight 2. Suppose that λ is a prime of E_π above l. Suppose that $\wp$ is an unramified prime of K above l which does not divide $N(\pi)$. Let I denote the inertia group at $\wp$ and let ψ_i denote the natural character $I \twoheadrightarrow \mathbb{F}_{l^i}^\times$ for $i \in \mathbb{Z}_{\geq 0}$. Then over an algebraicly closed field the semisimplification of $\overline{\rho_{\pi,\lambda}}$ is either $\psi_0 \oplus \psi_1$ or $\psi_2 \oplus \psi_2^l$.*

From the classification of the simple finite flat group schemes over the maximal unramified extension of $\mathbb{Q}_l$, which extend to finite flat group schemes over the ring of integers of the maximal unramified extension of $\mathbb{Q}_l$ (see [**8**]), we know that the simple factors of $\overline{\rho_{\pi,\lambda}}^{ss}$ must be of the form $\psi_i^{l^j}$. Also the product must be ψ_1 because the determinant of $\overline{\rho_{\pi,\lambda}}$ is the product of a character unramified at $\wp$ with the cyclotomic character. This leaves only the two possibilities mentioned in the lemma.

Proposition 1 follows immediately from this lemma, proposition 2 and the following result, which is essentially due to Serre.

Proposition 3.3. *Let K and E be number fields and let Γ denote the set of embeddings of K into the algebraic closure of E. Let N be an ideal of $\mathcal{O}_K$.*

Let $\mathcal{L}$ be an infinite set of primes λ of $\mathcal{O}_E$. Let λ lie above the rational prime $l(\lambda)$ and have residue field k_λ. For $\lambda \in \mathcal{L}$ let $\rho_\lambda : \mathrm{Gal}(\overline{K}/K) \to GL_d(\mathcal{O}_{E,\lambda})$ be a continuous representation. If $\wp \not| l(\lambda)$ suppose that the conductor of $\rho_\lambda|_{\mathrm{Gal}(\overline{K_\wp}/K_\wp)}$ is less than or equal to the $\wp$-adic valuation of N. If $\wp \not| N$ suppose that there exists a polynomial $P_\wp(T) \in \mathcal{O}_E[T]$ such that $P_\wp(T)$ is the characteristic polynomial of $\rho_\lambda(\mathrm{Frob}_\wp)$ for all λ with residue characteristic distinct from that of $\wp$.

Also suppose that for each λ, $\overline{\rho_\lambda}^{ss}$ is the sum of d characters $\theta_{\lambda,i}$. Restricting these characters to the inertia groups at primes above $l(\lambda)$ and composing with the local reciprocity map we get maps $\theta^*_{\lambda,i} : \mathcal{O}^\times_{K,l(\lambda)} \to k^\times_\lambda$. Suppose finally that there are integers $n_{\lambda,i}(\sigma)$ for $\sigma \in \Gamma$ which are bounded independently of λ and such that $\theta^*_{\lambda,i}$ is the reduction of $\alpha \mapsto \prod_{\sigma\in\Gamma}(\sigma\alpha)^{n_{\lambda,i}(\sigma)}$.

Then for infinitely many $\lambda \in \mathcal{L}$ the representation ρ^{ss}_λ is abelian.

If $E = \mathbb{Q}$ this is theorem 1 of **[10]**. The same proof works in general. Replacing $\mathcal{L}$ by an infinite subset we may assume that the λ have distinct residue characteristics and that the $n_{\lambda,i}(\sigma)$ are independent of λ. Again replacing $\mathcal{L}$ by an infinite subset and applying proposition 20 of **[10]** we may assume that there are characters $\psi_1, ..., \psi_d \in X(S_N)$ over $\overline{E}$ with $(\widetilde{\psi_i})_{l(\lambda)} = \overline{\theta_{\lambda,i}}$. Here and in the rest of this paragraph we use the notation of **[10]** and we suppose the prime of $\overline{E}$ above $l(\lambda)$ implicit in the notation is in fact chosen to be above λ. Let $\psi = \bigoplus_{i=1}^d \psi_i$. If $\wp$ is a prime of K not dividing N then as on page 292 of **[10]** one sees that $\psi(F_\wp)$ has characteristic polynomial $P_\wp(T)$. Let ψ_λ be the corresponding $l(\lambda)$-adic representation of $\mathrm{Gal}(\overline{K}/K)$ as defined in section 3.4 of **[10]**. If $\wp \not| Nl(\lambda)$ then ψ_λ and ρ_λ are both unramified at $\wp$ and $\psi_\lambda(\mathrm{Frob}_\wp)$ and $\rho_\lambda(\mathrm{Frob}_\wp)$ have the same characteristic polynomial $(P_\wp(T))$. The proposition follows.

References

[1] D.Blasius and J.Rogawski, *Motives for Hilbert modular forms*, Invent. Math. 114 (1993) 55-87.

[2] H.Carayol, *Sur les représentations p-adiques associées aux formes modulaires de Hilbert*, Ann. Sci. ENS (4) 19 (1986) 409-468.

[3] H.Carayol, *Formes modulaires et représentations Galoisiennes à valeurs dans un anneau local complet*, in "p-adic monodromy and the Birch-Swinnerton-Ryer conjecture" (eds. B.Mazur and G.Skerers), Contemporary Math. 165 (1994).

[4] G.Faltings, *Crystalline cohomology and p-adic Galois representations*, Algebraic Analysis, Geometry and Number Theory, Proc. JAMI Inaugural Conference, Johns-Hopkins Univ. Press (1989), 25-79.

[5] G.Henniart, *Représentations l-adiques abéliennes*, in Séminaires de Theorie des Nombres, Paris 1980-81 (ed. M.-J.Bertin), Birkhäuser (1982), 107-126.

[6] H.Jacquet and R.P.Langlands, *Automorphic forms on GL_2*, LNM 114, Springer 1970.

[7] H.Jacquet and J.Shalika, *On Euler products and the classification of automorphic representations I*, Am. J. Math. 103(2) (1981), 499-558.

[8] M.Raynaud, *Schémas en groupes de type $(p, ..., p)$*, Bull. Soc. Math. France 102 (1974) 241-280.

[9] K.Ribet, *Galois representations attached to eigenforms with nebentypus*, in modular Forms of One Variable V (ed. J.-P.Serre and D.Zagier), LNM 601 Springer 1977, 17-51.

[10] J.-P.Serre, *Propriétés galoisiennes des points d'ordre fini des courbes elliptiques*, Invent. Math. 15 (1972), 259-331.

[11] J.Tate, *Number theoretic background*, in Automorphic Forms, Representations and L-functions (ed. A.Borel and W.Casselman), Proc. Symp. Pure Math. XXXIII (2), AMS 1979, 3-26.

[12] R.Taylor, *On Galois representations associated to Hilbert modular forms*, Invent. Math. 98 (1989), 265-280.

Made in the USA
Coppell, TX
10 November 2026

11835566R00190